Handbook of
Behavioral Neurobiology

Volume 11
Sexual
Differentiation

HANDBOOK OF BEHAVIORAL NEUROBIOLOGY

General Editor:
Norman T. Adler
University of Pennsylvania, Philadelphia, Pennsylvania

Volume 1   Sensory Integration
Edited by R. Bruce Masterton

Volume 2   Neuropsychology
Edited by Michael S. Gazzaniga

Volume 3   Social Behavior and Communication
Edited by Peter Marler and J. G. Vandenbergh

Volume 4   Biological Rhythms
Edited by Jürgen Aschoff

Volume 5   Motor Coordination
Edited by Arnold L. Towe and Erich S. Luschei

Volume 6   Motivation
Edited by Evelyn Satinoff and Philip Teitelbaum

Volume 7   Reproduction
Edited by Norman T. Adler, Donald Pfaff, and Robert W. Goy

Volume 8   Developmental Psychobiology and Developmental Neurobiology
Edited by Elliott M. Blass

Volume 9   Developmental Psychobiology and Behavioral Ecology
Edited by Elliott M. Blass

Volume 10   Neurobiology of Food and Fluid Intake
Edited by Edward M. Stricker

Volume 11   Sexual Differentiation
Edited by Arnold A. Gerall, Howard Moltz, and Ingeborg L. Ward

---

Handbook of
Behavioral Neurobiology

Volume 11
Sexual
Differentiation

Edited by

Arnold A. Gerall
Tulane University
New Orleans, Louisiana

Howard Moltz
The University of Chicago
Chicago, Illinois

and

Ingeborg L. Ward
Villanova University
Villanova, Pennsylvania

PLENUM PRESS • NEW YORK AND LONDON

Library of Congress Cataloging-in-Publication Data

Sexual differentiation / edited by Arnold A. Gerall, Howard Moltz, and
Ingeborg L. Ward.
        p.   cm. -- (Handbook of behavioral neurobiology ; v. 11)
    Includes bibliographical references and index.
    ISBN 0-306-43983-2
    1. Sex differentiation. 2. Sex differences.   I. Gerall, Arnold
A. II. Moltz, Howard. III. Ward, Ingeborg L. IV. Series.
QP278.S48 1992
599'.036--dc20   92-4962
                                                           CIP

ISBN 0-306-43983-2

© 1992 Plenum Press, New York
A Division of Plenum Publishing Corporation
233 Spring Street, New York, N.Y. 10013

Printed in the United States of America

Contributors

WILLIAM W. BEATTY, *Department of Psychiatry and Behavioral Sciences, University of Oklahoma Health Sciences Center, Oklahoma City, Oklahoma 73190*

T. O. FOX, *Program in Neuroscience, Harvard Medical School, and Department of Biochemistry, E.K. Shriver Center, Waltham, Massachusetts 02254*

ARNOLD A. GERALL, *Department of Psychology, Tulane University, New Orleans, Louisiana 70112*

LIOR GIVON, *Department of Psychology, Tulane University, New Orleans, Louisiana 70112*

WENDY HELLER, *Department of Psychology, University of Illinois, Urbana, Illinois 61820*

SHELTON E. HENDRICKS, *Department of Psychology, University of Nebraska at Omaha, Omaha, Nebraska 68182*

JERRE LEVY, *Department of Psychology, University of Chicago, Chicago, Illinois 60637*

KATHIE L. OLSEN, *Behavioral Neuroendocrinology Program, National Science Foundation, Washington, D.C. 20550*

JUNE MACHOVER REINISCH, *The Kinsey Institute for Research in Sex, Gender, and Reproduction, and Department of Psychology, Indiana University, Bloomington, Indiana 47405; and Department of Psychiatry, Indiana University School of Medicine, Indianapolis, Indiana 46202*

STEPHANIE A. SANDERS, *The Kinsey Institute for Research in Sex, Gender, and Reproduction, Indiana University, Bloomington, Indiana 47405*

S. A. TOBET, *Program in Neuroscience, Harvard Medical School, and Department of Biochemistry, E.K. Shriver Center, Waltham, Massachusetts 02254*

INGEBORG L. WARD, *Department of Psychology, Villanova University, Villanova, Pennsylvania 19085*

O. BYRON WARD, *Department of Psychology, Villanova University, Villanova, Pennsylvania 19085*

MARK E. WILSON, *Yerkes Regional Primate Research Center, Field Station, Emory University, Lawrenceville, Georgia 30243*

Preface

The contributors to this volume integrate results and discuss conceptual issues arising from empirical research on differentiation of sexually dimorphic neural and behavioral characteristics in several species, including higher primates and human beings. The organizational theory of sexual differentiation proposed by W. C. Young and his colleagues served as a useful frame of reference or starting point for the presentation of several authors. Each of these chapters presents information and hypotheses that extend or modify this theory. Accordingly, genetic and experiential as well as gonadal hormonal effects on the expression of a variety of sex-related processes are discussed. Additionally, the organizing consequences of hormones are considered over more than one period in the life span of the organism. Also discussed are issues and data that relate to infantile and prepubertal as well as pubertal and senescent stages.

In the first chapter, Olsen comprehensively analyzes the contribution of genetic factors to mechanisms guiding normal development of sexual behavior. Then, she discusses how the experimental use of genetic mutations and strain differences can enhance understanding of fundamental developmental processes.

The next three chapters focus on gonadal hormone influence on sexual differentiation, mostly, but not entirely, in subprimate species. Tobet and Fox discuss how the development of sexual differences in central nervous system morphology and neurochemical distribution is influenced by prenatal and neonatal steroid hormones. The authors highlight the importance of considering methodological and species-related differences when integrating results from histochemical and neuroanatomical studies. Such studies, being conducted in large numbers, are revealing significant structural alterations in the nervous system that are induced by gonadal hormone manipulations. Beatty provides an overview and demonstrates the theoretical relevance of studies examining gonadal hormone involvement in the development and expression of adult male and female nonreproductive behaviors. Hendricks tackles the task of evaluating evidence indicating that the ovarian steroidal hormones, estrogen and progesterone, contribute to differentiation of female behavioral characteristics.

In the next four chapters, early hormonal modification of sexually dimorphic behaviors in animals, and gender-related behaviors in humans, are discussed in the broader context of social factors encountered at various ages. I. L. Ward provides experimental evidence that the characteristics of adult reproductive behaviors in rodents are determined not only by prenatal hormonal conditions but also by pre-pubertal social experiences. Data to support this hypothesis are drawn from a variety of experiments, including those involving prenatal stress. O. B. Ward draws together experimental data from animal research and emerging observational data from child studies, describing the consequences of fetal exposure to several common drugs for steroidogenesis and differentiation of sexually dimorphic and gender-related behaviors. Reinisch and Sanders discuss the development of cognitive and personality characteristics in children whose prenatal circulating androgen levels may have been altered because of a specific clinical endocrine syndrome or exposure to drugs. The authors point to male vulnerability resulting from exposure to subasymptotic prenatal androgen levels. They suggest that this condition can predispose such males to respond to socioenvironmental conditions in a manner that permanently sets cognitive style. Levy and Heller explore the effect of steroidal hormone conditions during infancy on the development of human cortical hemispheric asymmetries and the expression of the specialized cognitive abilities associated with each hemisphere. The hypothesis that early appearance of androgen retards the maturation rate of the left hemisphere is related both to gender differences in cognitive abilities and to the occurrence of genetic diseases that result in fetal and infantile steroid hormone deficiencies.

The last two chapters concentrate on specific age-related changes in reproductive processes. Wilson summarizes recent research on hormonal factors acting at the time of puberty in primates. Gerall and Givon outline how perinatal steroidal hormones may determine quantitative characteristics of reproductive processes, particularly ovulation and lordosis, throughout the life span of the laboratory rat.

Throughout the volume, the authors discuss results and concepts that directly relate to the origin and nature of sex differences in behavior and the underlying neural processes. The understanding of differences between male and female characteristics has clear relevance to many scientific and societal concerns. It should also be noted, however, that research into the mechanisms underlying sex differences in a particular behavior or psychological capacity also contributes to the scientific knowledge concerning that very behavior or capacity. Accordingly, if perinatal androgen affects the development of intrinsic differences between male and female behavioral patterns (e.g., those involved in expressing aggression or performing quantitative or spatial tasks), then this hormone also has a high probability of influencing the neural substrates or behavioral systems specifically mediating aggression or cognitive functions. Thus, it is hoped that this volume will be useful both to readers with specific interests in the topic of sexual differentiation and to those concerned with fundamental principles governing development and the expression of behavior over time.

Appreciation is expressed to Mr. Robert E. Flowerree, who generously supported a conference held at Tulane University, at which many of the contributors to this volume initially discussed their work.

Contents

CHAPTER 3

Gonadal Hormones and Sex Differences in Nonreproductive Behaviors

William W. Beatty

CHAPTER 4

Role of Estrogens and Progestins in the Development of Female Sexual
      Behavior Potential

Shelton E. Hendricks

CHAPTER 5

Sexual Behavior: The Product of Perinatal Hormonal and Prepubertal
      Social Factors

Ingeborg L. Ward

CHAPTER 8

Gender Differences in Human Neuropsychological Function

Jerre Levy and Wendy Heller

CHAPTER 9

Factors Determining the Onset of Puberty

Mark E. Wilson

CHAPTER 10

Early Androgen and Age-Related Modifications in Female Rat Reproduction

Arnold A. Gerall and Lior Givon

Genetic Influences on Sexual Behavior Differentiation

KATHIE L. OLSEN

INTRODUCTION

The importance of perinatal hormonal stimulation for sexual behavior differentiation is well established in mammals. Hormones secreted by the testes during a critical developmental period have masculinizing and defeminizing effects on sexual behaviors. Moreover, these differentiating actions of testicular hormones on behavior are independent of genetic sex. Females exposed perinatally to testosterone will show more male mating responses and fewer female mating responses in adulthood. Similarly, males deprived of perinatal testicular stimulation display limited male mating behavior but high levels of female mating behavior following the appropriate hormone treatment in adulthood. Although these data demonstrate that the development of sexual behaviors can be independent of genetic sex, the genotype does play an important role in normal sexual differentiation.

Grumbach and Conte (1985) estimate that at least 30 specific genes located on sex chromosomes and on autosomes are involved in sexual differentiation. Table 1 lists some of the ways that genes are involved in normal sexual development. For example, the sex chromosomes are responsible for formation of the gonads. In the presence of the Y chromosome, testes develop, which, in turn, secrete humoral substances (e.g., testosterone) responsible for the differentiation of sexual behaviors. The synthesis of testosterone and other steroids requires specific enzymes controlled by genes located on the autosomes. Furthermore, genes on the X chromosome code for specific protein receptors necessary for androgen action. Since

KATHIE L. OLSEN   Behavioral Neuroendocrinology Program, National Science Foundation, Washington, D.C. 20550.

Sexual Differentiation, Volume 11 of *Handbook of Behavioral Neurobiology*, edited by Arnold A. Gerall, Howard Moltz, and Ingeborg L. Ward, Plenum Press, New York, 1992.

TABLE 1. SOME GENETIC COMPONENTS INVOLVED IN SEXUAL DIFFERENTIATION[a]

Location of genes	Gene function	Result
Y chromosome (H-Y antigen?)	Testicular differentiation	Functional testes
X chromosome	Specific receptor proteins for steroid hormones	Target tissues respond to androgens
Autosomes	Enzymes in steroid biosynthetic pathways	Testosterone biosynthesis (as well as other steroids)
	5α-Reductase	Dihydrotestosterone is formed from testosterone
(Autosomal or X-linked)	(Receptors for anti-Müllerian hormones) (Biosynthesis of Müllerian inhibiting factor)	Müllerian structures involute

[a] Adapted from Grumbach and Conte (1985), Imperato-McGinley (1983), and Jost (1985).

the genotype influences the type and amount of hormone as well as the responsiveness of the target tissues, it plays a critical role in events underlying sexual behavior differentiation.

In this chapter, the various contributions the genotype makes to normal mechanisms underlying the development of sexual behavior are examined. Normal sexual differentiation is briefly reviewed, and then the use of genetic mutations (Tables 2 and 3) and strain comparisons as tools to investigate the normal developmental mechanisms is examined.

NORMAL SEXUAL DEVELOPMENT

Sexual differentiation of both peripheral and neural systems involves a series of sequential and orderly events. Disruption of the series at any stage by genetic or environmental factors can profoundly influence later developmental events. As illustrated in Figure 1, there are four major aspects of normal development: (1) genetic sex, (2) gonadal sex, (3) phenotypic or body sex, and (4) brain sex. More detailed descriptions of these events are found in several recent reviews (Crews & Moore, 1986; Grumbach & Conte, 1985; Jost, 1983, 1985; Martin, 1985; Wilson, George, & Griffin, 1981; Wilson, Griffin, George, & Leshin, 1981).

GENETIC SEX

Genotypic influence begins at the moment of conception with the establishment of genetic sex. Among mammals, each individual has a set number of paired autosomes and two sex chromosomes. The sex chromosomes determine whether the bipotential gonads develop into testes (XY, heterogametic) or ovaries (XX, homogametic). For ovarian development, two intact X chromosomes are important; individuals with XO karyotype (Turner syndrome) have incomplete ovarian development. For testicular development, the presence of the Y chromosome is necessary. Individuals with two or more X chromosomes existing along with the Y chromosome (Klinefelter syndrome or one of its variant forms) have functioning testes.

GONADAL SEX

In the presence of the Y chromosome, testes begin to develop. Initially, the seminiferous tubules (Sertoli cells) and then interstitial (Leydig) cells are formed. In the absence of the Y chromosome, the bipotential gonads develop into ovaries. Although the testes and ovaries form from a common anlage, the testis begins differentiating earlier than the ovary.

Until recently, it is not clear what induces the anlage to differentiate into testes. Jost, Vigier, Prepin, and Perchellet (1973) proposed that a nonandrogenic substance controlled by the Y chromosome and produced locally within the gonadal cells serves as the testicular organizer. H-Y antigen was suggested to be that inducer substance (Wachtel, Ohno, Koo, & Boyse, 1975). Evidence to support H-Y antigen's involvement in testicular differentiation has been extensively reviewed (see reviews: Müller, 1985; Ohno, 1979; Wachtel, 1983) and, therefore, is only briefly summarized. Historically, the term H-Y antigen was used by Eichwald and Silmser (1955) to describe the presence of a histocompatibility antigen in inbred male mice that led to the rejection of skin grafts by female mice of the same strain. Following the development of an assay to measure this histocompatibility factor serologically (Goldberg, Boyse, Bennett, Scheid, & Carswell, 1971), H-Y antigen was shown to be present on all cell surfaces of XY males except those of immature germ cells; it is absent from cells of females. Moreover, it was found not only in inbred mouse strains but also in the heterogametic sex of a wide range of vertebrates, indicating phylogenetic conservation of the H-Y gene (Wachtel, Koo, & Boyse, 1975). The H-Y antigen is present prior to testicular differentiation, and antibodies against it prevent testicular formation. Finally, its serologically detected levels correlate with the number of Y chromosomes. These observations suggest that H-Y antigen is the testicular differentiating substance (Ohno, 1979; Wachtel, 1983).

Although a considerable amount of data supported the testicular-organizing role of H-Y antigen, most of this evidence was indirect. Conclusive findings were not forthcoming. Consequently, a number of investigators began questioning whether H-Y antigen is necessary for testicular development or whether its presence results from its linkage with the Y chromosome (Andersson, Page, & de la Chapelle, 1986; Eicher, 1982; Johnson, Sargent, Washburn, & Eicher, 1982; Page, de la Chapelle, & Weissenbach, 1985; Page, Mosher, *et al.*, 1987; Silvers, Gasser, & Eicher, 1982; Simpson *et al.*, 1987). For instance, testicular-feminized (*Tfm/Y*) mice are resistant to their testicular androgens and consequently develop an external female phenotype. H-Y antigen is present in these mutants, indicating that the antigen is not associated with androgen responsiveness (Bennett *et al.*, 1975). However, these data do not distinguish between Y linkage of the H-Y gene and its possible role as the inducer substance. The sex-reversed (*Sxr*) mutation in mice initially provided powerful support for the H-Y antigen hypothesis, since XX,*Sxr* and XO,*Sxr* mice have masculine phenotypes, testes, and detectable levels of H-Y antigen (Bennett *et al.*, 1977; Wachtel, 1983). However, it is now clear that a fragment of the Y chromosome originating in the sex-determining region is attached to the X chromosome of the XX,*Sxr* male mice (Evans, Burtenshaw, & Cattanach, 1982; McLaren, 1983; Singh & Jones, 1982). Thus, much of the evidence for H-Y antigen being an inducer

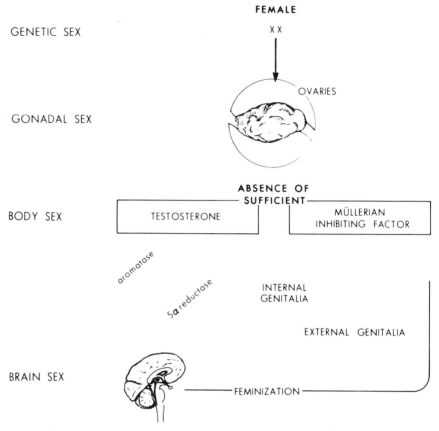

Figure 1. Schematic view of the sequential events in sexual differentiation.

was only correlational and, therefore, did not separate the possibility that its presence is only circumstantial, resulting from its Y linkage.

Another controversial area was the relationship between XY females and ovarian tissue. Support for H-Y antigen as the inducer comes from the findings that within a population of Scandinavian wood lemming, *Myopus schisticolor,* certain XY individuals develop ovaries and are H-Y antigen-negative (Haseltine & Ohno, 1981). However, its inducer role became questionable in view of other findings that demonstrate the existence of a mutant strain of XY female mice that have both H-Y antigen and ovarian tissue (Eicher, 1982; Johnson *et al.,* 1982). Another equivocal set of data is derived from patients with certain abnormal sexual differentiation syndromes. Earlier work indicated a strong correlation between the presence of detectable levels of serological H-Y antigen and testicular tissue, even in the absence of a "detectable" Y chromosome or its fragment (Haseltine & Ohno, 1981; Müller, 1985; Petit *et al.,* 1987; Wachtel, 1983). With the advent of more sophisticated molecular techniques involving Y-chromosomal DNA probes, Page and colleagues report that a small translocation of the Y-chromosomal material has been detected in "most" human XX males (Andersson *et al.,* 1986; de la Chapelle, Tippett, Wetterstrand, & Page, 1984; Page *et al.,* 1985; Page, Brown, & de la Chapelle, 1987). Others have also detected Y-chromosomal fragments attached to XX chromosomes of males (Müller *et al.,* 1986). Recently, Page and his associates (Page, Brown *et al.,* 1987; Simpson *et al.,* 1987) found that the Y-chromosome fragment present in XX

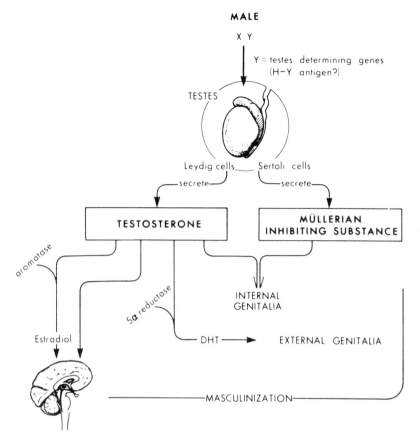

Figure 1. (*continued*)

men and lacking in XY females is in the deletion interval 1A. This fragment associated with the testes-determining factor is located on the distal short arm (in intervals 1A1 and 1A2), whereas the gene for the H-Y antigen is found near the centromere or on the long arm of the Y chromosome. They were able to clone the gene(s?) within this region that contains the testis-determining factor. Based on the identified DNA sequences, it appears that this factor is a protein, and Page, Mosher *et al.* (1987) propose that the encoded protein binds to DNA or RNA and regulates their transcription.

The exciting findings by Page and his colleagues provided the additional support against the role of H-Y antigen and started the search for the gene that might serve as the testicular differentiating factor. A group of scientists at the MRC National Institute for Medical Research and the Imperial Cancer Research Fund in London carried out a series of elegant studies that initially identified a sex-determining gene on the Y chromosome referred to as *SRY* in humans and *Sry* in mice (Berta, Hawkins, Sinclair, Taylor, Griffiths, Goodfellow, and Fellous, 1990; Gubbay, Collignon, Koopman, Capel, Economou, Munsterberg, Vivian, Goodfellow, and Lovell-Badge, 1990; Koopman, Münsterberg, Capel, Vivian, and Lovell-Badge, 1990; Sinclair, Berta, Palmer, Hawkins, Griffiths, Smith, Foster, Frischauf, Lovell-Badge, and Goodfellow, 1990) and then, by preparing *Sry* transgenic mice, demonstrated that this 14-kb genomic fragment induces the indifferent gonad to develop into the testes (Koopman, Gubbay, Vivian, Goodfellow, and Lovell-Badge, 1991). Based on the accumulated evidence, these investigators suggest that the *Sry* is the

Tdy (testis-determining gene on the Y chromosome). Work is now underway to identify other genes that are vital to the role that *Sry* plays in the differentiation of the testes.

PHENOTYPIC OR BODY SEX

DIFFERENTIATION OF INTERNAL GENITALIA.   In both sexes, the anlagen for the Müllerian duct and the Wolffian duct systems are present during early development. The Müllerian ducts develop into the uterus, oviducts, and the upper portion of the vagina, whereas the Wolffian ducts develop into the vas deferens, epididymis, and seminal vesicles.

Our understanding of the role of hormones in differentiation was strongly influenced by the elegant studies carried out by a number of scientists including Burns, Lillie, Jost, and Witschi (see reviews: Burns, 1961; Jost, 1953, 1985; Jost *et al.*, 1973; Jost & Magre, 1984). In Jost's early work, the gonads were removed from intrauterine rabbit fetuses at various developmental stages. If gonadectomies were performed prior to the onset of differentiation of the internal and external genitalia, female phenotypes developed regardless of the genetic sex. Based on these findings, Jost (1953) proposed that testicular secretions induced masculinization of the genitalia and that female differentiation did not depend on the presence of ovaries. In another experiment, testes were removed unilaterally from rabbit fetuses. On the side without the testis, an oviduct developed, whereas on the side with the testis, normal regression of the Müllerian system occurred; the Wolffian duct system was stimulated on both sides. Moreover, normal testicular grafts implanted in female rabbit fetuses promoted masculinization of the internal genitalia, whereas crystalline testosterone propionate (TP) placed in the same location as the grafts did not induce reabsorption of the Müllerian duct structures.

Jost deduced from these studies that two secretions from the fetal testes are critical for male development: a Müllerian-inhibiting substance and an androgen. Indeed, it has been shown that the Sertoli cells within the seminiferous tubules secrete a Müllerian-inhibiting substance that is a nonsteroidal, high-molecular-weight polypeptide. The substance acts locally to cause the regression of the Müllerian duct primordia (Josso, Picard, & Tran, 1977). The reabsorption of the Müllerian ducts occurs prior to stimulation of the Wolffian duct system and requires the presence of high levels of testosterone and its androgen receptor. Testosterone is synthesized by and secreted from the Leydig cells, which are activated initially by chorionic gonadotropin of placental origin and later by luteinizing hormone (LH) emanating from the fetal hypothalamic–pituitary axis. The ovaries in females do not secrete these substances, thereby promoting the development of the Müllerian duct system and regression of the Wolffian duct system.

DIFFERENTIATION OF EXTERNAL GENITALIA.   The embryonic anlagen of the external genitalia of both sexes are initially identical and have the capacity to differentiate into structures normal for either sex. The primordial structures consist of (1) the genital tubercle, which becomes the penis in males and the clitoris in females; (2) the urethral genital folds, which become the corpus spongiosum (penile shaft) in males and the labia minor in females; and (3) the labioscrotal (genital) swellings, which become the scrotum in males and labia major in females.

Masculinization of the external genitalia requires the presence of dihydrotestosterone (DHT), which is formed enzymatically from testosterone within the target tissue. Unlike the primordial structures of the internal genitalia, the anlagen for the external genitalia possess high levels of 5α-reductase. This enzyme is present even before the testes begin to secrete testosterone (Siiteri & Wilson, 1974). Masculinization of the external genitalia as well as the Wolffian-derived internal structures requires the presence of specific androgen receptors, which are large proteins found in the cytoplasm and nucleus of the target cells (Wilson, 1978; Wilson, Griffin, Leshin, & MacDonald, 1983). In the accepted model of steroid action, the binding of the hormone to its receptor is the initial step. This hormone–receptor complex is then "activated" and binds to acceptor sites on the DNA (although binding can also occur to the nuclear matrix and nuclear membranes). This interaction of the hormone–receptor complex with the chromatin material initiates mRNA transcription that ultimately leads to translation of specific proteins. Thus, steroid hormones exert their effects through an alteration of gene expression (see Clark, Schrader, & O'Malley, 1985). In this case, testosterone and DHT bind to specific receptors to induce the differentiation of primordial structures in the masculine direction. In the presence of ovaries, a condition associated with a lack of DHT stimulation, female external genitalia develop from the primordial structures.

BRAIN SEX

Initially, the neural regions controlling reproductive and pituitary function are bipotential. During a critical developmental period, humoral substances secreted by the perinatal testes act on the developing CNS to permanently alter its function (e.g., Barraclough & Gorski, 1962; Gerall & Ward, 1966; Harris, 1964; Phoenix, Goy, Gerall, & Young, 1959). The presence of testicular hormones causes masculinization (enhancement of male-type sexual responses, e.g., mounts, intromissions, ejaculation) and defeminization (suppression of female-type sexual responses, e.g., lordosis, ear wiggle, hop–dart) of CNS tissue mediating sexual behavior. Thus, normal adult male rats display the complete pattern of male sexual behavior and little or no female sexual behavior. Females treated with testosterone or TP during perinatal development show elevated levels of masculine sexual responses and reduced levels of feminine sexual responses in adulthood (e.g., Phoenix *et al.*, 1959; Whalen & Edwards, 1967). The absence of testicular secretions causes the converse —demasculinization and feminization. Females and neonatally castrated males exhibit limited male copulatory behaviors after TP treatment and display lordosis after estradiol benzoate (EB) plus progesterone treatment in adulthood (e.g., Grady, Phoenix, & Young, 1965; Whalen & Edwards, 1967). A critical point in understanding sexual differentiation is that masculinization and defeminization are viewed as two independent, although correlated, processes (Goy & Goldfoot, 1975; Reinisch, 1976; Whalen, 1974, 1982).

Although it is clear that testosterone induces the masculinization of internal genitalia and DHT promotes masculinization of the external genitalia, the testicular substance involved in masculinization and defeminization of the CNS is still speculative. Originally, it was assumed that testicular androgens were the active agents. Indeed, androgen receptors are present in the perinatal brain of a number of species including hamsters, ferrets, mice, rats, and monkeys (e.g., Attardi & Ohno, 1976; Fox, Vito, & Wieland, 1978; Kato, 1976; Pomerantz, Fox, Sholl, Vito, & Goy, 1985; Sheridan, Sar, & Stumpf, 1975; Vito, Baum, Bloom, & Fox, 1985; Vito, Wieland, & Fox, 1979). However, the primacy of androgens was questioned when it became apparent that testosterone can be aromatized into estrogens within neural regions critical for reproductive behaviors (McDonald *et al.*, 1970; Naftolin *et al.*, 1975). Not only can estrogen be formed from androgen in the perinatal CNS (Reddy, Naftolin, & Ryan, 1974; Weisz & Gibbs, 1974), but the putative estrogen receptors are also present (e.g., Barley, Ginsburg, Greenstein, MacLusky, & Thomas, 1974; Kato, Atsumi, & Inaba, 1971; MacLusky, Chaptal, Lieberburg, & McEwen, 1976; Plapinger & McEwen, 1973; Pomerantz *et al.*, 1985; Vito & Fox, 1979, 1982). Our current understanding is that androgens *per se* may have a role in masculinization, but their conversion into estrogen is required for defeminization (Baum, 1979; Feder, 1984; Olsen, 1983, 1985; Whalen, 1982). However, the role of androgens and their estrogenic metabolites in CNS differentiation may differ depending on the species, the strain, or the specific sexual dimorphism being studied, and this is discussed fully in another chapter.

USE OF GENETIC DISORDERS OF SEXUAL DIFFERENTIATION

Studying genetic disorders of sexual differentiation has revealed that at least 30 specific genes found on the X and Y chromosomes and on the autosomes are involved in sexual differentiation (Grumbach & Conte, 1985; also see Table 1). A genetic defect at any stage of the process can have a profound effect on peripheral organs and possibly neural development. Some of the genetic disorders found in humans are particularly suited to evaluate the effects of prenatal hormones on psychosexual differentiation, since many of these defects influence levels of androgens during fetal development (Table 2).

As pointed out in Table 1, testosterone synthesis is under the control of genes

TABLE 2. CLASSIFICATION OF GENETIC DISORDERS DISCUSSED
WITHIN TEXT[a]

Disorder	Example
1. Chromosomal and/or H-Y antigen defects	
A. Genetic recombinations	Sex-reversed *Sxr* mice
B. Too few chromosomes	Turner's syndrome, XO
C. Too many chromosomes	XXY
D. XY individual with ovaries	Scandinavian wood lemmings, mutant mouse strains
2. Excessive androgen stimulation	
A. Enzymatic defects resulting in overproduction of adrenal androgens	Congenital adrenal hyperplasia
3. Deficient androgen stimulation	
A. Enzymatic defects resulting in decrease of testosterone synthesis	3β-hydroxysteroid dehydrogenase deficiencies
B. Enzymatic defects resulting in inadequate formation of DHT from testosterone	5α-reductase deficiencies
C. Target organs unresponsive to androgens	Testicular feminization syndrome, *Tfm*, and its variant forms
D. Hypothalamic or pituitary defects that influence testicular androgen levels	Hypogonadism, *hpg*; obese *ob/ob* mice
E. Deficiency in gonadotropins	

[a] Adapted from Blizzard (1979) and Grumbach and Conte (1985).

thought to be located on the autosomes (Forest, 1981). These genes regulate enzymes in the steroid biosynthetic pathway such that a defect in one particular enzyme can influence the synthesis of several hormones, resulting in decreases or increases in their circulating plasma levels. For example, congenital adrenal hyperplasia is a family of disorders inherited as an autosomal recessive trait that is caused by deficiencies in adrenal steroidogenesis (New, Dupont, Grumbach, & Levine, 1983). Depending on which enzyme is defective, individuals can have impaired cortisol secretion and an overproduction of adrenal androgens and androgen precursors. A secondary consequence is that affected females are exposed to high levels of androgens *in utero*. Since excessive levels of testosterone have major effects on brain development and behavior in lower species, the psychosexual development of these individuals is critically important and has been extensively studied (Ehrhardt and Meyer-Bahlburg, 1981).

In general, the gender identity corresponds with sex of rearing. Money and Daléry (1976) reported that 46,XX individuals reared as boys adopt a male gender identity and, although sterile, marry and maintain a heterosexual life as a man. In contrast, the gender roles of those 46,XX individuals reared as females differ from those of matched controls in that these females exhibit behavioral patterns referred to as tomboyism, which is characterized by high energy expenditure (Ehrhardt, Epstein, & Money, 1968). More recently, Money, Schwartz, and Lewis (1984) reported that prenatal hormonal masculinization may influence erotosexual imagery or experience, since a higher percentage of these affected females were bisexual or predominantly homosexual.

Insufficient androgenization has been examined in individuals with the androgen insensitivity (testicular feminization) syndrome. This syndrome is discussed in more detail in the next section. Briefly, these 46,XY males have testes, normal testosterone production, and a female external phenotype. They are typically reared as females and adopt a feminine gender identity and role, resulting in a heterosexual life as a woman (Masica, Money, & Ehrhardt, 1977; Money, Ehrhardt, & Masica, 1968; Money *et al.*, 1984). In addition, there are also genetic males with enzymatic defects (e.g., deficiency in 3β-hydroxysteroid dehydrogenase, 17α-hydroxylase, or 17β-hydroxysteroid dehydrogenase) that lead to a decrease in the synthesis of testosterone *in utero* (Forest, 1981; Grumbach & Conte, 1985; Peterson & Imperato-McGinley, 1984). Possibly, the effects of insufficient androgenization on psychosexual development could also be examined in these affected individuals.

Genetic defects have also been detected in 5α-reductase, the enzyme that converts testosterone into DHT. Males having this autosomal recessive trait are exposed *in utero* to testosterone but, because of inadequate formation of DHT, develop a female external phenotype. Imperato-McGinley and colleagues (Imperato-McGinley, 1983; Imperato-McGinley, Peterson, Gautier, & Sturla, 1979) have studied a group of 5α-reductase-deficient males who were raised as girls until puberty but then were virilized and adopted a male gender identity and male gender role. These researchers have used this genetic mutation found among families in the Dominican Republic to argue that exposure of the brain to testosterone *in utero*, in combination with androgenic exposure at puberty and in adulthood, is important for normal psychosexual development.

Although the evidence accumulating from various clinical syndromes suggests that prenatal exposure to gonadal hormones influences psychosexual development, postnatal rearing history as well as other social factors are also extremely important. However, it is inherently difficult to dissociate endocrine and social factors in human behavior. One strategy has been to study similar genetic mutations in animals, in which the confound of social rearing presumably is not an issue. Moreover, this provides a means to directly study neural mechanisms that may prove important. Several mutations characterized in lower animal species (Table 3) are described in the following section.

TESTICULAR FEMINIZATION SYNDROME (*Tfm* AND *tfm*)

The *Tfm* mutation has been invaluable for elucidating the molecular basis of sexual dimorphism (Bardin & Catterall, 1981) as well as for identifying the hormones mediating sexual differentiation of brain and behavior (Arnold & Gorski, 1984; Olsen, 1979a). The testicular feminization syndrome was initially described in humans (Morris, 1953) and later in rats (Stanley & Gumbreck, 1964; Stanley, Gumbreck, Allison, & Easley, 1973), mice (Lyon & Hawkes, 1970), cattle (Nes, 1966), and chimpanzees (Eil *et al.*, 1980). It is caused by an X-linked recessive gene defect. Since males have only one X chromosome, the mutation is present in all cells. Females have two X chromosomes, one of which is randomly inactivated during development, resulting in normal cells and some cells that express the mutation (Lyon, 1961, 1972). Differences in the number of normal (wild-type) versus mutant cells influence the expression of the mutation in affected females.

Males with this inherited X-linked mutation are insensitive to androgens secreted by their testes and develop phenotypically as females. These individuals have a blind vaginal pouch and a clitoris instead of a scrotum and a penis. The testes are

TABLE 3. SUMMARY OF GENETIC DISORDERS

Genetic disorders	Genotype	Gonads	External phenotype	Description
Testicular feminization (*Tfm*) mutation Humans: Androgen-insensitivity syndrome Rats: *tfm* or Stanley–Gumbreck androgen-insensitive Mice: *Tfm* or androgen-insensitive	XY	Small testes	Feminine	Deficiency in androgen receptor; some cases, receptor protein abnormal
Sex-reversed (*Sxr*) mutation XX*Sxr* mice XO,*Sxr* mice	XX XO	Small testes Normal-size testes	Masculine Masculine	Recombination between X and abnormal Y chromosome with *Sxr* genes
Hypogonadism (*hpg*) XY, males (e.g., mice, humans) XX females (e.g., mice)	XY XX	Small, infantile testes Infantile ovaries	Masculine Feminine	Deficiency in gonadotropin-releasing hormones with secondary effects on pituitary LH and FS and, thus, gonadal hormones

located in the inguinal canal, but since these males are not insensitive to Müllerian inhibiting substance, both the Müllerian and Wolffian structures fail to develop. The adult hormonal profile is characterized by high levels of plasma LH, testosterone, and estradiol (Lacroix, McKenna, & Rabinowitz, 1979; Naftolin *et al.*, 1983; Purvis, Haug, Clausen, Naess, & Hansson, 1977). This may be species dependent, since, unlike the other species, *Tfm* mice have lower plasma testosterone levels than wild-type male littermates (Amador, Parkening, Beamer, Bartke, & Collins, 1986; Goldstein & Wilson, 1972; K. L. Olsen & B. A. Gladue, unpublished observations).

It is generally believed that the inherited nonresponsiveness to androgens results from a severe deficiency in the number of androgen receptors. Androgen binding in the brain is reduced by 80–90% in both *tfm* rats (Fox, Olsen, Vito, & Wieland, 1982; Krey, Lieberburg, MacLusky, Davis, & Robbins, 1982; Naess *et al.*, 1976) and *Tfm* mice (Attardi, Geller, & Ohno, 1976; Fox, 1975; Wieland & Fox, 1979; Wieland, Fox, & Savakis, 1978; MacLusky, Luine, Gerlach, Fischette, Naftolin, and McEwen, 1988) as compared to wild-type male siblings. A reduction in androgen binding associated with the *Tfm* mutation has also been detected in other organs, such as kidney (e.g., Bardin, Bullock, Sherins, Mowszowicz, & Blackburn, 1973; Bullock & Bardin, 1974; Wieland & Fox, 1981), submaxillary gland (e.g., Verhoeven & Wilson, 1976), and muscle (Max, 1981). Carrier females show intermediate levels of receptor activity and androgen responsiveness resulting from increased variability in X inactivation (Bullock, Mainwaring, & Bardin, 1975; Ohno & Lyon, 1970; Tettenborn, Dofuku, & Ohno, 1971). Cultured skin fibroblasts from human patients with complete androgen insensitivity also have reduced androgen binding (Keenan, Meyer, Hadjian, Jones, & Migeon, 1974; Migeon, Amrhein, Keenan, Meyer, & Migeon, 1979; Migeon, Brown, Axelman and Migeon, 1981; Pinsky, 1981; Wilson *et al.*, 1983), whereas carrier females show intermediate levels of activity (Meyer, Migeon, & Migeon, 1975). The *Tfm* mutation is selective for androgen putative receptors, since estrogen binding activity appears to be normal (Fox, 1975; Fox *et al.*, 1982; Krey *et al.*, 1982; MacLusky *et al.*, 1988; Olsen & Fox, 1981; Olsen & Whalen, 1982; Sheridan, 1978).

It has been concluded that the *tfm* locus on the X chromosome is either the structural gene for the receptor or a regulatory gene that controls its activity (Bardin & Catterall, 1981; Ohno, 1979). The gene is located between Xq13 and Xp11 on the human X chromosome (Migeon *et al.*, 1981). Recently, the complementary DNAs (cDNAs) encoding the androgen receptors have been identified (Chang, Kokontis, and Liao, 1988; Lubahn, Joseph, Sullivan, Willard, French, and Wilson, 1988). Using this information, Lubahn and colleagues found that it is the structural gene for the androgen receptor that is located near the centromere between Xq13 and Xp11. Moreover, cloning of the androgen receptor cDNAs has provided a means to examine the molecular defects associated with this mutation. In families with complete androgen insensitivity, both a partial deletion (Brown, Lubahn, Wilson, Joseph, French, and Migeon, 1988) and a single base mutation (Lubahn, Brown, Simental, Higgs, Migeon, Wilson, and French, 1989) in the androgen receptor domain have been described. Comparison of *tfm* rats with wild-type males revealed a single base change in the steroid binding domain of the androgen receptor (Yarbrough, Quarmby, Simental, Joseph, Sar, Lubahn, Olsen, French, and Wilson, 1990). Specifically, a single transition mutation changed arginine 734, which is highly conserved within the family of nuclear receptors, to glutamine. In *Tfm* mice, a single base deletion in the N-terminal region of the gene for the androgen receptor is likely to be responsible for the androgen-insensitivity (He, Young, and Tindall,

1990; Charest, Zhou, Lubahn, Olsen, Wilson, and French, 1991). Consequently, there is a shift in the reading frame, and both premature termination of transcription and reduced androgen receptor mRNA levels. Thus, the *Tfm* mutation has proven invaluable for determining the locus of the androgen receptor and for demonstrating the critical role that genes on the X chromosome play in androgen action.

The *tfm* mutation also provides an excellent means to separate androgen- and estrogen-regulated functions. Indeed, this mutation has been utilized to identify which steroid is involved in the differentiation of sexually dimorphic brain structures (Arnold & Gorski, 1984; Breedlove, 1986), the differentiation of sexually dimorphic nonreproductive behaviors (Meaney, Stewart, Poulin, & McEwen, 1983; Shapiro & Goldman, 1973), as well as the induction of various hormone-regulated enzymes and proteins in adults (Bardin & Catterall, 1981; Luine, MacLusky, & McEwen, 1979; MacLusky *et al.*, 1988; Olsen, Heydorn, Rodriguez-Sierra, and Jacobowitz, 1989). Probably the most extensive use of the *Tfm* mutation has been in examining the role that androgens and estrogens play in the development of sexual behavior. The rest of this section is devoted to summarizing the latter topic.

SEXUAL BEHAVIOR: ANDROGEN-INSENSITIVE STANLEY–GUMBRECK (*tfm*) RATS.   Shapiro, Goldman, Steinbeck, and Neumann (1976) reported that Stanley–Gumbreck androgen-insensitive *tfm* rats (herein referred to as *tfm*) were asexual. Although no quantitative data were presented, the investigators stated that gonadectomized *tfm* rats did not mount when injected daily with TP (1,000 µg) and did not exhibit lordosis when administered estradiol benzoate (EB) (200 µg) and progesterone (1,000 µg).

Beach and Buehler (1977) reported that these *tfm* rats showed less masculine sexual behavior than normal wild-type males. However, contrary to Shapiro *et al.* (1976), all their mutant males investigated the genital region of the receptive females, and 91% mounted them. The mutant males were highly variable, with some exhibiting no behavior and others showing the ejaculation pattern. Injections of TP did not facilitate sexual behaviors. Neither genotype readily displayed feminine behavior in response to EB and progesterone injections. Lordosis levels were low, and no proceptive behaviors were observed. Based on these data, Beach and Buehler (1977) postulated that "central neural mechanisms involved in mediation of male behavior are normally organized but markedly insensitive to testosterone; and that mechanisms responsible for manifestations of feminine behavior are like those of the normal male in their relative lack of responsiveness to estrogen and progesterone."

We carried out a series of studies (Olsen, 1979a, 1979b; Olsen & Whalen, 1981) that directly tested this hypothesis. We postulated that the deficiency in copulatory responses exhibited by *tfm* rats either results from an insufficient response of the neural tissues to androgens during adulthood (Beach and Buehler's hypothesis) or suggests that the CNS did not differentiate in the masculine direction during development. We assumed that estrogen could distinguish between the two hypotheses, since it induces male copulatory behavior in adult rats (Gorzalka, Rezek, & Whalen, 1975; Södersten, 1973) and the *tfm* mutation does not appear to interfere with its receptors (Olsen & Whalen, 1982). Therefore, if the neural systems underlying male mating behavior are masculinized during development but insensitive to androgens in adulthood, estrogen should stimulate copulatory responses in *tfm* rats. On the other hand, if the neural systems underlying masculine behavior are not differentiated during development, estrogen will not induce these mutants to mount receptive females.

We found that castrated androgen-insensitive *tfm* males exhibited mounting and intromission patterns in response to EB (Olsen, 1979b). Testosterone propionate, an androgen that can be converted to estrogen, was also effective in stimulating masculine behaviors. Dihydrotestosterone, which is not aromatized in the CNS, did not induce mating responses in the mutants, although wild-type males exhibited all aspects of copulatory behavior including the ejaculation pattern (Olsen, 1979b; Olsen & Whalen, 1984). Others have also reported that EB treatment activates mounting behavior in these mutant rats (Shapiro, Levine, & Adler, 1980). Thus, as hypothesized by Beach and Buehler (1977), the neural substrate underlying male mating behavior is masculinized in the *tfm* rats. However, these androgen-insensitive rats show less masculine behavior than wild-type males. This deficiency, as measured by frequency of intromission and ejaculatory patterns, may be a consequence of incomplete masculinization during development. For example, if one assumes that masculinization of the mounting system is independent of masculinization of the ejaculatory component (Feder, 1984; Hart, 1972, 1977; van der Schoot, 1980; Whalen, Yahr, & Luttge, 1985), then it is possible that *tfm* rats are only partially masculinized during development. If true, then the two indices of masculinization may require either different degrees of hormonal stimulation or different hormones.

Although the mutant and wild-type males differ in their expression of male sexual behavior, neither genotype shows lordotic behavior in response to ovarian hormones (Beach & Buehler, 1977; Olsen, 1979a; Olsen & Whalen, 1981; Shapiro *et al.*, 1976). However when *tfm* and wild-type males are castrated at birth, they will display female mating responses in adulthood following treatment with EB and progesterone (Olsen, 1979a) or testosterone and progesterone (Krey *et al.*, 1982). Lordosis is not readily induced in either genotype when they are castrated at birth and immediately exposed to high levels of TP (500 μg) or EB (25 or 250 μg) (Olsen & Whalen, 1981). Thus, like wild-type males, androgen-insensitive *tfm* rats are defeminized by the presence of testes.

SEXUAL BEHAVIOR: *Tfm* MICE. The sexual behavior of *Tfm* mice has also contributed to our understanding of the hormones involved in the sexual differentiation process. Ohno, Geller, and YoungLai (1974) reported that *Tfm* mice showed no male-typical or female-typical behaviors. However, the testing and hormonal paradigm utilized was not exemplary of methods normally used in measuring sexual behavior. Since their findings were very intriguing, we decided to replicate and extend their work using the procedures outlined by McGill (1962).

Tfm mice and their littermates were tested for their ability to display masculine behavior when gonadally intact or following castration and hormone replacement (Olsen, 1987). Comparisons were made among *Tfm*/Y (mutant males), *Ta+*/Y males (wild-type males), *Tfm*/+*Ta* (carrier females), and *Ta+*/+*Ta* (wild-type females). These genotypes are easily identified since the *Tfm* mutation is closely linked to the coat color *Ta*. In our paradigm, mice were given eight weekly 1-hour mating tests beginning at 90 days of age. After the second mating test, the mice were gonadectomized and injected daily with either TP (200 μg), EB (1 μg), DHT (200 μg), DHT + EB (200 μg + 1 μg), or oil vehicle. Eight mice were in each hormone treatment group.

The most striking finding was that gonadally intact *Tfm* mice did not mount. In contrast, 85% of the gonadally intact $Ta+/Y$ males mounted, and 50% of these males ejaculated during one of the two mating tests. Even 20% of the females displayed some mounting and intromitting responses toward receptive stimulus mice, even though no attempt was made to test at a particular stage of the estrous cycle. Given these data, two additional studies were conducted with only the Tfm mice. To examine whether experience would influence the probability of displaying masculine behaviors, we tested gonadally intact mutant mice either at weekly intervals for 5 weeks or for five consecutive days. None of these Tfm mice mounted. Thus, gonadally intact Tfm mice do not readily exhibit male-typical mating behaviors.

Table 4 summarizes the masculine behavior exhibited by the gonadectomized mice given the various hormones. As can be seen, castrated Tfm mice showed little masculine behavior regardless of the hormone treatment. For example, TP induced only one Tfm mouse to mount and intromit, and this mouse only displayed the behavior on one of six mating tests. In contrast, 100% of the wild-type males ($Ta+/Y$), 50% of the wild-type ($Ta+/+Ta$) females, and 75% of the carrier females ($Tfm/$

TABLE 4. HORMONAL INDUCTION OF MASCULINE SEXUAL BEHAVIOR IN GONADECTOMIZED WILD-TYPE (Ta/Y) AND ANDROGEN-INSENSITIVE (Tfm/Y) MALE MICE AND THEIR FEMALE LITTERMATES

| | | Masculine behaviors[b] | | | | | |
| | | Mounts | | Intromissions | | Ejaculations | |
Genotype	Hormone treatment[a]	Percentage displaying behavior	Tests[c]	Percentage displaying behavior	Tests[c]	Percentage displaying behavior	Tests[c]
Tfm/Y (mutant male)	TP	12%	2%	12%	2%	0	0
	EB	37%	19%	25%	16%	25%	6%
	DHT	0	0	0	0	0	0
	EB + DHT	37%	25%	37%	25%	25%	6%
	Oil	0	0	0	0	0	0
Ta/Y (wild-type male)	TP	100%	100%	100%	98%	100%	77%
	EB	100%	89%	100%	79%	75%	25%
	DHT	100%	85%	100%	77%	100%	50%
	EB + DHT	87%	85%	87%	85%	87%	71%
	Oil	75%	67%	75%	52%	62%	17%
Tfm/Ta (carrier female)	TP	75%	64%	75%	56%	0	0
	EB	75%	52%	75%	46%	0	0
	DHT	0	0	0	0	0	0
	EB + DHT	87%	69%	87%	62%	12%	2%
	Oil	12%	2%	0	0	0	0
Ta/Ta (wild-type female)	TP	50%	44%	50%	42%	0	0
	EB	62%	31%	62%	31%	0	0
	DHT	25%	4%	12%	2%	0	0
	EB + DHT	87%	71%	87%	69%	0	0
	Oil	0	0	0	0	0	0

[a] Mice were injected daily for 42 days with one of the following hormones or combinations: TP, 200 μg of testosterone propionate; EB, 1 μg of estradiol benzoate; DHT + EB, 200 μg of dihydrotestosterone in combination with 1 μg of estradiol benzoate; DHT, 200 μg of dihydrotestosterone; Oil, 0.1 cc of peanut oil vehicle.
[b] Six mating tests with estrous females were given at weekly intervals. The test was ended after 1 hour or following an ejaculation.
[c] Percentage of mating tests in which mounts, intromission, or ejaculations occurred for all animals. Total number of tests positive for behavior for each treatment = 48 (8 animals/group × 6 tests).

+Ta) intromitted in response to TP during the mating tests. Not surprisingly, the nonaromatizable androgen DHT was ineffective in stimulating mounting behavior in any of the androgen-insensitive mice. Although the lack of behavior displayed by *Tfm* males may result from insensitivity to androgens in adulthood, only three of eight mice mounted following EB or EB + DHT treatment. Thus, as illustrated in Table 4, *Tfm* mice showed less mating in comparison with their male and female littermates under all hormonal conditions.

Similar to wild-type males, *Tfm* mice showed low levels of female mating responses when castrated as adults and given estrogen alone (Ohno *et al.*, 1974) or estrogen and progesterone (K. L. Olsen, unpublished data). In contrast to the genetic males, ovariectomized $Ta+/+Ta$ and $Tfm/+Ta$ females given estrogen and progesterone displayed lordosis under the same testing conditions (K. L. Olsen, unpublished data). Analyzing fertility, Lyon and Glenister (1980) found no impairment of female reproductive behaviors in the heterozygous $Tfm/+Ta$ female mice.

COMPARISON OF *tfm* RAT AND *Tfm* MOUSE MUTANT: MODEL FOR IDENTIFYING THE DIFFERENTIATING HORMONES. As summarized in Table 5, masculine behavior exhibited by *tfm* rat and *Tfm* mouse mutants is clearly inferior compared with their respective wild-type males. Differences are also apparent between rat and mouse mutants in their potential to display mounting behavior in response to various hormones. For example, 100% of the *tfm* rats mounted and intromitted following treatment with TP, EB, or DHT + EB (Olsen, 1979b), whereas only a few of the *Tfm* mice displayed masculine behavior in response to these hormones (see Table 2). This distinction between the mutants from these two species is even more apparent when one compares the masculine behavior of the gonadally intact *tfm* rats and *Tfm* mice. Unlike rats that exhibit some mounting and intromission responses, *Tfm* mice with testes do not mount. Besides masculine behavior, other differences have been noted between *tfm* rat and mice, leading Bardin and Catterall (1981) to postulate that *tfm* rats are more responsive to androgen than *Tfm* mice. For exam-

TABLE 5. SUMMARY OF THE POTENTIAL OF ADULT *Tfm* AND WILD-TYPE MICE AND RATS TO DISPLAY SEXUAL BEHAVIOR

	Sexual behavior		
	Male[a]		Female[b]
	Intact	Castrate + hormone	Castrate + EB + prog
Mice			
Tfm/Y	——	—— > +	——
$Ta +/Y$	++−	++−	——
Rats			
tfm/Y	++	++	——
Wild type/Y			
(King × Holtzman)	+++	+++	——

[a] Male mating behavior was assessed by measuring the latency and frequency of mounts, mounts with intromissions, and ejaculations. The rodents were tested either with their own gonads (Intact) or following castration and hormone replacement. Hormones administered were testosterone propionate (TP), estradiol benzoate (EB), dihydrotestosterone (DHT), and EB + DHT. —— indicates that rodents rarely displayed the behavior; + indicates that only a few rodents display limited behavior; ++ indicates that rodents mount and intromit but do not ejaculate; ++− indicates some ejaculation; +++ indicates that rodents display full copulatory sequence, including ejaculation.
[b] Lordosis quotient was used to assess female mating behavior in castrated rodents given EB followed by progesterone.

ple, *tfm* rats show some androgen-induced enzyme synthesis (Sherins & Bardin, 1971) and reduced LH levels (Naess *et al.*, 1976) when given large doses of TP or DHT, whereas *Tfm* mice are considered very resistant, since these steroids have only minimal or no effects on androgen-regulated functions (Bardin & Catterall, 1981; Lyon, Hendry, & Short, 1973). The difference in responsiveness is not correlated with the level of androgen binding, since both mutants are severely deficient in the number of androgen receptors.

Analysis of the small number (residual) of androgen receptors in *tfm* target organs has led to the hypothesis that both the quantity and quality of the receptor are important for androgen responsiveness (Fox, Blank, & Politch, 1983; Fox *et al.*, 1982; Olsen & Fox, 1981; Wieland & Fox, 1981). For example, DNA-cellulose elution patterns of putative androgen receptors from CNS can be used to distinguish rat and mouse mutants from one another and from their wild-type controls (Fox *et al.*, 1982, 1983; Olsen & Fox, 1981; Wieland *et al.*, 1978). The elution profile of androgen receptors consists of a major (lower-salt-eluting peak) and a minor (higher-salt-eluting peak) form. In our study (Olsen & Fox, 1981), androgen receptors from *tfm* rat brain eluted with the same profile as, but at a uniformly lower level than, that of wild-type rats. In contrast, the few androgen receptors from the CNS of *Tfm* mice eluted with an altered pattern as compared with wild-type mice. The major form found in wild-type male mice was not detectable by our procedures in *Tfm* mice, and the minor form was reduced in the mutants by 80–90%. Others have also reported a qualitative defect in the androgen receptor protein (Young, Johnson, Prescott, and Tindall, 1989). Moreover, Fox, *et al.* (1983) examined the androgen-binding activities in several androgen-resistant mutants that differ in degree of androgen responsiveness and have found at least three different types of residual receptors.

These data suggest that both the quantity and the quality of the androgen receptor may be critical for the action of the hormone. Qualitative defects in androgen receptors from skin fibroblasts of affected patients have also been demonstrated (Eil, 1983; Fox *et al.*, 1983; Griffin, 1979; Kaufman, Pinsky, Simard, & Wong, 1982). The data indicate that in some cases the *tfm* mutation affects the stability of the androgen receptor. Defects in stability may prevent a functional interaction among the androgen, its receptor, and DNA binding components within the genome, thus preventing androgen action. Such qualitative differences in androgen receptor could explain why patients with similar levels of binding activity or with no apparent reduction in the number of receptors may differ in their responsiveness to androgens (e.g., Pinsky, 1981; Wilson *et al.*, 1983).

Quantitative defects could also explain the differences in masculine behaviors exhibited by *tfm* rats and *Tfm* mice. In this case, the mutations differentially affect specific characteristics of androgen receptors, which, in turn, cause differential androgen responsiveness. For example, the neural system of rats may be partially masculinized by androgens acting via the few receptors found in the CNS. The masculinization is not complete because of the deficiency in the overall number of receptors. Thus, low levels of androgen stimulation cause masculinization of the mounting and intromission systems, but higher levels of androgen exposure may be required for the development of the ejaculation responses. In contrast, the qualitative defect superimposed on the quantitative defect may render *Tfm* mice almost totally insensitive to androgens, and, thus, they are not masculinized during development. This interpretation predicts that androgens mediate masculinization of neural systems. An alternative explanation is that the *tfm* mutations are the same in both species, but different steroids mediate masculinization in rats and mice. Estrogen derived from circulating testicular androgens causes masculinization of the neural systems controlling mounting and intromission patterns in rats but not in mice. The behavioral deficiency observed in *tfm* rats as compared with wild-type males suggests that both androgens and estrogens are necessary for complete masculinization. Regardless of which hypothesis proves to be correct, androgens are involved in the masculinization process.

The *Tfm* mice and rats do not show lordosis. Although comparable studies have not been completed in mice, *tfm* rats are defeminized by the presence of the testes. Therefore, this process is similar to that of wild-type or normal males. Aromatase activity has not been measured directly in perinatal *tfm* males; however, older *Tfm* mice metabolize androgens into estrogens within hypothalamic tissue (Naftolin *et al.*, 1975; Rosenfeld, Daley, Ohno, & YoungLai, 1977). Adult *tfm* rats can convert androgens into estrogens, since testosterone injections increase measurable estrogen receptor levels in hypothalamic nuclei, and neonatally castrated *tfm* rats given TP in combination with progesterone in adulthood display lordosis (Krey *et al.*, 1982). These mutants provide strong support for the hypothesis that estrogens, derived intracellularly from circulating androgens, interact within CNS target tissues to suppress the development of female reproductive behaviors in rats and mice.

In summary, the *Tfm* mutation has demonstrated the importance of genes on the X chromosome for sexual differentiation. Moreover, the genetic defect has provided a powerful means to examine the contribution of androgens to this process. Since testosterone, the major testicular hormone, is readily converted into estrogen within CNS neurons, it is difficult to distinguish androgen-regulated from estrogen-regulated functions in normal animals. The *Tfm* mutation allows the action of these two hormones to be separated. Identifying the hormone or combination of hormones that bring about behavioral masculinization and behavioral defeminization is essential for our eventual understanding of the neural mechanism underlying sexual differentiation.

SEX-REVERSED (*Sxr*) MUTATION

Another abnormal genetic syndrome that has contributed to our understanding of the normal events underlying sexual differentiation is the sex-reversed *Sxr* mutation. First described by Cattanach, Pollard, and Hawkes (1971), this mutation is thought to be a dominant autosomal gene that causes genetic females to develop as phenotypic males. The XX sex-reversed (*Sxr*) mice are normally masculinized except for small testes that lack germ cells. Another variant, the XO *Sxr* males, are masculinized and have normal-size testes capable of producing sperm; however, they are sterile because the sperm are morphologically abnormal and immobile. The mutation is caused by a recombination between the X and an abnormal Y chromosome in male mice carrying the *Sxr* gene. As a result a fragment of the Y chromosome is transferred to the distal region of the X chromosome (Evans *et al.*, 1982; McLaren, 1983; Singh & Jones, 1982). Eicher and Washburn (1986) suggest that these mice may share some similarities with the human clinical syndrome of XXY males.

Both XX and XO male mice display copulatory behavior toward females but are sterile (Cattanach *et al.*, 1971). Even though circulating testosterone levels are lower in adult *Sxr* males than normal males derived from the same stock, these mutant mice exhibited similar masculine mating behavior as compared with normal male littermates (Ohno *et al.*, 1974). In addition, a massive single injection of testosterone or estradiol can maintain and induce male behavior in XX,*Sxr*/+ males (Ohno *et al.*, 1974). Some behavioral differences between the mutant and control males were noted when the investigators measured aggressiveness toward an intruder. Unlike normal male siblings, XX,*Sxr*/+ mice were mounted instead of being attacked by the resident males, a response that may be related to their low levels of circulating testosterone. It is not clear whether the mutant mice exhibited lordosis when mounted. Careful analyses of the potential of XX and XO *Sxr* males to display male and female mating responses have not been carried out, but the available data suggest that *Sxr* males are normally masculinized during perinatal development.

The *Sxr* mutation has been combined with the *Tfm* mutation to address questions about androgen action during development (Ohno *et al.*, 1974; Breedlove, 1986). The *Sxr* carrier males and *Tfm* carrier females were mated to obtain offspring that carried both traits. In the study by Ohno and colleagues (Ohno *et al.*, 1974), the carrier mice also possess the (O^{hv}) and (O^+) alleles, which in (O^{hv})/(O^+) heterozygotes inhibit the random X inactivation, so that the (O^{hv})-carrying X is preferentially activated. Thus, their mating scheme provided a mechanism for maximizing the number of cells containing the X-linked mutation. Ohno *et al.* (1974) found that, unlike the *Tfm* mice in their study (see also Tables 2 and 3), some of the gonadally intact *Tfm*(O^{hv})/+(O^+),*Sxr*/+ males were capable of showing masculine sexual behavior. Moreover, administering TP induced copulatory behavior in all *Tfm/Sxr* XX mice. These data were not predicted, since the genetic mating scheme combined with the predominately female phenotype predicts that the X-linked mutation would be expressed in the majority of the cells. Based on these data, it was suggested that the "centers for neuronal control of sexual behavior contain sufficient redundancy, so that the control cannot be abolished until a center involved becomes effectively *Tfm* monoclonal" (Ohno *et al.*, 1974). This idea proposes that only a few androgen-responsive cells are necessary for sufficient masculinization to occur.

HYPOGONADAL (*hpg*) MUTATION

Whereas the *Tfm* mutation affects responsiveness to androgens and the *Sxr* mutation causes sex reversal, other genetic defects have been described that influence the production of hormones (Forest, 1981; Grumbach & Conte, 1985; New *et al.*, 1983). One such mutation is hypogonadism (*hpg*), a genetic defect (autosomal recessive trait) that was initially described in humans (Kallman, Schoenfeld, & Barrera, 1944; Money & Mazur, 1978; Naftolin, Harris, & Bobrow, 1971; and others) and later in mice (Cattanach, Iddon, Charlton, Chiappa, & Fink, 1977). Since this mutation has been described in mice, the neural mechanisms underlying this defect can be directly studied. It has been found that *hpg* is characterized by a severe deficiency in gonadotropin-releasing hormone (GnRH). This hypothalamic hormone controls the release of the pituitary hormones LH and follicle-stimulating hormone (FSH), and, consequently, their levels are also deficient (Cattanach *et al.*, 1977). These pituitary hormones are essential for normal gonadal functions.

The GnRH deficiency does not affect the differentiation of the gonads and accessory organs but interferes with their further development. Thus, *hpg* mice do not mature sexually beyond an early juvenile stage. Since this neuroendocrine defect ultimately leads to abnormally low levels of testicular and ovarian steroids, it is not surprising that gonadally intact male and female *hpg* mice do not readily show sexual behavior.

Female mutants could be induced to display lordosis by injecting estrogen and progesterone or estrogen and synthetic GnRH in adulthood (Ward & Charlton, 1981). Adult *hpg* males given TP implants, however, did not mate successfully with normal females even though spermatogenesis was stimulated; *hpg* males sired off-spring only when they were exposed to testosterone on the day of birth in addition to receiving the TP implant as adults (Krieger & Gibson, 1984). If exposure to perinatal testosterone is necessary for complete copulatory behavior in males, these data suggest that the mutation may also be interfering with the organization of male sexual behavior and thus could prove useful in defining the role of the hypothalamic gonadotropin system in sexual differentiation.

The *hpg* mutation has already been shown to be an excellent model for the identification of the neural and genetic mechanisms regulating gonadal function. To understand the neural factors, Krieger and colleagues (Krieger & Gibson, 1984; Krieger *et al.*, 1982) transplanted the preoptic area, the major site of GnRH production, from normal fetal mice into the third ventricle of *hpg* mice. Male mutants with successful preoptic area implants have higher concentrations of hypothalamic GnRH, pituitary and serum gonadotropins (LH and FSH), and serum testosterone than untreated *hpg* males, although the levels were not normal. No data assessing the sexual behavior of *hpg* mice with preoptic transplants were presented. Based on the endocrine assays, it is clear that the preoptic area brain grafts partially corrected the CNS brain defect.

Unlike the *hpg* males, females receiving successful preoptic area brain grafts have normal levels of pituitary gonadotropins. Ovarian and uterine development was detected, in addition to the cornified vaginal cells that signal estrus (Gibson, Charlton, *et al.*, 1984; Krieger & Gibson, 1984). Gibson and colleagues found that *hpg* females with successful implants of normal fetal preoptic area tissue were now capable of mating, achieving pregnancy, and delivering young (Gibson, Krieger, Zimmerman, Silverman, & Perlow, 1984). An analysis of their mating behavior under defined laboratory conditions revealed that *hpg* mice with preoptic area implants showed normal or increased receptivity, since normal males mounted and ejaculated sooner with the *hpg* females than with normal females (Gibson, Korovis, Silverman, & Zimmerman, 1985). Krieger and Gibson (1984) conclude that these mutant mice with brain transplants can serve as a model to examine neural mechanisms involved in the neuroendocrinology of gonadal function and to help delineate the role of sexually dimorphic CNS structures in behavior and cyclicity.

More recently, investigators have restored the reproductive capacity of *hpg* mice with gene therapy (Mason, Hayflick, *et al.*, 1986; Mason, Pitts, *et al.*, 1986). Mason and colleagues (Mason, Hayflick, *et al.*, 1986) compared the GnRH gene from normal and mutant mice and identified a deletional mutation of at least 33.5 kb in the GnRH gene of *hpg* mice. This region consists of two exons of the GnRH gene that encode most of the GnRH-associated peptide (GAP). The mutation is thought to affect the translational activity of the gene. Using *in situ* hybridization histochemistry, Mason, Hayflick, *et al.* (1986) determined that the gene has transcriptional activity in *hpg* hypothalamic tissue and that the number and location of GnRH neurons appear not to be affected by the mutation. In view of these findings, an intact GnRH gene was introduced into the genome of *hpg* mice by a series of matings of transgenic mice and *hpg* mutants (Mason, Pitts, *et al.*, 1986). The transgenic *hpg/hpg* male and female offspring had normal levels of pituitary and serum LH and FSH. In contrast, the preoptic area implants in *hpg* males were only partially effective in restoring normal hormone levels. In addition, the deficiency of GnRH in hypothalamic neurons was corrected, which, according to Mason and colleagues, indicates a neural-specific expression of the introduced gene. Both *hpg* male and female mice receiving the intact GnRH gene were capable of mating and siring offspring. These studies employing brain transplants and gene therapy illustrate some of the clever ways that genetic mutations can be used to examine normal mechanisms.

In summary, this section described work that utilized the *Tfm*, *Sxr*, and *hpg* mutations as a means of elucidating normal mechanisms underlying sexual differentiation (Table 3). Additional genetic defects have been identified that affect testis determination and the endocrine system, and these, too, could prove valuable in examining the development of sexual behavior. For instance, Eicher (1982) describes several inherited mutations, including the *Sxr*, that cause sex reversal in mice. Some of these Y-linked traits or autosomal defects result in partial sex reversal (both ovarian and testicular tissue), and others involve complete sex reversal (XY with only ovaries and XX with only testes). These mutants have proven invaluable for studying events underlying gonadal differentiation but have not yet been utilized to address questions about sexual behavior differentiation.

In addition to the genetic defects already briefly discussed (e.g., congenital adrenal hyperplasia, 5α-reductase, *hpg;* see Tables 2 and 3), a number of other mutations have a direct consequence on various endocrine systems. Shire (1981) summarizes reports that indicate 21 and 35 different loci affecting the endocrine system but feels that this is an underestimate. Many of these mutations, like the *hpg* mice, have their primary effect on hypothalamic and pituitary hormones, which, in turn, may result in a disruption of sexual behavior (Bartke, 1979; Hall, Greenspan, & Harris, 1982; Shire, 1981). These effects on sexual behavior may be secondary to CNS defects and not have any direct consequences on sexual differentiation. Alternatively, some of these genetic defects that produce a deficit in adult sexual behavior may have been interfering with normal brain development. For example, reproductive behavior is impaired in both genetically diabetic (*db/db*) mice (Johnson & Sidman, 1979) and genetically obese (*ob/ob*) mice (Swerdloff, Batt, & Bray, 1976). These two mutations are found on different chromosomes but manifest similar syndromes of insulin-resistant diabetes and obesity. Closer examination of the hormonal profile revealed deficiencies in the level of LH, FSH, and testosterone in obese (*ob/ob*) male mice but not in the diabetic (*db/db*) male mice. Johnson and Sidman (1979) did not find a reduction in the gonadotropins that would explain the lack of sexual behavior in the (*db/db*) males. Possibly, the *db/db* mutation could also be interfering with normal events underlying sexual differentiation of the brain. There is some precedence for this idea, since sexual differentiation of the brain is thought to be deficient in the genetically obese male Zucker rats (Doherty, Baum, & Finkelstein, 1985; Withyachumnarnkul & Edmonds, 1982; Young, Fleming, & Matsumoto, 1986).

Thus, this section was devoted to the use of genetic mutations as a means of understanding sexual behavior differentiation. Although this research strategy has not been fully exploited in this field, other disciplines have recognized the value of such an approach (Eicher, 1982; Eicher & Washburn, 1986; Hall *et al.*, 1982; Sidman, 1982). Finding and analyzing mutations in animals that interfere with sexual differentiation provide novel means of elucidating normal mechanisms while at the same time providing information about genetic influences on the development of sexual behavior.

Strain comparisons provide another major approach used to examine genetic influences on sexual behavioral differentiation. Before we review this literature, it is important to define several terms that serve as the rubric for behavioral genetics (for more details, see Bailey, 1981; Festing, 1979; Plomin, DeFries, & McClearn, 1980; Wimer & Wimer, 1985). A *line* or *stock* of animals refers to a group that has been selected and maintained for some specific purpose. These animals can be inbred, but inbreeding does not necessarily need to be specific for any one characteristic. *Inbreeding* involves nonrandom mating between individuals that are related to each other, and an *inbred strain* results from inbreeding of 20 generations of full-brother-and-sister matings (full-sib mating). The requirement of 20 generations was established by the Committee on Standardized Genetic Nomenclature for Mice in 1952. No strains are ever fully inbred (coefficient of inbreeding = 100%), but after 20 generations the probability that a pair of genes present at any given locus is homozygous is 98.6% (Festing, 1979). Individuals within an inbred strain are genetically identical or *isogenic*.

The nomenclature for inbred strains is very specific (see Staats, 1976; Festing, 1979). An inbred strain is designated by one to four capital letters, for example, the mouse strains A and BALB. Inbred strains can contain numbers (e.g., C57BL or C3H) if named prior to the establishment of the guidelines in 1952.

Substrains develop from the strains when (1) inbreeding is halted after 8–19 matings (20 generations = inbred strain) and these animals are not mated together again for at least 12 or more generations, (2) a genetic difference is observed, or (3) the strain is maintained in a new laboratory. Substrains are used more frequently in research than original strains. Substrains are designated by the name of the parent strain followed by a virgule and the substrain symbol, usually representing the name of the person or laboratory maintaining the animals, although numbers are also used. Some popular mouse substrains are C57BL/6J and C57BL/10J (J = Jackson Laboratory), DBA/1, and DBA/2.

When two inbred strains are mated, an F_1 generation is produced. This cross reinitiates heterozygosity at all loci in which the two strains differ. The F_1 offspring are isogenic but not homozygous. Whereas *inbreeding depression* can result from increasing homozygosity, *hybrid vigor or heterosis*, which is an increase in fitness or viability, is thought to occur in the F_1 generation. The accepted nomenclature is mother's strain × father's strain, followed by F_1. Since this can be quite long, abbreviated strain names are more commonly used. For example, McGill and Manning (1976) mated C57BL/6 females and DBA/2 males to produce C57BL/6 × DBA/2 F_1, which is shortened to B6D2F$_1$ or BDF$_1$.

A number of strategies have been employed to dissect out the genetic component of behavior. This is critical, since strain differences may be attributed not only to genetic differences but also to maternal–paternal and/or environmental factors (Fulker, 1970). Such variables as temperature, noise, caging and bedding, social interactions, lighting, handling, etc. can have a profound effect on behavior, either independent of or by interacting with the genotype. To identify whether any of these environmental factors are contributing to differences in behavior, the investigator can examine a single inbred strain under different rearing conditions. *Reciprocal crosses* between two strains are used to determine whether intrauterine or postnatal factors promote differences in behavior. This breeding scheme produces F_1

offspring with identical genotypes and genotypically different mothers. If the behaviors of the hybrids differ, then maternal factors are involved. Two techniques have been employed to separate prenatal from postnatal maternal influences. First, the pups can be *cross-fostered,* which involves transferring the newborn offspring to a mother of a different genotype. This procedure provides a means of determining whether differences in maternal care of the young or milk yield contribute to behavioral differences. Second, *ovary transplants* are used to examine intrauterine influences or changes that may result from the origin of maternal egg cell cytoplasm. Ovaries from two inbred strains are transplanted into their F_1 hybrids, and these females are mated with males of the same strain as their newly transplanted ovaries.

An analysis of strain differences ultimately leads to producing a number of crosses between the strains. These crosses, which include the F_1 generation, F_2 *hybrids* (obtained by intercrossing F_1 animals), and *backcrosses* (obtained by crossing the F_1 hybrids to the parental lines), provide critical information about the mode of inheritance (polygenic or single-gene) between the inbred strains (Festing, 1979; Plomin *et al.*, 1980). A *diallel design* involves all possible crosses between a set of inbred strains and uses the F_1 hybrids as a tool to dissect genetic factors, heterosis, and maternal effects. These breeding paradigms are essential in a genetic analysis of differences between inbred strains.

The *recombinant inbred strains* method is particularly suited for amplifying gene effects above environmental factors. First described by Bailey (1971), it has been used successfully to carry out genetic analyses of a number of behaviors. For instance, this method has also been applied in dissecting drug–behavior–genetic interactions (Eleftheriou & Elias, 1975) and neural–behavior–genetic interactions (Vadász, Kobor, & Lajtha, 1982). Recombinant inbred strains are derived by crossing two unrelated inbred strains to produce both the F_1 and F_2 generations. The progeny of the F_2 generation are then inbred for at least 20 generations, resulting in a group of inbred strains that have assorted and replicable genotypes. Recombinant inbred strains provide a means of assessing whether the behavior is determined by a single locus or a polygenic mode of inheritance. Besides estimating the number of loci, this strategy allows examination of differences between the sexes, estimating the size and direction of gene effects, and genetic linkage analysis (Bailey, 1981).

Congenic strains provide another tool for identifying and examining genes with subtle effects on behavior. Congenic strains are created by introducing a mutation or gene of interest into an inbred strain. The carrier of the gene of interest is backcrossed to the inbred strain for at least seven generations. Both the methods of recombinant inbred strains and congenic strains would be particularly suited for identifying genetic components underlying development of sexual behavior.

STRAIN COMPARISONS: ADULT SEXUAL BEHAVIOR

Comparing a number of strains within a species is a powerful procedure to identify genotypic influences on sexual behavior differentiation. This type of research has proven to be very successful in isolating genetic factors in the development and expression of aggressive behaviors (see reviews: Maxson, 1981; Selmanoff & Ginsburg, 1981; Simon, 1979), open-field activity (Plomin *et al.*, 1980), and learning (Wimer & Wimer, 1985). However, very few systematic studies have examined the development of sexual behavior (see reviews: McGill, 1978a, 1978b; Olsen, 1983; Shrenker & Maxson, 1983). Instead, most research has been limited to comparing the ability of adult animals of various strains to display sexual behavior with or without exogenous hormone treatment. Although there is evidence to suggest genetic components in the development and expression of sexual behavior, as is evident in later discussion, additional research is required to establish the validity of the conclusions.

GENETIC–BEHAVIOR INTERACTIONS. Quantitative differences in mating behavior are found among strains of mice (Batty, 1978; McGill, 1962, 1978a, 1978b; Shrenker & Maxson, 1983) and rats (Blizard & Fulker, 1979; Moralí, Acatécatl, & Figueroa, 1986), which then can be studied for genetic factors. This is exemplified by the work of McGill (1962), who compared the copulatory behavior displayed by three inbred strains of mice and reported that (1) C57BL/6 males exhibited copulatory patterns characterized by short intromission and ejaculation latencies, (2) DBA/2 males took a longer time to ejaculate but required fewer intromissions than the other strains, and (3) BALB/c males were slow maters, exhibiting a large number of thrusts with each mount over a relatively long time period before ejaculating. Because qualitative assessment of sexual performance is usually based on the frequency and latencies of mounts, intromissions, and ejaculations, experimental paradigms utilizing short mating tests could exaggerate the behavioral differences among these three strains. Indeed, in standardized mating tests, BALB/c and DBA/2 males were labeled as "poor maters" because very few of these mice achieved ejaculation (Batty, 1978). Along with the three inbred strains discussed above, Batty (1978) also examined the male mating responses of five additional genotypes and reported significant differences among the strains in the number of tests in which mount, intromission, and ejaculation patterns occurred and the latencies to show these responses.

The discovery of quantitative differences among inbred strains suggests that genetic factors are important, although other variables may also contribute. Therefore, it becomes essential to establish breeding schedules that can begin to identify the genetic component. For example, an analysis of the F_1 progeny derived from inbred strains suggests the possibility of inheritance of a behavioral trait from a dominance of one parental genotype (exhibited behavior similar to only one parent), a mixture (exhibited behavior is between those of the parents), or heterosis (exhibited behavior is superior to both parents).

McGill (1978a, 1978b) utilized this approach to examine further the behavioral differences among three inbred strains and reported that the genotype has a major effect on the retention of ejaculation following castration. The $CD2F_1$ males, resulting from a cross between inbred BALB/c females and DBA/2 males, lost their ejaculatory reflex within a week after castration (Champlin, Blight, & McGill, 1963), whereas $B6D2F_1$ hybrid males, derived from a cross between C56BL/6 females and DBA/2 males, continued to ejaculate for a considerably longer time after castration than either of the parental strains (McGill & Tucker, 1964; McGill & Haynes, 1973; McGill & Manning, 1976). Indeed, McGill (1978a) reports that although these mice may show an initial reduction in behavior immediately after castration, some of these $B6D2F_1$ hybrids were still ejaculating 2 years following castration, and Clemens and colleagues (Wee, Weaver, Goldman, & Clemens, 1986) have observed this behavior up to 1 year after the operation. Thus, these animals exhibit the entire copulatory pattern independent of androgen stimulation. These data are contrary to our understanding of the relationship between hormones and male sexual behavior and thus raise a number of questions regarding the nature of this genotypic difference.

Based on these studies with the F_1 hybrid males, McGill and Manning (1976) hypothesized that retention of the ejaculatory response following gonadectomy may relate to heterozygosity at particular genetic loci. Using a diallelic design involving the three inbred strains (C57BL/6, DBA/2, and BALB/c) and their six progenies, they found that manipulation of the genotype within a species can influence the retention of ejaculation after castration, but "superior" retention was not associated with heterosis. Instead, this behavioral trait is unique for the B6D2F_1 hybrid males, suggesting that the characteristic may be sex-linked and/or affected by maternal factors.

Another line of genetic research has generated data indicating that genes on Y chromosomes influence mounting behavior in mice (Shrenker & Maxson, 1983, 1984). Capitalizing on congenic strains (DBA/1Bg, DBA/2BG, and DBA/2.DBA/1-YBg) that were previously developed to substantiate the relationship between Y chromosomes and aggression (Maxson, Ginsburg, & Trattner, 1979; Selmanoff, Jumonville, Maxson, & Ginsburg, 1975; Selmanoff, Maxson, & Ginsburg, 1976), Shrenker and Maxson (1983, 1984) reported strain differences in the proportion of males that mounted (50% of DBA/2BG versus 13% of DBA/2.DBA/1-YBg). Since the only conceivable difference between these two genotypes would reside in the Y chromosome, these data provide strong evidence for this relationship. However, Stewart, Manning, and Batty (1980) did not find this correlation when comparing offspring that were developed from the parental lines, C57Bl/6 and CBA, two strains that differ both behaviorally and in the ability of the Y chromosome to affect testicular weights. Genotypes differing in the origin of the Y chromosome were not distinguishable on any of the standard measures used to define male sexual behavior, including mounting. However, similar to previous reports (Maxson *et al.*, 1979; Selmanoff & Ginsburg, 1981; Selmanoff *et al.*, 1975, 1976), a relationship was found between Y chromosomes and aggression (Steward *et al.*, 1980). Taken together, these data are consistent with the hypothesis that different genes located on Y chromosomes influence male sexual behavior and aggression (Shrenker & Maxson, 1983).

GENETIC–HORMONE–BEHAVIOR INTERACTIONS.   Genetic diversity has also been reported in the ability of exogenous hormones to activate sexual behavior in adult gonadectomized animals. Comparison of strains of mice (Gorzalka & Whalen, 1974, 1976; Thompson & Edwards, 1971), of rats (Whalen, Gladue, & Olsen, 1986), and of guinea pigs (Goy & Young, 1957; Thornton, Wallen, & Goy, 1987) revealed striking differences in the sensitivity to estrogens and progestins as indicated by the amount of lordosis behavior that could be activated. Differences in sensitivity to androgens, as measured by activation of copulatory responses (Luttge & Hall, 1973; Olsen, 1979b) or their retention following castration (McGill, 1978a, 1978b), have also been reported to be influenced by genotype.

Although these initial observations of genetic diversity are exciting, further research incorporating selective breeding paradigms is essential to eliminate the possibility that they are not the result of other influences (e.g., accidental gene associations, maternal effects) that affect the observed hormone–behavior relationships (Festing, 1979; Fulker, 1970; Plomin *et al.*, 1980). A few investigators have begun to use selective breeding to extend their initial work (Gorzalka & Whalen, 1976; McGill, 1978a, 1978b). Gorzalka and Whalen (1976) bred CD-1 and Swiss–Webster mice to produce their reciprocal F_1 hybrids. In their original study, differences between these two genetically heterogeneous outbred strains were found in the ability of progesterone and dihydroprogesterone to stimulate female receptivity (Gorzalka & Whalen, 1974). Lordotic levels were higher in ovariectomized CD-1 mice given EB plus progesterone than in Swiss–Webster mice, and only the CD-1 strain displayed lordosis when dihydroprogesterone was substituted for the progesterone. In an extension of this work, the F_1 hybrid groups behaved like the CD-1 mice, suggesting that sensitivity to dihydroprogesterone is a dominant trait. However, significant variability was found within the hybrid groups in their response to dihydroprogesterone, suggesting that additional breeding may possibly select for behavioral sensitivity to this progestin metabolite.

STRAIN COMPARISONS: DEVELOPMENT OF SEXUAL BEHAVIOR

To date, only a few studies have utilized the unique properties of inbred strains to examine the interaction of genotype and perinatal hormones on the differentiation of sexual behavior. Vale, Ray, and Vale (1974) injected 3-day-old males of the inbred strains A, BALB/c, and C57BL/6 with either 1 mg of TP or oil. As adults, these gonadally intact males were tested for the display of male mating responses. Neonatal androgen treatment enhanced male copulatory responses in mice of the BALB/c and C57BL/6 strains. The TP-treated males of these two strains tended to mount and intromit more quickly and more frequently than oil-treated males. In contrast, the A strain was relatively unresponsive to this perinatal treatment.

Batty (1979) also examined the effects of perinatal androgen treatment on the development of masculine behavior among four mouse strains selected on the basis of their adult performance (i.e., adult BALB/c and DBA/2J males display fewer ejaculations, suggesting less masculine behavior than either the C57BL/6Fa or BDF_1 hybrids) (Batty, 1978; McGill, 1962). Neonatal males were injected 4 days after birth with TP (100 μg) and tested for male sexual behavior without exogenous androgen treatment as adults. Neonatal TP treatment significantly increased mount, intromission, and ejaculation frequency in all four strains. In addition, exposure to TP during early development enhanced the number of BALB/c and DBA/2J mice that ejaculated.

Campbell and McGill (1970) exposed DBA/2 male mice to 100 μg of TP, 50 μg of EB, or oil at birth and tested them under the influence of their own testicular hormones as adults. Although all mice, regardless of neonatal hormone treatment, eventually ejaculated, the groups differed in that TP-treated males ejaculated sooner and EB-treated males began to ejaculate later than oil-treated males. In addition, EB-treated males had longer ejaculation latencies. Similar to the previous reports, neonatal TP treatment had an enhancing effect on the development of masculine behavior in DBA/2J males.

The interaction of genotype and perinatal hormones on sexual behavior differentiation has also been examined in females. Vale, Ray, and Vale (1973) injected 3-day-old females of the inbred strains A, BALB/c, and C57BL/6 with 1 mg of TP, 500 μg of EB, or oil. In adulthood, the females were tested for male mating responses, independent of any exogenous hormones. Following these tests the gonadally intact females were injected with EB followed by progesterone, and the behavior of males toward them was recorded. Lordosis scores were not presented. Similar to the males, neonatal exposure to TP enhanced the ability of BALB/c and C57BL/6 females, but not of A females, to display male mating responses. Neonatal estrogen treatment was ineffective in augmenting mounting in any strain. Although no quantitative scores were presented for female behavior, males readily intromitted with the oil-treated but not with the TP- or EB-treated females of the A and BALB/c strains. In contrast, C57BL/6 females treated with androgen or oil received a similar number of intromissions by males. Neonatal estrogen, however, modified female receptivity in that males failed to achieve intromissions. Thus, estrogens appear to have defeminizing activities in all three strains.

Lordosis behavior was quantified in Swiss–Webster females exposed neonatally to the aromatizable androgens, TP, testosterone, or androstenedione (Edwards & Burge, 1971; Edwards, 1971). Unlike oil-treated mice, androgenized females from this outbred strain rarely showed lordosis in adulthood following EB plus progesterone treatment. Interestingly, Swiss–Webster females were similar to A-strain females (Vale *et al.*, 1973) in that neonatal androgen treatment did not enhance the development of male sexual behavior but suppressed female sexual behavior (Edwards, 1971; Edwards & Burge, 1971). Swiss–Webster females given 100 μg of TP or oil at birth did not differ in their ability to display mounting behavior in adulthood in response to exogenous androgens. On the other hand, genetically homogeneous BDF_1 females given 100 μg of TP at birth exhibited the full copulatory sequence including the ejaculatory pattern in adulthood. Moreover, these androgenized females ejaculated in adulthood with or without exogenous TP treatment (Manning & McGill, 1974). Since these two strains (Swiss–Webster and BDF_1) were not directly compared within the same testing paradigms, nongenetic factors may have contributed or interacted with the genotype. This seems possible, since Edwards and Burge (1971) never observed ejaculatory behavior in normal Swiss–Webster males during the 10-minute mating test. Possibly, the shortened test period could have underestimated the potential of the androgenized females to display masculine behavior.

Although genetic and nongenetic factors cannot be distinguished in the studies described above, Vale *et al.* (1973) found that neonatal TP treatment enhanced male mating responses in the females of some strains but not others. In addition, in comparing two strains selectively bred for aggressiveness (TA) and nonaggressiveness (TNA), Lagerspetz and Lagerspetz (1975) observed augmented mounting behavior in females exposed to TP on day 2 post-partum and/or in adulthood. Interestingly, females of the nonaggressive TNA strain rarely mounted (28%) unless they had also been exposed to TP neonatally (75%). On the other hand, 80% of TA females that only received TP as adults attempted to mount. Thus, there is a significant strain difference in adult copulatory behavior that can be overcome by neonatal exposure to androgens. Interestingly, neonatal TP treatment had no effect on strain differences in aggression.

Most studies comparing strains have used mice. This is not surprising since over 200 inbred strains have been identified (Festing, 1979). Although there are a few inbred strains of rats, comparisons within this species usually involve Long–Evans, Sprague–Dawley, or Wistar rats. As pointed out by Festing (1979), the various colonies of such stocks with the same generic name can differ markedly in their properties; e.g., Sprague–Dawley rats obtained from one supplier may differ from those bred by another. With this caveat in mind, behavioral differences have been found between Long–Evans (Charles River) and Sprague–Dawley (Simonsen Laboratories) rats that most likely result from strain differences in hormonal sensitivity (Whalen *et al.*, 1986). Clemens and Gladue (1978) reported that Long–Evans male rats exposed *in utero* to the aromatase inhibitor 1,4,6-androstatriene-3,17-dione (ATD) displayed high levels of female sexual behavior in adulthood following EB plus progesterone treatment, whereas Whalen and Olsen (1981) observed only a slight enhancement of adult lordotic behavior in ATD-treated Sprague–Dawley males. A major difference was that Long–Evans rats littered after a 22-day gestation period, whereas Sprague–Dawley rats littered after 21 days. Therefore, the fetuses had received one additional day of ATD treatment in the Clemens and Gladue study. It is conceivable that the behavioral differences could reflect (1) strain differences in gestation length and, therefore, differences in the timing of the sensitive period of sexual differentiation; (2) strain differences in sensitivity to ATD treatment; or (3) strain differences in sensitivity to the defeminizing action of gonadal hormones. To examine these possibilities, we compared the effects of perinatal ATD on the development of female mating behaviors between Long–Evans and Sprague–Dawley males maintained under identical treatment and testing conditions. The major finding was that the strains differed in their sensitivity to the lordosis-inducing properties of estrogen (Whalen *et al.*, 1986).

STRAIN COMPARISONS: TOOLS TO IDENTIFY GENETIC INFLUENCES ON SEXUAL BEHAVIOR DIFFERENTIATION

In summary, strain differences have been found both in the patterns of behavior normally displayed during copulation and in responsiveness to the steroids that activate these patterns. How can these strain differences be used to understand mechanisms underlying the development of behavior?

First, comparing similarities as well as differences among strains provides a powerful means of examining the generality of a given phenomenon across species. Results obtained from a single inbred strain (a single genotype) or an outbred strain (unknown variability) may lead to wrong conclusions. For example, Vale *et al.* (1973) reported that C57BL/6 females neonatally treated with TP or oil received a similar number of mounts with intromissions from normal males in adulthood. This finding could have led to the erroneous conclusion that androgens are not involved in defeminization in mice. Fortunately, a number of other strains examined under identical conditions indicated that C57BL/6 mice may be differentially sensitive to androgens.

Second, identifying strain differences leads to further genetic analyses, which, in turn, provide information about nongenetic environmental and maternal factors that may contribute to the variability. For example, we already are aware of the major effect that the uterine environment can have on the display of adult masculine behavior by females (Clemens, 1974; Clemens, Gladue, & Coniglio, 1978; Meisel & Ward, 1981). Gestational stress is another major factor influencing the development of sexual behavior in males (Ward, 1972, 1983), possibly by altering circulating levels of testosterone in the fetuses (Ward, Orth, & Weisz, 1983; Ward & Weisz, 1984; Weisz & Ward, 1980). Moreover, it is becoming more apparent that mother–offspring interactions can influence adult sexual behaviors (Moore, 1985). Thus, analysis of strain differences can examine the extent to which a particular finding is dependent on genetic or nongenetic factors. Finally, understanding strain differences provides a means of identifying some characteristic that may be related to or contribute directly to the development of sexual behavior.

How do genes mediate their effects on adult sexual behavior? Do the strain differences result from genotypic influences on (1) developmental processes that subsequently affect adult sexual behavior, (2) activational processes in adulthood, or (3) a combination of the two events? If we consider that strain comparisons provide a means of identifying possible traits that may be related to or contribute directly to the observed strain difference, then based on our current understanding of the normal events underlying sexual behavior differentiation (Figure 1 and Table 1), two possibilities come immediately to mind: testosterone and aromatase activity. Each has major effects on development, and both the expression of sexual behavior and the steroidogenic pathways are under genetic control.

Let's consider the possibility that genes affect circulating testosterone levels throughout the life span of the animals. During perinatal development, differences in testosterone levels result in differences in the degree of masculinization and defeminization of the neural systems underlying sexual behavior. In adulthood, differences in circulating levels of testosterone may influence the activation of copulatory behavior. Consistent with this hypothesis are the findings that genes on the autosomes are known to control steroid biosynthesis, including testosterone, and that strain differences have been reported in testes weight, circulating plasma testosterone, and LH levels (Bartke, 1974; Bartke & Shire, 1972; Batty, 1978; Selmanoff, Goldman, & Ginsburg, 1977; and others).

Since it is readily accepted that testosterone activates male sexual behavior in most mammalian species, it would be expected that levels of testosterone and sexual behavior are correlated. Although this may be true in some species, a comparison of a number of mouse strains indicates the opposite. Batty (1978) measured male copulatory behaviors and plasma testosterone levels in adult male mice from eight strains. The genotypes (BALB/c and CBA/H) ranked as the poorest maters had the highest levels of plasma testosterone, and males of the BDF_1 and DBF_1 strains, which ejaculated on almost all of the mating tests, had the lowest levels. The strain differences in testosterone levels were apparent even when blood was measured from mice with no previous heterosexual experience. Interestingly, these strains also responded differently to neonatal androgen treatment (Batty, 1979). In BALB/c males, sexual behavior but not plasma testosterone levels was enhanced in adulthood by injecting TP (100 μg) 4 days after birth. In contrast, neonatal androgen treatment increased testosterone levels but not sexual behavior in BDF_1 males. Since BDF_1 males routinely ejaculate during mating tests and BALB/c males have high plasma testosterone levels, it is possible that the experimental procedures (e.g., ceiling effects) contributed to these findings.

Batty (1978) further analyzed the gene–hormone–behavior relationship in adult mice to determine whether it had occurred by chance or was a reliable association. Testosterone levels and male sexual behavior were measured in two F_2 populations derived from the BDF_1 and CBF_1 hybrids. The parental lines included BALB/c and DBA(F), strains characterized by high testosterone and low sexual behavior, and C57BL/6FA, a strain with lower testosterone levels and higher behavioral scores. The BDF_2 males did not show the predicted correlation. However, the F_2 hybrids derived from the BDF_1 males were characterized by low circulating testosterone levels, high sexual activity, and an unusually long retention of the ejaculation reflex following castration (McGill & Manning, 1976; McGill, 1978a, 1978b). On the other hand, the negative correlation between testosterone levels and some measures of sexual behavior was apparent for the CBF_2 males. Although this hormone–behavior relationship would not have been predicted, others also have observed high circulating testosterone levels associated with low levels of male sexual behavior (Shrenker & Maxson, 1983). In addition, McGill (1978b) reported that exogenous TP given to castrated mice from several different strains tends to increase, not decrease, ejaculation latencies.

Genes also could influence sexual behavior differentiation by coding for aromatase. Similar to testosterone, aromatase levels are strain dependent (Sheridan & Melgosa, 1983). If behavioral defeminization is mediated by estrogens derived from circulating androgens, then one prediction is that differences in aromatase activity may be correlated with differences in adult female sexual behavior. For instance, Vale *et al.* (1973) reported that females of the inbred strains A, BALB/c, and C57BL/6 exposed neonatally to EB were not mounted by males as adults, suggesting that they were defeminized by this hormone treatment. On the other hand, unlike the other two strains, C57BL/6 females given TP on day 4 were mounted in adulthood. It was suggested that the C57BL/6 strain may be relatively insensitive to androgens, but this does not seem likely since neonatal TP was capable of enhancing male sexual behaviors, especially in C57BL/6 mice (Vale *et al.*, 1973, 1974). An alternative explanation is that C57BL/6 mice have a decreased ability to aromatize androgens within the CNS during development, which in turn results in higher levels of female sexual behavior as adults.

It is not known how genes mediate their effects on sexual behavior differentiation. The possibility that testosterone and aromatase are gene products was only proposed because of their already established importance in the sexual differentiation process. The idea that genetic differences result in differences in their activity that in turn influence neural and behavioral development is speculative.

We do know that behavior is controlled by neural circuits or systems and that genes code for the necessary proteins and enzymes involved in the formation of these systems. A wide range of neuroanatomical parameters differ among mouse strains (Ingram & Corfman, 1980), including the organization of the medial preoptic area, which is critical for male sexual behavior and cyclicity (Robinson, Fox, & Sidman, 1985). Moreover, a number of specific genetic defects interfere with normal brain development (Sidman, 1982). We also know that genes code for proteins and enzymes involved in maintaining the integrity of these neural systems and in regulating their activity and, thus, behavior. For instance, neurotransmitter activity is thought to be under genetic control. Strain differences have been reported in tyrosine hydroxylase activity (Vadász, Baker, Fink, & Reis, 1985) and in the number of dopamine and serotonin receptors found in several brain regions including the hypothalamus (Boehme & Ciaranello, 1981, 1982; Reis, Baker, Fink, & Joh, 1981). Furthermore, genes code for peptide hormones and other neuromodulators. Thus, in addition to what we already know about the genetic control of sexual differentiation (see Table 1), it is also important to consider that the systems (e.g., neuroanatomical, neurochemical, neuropharmacological) within which steroids interact are also influenced by the genotype. Indeed, many genetic as well as nongenetic factors are essential for the development and expression of sexual behavior.

Summary and Conclusions

The overpowering effects of testicular hormones on sexual behavior differentiation make it easy to ignore genetic influences. Subtle differences in copulatory behavior among strains suggest a genetic component, but the effects are not striking. Only when dramatic genetic mutations, such as testicular feminization syndrome or 5α-reductase deficiency, are described do we begin to acknowledge and appreciate the importance of genes in normal sexual differentiation.

This chapter examined the use of gene defects and strain comparisons as a means of identifying genotypic influences and elucidating normal mechanisms underlying sexual behavior differentiation. Three major genetic defects, testicular feminization (*Tfm*), sex-reversal (*Sxr*), and hypogonadism (*hpg*) were described. These mutations are found in lower animal species as well as variants in humans and, therefore, can serve as models for investigating the neural mechanisms underlying the clinical disorder. Finding and analyzing mutations in animals that interfere with sexual differentiation is an excellent strategy for elucidating normal processes. Although still at an early stage, this line of research has already made many important contributions.

Strain comparisons have revealed a number of differences in copulatory behavior and in behavioral responsiveness to exogenous hormones. However, many of these studies have not progressed beyond the initial observations to determine whether the differences are caused by genetic or nongenetic factors. We need to progress beyond the mere recognition that strain differences exist or that genes influence sexual behavior and begin to examine the mechanisms whereby genes mediate these differences. Indeed, techniques are now available to isolate, identify, and map individual genes. Possibly a number of the tools used by behavioral geneticists and molecular biologists could be applied to this research. Identifying the genes and elucidating their mechanism of action can lead to major advances in our understanding of sexual behavior differentiation.

Acknowledgments

This research was supported by a grant from NIH (HD-18893).

REFERENCES

Amador, A. G., Parkening, T. A., Beamer, W. G., Bartke, A., & Collins, T. J. (1986). Testicular LH receptors and circulating hormone levels in three mouse models for inherited diseases (*Tfm/y lit/lit* and *hyt/hyt*). *Endocrinologica Experimentalis, 20*, 349–358.

Andersson, M., Page, D. C., & de la Chapelle, A. (1986). Chromosome Y-specific DNA is transferred to the short arm of X chromosome in human XX males. *Science, 233*, 786–788.

Arnold, A. P., & Gorski, R. A. (1984). Gonadal steroid induction of structural sex differences in the central nervous system. *Annual Review of Neuroscience, 7*, 413–442.

Attardi, B., & Ohno, S. (1976). Androgen and estrogen receptors in the developing mouse brain. *Endocrinology, 99*, 1279–1290.

Attardi, B., Geller, L. N., & Ohno, S. (1976). Androgen and estrogen receptors in brain cytosol from male, female and testicular feminized (*Tfm/ₒy*) mice. *Endocrinology, 98*, 864–874.

Bailey, D. W. (1971). Recombinant-inbred strains. An aid to finding identity, linkage, and function of histocompatibility and other genes. *Transplantation, 11*, 325–327.

Bailey, D. W. (1981). Strategic uses of recombinant inbred, and congenic strains in behavior genetics research. In E. S. Gershon, S. Matthysse, X. O. Breakefield, & R. D. Ciaranello (Eds.), *Genetic research strategies for psychobiology and psychiatry* (pp. 189–198). Pacific Grove: The Boxwood Press.

Bardin, C. W., & Catterall, J. F. (1981). Testosterone: A major determinant of extragenital sexual dimorphism. *Science, 211*, 1285–1294.

Bardin, C. W., Bullock, L. P., Sherins, R. J., Mowszowicz, I., & Blackburn, W. R. (1973). Part II. Androgen metabolism and mechanism of action in male pseudohermaphroditism: A study of testicular feminization. *Recent Progress in Hormone Research, 29*, 65–107.

Barley, J., Ginsburg, M., Greenstein, B. D., MacLusky, N. J., & Thomas, P. J. (1974). A receptor mediating sexual differentiation. *Nature, 252*, 259–260.

Barraclough, C. A., & Gorski, R. A. (1962). Studies on mating behavior in the androgen-sterilized rat and their relation to the hypothalamic regulation of sexual behavior in the female rat. *Journal of Endocrinology, 25*, 175–182.

Bartke, A. (1974). Increased sensitivity of seminal vesicles to testosterone in a mouse strain with low plasma testosterone level. *Journal of Endocrinology, 60,* 145–148.

Bartke, A. (1979). Genetic models in the study of anterior pituitary hormones. In J. G. M. Shire (Ed.), *Genetic variation in hormone systems, Vol. I* (pp. 113–126). Boca Raton: CRC Press.

Bartke, A., & Shire, J. G. M. (1972). Differences between mouse strains in testicular cholesterol levels and androgen target organs. *Journal of Endocrinology, 55,* 173–184.

Batty, J. (1978). Plasma levels of testosterone and male sexual behaviour in strains of the house mouse (*Mus musculus*). *Animal Behaviour, 26,* 339–348.

Batty, J. (1979). Influence of neonatal injections of testosterone propionate on sexual behavior and plasma testosterone levels in the male house mouse. *Developmental Psychobiology, 12,* 231–238.

Baum, M. J. (1979). Differentiation of coital behavior in mammals: A comparative analysis. *Neuroscience and Biobehavioral Reviews, 3,* 265–284.

Beach, F. A., & Buehler, M. G. (1977). Male rats with inherited insensitivity to androgen show reduced sexual behavior. *Endocrinology, 100,* 197–200.

Bennett, D., Boyse, E. A., Lyon, M. F., Mathieson, B. J., Schied, M., & Yanagisawa, K. (1975). Expression of H-Y (male) antigen in phenotypically female *Tfm/Y* mice. *Nature, 257,* 236–238.

Bennett, D., Mathieson, B. J., Schied, M., Yanagisawa, K., Boyse, E. A., Wachtel, S. S., & Cattanach, B. M. (1977). Serological evidence for H-Y antigen in *Sxr,*XX sex-reversed phenotypic males. *Nature, 265,* 255–257.

Berta, P., Hawkins, J. R., Sinclair, A. H., Taylor, A., Griffiths, B. L., Goodfellow, P. N., & Fellous, M. (1990). Genetic evidence equating *SRY* and the tenths-determining factor. *Nature, 348,* 448–450.

Blizard, R. A., & Fulker, D. W. (1979). Interactions of strain of male and female on the mating behavior of the rat. *Behavior and Neural Biology, 25,* 99–114.

Blizzard, R. M. (1979). Overview. In: H. L. Vallet & I. H. Porter (Eds.), *Genetic mechanisms of sexual development* (pp. 485–490). New York: Academic Press.

Boehme, R. E., & Ciaranello, R. D. (1981). Strain differences in mouse brain dopamine receptors. In E. S. Gershon, S. Matthysse, X. O. Breakefield, & R. D. Ciaranello (Eds.), *Genetic research strategies in psychobiology and psychiatry* (pp. 231–240). Pacific Grove: The Boxwood Press.

Boehme, R. E., & Ciaranello, R. D. (1982). Genetic control of dopamine and serotonin receptors in brain regions of inbred mice. *Brain Research, 266,* 51–65.

Breedlove, S. M. (1986). Cellular analyses of hormone influence on motoneuronal development and function. *Journal of Neurobiology, 17,* 157–176.

Bullock, L. P., & Bardin, C. W. (1974). Androgen receptors in mouse kidney: A study of male, female and androgen-insensitive (*tfm/y*) mice. *Endocrinology, 94,* 746–756.

Bullock, L. P., Mainwaring, W. I. P., & Bardin, C. W. (1975). The physicochemical properties of the cytoplasmic androgen receptor in the kidneys of normal, carrier female (*tfm/+*) and androgen-insensitive (*tfm/y*) mice. *Endocrine Research Communications, 2,* 25–45.

Burns, R. K. (1961). The role of hormones in the differentiation of sex. In W. C. Young (Ed.), *Sex and internal secretions* (3rd ed.) (pp. 76–158). New York: Robert E. Krieger.

Campbell, A. B., & McGill, T. E. (1970). Neonatal hormone treatment and sexual behavior in male mice. *Hormones and Behavior, 1,* 145–150.

Cattanach, B. M., Pollard, C. E., & Hawkes, S. G. (1971). Sex-reversed mice: XX and XO males. *Cytogenetics, 10,* 318–337.

Cattanach, B. M., Iddon, C. A., Charlton, H. M., Chiappa, S. A., & Fink, G. (1977). Gonadotrophin-releasing hormone deficiency in a mutant mouse with hypogonadism. *Nature, 269,* 338–340.

Champlin, A. K., Blight, W. C., & McGill, T. E. (1963). The effects of varying levels of testosterone on the sexual behaviour of the male mouse. *Animal Behavior, 11,* 244–245.

Chang, C., Kokontis, J., & Liao, S. (1988). Molecular cloning of human and rat complementary DNA encoding androgen receptors. *Science, 240,* 324–326.

Charest, N. J., Zhou, Z.-X., Lubahn, D. B., Olsen, K. L., Wilson, E. M., & French, F. S. (1991). A frameshift mutation destabilizes androgen receptor messenger RNA in the *Tfm* mouse. *Molecular Endocrinology, 5,* 573–581.

Clark, J. H., Schrader, W. T., & O'Malley, B. W. (1985). Mechanisms of steroid hormone action. In J. D. Wilson & D. W. Foster (Eds.), *Williams textbook of endocrinology* (pp. 33–75). Philadelphia: W. B. Saunders.

Clemens, L. G. (1974). Neurohormonal control of male sexual behavior. In W. Montagna & W. A. Sadler (Eds.), *Reproductive behavior* (pp. 23–53). New York: Plenum Press.

Clemens, L. G., & Gladue, B. A. (1978). Feminine sexual behavior in rats enhanced by prenatal inhibition of androgen aromatization. *Hormones and Behavior, 11*, 190–201.

Clemens, L. G., Gladue, B. A., & Coniglio, L. P. (1978). Prenatal endogenous androgenic influences on masculine sexual behavior and genital morphology in male and female rats. *Hormones and Behavior, 10*, 40–53.

Crews, D., & Moore, M. C. (1986). Evolution of mechanisms controlling mating behavior. *Science, 231*, 121–125.

de la Chapelle, A., Tippett, P. A., Wetterstrand, G., & Page, D. (1984). Genetic evidence of X–Y interchange in a human XX male. *Nature, 307*, 170–171.

Doherty, P. C., Baum, M. J., & Finkelstein, J. A. (1985). Evidence of incomplete behavioral sexual differentiation in obese male Zucker rats. *Physiology and Behavior, 34*, 177–179.

Edwards, D. A. (1971). Neonatal administration of androstenedione, testosterone or testosterone propionate: Effects on ovulation, sexual receptivity and aggressive behavior in female mice. *Physiology and Behavior, 6*, 223–228.

Edwards, D. A., & Burge, K. G. (1971). Early androgen treatment and male and female sexual behavior in mice. *Hormones and Behavior, 2*, 49–58.

Ehrhardt, A. A., & Meyer-Bahlburg, H. F. L. (1981). Effects of prenatal sex hormones on gender-related behavior. *Science, 211*, 1312–1318.

Ehrhardt, A. A., Epstein, R., & Money, J. (1968). Fetal androgens and female gender identity in the early-treated adrenogenital syndrome. *Johns Hopkins Medical Journal, 123*, 160–167.

Eicher, E. M. (1982). Primary sex determining genes in mice. In R. P. Amann & G. E. Seidel, Jr. (Eds.), *Prospects for sexing mammalian sperm* (pp. 121–135). Boulder: Colorado Associated University Press.

Eicher, E. M., & Washburn, L. L. (1986). Genetic control of primary sex determination in mice. *Annual Review of Genetics, 20*, 377–360.

Eichwald, E. J., & Silmser, C. R. (1955). Untitled communication. *Transplantation Bulletin, 2*, 148–149.

Eil, C. (1983). Familial incomplete male pseudohermaphroditism associated with impaired androgen retention: Studies in cultured skin fibroblasts. *Journal of Clinical Investigations, 71*, 850–858.

Eil, C., Merriam, G. R., Bowen, J., Ebert, J., Tabor, E., White, B., Douglass, E. C., & Loriaux, D. L. (1980). Testicular feminization in the chimpanzee. *Clinical Research, 28*, 624A.

Eleftheriou, B. E., & Elias, P. K. (1975). Recombinant inbred strains: A novel genetic approach for psychopharmacogeneticists. In B. E. Eleftheriou (Ed.), *Psychopharmacogenetics* (pp. 43–71). New York: Plenum Press.

Evans, E. P., Burtenshaw, M. D., & Cattanach, B. M. (1982). Meiotic crossing-over between the X and Y chromosomes of male mice carrying the sex-reversed (*Sxr*) factor. *Nature, 300*, 443–445.

Feder, H. H. (1984). Hormones and sexual behavior. *Annual Review of Psychology, 35*, 165–200.

Festing, M. F. W. (1979). *Inbred strains in biomedical research.* New York: Oxford University Press.

Forest, M. G. (1981). Inborn errors of testosterone biosynthesis. In Z. Laron & P. Tikva (Eds.), *Pediatric and adolescent endocrinology 8* (pp. 133–155). Basel: S. Karger.

Fox, T. O. (1975). Androgen- and estrogen-binding macromolecules in developing brain: Biochemical and genetic evidence. *Proceedings of the National Academy of Sciences U.S.A., 72*, 4303–4307.

Fox, T. O., Vito, C. C., & Wieland, S. J. (1978). Estrogen and androgen receptor proteins in embryonic and neonatal brain: Hypotheses for roles in sexual differentiation and behavior. *American Zoologist, 18*, 525–537.

Fox, T. O., Olsen, K. L., Vito, C. C., & Wieland, S. J. (1982). Putative steroid receptors: Genetics and development. In F. O. Schmitt, S. Bird, & F. E. Bloom (Eds.), *Molecular genetics and neurosciences: A new hybrid* (pp. 289–306). New York: Raven Press.

Fox, T. O., Blank, D., & Politch, J. A. (1983). Residual androgen binding in testicular feminization (*Tfm*). *Journal of Steroid Biochemistry, 19*, 577–581.

Fulker, D. W. (1970). Maternal buffering of rodent genotypic responses to stress: A complex genotype–environment interaction. *Behavioral Genetics, 1*, 119–124.

Gerall, A. A., & Ward, I. L. (1966). Effects of prenatal exogenous androgen on the sexual behavior of the female albino rat. *Journal of Comparative and Physiological Psychology, 62*, 370–375.

Gibson, M. J., Charlton, H. M., Perlow, M. J., Zimmerman, E. A., Davies, T. F., & Krieger, D. T. (1984). Preoptic area brain grafts in hypogonadal (*hpg*) female mice abolish effects of congenital hypothalamic gonadotropin-releasing hormone (GnRH) deficiency. *Endocrinology, 114*, 1938–1940.

Gibson, M. J., Krieger, D. T., Zimmerman, E. A., Silverman, A. J., & Perlow, M. J. (1984). Mating and pregnancy can occur in genetically hypogonadal mice with preoptic area brain grafts. *Science, 225*, 949–951.

Gibson, M. J., Korovis, G., Silverman, A. J., & Zimmerman, E. A. (1985). Mating behavior in hypogonadal female mice with GnRH cell-containing brain grafts. *Society for Neuroscience, 11,* 528.

Gubbay, J., Collignon, J., Koopman, P., Capel, B., Economou, A., Münsterberg, A., Vivian, N., Goodfellow, P., & Lovell-Badge, R. (1990). A gene mapping to the sex-determining region of the mouse Y chromosome is a member of a novel family of embryonically expressed genes. *Nature, 346,* 245–250.

Goldberg, E. H., Boyse, E. A., Bennett, D., Scheid, M., & Carswell, E. A. (1971). Serological demonstration of H-Y (male) antigen on mouse sperm. *Nature, 232,* 478–480.

Goldstein, J. I., & Wilson, J. D. (1972). Studies on the pathogenesis of the pseudohermaphroditism in the mouse with testicular feminization. *Journal of Clinical Investigations, 51,* 1647–1658.

Gorzalka, B. B., & Whalen, R. E. (1974). Genetic regulation of hormone action: Selective effects of progesterone and dihydroprogesterone (5α-pregnane-3,20-dione) on sexual receptivity in mice. *Steroids, 23,* 499–505.

Gorzalka, B. B., & Whalen, R. E. (1976). Effects of genotype on differential behavioral responsiveness to progesterone and 5-α-dihydroprogesterone in mice. *Behavioral Genetics, 6,* 7–15.

Gorzalka, B. B., Rezek, D. L., & Whalen, R. E. (1975). Adrenal mediation of estrogen-induced ejaculatory behavior in the male rat. *Physiology and Behavior 14,* 373–376.

Goy, R. W., & Goldfoot, D. A. (1975). Neuroendocrinology: Animal models and problems of human sexuality. *Archives of Sexual Behavior, 4,* 405–420.

Goy, R. W., & Young, W. C. (1957). Strain differences in the behavioral responses of female guinea pigs to alpha-estradiol benzoate and progesterone. *Behaviour, 10,* 340–354.

Grady, K. L., Phoenix, C. H., & Young, W. C. (1965). Role of the developing rat testis in differentiation of the neural tissues mediating mating behavior. *Journal of Comparative and Physiological Psychology, 59,* 176–182.

Griffin, J. E. (1979). Testicular feminization associated with a thermolabile androgen receptor in cultured fibroblasts. *Journal of Clinical Investigations, 64,* 1624–1631.

Grumbach, M. M., & Conte, F. A. (1985). Disorders of sexual differentiation. In J. D. Wilson & D. W. Foster (Eds.), *Williams textbook of endocrinology* (pp. 321–401). Philadelphia: W. B. Saunders.

Hall, J. C., Greenspan, R. J., & Harris, W. A. (1982). *Genetic neurobiology,* Cambridge: MIT Press.

Harris, G. W. (1964). Sex hormones, brain development and brain function. *Endocrinology, 75,* 627–648.

Hart, B. L. (1972). Manipulation of neonatal androgen: Effects on sexual responses and penile development in male rats. *Physiology and Behavior, 8,* 841–845.

Hart, B. L. (1977). Neonatal dihydrotestosterone and estrogen stimulation: Effects on sexual behavior of male rats. *Hormones and Behavior, 8,* 193–200.

Haseltine, F. P., & Ohno, S. (1981). Mechanisms of gonadal differentiation. *Science, 211,* 1272–1278.

He, W. W., Young, C. Y.-F., & Tindall, D. J. (1990). The molecular basis of the mouse testicular feminization (*Tfm*) mutation: a frameshift mutation. *72nd Annual Meeting of the Endocrine Society,* Atlanta, GA 1990, p. 240.

Imperato-McGinley, J. (1983). Sexual differentiation: Normal and abnormal. In L. Martini & V. H. T. James (Eds.), *Current topics in experimental endocrinology* (pp. 231–307). New York: Academic Press.

Imperato-McGinley, J., Peterson, R. E., Gautier, T., & Sturla, E. (1979). Androgens and the evolution of male-gender identity among male pseudohermaphrodites with 5α-reductase deficiency. *New England Journal of Medicine, 300,* 1233–1237.

Ingram, D. K., & Corfman, T. P. (1980). An overview of neurobiological comparisons in mouse strains. *Neuroscience and Biobehavioral Reviews, 4,* 421–435.

Johnson, L. L., Sargent, E. L., Washburn, L. L., & Eicher, E. M. (1982). XY female mice express H-Y antigen. *Developmental Genetics, 3,* 247–253.

Johnson, L. M., & Sidman, R. L. (1979). A reproductive endocrine profile in the diabetes (*db*) mutant mouse. *Biology of Reproduction, 20,* 552–559.

Josso, N., Picard, J. Y., & Tran, D. (1977). The anti-Müllerian hormone. *Recent Progress in Hormone Research, 33,* 117–160.

Jost, A. (1953). Problems of fetal endocrinology: The gonadal and hypophyseal hormones. *Recent Progress in Hormone Research, 8,* 379–418.

Jost, A. (1983). Genetic and hormonal factors in sex differentiation of the brain. *Psychoneuroendocrinology, 8,* 183–193.

Jost, A. (1985). Sexual organogenesis. In N. Adler, D. Pfaff, & R. W. Goy (Eds.), *Handbook of behavioral neurobiology, Vol. 7* (pp. 3–19). New York: Plenum Press.

Jost, A., & Magre, S. (1984). Testicular development phases and dual hormonal control of sexual organogenesis. In M. Serio, M. Motta, M. Zanisi, & R. L. Martin (Eds.), *Sexual differentiation: Basic and clinical aspects, Serono Symposium 11* (pp. 1–15). New York: Raven Press.

Jost, A., Vigier, B., Prepin, J., & Perchellet, J. P. (1973). Studies on sex differentiation in mammals. *Recent Progress in Hormone Research, 29,* 1–41.

Kallman, F. S., Schoenfeld, W. A., & Barrera, S. E. (1944). The genetic aspects of pituitary eunuchoidism. *American Journal of Mental Deficiency, 48,* 203–236.

Kato, J. (1976). Cytosol and nuclear receptors for 5α-DHT and testosterone in the hypothalamus and hypophysis, and testosterone receptors isolated from neonatal female rat hypothalamus. *Journal of Steroid Biochemistry, 17,* 1179–1187.

Kato, J., Atsumi, Y., & Inaba, M. (1971). Development of estrogen receptors in the rat hypothalamus. *Journal of Biochemistry, 70,* 1051–1053.

Kaufman, M., Pinsky, L., Simard, L., & Wong, S. C. (1982). Defective activation of androgen–receptor complexes: A marker of androgen insensitivity. *Molecular and Cellular Endocrinology, 25,* 151–162.

Keenan, B. S., Meyer, W. J. III, Hadjian, A. J., Jones, H. W., & Migeon, C. J. (1974). Syndrome of androgen insensitivity in man: Absence of 5α-dihydrotestosterone binding protein in skin fibroblasts. *Journal of Clinical Endocrinology and Metabolism, 38,* 1143–1146.

Koopman, P., Münsterberg, A., Capel, B., Vivian, N., & Lovell-Badge, R. (1990). Expression of a candidate sex-determining gene during mouse testis differentiation. *Nature, 348,* 450–452.

Koopman, P., Gubbay, J., Vivian, N., Goodfellow, P., & Lovell-Badge, R. (1991). Male development of chromosomally female mice transgenic for *Sry. Nature, 351,* 117–121.

Krey, L. C., Lieberburg, I., MacLusky, N. J., Davis, P. G., & Robbins, R. (1982). Testosterone increases cell nuclear estrogen receptor levels in the brain of the Stanley–Gumbreck pseudohermaphrodite male rat: Implications for testosterone modulation of neuroendocrine activity. *Endocrinology, 110,* 2168–2176.

Krieger, D. T., & Gibson, M. J. (1984). Correction of genetic gonadotropin hormone-releasing hormone deficiency by preoptic area transplants. In J. R. Sladek, Jr., & D. M. Gash (Eds.), *Neural transplants, development and function* (pp. 187–203). New York: Plenum Press.

Krieger, D. T., Perlow, M. J., Gibson, M. J., Davies, T. F., Zimmerman, E. A., Ferin, M., & Charlton, H. M. (1982). Brain grafts reverse hypogonadism of gonadotropin releasing hormone deficiency. *Nature, 298,* 468–471.

Lacroix, A., McKenna, T. J., & Rabinowitz, D. (1979). Sex steroid modulation of gonadotropins in normal men and in androgen insensitivity syndrome. *Journal of Clinical Endocrinology and Metabolism, 48,* 235–240.

Lagerspetz, K. M. J., & Lagerspetz, K. Y. N. (1975). The expression of the genes of aggressiveness in mice: The effect of androgen on aggression and sexual behavior in females. *Aggressive Behavior, 1,* 291–296.

Lubahn, D. B., Joseph, D. R., Sullivan, P. M., Willard, H. F., French, F. S., & Wilson, E. M. (1988). Cloning of human androgen receptor complementary DNA and localization to the X chromosome. *Science, 240,* 327–330.

Luine, V. N., MacLusky, N. J., & McEwen, B. S. (1979). Testosterone effects on enzymes in central and peripheral target sites in *Tfm* mutant mice. *Society for Neuroscience, 5,* 451.

Luttge, W. G., & Hall, M. R. (1973). Differential effectiveness of testosterone and its metabolites in the induction of male sexual behavior in two strains of albino mice. *Hormones and Behavior, 4,* 31–43.

Lyon, M. F. (1961). Gene action in the X-chromosome of the mouse (*Mus musculus L.*). *Nature, 190,* 372–373.

Lyon, M. F. (1972). X-chromosome inactivation and developmental patterns in mammals. *Biology Reviews, 47,* 1–35.

Lyon, M. F., & Glenister, P. H. (1980). Reduced reproductive performance in androgen-resistant *Tfm/ Tfm* female mice. *Proceedings of the Royal Society, London, B208,* 1–12.

Lyon, M. F., & Hawkes, S. G. (1970). X-linked gene for testicular feminization in the mouse. *Nature, 227,* 1217–1219.

Lyon, M. F., Hendry, I., & Short, R. V. (1973). The submaxillary salivary glands as test organs for response to androgen in mice with testicular feminization. *Journal of Endocrinology, 58,* 357–362.

MacLusky, N. J., Chaptal, C., Lieberburg, I., & McEwen, B. S. (1976). Properties and subcellular interrelationships of presumptive estrogen receptor macromolecules in the brains of neonatal and prepubertal female rats. *Brain Research, 114,* 158–165.

MacLusky, N. J., Luine, V. N., Gerlach, J. L., Fischette, C., Naftolin, F., & McEwen, B. S. (1988). The role of androgen receptors in sexual differentiation of the brain: Effects of the testicular feminization (*Tfm*) gene on androgen metabolism, binding, and action in the mouse. *Psychobiology, 16,* 381–397.

Manning, A., & McGill, T. E. (1974). Neonatal androgen and sexual behavior in female house mice. *Hormones and Behavior, 5,* 19–31.

Martin, C. R. (1985). *Endocrine physiology* (pp. 499–569). New York: Oxford University Press.

Mascia, D. N., Money, J., & Ehrhardt, A. A. (1977). Fetal feminization and female gender identity in the testicular feminizing syndrome of androgen insensitivity. *Archives of Sexual Behavior, 1,* 131–142.

Mason, A. J., Hayflick, J. S., Zoeller, R. T., Young, W. S. III, Phillips, H. S., Nikolics, K., & Seeburg, P. H. (1986). A deletion truncating the gonadotropin-releasing hormone gene is responsible for hypogonadism in the hpg mouse. *Science, 234,* 1366–1371.

Mason, A. J., Pitts, S. L., Nikolics, K., Szonyi, E., Wilcox, J. N., Seeburg, P. H., & Stewart, T. A. (1986). The hypogonadal mouse: Reproductive functions restored by gene therapy. *Science, 234,* 1372–1378.

Max, S. R. (1981). Cytosolic androgen receptor in skeletal muscle from normal and testicular feminization mutant (*Tfm*) rats. *Biochemical and Biophysical Research Communications, 101,* 792–799.

Maxson, S. C. (1981). The genetics of aggression in vertebrates. In P. F. Brain & D. Benton (Eds.), *The biology of aggression* (pp. 69–104). Alphen aan den Rijn: Sythoff & Noordhoff.

Maxson, S. C., Ginsburg, B. E., & Trattner, A. (1979). Interaction of Y-chromosomal gene(s) in the development of intermale aggression in mice. *Behavior Genetics, 9,* 219–226.

McDonald, P., Beyer, C., Newton, F. Brien, B., Baker, R., Tan, H. S., Sampson, C., Kitching, P., Greenhill, R., & Pritchard, D. (1970). Failure of 5α-dihydrotestosterone to initiate sexual behavior in the castrated male rat. *Nature, 277,* 964–965.

McGill, T. E. (1962). Sexual behavior in three inbred strains of mice. *Behaviour, 19,* 341–350.

McGill, T. E. (1978a). Genotype–hormone interactions. In T. E. McGill, D. A. Dewsbury, & B. D. Sachs (Eds.), *Sex and behavior* (pp. 161–187). New York: Plenum Press.

McGill, T. E. (1978b). Genetic factors influencing the action of hormones on sexual behavior. In J. B. Hutchison (Ed.), *Biological determinants of sexual behavior* (pp. 7–28). New York: John Wiley & Sons.

McGill, T. E., & Haynes, C. M. (1973). Heterozygosity and retention of ejaculatory reflex after castration in male mice. *Journal of Comparative and Physiological Psychology, 84,* 423–429.

McGill, T. E., & Manning, A. (1976). Genotype and retention of the ejaculatory reflex in castrated male mice. *Animal Behavior, 24,* 507–518.

McGill, T. E., & Tucker, G. R. (1964). Genotype and sex drive in intact and in castrated male mice. *Science, 145,* 514–515.

McLaren, A. (1983). Sex reversal in the mouse. *Differentiation, 23*(Suppl.), S93–S98.

Meaney, M. J., Stewart, J., Poulin, P., & McEwen, B. S. (1983). Sexual differentiation of social play in rat pups is mediated by the neonatal androgen–receptor system. *Neuroendocrinology, 37,* 85–90.

Meisel, R. L., & Ward, I. L. (1981). Fetal female rats are masculinized by male littermates located caudally in the uterus. *Science, 213,* 239–242.

Meyer, W. J. III, Migeon, B. R., & Migeon, C. J. (1975). A locus on human X-chromosome for dihydrotestosterone receptor and androgen-insensitivity. *Proceedings of the National Academy of Sciences, U.S.A., 72,* 1469–1472.

Migeon, C. J., Amrhein, J. A., Keenan, B. S., Meyer, W. J., & Migeon, B. R. (1979). The syndrome of androgen insensitivity in man: Its relation to our understanding of male sexual differentiation. In H. L. Vallet & I. H. Porter (Eds.), *Genetic mechanisms of sexual development* (pp. 93–128). New York: Academic Press.

Migeon, B. R., Brown, T. R., Axelman, J., & Migeon, C. J. (1981). Studies of the locus for androgen receptor: Localization on the human X chromosome and evidence for homology with the *Tfm* locus in the mouse. *Proceedings National Academy of Sciences, U.S.A., 78,* 6339–6343.

Money, J., & Daléry, J. (1976). Iatrogenic homosexuality: Gender identity in seven 46,XX chromosomal females with hyperadrenocortical hermaphroditism born with a penis, three reared as boys, four reared as girls. *Journal of Homosexuality, 1,* 357–371.

Money, J., & Mazur, T. (1978). Endocrine abnormalities and sexual behavior in man. In J. Money & H. Musaph (Eds.), *Handbook of sexology* (pp. 485–492). New York: Elsevier.

Money, J., Ehrhardt, A. A., & Masica, D. N. (1968). Fetal feminization induced by androgen insensitivity in the testicular feminizing syndrome: Effect on marriage and maternalism. *Johns Hopkins Medical Journal, 123,* 105–114.

Money, J., Schwartz, M., & Lewis, V. G. (1984). Adult erotosexual status and fetal hormonal masculinization and demasculinization: 46,XX congenital virilizing adrenal hyperplasia and 46,XY androgen-insensitivity syndrome compared. *Psychoneuroendocrinology, 9,* 405–414.

Moore, C. L. (1985). Another psychobiological view of sexual differentiation. *Developmental Review, 5,* 18–55.

Moralí, G., Acatécatl, M., & Figueroa, M. T. (1986). Comparative study of the masculine copulatory motor pattern of three strains of rats. *Conference on Reproductive Behavior, 18,* 103.

Morris, J. M. (1953). The syndrome of testicular feminization in male pseudohermaphrodites. *American Journal of Obstetrics and Gynecology, 65,* 1192–1211.

Müller, U. (1985). The H-Y antigen: Identification, function, and role in sexuality. In A. A. Sandberg (Ed.), *The Y chromosome, part A: Basic characteristics of the Y chromosome* (pp. 63–80). New York: Alan R. Liss.

Müller, U., Donlon, T., Schmid, M., Fitch, N., Richer, C.-L., Lalande, M., & Latt, S. A. (1986). Deletion mapping of the testis determining locus with DNA probes in 46,XX males and in 46,XY and 46,X,dic(Y) females. *Nucleic Acids Research, 14,* 6489–6505.

Naess, O., Haug, E., Attramadal, A., Aakvaag, A., Hansson, V., & French, F. (1976). Androgen receptors in the anterior pituitary and central nervous system of the androgen "insensitive" (*Tfm*) rat: Correlation between receptor binding and effects of androgens on gonadotropin secretion. *Endocrinology, 99,* 1295–1303.

Naftolin, F., Harris, G. W., & Bobrow, M. (1971). Effect of purified luteinizing hormone releasing factor on normal and hypogonadotrophic anosmic men. *Nature, 232,* 496–497.

Naftolin, F., Ryan, K. S., Davies, I. J., Reddy, V. V., Flores, F., Petro, Z. Kuhn, M., White, R. S., Takaoka, Y., & Wolin, L. (1975). The formation of estrogens by central neuroendocrine tissues. In R. O. Greep (Ed.), *Recent progress in hormone research* (pp. 295–319). New York: Academic Press.

Naftolin, F., Pujol-Amat, P., Corker, C. S., Shane, J. M., Polani, P. E., Kohlinsky, S., Yen, S. S. C., & Bobrow, M. (1983). Gonadotropins and gonadal steroids in androgen insensitivity (testicular feminization) syndrome: Effects of castration and sex steroid administration. *American Journal of Obstetrics and Gynecology, 147,* 491–496.

Nes, N. (1966). Testikulaer feminisering hos storfe. *Norwegian Veternarian Medicine, 18,* 19–29.

New, M. I., Dupont, B., Grumbach, K., & Levine, L. S. (1983). Congenital adrenal hyperplasia and related conditions. In J. B. Stanbury (Ed.), *The Metabolic Basis of Inherited Disease* (pp. 973–1000). New York: McGraw-Hill.

Ohno, S. (1979). *Major sex-determining genes.* New York: Springer Press.

Ohno, S., & Lyon, M. F. (1970). X-linked testicular feminization in the mouse as a non-inducible regulatory mutation of the Jacob–Monod type. *Clinical Genetics, 1,* 121–127.

Ohno, S., Geller, L. N., & YoungLai, E. V. (1974). *Tfm* mutation and masculinization versus feminization of the mouse central nervous system. *Cell, 3,* 235–242.

Olsen, K. L. (1979a). Androgen-insensitive rats are defeminized by their testes. *Nature, 279,* 238–239.

Olsen, K. L. (1979b). Induction of male mating behavior in androgen-insensitive (*tfm*) and normal (King–Holtzman) male rats: Effect of testosterone propionate, estradiol benzoate and dihydrotestosterone. *Hormones and Behavior, 13,* 66–84.

Olsen, K. L. (1983). Genetic determinants of sexual differentiation. In J. Balthazart, E. Pröve, & R. Gilles (Eds.), *Hormones and behavior in higher vertebrates* (pp. 138–158). Heidelberg: Springer-Verlag.

Olsen, K. L. (1985). Aromatization: Is it critical for differentiation of sexually dimorphic behaviors? In R. Gilles & J. Balthazart (Eds.), *Neurobiology* (pp. 149–164). Heidelberg: Springer-Verlag.

Olsen, K. L. (1987). Androgen-insensitive (*Tfm*) mice are defeminized but not masculinized. In *Second World Congress of Neuroscience, Budapest, 22,* S155.

Olsen, K. L., & Fox, T. O. (1981). Differences between androgen-resistant rat and mouse mutants. *Society for Neuroscience, 7,* 219.

Olsen, K. L., & Whalen, R. E. (1981). Hormonal control of the development of sexual behavior in androgen-insensitive (*tfm*) rats. *Physiology and Behavior, 27,* 883–886.

Olsen, K. L., & Whalen, R. E. (1982). Estrogen binds to hypothalamic nuclei of androgen-insensitive (*tfm*) rats. *Experientia, 38,* 139–140.

Olsen, K. L., & Whalen, R. E. (1984). Dihydrotestosterone activates male mating behavior in castrated King–Holtzman rats. *Hormones and Behavior, 18,* 380–392.

Olsen, K. L., Heydorn, W. E., Rodriguez-Sierra, J. F., & Jacobowitz, D. M. (1989). Quantification of proteins in discrete brain regions of androgen-insensitive testicular feminized (*Tfm*) mice, *Neuroendocrinology, 50,* 392–399.

Page, D. C., de la Chapelle, A., & Weissenbach, J. (1985). Chromosome Y-specific DNA in related human XX males. *Nature, 315,* 224–226.

Page, D. C., Brown, L. G., & de la Chapelle, A. (1987). Exchange of terminal portions of X- and Y-chromosomal short arms in human XX males. *Nature, 328,* 437–440.

Page, D. C., Mosher, R., Simpson, E. M., Fisher, E. M. C., Mardon, G., Pollack, J., McGillivray, B., de la Chapelle, A., & Brown, L. G. (1987). The sex-determining region of the human Y chromosome encodes a finger protein. *Cell, 51,* 1091–1104.

Peterson, R. E., & Imperato-McGinley, J. (1984). Male pseudohermaphroditism due to inherited deficiencies of testosterone biosynthesis. In M. Serio, M. Motta, M. Zanisi, & L. Martini (Eds.), *Sexual differentiation: Basic and clinical aspects* (pp. 301–319). New York: Raven Press.

Petit, C., de la Chapelle, A., Levilliers, J., Castillo, S., Noël, B., & Weissenbach, J. (1987). An abnormal terminal X-Y interchange accounts for most but not all cases of human XX maleness. *Cell, 49,* 595–602.

Phoenix, C. H., Goy, R. W., Gerall, A. A., & Young, W. C. (1959). Organizing action of prenatally administered testosterone propionate on the tissues mediating mating behavior in the female guinea pig. *Endocrinology, 65,* 369–382.

Pinsky, L. (1981). Sexual differentiation. In R. Collu, J. R. Ducharme, & H. Guyda (Eds.), *Pediatric endocrinology* (pp. 231–239). New York: Raven Press.

Plapinger, L., & McEwen, B. S. (1973). Ontogeny of estradiol-binding sites in rat brain. I. Appearance of presumptive adult receptors in cytosol and nuclei. *Endocrinology, 93,* 1119–1128.

Plomin, R., DeFries, J. C., & McClearn, G. E. (1980). *Behavioral genetics: A primer.* San Francisco: W. H. Freeman.

Pomerantz, S. M., Fox, T. O., Sholl, S. A., Vito, C. C., & Goy, R. W. (1985). Androgen and estrogen receptors in fetal rhesus monkey brain and anterior pituitary. *Endocrinology, 116,* 83–89.

Purvis, K., Haug, E., Clausen, O. P. F., Naess, O., & Hansson, V. (1977). Endocrine status of the testicular feminized male (*TFM*) rat. *Molecular Cellular Endocrinology, 8,* 317–334.

Reddy, V. V. R., Naftolin, F., & Ryan, K. J. (1974). Conversion of androstenedione to estrone by neural tissues from fetal and neonatal rats. *Endocrinology, 94,* 117–121.

Reinisch, J. M. (1976). Effects of prenatal hormone exposure on physical and psychological development in humans and animals: With a note on the state of the field. In E. J. Sachar (Ed.), *Hormones, behavior and psychopathology* (pp. 69–93). New York: Raven Press.

Reis, D. J., Baker, H., Fink, J. S., & Joh, T. H. (1981). A genetic control of the number of dopamine neurons in mouse brain: Its relationship to brain morphology, chemistry and behavior. In E. S. Gershon, S. Matthysse, X. O. Breakefield, & R. D. Ciaranello (Eds.), *Genetic research strategies in psychobiology and psychiatry* (pp. 215–230). Pacific Grove: The Boxwood Press.

Robinson, S. M., Fox, T. O., & Sidman, R. L. (1985). A genetic variant in the morphology of the medial preoptic area in mice. *Journal of Neurogenetics, 2,* 381–388.

Rosenfeld, J. M., Daley, J. D., Ohno, S., & YoungLai, E. V. (1977). Central aromatization of testosterone in testicular feminized mice. *Experientia, 33,* 1392–1393.

Selmanoff, M. K., & Ginsburg, B. E. (1981). Genetic variability in aggression and endocrine function in inbred strains of mice. In P. F. Brain & D. Benton (Eds.), *Multidisciplinary approaches to aggressive research* (pp. 247–268). Amsterdam: Elsevier/North-Holland Biomedical Press.

Selmanoff, M. K., Jumonville, J. E., Maxson, S. C., & Ginsburg, B. E. (1975). Evidence for a Y chromosomal contribution to an aggressive phenotype in inbred mice. *Nature, 253,* 529–530.

Selmanoff, M. K., Maxson, S. C., & Ginsburg, B. E. (1976). Chromosomal determinants of intermale aggressive behavior in inbred mice. *Behavior Genetics, 6,* 53–69.

Selmanoff, M. K., Goldman, B. D., & Ginsburg, B. E. (1977). Developmental changes in serum luteinizing hormone, follicle stimulating hormone and androgen levels in males of two inbred mouse strains. *Endocrinology, 100,* 122–127.

Shapiro, B. H., & Goldman, A. S. (1973). Feminine saccharin preference in the genetically androgen insensitive male rat pseudohermaphrodite. *Hormones and Behavior, 4,* 371–375.

Shapiro, B. H., Goldman, A. S., Steinbeck, H. F., & Neumann, F. (1976). Is feminine differentiation of the brain hormonally determined? *Experientia, 32,* 650–651.

Shapiro, B. H., Levine, D. C., & Adler, N. T. (1980). The testicular feminized rat: A naturally occurring model of androgen independent brain masculinization. *Science, 209,* 418–420.

Sheridan, P. J. (1978). Localization of androgen- and estrogen-concentrating neurons in the diencephalon and telencephalon of the mouse. *Endocrinology, 103,* 1328–1334.

Sheridan, P. J., & Melgosa, R. T. (1983). Aromatization of testosterone to estrogen varies between strains of mice. *Brain Research, 273,* 285–289.

Sheridan, P. J., Sar, M., & Stumpf, W. (1975). Estrogen and androgen distribution in the brain of neonatal rats. In W. E. Stumpf & L. D. Grant (Eds.), *Anatomical neuroendocrinology* (pp. 134–141). Basel: S. Karger.

Sherins, R. J., & Bardin, C. W. (1971). Preputial gland growth and protein synthesis in the androgen-insensitive male pseudohermaphroditic rat. *Endocrinology, 89,* 835–841.

Shire, J. G. M. (1981). Genes and hormones in mice. In R. J. Berry (Ed.), *Biology of the house mouse* (pp. 547–574). New York: Academic Press.

Shrenker, P., & Maxson, S. C. (1983). The genetics of hormonal influences on male sexual behavior of mice and rats. *Neuroscience and Biobehavioral Reviews, 7,* 349–359.

Shrenker, P., & Maxson, S. C. (1984). The DBA/1Bg and DBA/2Bg Y chromosomes compared for their effects on male sexual behavior. *Behavioral and Neural Biology, 42,* 33–37.

Sidman, R. L. (1982). Mutations affecting the central nervous system in the mouse. In F. O. Schmitt, S. J. Bird, & F. E. Bloom (Eds.), *Molecular genetic neuroscience* (pp. 389–400). New York: Raven Press.

Siiteri, P. K., & Wilson, J. D. (1974). Testosterone formation and metabolism during male sexual differentiation in the human embryo. *Journal of Clinical Endocrinology and Metabolism, 38,* 113–125.

Silvers, W. K., Gasser, D. L., & Eicher, E. M. (1982). H-Y antigen, serologically detectable male antigen and sex determination. *Cell, 28,* 439–440.

Simon, N. G. (1979). The genetics of intermale aggressive behavior in mice: Recent research and alternative strategies. *Neuroscience and Biobehavioral Reviews, 3,* 97–106.

Simpson, E., Chandler, P., Goulmy, E., Disteche, C. M., Ferguson-Smith, M. A. & Page, D. C. (1987). Separation of the genetic loci for the H-Y antigen and for testis determination on human Y chromosome. *Nature, 326,* 876–878.

Sinclair, A. H., Berta, P., Palmer, M. S., Hawkins, J. R., Griffiths, B. L., Smith, M. J., Foster, J. W., Frischauf, A.-M., Lovell-Badge, R., & Goodfellow, P. N. (1990). A gene from the human sex-determining region encodes a protein with homology to a conserved DNA-binding motif. *Nature, 346,* 240–244.

Singh, L., & Jones, K. W. (1982). Sex reversal in the mouse (*Mus musculus*) is caused by a recurrent nonreciprocal crossover involving the X and an aberrant Y chromosome. *Cell, 28,* 205–216.

Södersten, P. (1973). Estrogen-activated sexual behavior in male rats. *Hormones and Behavior, 4,* 247–256.

Staats, J. (1976). Standardized nomenclature for inbred strains of mice: Sixth listing. *Cancer Research, 36,* 4333–4377.

Stanley, A. J., & Gumbreck, L. G. (1964). Male pseudohermaphroditism with feminizing testes in the male rat—a sex-linked recessive character. *The Endocrine Society, 46,* 40.

Stanley, A. J., Gumbreck, L. G., Allison, J. E., & Easley, R. B. (1973). Part I. Male pseudohermaphroditism in the laboratory Norway rat. *Recent Progress in Hormone Research, 29,* 43–64.

Stewart, A. D., Manning, A., & Batty, J. (1980). Effects of Y-chromosome variants on the male behaviour of the mouse *Mus musculus. Genetic Research, 35,* 261–268.

Swerdloff, R. S., Batt, R. A., & Bray, G. A. (1976). Reproductive hormonal function in the genetically obese (*ob/ob*) mouse. *Endocrinology, 98,* 1359–1364.

Tettenborn, U., Dofuku, R., & Ohno, S. (1971). Noninducible phenotype exhibited by a proportion of female mice heterozygous for the X-linked testicular feminization mutation. *Nature New Biology, 234,* 37–40.

Thompson, M. L., & Edwards, D. A. (1971). Experimental and strain determinants of the estrogen-progesterone induction of sexual receptivity in spayed female mice. *Hormones and Behavior, 2,* 299–305.

Thorton, J. E., Wallen, K., & Goy, R. W. (1987). Lordosis behavior in males of two inbred strains of guinea pig. *Physiology and Behavior, 40,* 703–709.

Vadász, C., Kobor, G., & Lajtha, A. (1982). Neuro-behavioral genetic analysis in recombinant inbred strains. In I. Lieblich (Ed.), *Genetics of the brain* (pp. 127–154). Amsterdam: Elsevier Biomedical Press.

Vadász, C., Baker, H., Fink, S. J., & Reis, D. J. (1985). Genetic effects and sexual dimorphism in tyrosine hydroxylase activity in two mouse strains and their reciprocal F_1 hybrids. *Journal of Neurogenetics, 2,* 219–230.

Vale, J. R., Ray, D., & Vale, C. A. (1973). The interaction of genotype and exogenous neonatal androgen and estrogen: Sex behavior in female mice. *Developmental Psychobiology, 6,* 319–327.

Vale, J. R., Ray, D., & Vale, C. A. (1974). Neonatal androgen treatment and sexual behavior in males of three inbred strains of mice. *Developmental Psychobiology, 7,* 483–488.

van der Schoot, P. (1980). Effects of dihydrotestosterone and oestradiol on sexual differentiation in male rats. *Journal of Endocrinology, 84,* 397–407.

Verhoeven, G., & Wilson, J. D. (1976). Cytosol androgen binding in submandibular gland and kidney of the normal mouse and the mouse with testicular feminization. *Endocrinology, 99,* 79–92.

Vito, C. C., & Fox, T. O. (1979). Embryonic rodent brain contains estrogen receptors. *Science, 204,* 517–519.

Vito, C. C., & Fox, T. O. (1982). Androgen and estrogen receptors in embryonic and neonatal rat brain. *Developmental Brain Research, 2,* 97–110.

Vito, C. C., Wieland, S. J., & Fox, T. O. (1979). Androgen receptors exist throughout the critical period of brain sexual differentiation. *Nature, 282,* 308–310.

Vito, C. C., Baum, M. J., Bloom, C., & Fox, T. O. (1985). Androgen and estrogen receptors in perinatal ferret brain. *Journal of Neuroscience, 5,* 268–274.

Wachtel, S. S. (1983). *H-Y antigen and the biology of sex determination.* New York: Grune & Stratton.

Wachtel, S. S., Koo, G. C., & Boyse, E. A. (1975). Evolutionary conservation of H-Y ("male") antigen. *Nature, 254,* 270–272.

Wachtel, S. S., Ohno, S., Koo, G. C., & Boyse, E. A. (1975). Possible role for H-Y antigen in the primary determination of sex. *Nature, 257,* 235–236.

Ward, B. J., & Charlton, H. M. (1981). Female sexual behaviour in the GnRH deficient, hypogonadal (hpg) mouse. *Physiology and Behavior, 27,* 1107–1109.

Ward, I. L. (1972). Prenatal stress feminizes and demasculinizes the behavior of males. *Science, 175,* 82–84.

Ward, I. L. (1983). Effects of maternal stress on the sexual behavior of male offspring. In M. Schlumpf & W. Lichtensteiger (Eds.), *Monograph in neural science, Vol. 9: Drugs and hormones in brain development* (pp. 169–175). Basel: S. Karger.

Ward, I. L., & Weisz, J. (1984). Differential effects of maternal stress on circulating levels of corticosterone, progesterone and testosterone in male and female rat fetuses and their mothers. *Endocrinology, 114,* 1635–1644.

Ward, O. B., Jr., Orth, J. M., & Weisz, J. (1983). A possible role of opiates in modifying sexual differentiation. In M. Schlumpf & W. Lichtensteiger (Eds.), *Monograph in neural science, Vol. 9: Drugs and hormones in brain development* (pp. 194–200). Basel: S. Karger.

Wee, B. E. F., Weaver, D. R., Goldman, B. D., & Clemens, L. G. (1986). Hormonal correlates of masculine sexual behavior after castration. *Conference on reproductive behavior, 18,* 24.

Weisz, J., & Gibbs, C. (1974). Metabolites of testosterone in the brain of the newborn female rat after an injection of tritiated testosterone. *Neuroendocrinology, 14,* 72–86.

Weisz, J., & Ward, I. L. (1980). Plasma testosterone and progesterone titers of pregnant rats, their male and female fetuses, and neonatal offspring. *Endocrinology, 106,* 306–316.

Whalen, R. E. (1974). Sexual differentiation: Models, methods and mechanisms. In R. C. Friedman, R. M. Richart, & R. L. Vande Wiele (Eds.), *Sex differences in behavior* (pp. 467–481). New York: John Wiley & Sons.

Whalen, R. E. (1982). Current issues in the neurobiology of sexual differentiation. In A. Vernadakis & P. S. Timiras (Eds.), *Hormones in development and aging* (pp. 273–304). New York: Spectrum Publications.

Whalen, R. E., & Edwards, D. A. (1967). Hormonal determinants of the development of masculine and feminine behavior in male and female rats. *Anatomical Records, 157,* 173–180.

Whalen, R. E., & Olsen, K. L. (1981). Role of aromatization in sexual differentiation: Effects of prenatal ATD treatment and neonatal castration. *Hormones and Behavior, 15,* 107–122.

Whalen, R. E., Yahr, P., & Luttge, W. G. (1985). The role of metabolism in hormonal control of sexual behavior. In N. Adler, D. Pfaff, and R. W. Goy (Eds.), *Handbook of behavioral endocrinology, Vol. 7* (pp. 609–993). New York: Plenum Press.

Whalen, R. E., Gladue, B. A., & Olsen, K. L. (1986). Lordotic behavior in male rats: Genetic and hormonal regulation of sexual differentiation. *Hormones and Behavior, 20,* 73–82.

Wieland, S. J., & Fox, T. O. (1979). Putative androgen receptors distinguished in wild-type and testicular-feminized (*Tfm*) mice. *Cell, 17,* 781–787.

Wieland, S. J., & Fox, T. O. (1981). Androgen receptors from rat kidney and brain; DNA-binding properties of wild-type and *tfm* mutant. *Journal of Steroid Biochemistry, 14,* 409–414.

Wieland, S. J., Fox, T. O., & Savakis, C. (1978). DNA-binding of androgen and estrogen receptors from mouse brain: Behavior of residual androgen receptor from *Tfm* mutant. *Brain Research, 140,* 159–164.

Wilson, J. D. (1978). Sexual differentiation. *Annual Review of Physiology, 40,* 279–306.

Wilson, J. D., George, F. W., & Griffin, J. E. (1981). The hormonal control of sexual development. *Science, 211*, 1278–1284.

Wilson, J. D., Griffin, J. E., George, F. W., & Leshin, M. (1981). The role of gonadal steroids in sexual differentiation. *Recent Progress in Hormone Research, 37*, 1–39.

Wilson, J. D., Griffin, J. E., Leshin, M., & MacDonald, P. C. (1983). The androgen resistance syndromes: 5α-Reductase deficiency, testicular feminization, and related disorders. In J. B. Stanbury (Ed.), *The metabolic basis of inherited disease* (pp. 1001–1026). New York: McGraw-Hill.

Wimer, R. E., & Wimer, C. C. (1985). Animal behavior genetics: A search for the biological foundations of behavior. In M. R. Rosenzweig & L. W. Porter (Eds.), *Annual Review of Psychology, 36*, 171–218.

Withyachumnarnkul, B., & Edmonds, E. S. (1982). Plasma testosterone levels and sexual performance of young obese male Zucker rats. *Physiology and Behavior, 29*, 773–777.

Yarbrough, W. G., Quarmby, V. E., Simental, J. A., Joseph, D. R., Sar, M., Lubahn, D. B., Olsen, K. L., French, F. S., & Wilson, E. M. (1990). A single base mutation in the androgen receptor gene causes androgen insensitivity in the testicular feminized rat. *The Journal of Biological Chemistry, 265*, 8893–8900.

Young, C. Y., Johnson, M. P., Prescott, J. L., & Tindall, D. (1989). The androgen receptor of the testicular-feminized (*Tfm*) mutant mouse is smaller than the wild-type receptor. *Endocrinology, 124*, 771–775.

Young, J. K., Fleming, M. W., & Matsumoto, D. E. (1986). Sex behavior and the sexually dimorphic hypothalamic nucleus in male Zucker rats. *Physiology and Behavior, 36*, 881–886.

2

Sex Differences in Neuronal Morphology Influenced Hormonally throughout Life

S. A. Tobet and T. O. Fox

Introduction

Sexual dimorphism is a phrase that has come to refer to sex differences in form, which includes qualitative or quantitative differences in structure or function (Goy & McEwen, 1980). This chapter deals exclusively with sexual dimorphisms in neural morphology and focuses on how age and hormone milieu interact to influence their appearance. We begin with a brief description of the types of measures (Table 1) that have been used in different model systems that show sex and hormone dependence. This is followed by a closer look at four exemplary systems (preoptic area/hypothalamus; amygdala; spinal cord motor neurons; and telencephalic song nuclei) that use these dependent measures to illustrate the emerging principles that govern the development and maintenance of sexual dimorphisms. We then examine two recently growing areas in sexual dimorphism research: sex differences visualized using immunocytochemistry and sex differences in nonneuronal cells such as glia or ependyma. Finally, general guidelines for studying sexual dimorphisms with a life-span approach are discussed.

Measures of Dimorphic Characteristics

The earliest attempts to delineate sex differences in brain morphology date to the mid-19th century, when macroscopic determinations of sizes of the whole brain

S. A. Tobet and T. O. Fox   Program in Neuroscience, Harvard Medical School, and Department of Biochemistry, E. K. Shriver Center, Waltham, Masschusetts 02254.

Sexual Differentiation, Volume 11 of *Handbook of Behavioral Neurobiology*, edited by Arnold A. Gerall, Howard Moltz, and Ingeborg L. Ward, Plenum Press, New York, 1992.

TABLE 1. SEXUALLY DIMORPHIC
CHARACTERISTICS

Gross size: regions or whole brain
Microscopic size: nuclei, nucleoli
  Perikaryal size
  Synaptic connections
  Dendritic morphology
Cell number, cytoarchitectonics
Chemically defined subsystems
  Peptides, proteins, etc.

and of specific brain regions were made (see Gould, 1981; Swaab & Hofman, 1984). These gross measures, which are still used, have revealed differences in the brains of several species (e.g., human, Swaab & Hofman, 1984; ferret, S. A. Tobet & M. J. Baum, unpublished observation; rat, Yanai, 1979). Two of the earliest morphological characteristics measured microscopically with respect to hormonal influence were the size of individual cell nuclei and of nucleoli. Experiments indicated that gonadectomy or estrogen treatment changed cell nuclear and nucleolar sizes (Szentagothai, Flerko, Mess, & Halasz, 1962; Lisk & Newlon, 1963; Ifft, 1964). Subsequent experiments investigated sex differences in these cellular characteristics (Pfaff, 1966; Dorner & Staudt, 1968).

By the early 1970s morphological correlates of hormone action in the brain were examined with increased resolution by electron microscopy. The ultrastructure of perikarya (Zambrano & de Robertis, 1968; Brawer, 1971) and synaptic connections (Raisman & Field, 1971) were found to be hormone dependent. Degeneration techniques were used in conjunction with electron microscopy to examine the sites of termination of the stria terminalis (Raisman & Field, 1971, 1973). By comparing numbers of dendritic shaft versus spine synapses, they revealed a sex difference in the distribution of nonamygdaloid terminals in one region of the preoptic area. Neonatal hormone exposure and sex were related to plasticity of ultrastructural features observed in neurons in adult rats' arcuate nuclei (Ratner & Adamo, 1971; King, 1972).

During the 1970s, investigators increasingly utilized multiple techniques to analyze sex-dependent neuronal connectivity. For example, electrophysiological experiments (Dyer, MacLeod, & Ellendorff, 1976) demonstrated sex differences in neural circuitry. By recording from cells in the rat's hypothalamus following stimulation of the amygdala, Dyer and colleagues (1976) identified more cells in males than females that were coupled. With Golgi impregnation, sex differences in the extent, density, and orientation of dendritic trees were described (Greenough, Carter, Steerman, & DeVoogd, 1977).

In 1971, Calaresu and Henry reported that the number of neurons in the intermediolateral cell column of the cat spinal cord was greater in males than females. Large histologically evident sex differences were noted by Nottebohm and Arnold (1976) in the volume of forebrain nuclear groupings in the brains of songbirds. This measure of size, which is the composite result of neuron number, packing density, and neuropil volume, sparked the search for similar histological sex differences in mammalian species. Subsequently, differences in volume of a region within the medial preoptic area were reported in rats (Gorski, Gordon, Shryne, & Southam, 1978). This finding spurred others to look for similar sexually dimorphic regions in other species (see below). Recently, immunocytochemical investigations have revealed sex- or hormone-dependent neural subsystems (e.g., King, Elkind, Gerall, & Millar, 1978; DeVries, Buijs, & Swaab, 1981). Specific systems may be constituted as neurotransmitter, peptide, second messenger, structural protein, or any other type of molecule that is not uniformly distributed throughout the brain. Although sex differences can be detected by quantification of various molecules in homogenized tissues, immunocytochemical techniques permit evaluation of the subtle heterogeneity within populations of cells that are lost in homogenates.

Questions

In characterizing the development and maintenance of sexual dimorphisms, several basic questions about specificity, location, timing, and cause can be asked of each system. The temporal dependence of sexual dimorphisms on hormone has been categorized in three main ways (Goy & McEwen, 1980). First, some dimorphisms only are apparent when appropriate gonadal steroids are in the circulation. Thus, the dimorphism may not be absolutely sex dependent but can be hormone dependent and demonstrable in both sexes. Second, some sexually dimorphic characteristics are independent of the hormonal environment and therefore can be detected in both intact and gonadectomized animals. In some cases sexually dimorphic characteristics may be demonstrable under hormone-independent conditions but amplified in the presence of hormones. Third, some sex differences only become apparent after both sexes are stressed, such as in the case of a lesion (Nance, Shryne, & Gorski, 1975; Loy & Milner, 1980). These categories raise an important question in examining any sexual dimorphism: *In what way is the sexual dimorphism dependent on the concurrent hormone environment?* (See Table 2.)

TABLE 2. SEXUAL DIMORPHISMS DEPENDENT ON CONCURRENT HORMONE ENVIRONMENT

Preoptic area/hypothalamus	The SDA in gerbils—a nuclear area that is enlarged in the presence of testosterone	Commins & Yahr, 1984b
Amygdala	Structural changes not characterized	
Spinal cord motor neurons	The size of neurons in the SNB in rats is dependent on circulating androgen	Breedlove & Arnold, 1981
Telencephalic nuclei that control avian vocal behavior	The size of cell groups and dendritic trees of individual neurons in canaries is dependent on testosterone exposure	Nottebohm, 1980; DeVoogd & Nottebohm, 1981a, 1981b
Neuropeptide and monoamine systems	Detectability of LHRH in neuronal perikarya is dependent on circulating steroid; castration transiently decreases the detectability of LHRH in males and females over different time courses.	King *et al.*, 1987
Glial cells	Testosterone may influence the number of glial cells in the adult canary telencephalon	Goldman & Nottebohm, 1983

Many morphological actions of gonadal hormones only occur during restricted, "critical" periods. One can ask how the appearance of a particular morphological characteristic depends on the *prior* hormonal history of the individual. The question raised is: *In what way is the appearance of a sexual dimorphism related to the action of gonadal steroids during earlier critical periods?*

The ultimate appearance of sexual dimorphisms can become apparent after a critical period has passed and elevated hormone levels are no longer present. Thus, in the amygdala, the ultimate number of synapses can be regulated by testosterone exposure postnatally, but the effect on synapse number is not observed until weeks after the hormone exposure (Nishizuka & Arai, 1981a). The temporal uncoupling can be envisioned in several ways. For instance, if the half-life of a hormone-induced protein is longer than the duration of steroid exposure, then a hormone-dependent event could occur after the animal has been exposed to hormone and before the disappearance of the protein. If several steps must occur between the initial direct hormone response and the eventual appearance of a structure, this lapse might be a long time (days, weeks, months). Another possibility is that steroid–receptor complexes alter chromatin structure such that particular genes are rendered inactive or are poised for future activation by hormones or other signals. The issue raises the question: *What is the time course for the appearance of sexual dimorphic phenotypes, and are sequential steps, including sequential hormonal signals, involved?* (See Table 3.)

Critical periods describe the limited stages during which the initiation of devel-

TABLE 3. FIRST APPEARANCE OF SEXUAL DIMORPHISMS AT AGES AFTER THE CRITICAL PERIOD OF HORMONE ACTION

Preoptic area/ hypothalamus	The rat SDN does not become sexually dimorphic until after birth, although prenatal hormone is important for the maximal increase in SDN volume	Jacobson *et al.*, 1981; Ito *et al.*, 1986
Amygdala	After early hormone exposure, sex differences in synapse number take 1 to 2 weeks to appear in the NMA	Nishizuka & Arai, 1981b
Spinal cord motor neurons	For the SNB, sex difference appears shortly after birth, but timing of hormone exposure to increase SNB size is flexible	Breedlove & Arnold, 1983b
Telencephalic nuclei that control avian vocal behavior	For soma areas in RA, the period of hormone sensitivity overlaps the period of morphological changes	Konishi & Akutagawa, 1988
Neuropeptide and monoamine systems	Immunoreactivity in Met-enkephalin fibers is sensitive to perinatal hormone(s) but not sexually dimorphic, phenotypically, until activational hormones are presented	Watson *et al.*, 1986
Glial cells	None characterized	

opment of sexually dimorphic features is permitted. Once formed, sexual dimorphisms may persist for differing lengths of time (Arnold & Breedlove, 1985). Longevity can be expected to be a function of the primary steroid action and of the intrinsic half-lives of the affected features. Thus, hormones might induce factors that are important for the formation of a synapse but that do not necessarily lend any characteristic longevity to the derived feature. Individual metabolites of gonadal steroids affect alternate aspects of neural morphology. Research over the last four decades has shown that in males, circulating androgens (testosterone in particular) are important for the differentiation of multiple sex-dependent functions in many different species (reviewed: Baum, 1979; Goy & McEwen, 1980; MacLusky & Naftolin, 1981). In the brain, androgens can be metabolized through two distinct pathways (Figure 1). Multiple hydroxylations of the androgen A-ring by the aromatase enzyme complex lead to "aromatization," resulting in estrogen formation. Ring A reduction leads to further androgenic metabolites (Callard, Petro, & Ryan, 1978; Martini, 1978; Jouan & Samperez, 1980; Tobet, Shim, Osiecki, Baum, & Canick, 1985). The resulting metabolites then interact with estrogen and androgen receptors in brain cells (Fox *et al.*, 1982). The heterogeneity of metabolites and receptors raises another important question to consider about sexual dimorphisms: *Which gonadal steroids and gonadal steroid receptors are important for the appearance of selective dimorphisms?*

After sexually dimorphic structures or regions are localized in a given species, consideration needs to be given to determining *whether the sexual dimorphism results from direct, primary actions of steroid hormones in the cells in which hormone is bound to specific receptors or from actions secondary to the initial steps in hormone response.* A morphological cascade (Toran-Allerand, 1984) involving connectivity does not re-

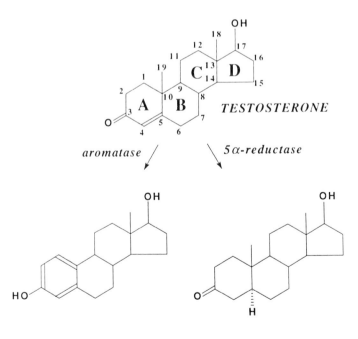

Figure 1.   Testosterone acts directly and as a metabolic precursor either to estradiol via aromatization or to 5α-dihydrotestosterone via reduction.

quire hormone involvement at every point. For instance, if the direct effect of a hormone is to produce a factor that causes dendritic outgrowth, the decision to form synapses might depend on responses to other influences. The resulting numbers and types of synapses would have been influenced by hormones, and the end result would therefore be secondarily hormone dependent: the action of steroid hormone could be necessary but not sufficient. Molecular cascades can be anticipated in which hormone-dependent products control a range of posttranslational modifications (e.g., glycosylation, phosphorylation, sulfation). Consequently, important endpoints would be modified substrates whose primary syntheses are not hormone dependent. By these interactions of hormone-dependent and hormone-independent activities, a spectrum of phenotypes can result.

Detailed descriptions of cellular and biochemical mechanisms that create sex differences in morphology will provide answers to the questions raised above. All mechanisms that act in the development of the nervous system are potential sites of hormone influence that could result in sexual dimorphisms, some of which may be transient and others permanent. At the cellular level, these include neuronal cell death, proliferation, and migration (Arnold & Gorski, 1984; Konishi & Akutagawa, 1985). Progress is being made in identifying potential mechanisms within the model systems considered below.

SEX DIFFERENCES IN NEURAL MORPHOLOGY: EXEMPLARY SYSTEMS

Sexual dimorphisms in various neural regions illustrate principles of sexual differentiation in the nervous system. Additional details have been reviewed by others (T. J. DeVoogd, S. M. Breedlove, & R. A. Gorski, cited by De Vries, De Bruin, Uylings, & Corner, 1984; Arnold, Bottjer, Brenowitz, Nordeen, & Nordeen, 1986).

PREOPTIC AREA/HYPOTHALAMUS

One easily identified sexual dimorphism in the mammalian forebrain is the difference in gross volume of a nuclear region in the preoptic/anterior hypothalamic area. In the rat, a nucleus has been labeled the sexually dimorphic nucleus of the preoptic area (SDN-POA) (Gorski, Harlan, Jacobson, Shryne, & Southam, 1980), and it lies between the midline crossing of the anterior commissure and the optic chiasm. It is embedded in a region important for the regulation of many homeostatic and/or sex-dependent functions. The region begins rostrally with the medial preoptic area and merges caudally with the medial anterior hypothalamic area and the medial division of the bed nucleus of the stria terminalis (Figure 2). Since the initial description of the SDN-POA in the rat, a number of sex-dependent cytoarchitectonic subdivisions have been recognized throughout this region in a number of species (Bleier, Byne, & Siggelkow, 1982; Commins & Yahr, 1984a, 1984b; Takami & Urano, 1984; Hines, Davis, Coquelin, Goy, & Gorski, 1985; Robinson, Fox, & Sidman, 1985; Tobet, Zahniser, & Baum, 1986a; Viglietti-Panzica *et al.*, 1986; Swaab & Fliers, 1985; Allen, Hines, Shryne, & Gorski, 1989).

Sex differences in POA/AH morphology have been examined in some detail with respect to the dependence on gonadal steroids in rodents and carnivores. For example, sex differences have been found in gross nuclear volume or shape of nuclei in the rat (SDN-POA, Gorski *et al.*, 1980), gerbil [the sexually dimorphic area (SDA) and its pars compacta (SDApc), Commins & Yahr, 1984b], guinea pig [sex-

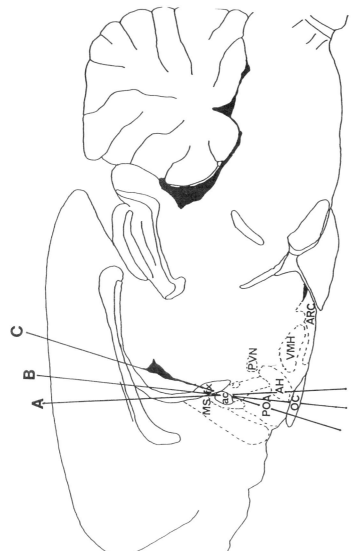

Figure 2. Parasagittal section of rat brain (adapted from Paxinos & Watson, 1982) illustrates different angles of sectioning (A, B, or C), which produce alternate dorsal/ventral alignments of neural structures. Thus, in coronal plane A, the preoptic area overlays different portions of the anterior hypothalamic area than in coronal plane C. MS, medial septum; ac, anterior commissure; POA, preoptic area; AH, anterior hypothalamic area; OC, optic chiasm; VMH, ventromedial nucleus of the hypothalamus; PVN, paraventricular nucleus of the hypothalamus; ARC, arcuate nucleus of the hypothalamus; fx, fornix.

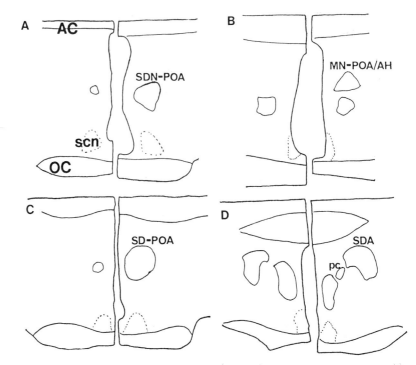

Figure 3. Schematic drawings are adapted from coronal sections through the border region of the preoptic area and anterior hypothalamus from (A) rats (Dohler, Coquelin *et al.,* 1984), (B) ferrets (Tobet *et al.,* 1986a), (C) guinea pigs (Hines *et al.,* 1985), and (D) gerbils (Commins & Yahr, 1984a). In each panel the left half represents the nuclear configuration of a female, and the right half the configuration of a male. [More detailed parcellations of the rat sexually dimorphic nucleus of the preoptic area (SDN-POA) can be found in Simerly *et al.* (1984) and Bloch & Gorski (1988).] Male nucleus of the preoptic/anterior hypothalamic area, MN-POA/AH; sexually dimorphic preoptic area, SD-POA; sexually dimorphic area, SDA; pars compacta, pc; optic chiasm, OC; anterior commissure, AC; suprachiasmatic nucleus, scn.

ually dimorphic preoptic area (SD-POA), Hines *et al.,* 1985; Byne & Bleier, 1987], and ferret [the male nucleus of the POA/AH (MN-POA/AH), Tobet *et al.,* 1986a, 1986b] (Figure 3). In these species sex differences are apparent regardless of the adult hormonal milieu. The volumes of the SDN-POA or MN-POA/AH are not affected by adult hormonal status (Gorski *et al.,* 1978; Hines *et al.,* 1985; Tobet *et al.,* 1986a), whereas in the gerbil the volume of the SDA is significantly increased in the presence of testosterone (Commins & Yahr, 1984b). Although the volume of the MN-POA/AH is not measurably affected in the ferret, the cells in the MN-POA/AH are 20% larger in the presence of gonadal steroids (Tobet *et al.,* 1986a). An effect of gonadal steroids on the size of perikarya or cell nuclei is a characteristic common to a number of systems that are hormone responsive, such as certain neurons in the amygdala, spinal cord motoneurons, and songbird telencephalic neurons.

Within the many reports on sex differences in nuclear volumes, a wide range of methods and criteria have been used to define and quantitate these differences. Early reports on differences in nuclear volume utilized multiple tracings of particular nuclei, which were cut out, weighed, and averaged for several investigators. Even with precautions to conceal specification of the sex of the animal from which a tissue section was taken, investigators could inadvertently develop perceptual recognition of the distinctive appearance of each sex, making objective specification of the nuclear boundary difficult. Methods are being developed to objectify the characterization of sexually dimorphic nuclei. Optical and computer-assisted methods can be used to set density thresholds as criteria for measurements. For nuclei in which somal sizes differ between males and females, these methods can be used to count cells and measure their size (Tobet *et al.*, 1986a; see Figure 4).

Computer-assisted methods give highly quantitative and reproducible numbers, which, with clear operational definitions, can be used as reliable measurements after methods are validated. Comparison of males and females are made by examining sequential sections through entire nuclei. Publications, however, usually contain only photographs of selected pairs of sections, thereby limiting readers to a small portion of the available morphological information. Computer-assisted three-dimensional reconstructions can aid in conveying this information and have the added benefit of simplifying determination of nuclear shapes, volumes, and surface areas (Lydic, Albers, Tepper, & Moore-Ede, 1982; Robinson, Fox, Dikkes, & Pearlstein, 1986; see Figure 5).

Sex differences in POA/AH morphology depend on the action of gonadal steroids during critical periods early in development. Testosterone administered to neonatal female rats increases the volume of the SDN-POA in adulthood, and cas-

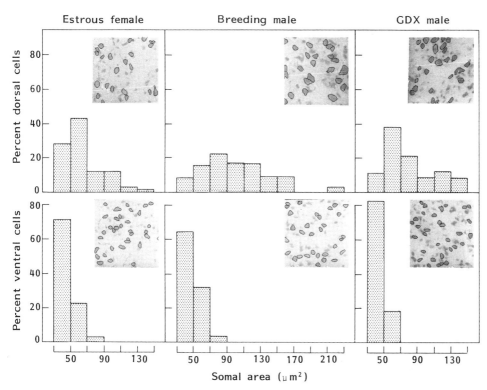

Figure 4.   Semiautomated process for delineating group differences in cell size. Histograms represent the frequency of specific cell sizes as measured using an image analysis system applied to specific windows of tissue. In this case each inset is one of several 128 × 128-μm panels used to generate the related frequency histogram. The data illustrate the significantly greater proportion of large cells in the MN-POA/AH (labeled dorsal cells) of the breeding male ferret as compared to the same nucleus in gonadectomized (GDX) males or the same region in estrous females. Also shown are cells in the ventral nucleus present in these animals. (From Tobet *et al.*, 1986a, with permission of Elsevier Science Publishers.)

S. A. TOBET AND
T. O. FOX

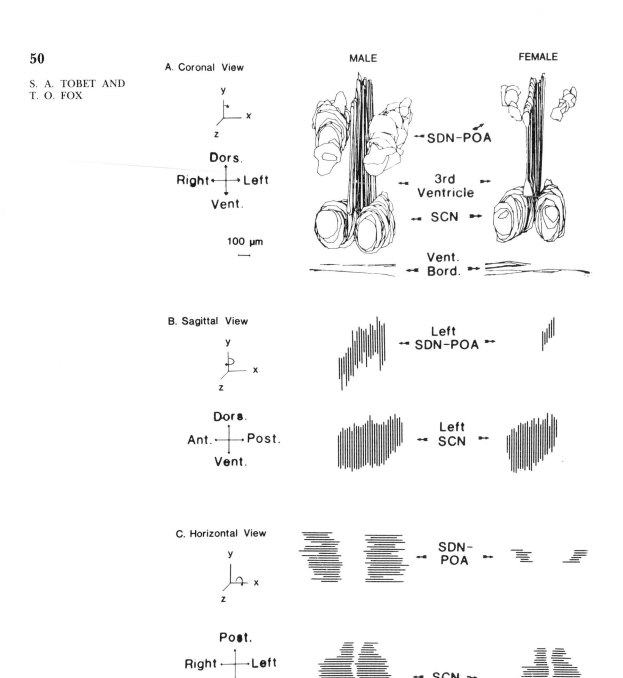

Figure 5. Three-dimensional computer reconstructions of the sexually dimorphic nucleus of the preoptic area (SDN-POA) and the suprachiasmatic nucleus (SCN) from adult male and female rats. (From Robinson *et al.* (1986), with permission of Elsevier Science Publishers.)

ration of neonatal males reduces the volume of the nucleus in adulthood (Gorski *et al.*, 1978; Jacobson, Csernus, Shryne, & Gorski, 1981; Handa, Corbier, Shryne, Schoonmaker, & Gorski, 1985). These neonatal manipulations, however, do not result in complete sex reversals of SDN-POA volume. Neonatally androgenized females have SDN-POA volumes that are larger than the volume in control females but smaller than the normal range for males. Measurements made after only prenatal testosterone administration indicate that exposure during this period also can contribute to SDN-POA development (Ito, Murakami, Yamanouchi, & Arai, 1986). Accordingly, when testosterone is given both pre- and postnatally, the volume of the SDN-POA in adult genetic females becomes indistinguishable from that in normal males (Dohler, Coquelin, *et al.*, 1982; Dohler, Coquelin, *et al.*, 1984). It is important to note that multiple steps in the differentiation process may act cumulatively or sequentially, inducing new states that are apparently more sensitive to the next hormone-dependent stimulus (Fox, 1987; Tobet & Baum, 1987). Thus, similar hormonal stimuli given to gonadectomized male and female rats after birth elicit a larger SDN-POA in males when examined in adulthood (Jacobson *et al.*, 1981). This is in keeping with the importance of combined prenatal and postnatal treatments described above.

The presence of the MN-POA/AH in ferrets, the SD-POA in guinea pigs, and the SDApc in gerbils is also dependent on exposure to gonadal hormones early in life. In ferrets and guinea pigs the critical period appears to be primarily prenatal. Thus, neonatal castration of male ferrets (Tobet *et al.*, 1986b) or guinea pigs (Byne & Bleier, 1987) does not disturb the normal organization of the MN-POA/AH or SD-POA examined at older ages. As expected, genetic females exposed *in utero* to exogenous testosterone have discernible MN-POA/AH or larger SDN-POA in adulthood. In gerbils, although neonatal castration of males did not disturb the neural organization of the SDApc, neonatal testosterone administration induced the organization of a masculine SDApc in females (Ulibarri & Yahr, 1988). For both rats and ferrets the age at which the preoptic area first appears sexually dimorphic is not concurrent with the critical period for hormone action. In the rat, although prenatal hormone exposure is needed for complete masculinization, the sex difference in SDN-POA volume is not detectable until after birth, when the number and size of cells that comprise the SDN-POA increase (Hsu, Chen, & Peng, 1980; Jacobson, Shryne, Shapiro, & Gorski, 1980; Hammer & Jacobson, 1984). In ferrets hormonal stimulation must be prenatal for the MN-POA/AH to form; however, the MN-POA/AH is not clearly discernible until after birth (S. A. Tobet, C. A. Gallagher, & M. J. Baum, unpublished observations) or only shortly before birth (Cherry *et al.*, 1990).

In rats (Jacobson & Gorski, 1981) and guinea pigs (Byne, Warren, & Bleier, 1987), many cells that comprise the SDN-POA are born later than most of the surrounding neurons in the POA/AH, as determined by tritiated-thymidine birthdate autoradiography. Thus, age-dependent changes in cells that come to comprise the SDN-POA can be followed. Sex differences in the number and temporal gradient of cell generation may account for part of the ultimate sex difference in SDN-POA volume in rats (Jacobson & Gorski, 1981). For the subpopulation of cells born only on embryonic day 18, hormones do not affect cell growth, migration, or number (Jacobson, Davis, & Gorski, 1985), leading to the two hypotheses that differential cell death may be hormone dependent and that cells born on days other than embryonic day 18 may be responsible for the sexual dimorphism in SDN-POA volume.

Specific gonadal steroid effects on the volume of the SDN-POA have been examined in adulthood and during development. Either testosterone or estradiol plus progesterone administered to animals gonadectomized in adulthood has no effect on SDN-POA volume in male or female rats (Gorski et al., 1978) or guinea pigs (Hines et al., 1985). However, perinatal treatment of females with testosterone and with the synthetic estrogen diethylstilbestrol results in significantly larger SDN-POA volumes in adulthood (Dohler, Hines, et al., 1982; Dohler, Coquelin, et al., 1984; Byne & Bleier, 1987; Hines, Alsum, Roy, Gorski, & Goy, 1987). These results suggest the importance of estrogenic metabolites for morphological differentiation. To delineate further the specificity of perinatal hormone action, researchers have treated animals both pre- and postnatally with the androgen receptor antagonist cyproterone acetate or with the estrogen receptor antagonist tamoxifen (Dohler, Srivastava, et al., 1984; Hines et al., 1987). If testosterone acts primarily by binding to androgen receptors, cyproterone acetate should diminish the effect. If, however, cyproterone acetate has no effect but tamoxifen does, then it is likely that estrogen converted from testosterone acts after binding to the estrogen receptor. Cyproterone acetate does not diminish the volume of the SDN-POA in males, whereas tamoxifen treatment does, thus supporting the hypothesis that the volume of the SDN-POA is estrogen dependent.

In line with the hypothesis that estrogens contribute to brain sexual differentiation in both sexes (Toran-Allerand, 1976, 1978, 1980; Dohler, 1978), tamoxifen-treated females have significantly smaller SDN-POA volumes than control females (Dohler, Srivastava, et al., 1984). It is harder, however, to interpret a result that decreases the volume of a nucleus to below the normal range, which raises the caution that possible toxic effects of synthetic hormones should be considered. In ferrets, treatment of mothers over the last 12 days of gestation with 1,4,6-androstatriene-3,17-dione (ATD), which inhibits aromatization of androgens to estrogens, inhibits the development of the MN-POA/AH in genetic male offspring. In contrast, timed treatment with the androgen receptor antagonist flutamide has no disorganizing effect (Tobet et al., 1986b). These results strongly support the hypothesis that estrogenic metabolites of androgen are important in diverse mammalian species for morphogenesis in this region of the HPOA (preoptic area/hypothalamus).

Cells that comprise sexually dimorphic features of the POA/AH are likely candidates for direct hormone action since, via autoradiography, estradiol binding has been noted in neonatal (Schoonmaker, Breedlove, Arnold, & Gorski, 1983) and adult (Jacobson, Arnold, & Gorski, 1987) rat SDN-POA cells and gerbil SDA cells (Commins & Yahr, 1985). Biochemical experiments also have indicated high levels of estradiol receptor binding throughout this region in rats (Rainbow, Parsons, & McEwen, 1982) and ferrets (Holbrook & Baum, 1983; Vito, Baum, Bloom, & Fox, 1985). Additionally, the enzyme responsible for the conversion of testosterone to estradiol, aromatase, is present in the POA/AH of rats (Tobet, Baum, Tang, Shim, & Canick, 1985) and ferrets (Tobet, Shim, et al., 1985) at higher levels in males than females at particular points during the critical period when estrogens are important for morphological differentiation. Higher aromatase substrate levels (i.e., plasma testosterone, Weisz & Ward, 1980; Erskine & Baum, 1982) with or without (Tobet, Shim, et al., 1985; Tobet & Baum, 1987) higher aromatase levels would combine to provide more estrogenic stimulation to developing male brains.

The morphogenetic actions of gonadal steroids may be mediated via a hormone that accumulates in the affected cell nuclei (Toran-Allerand, Gerlach, & McEwen,

1980). When estradiol, however, is implanted directly into the preoptic area of neonatal rats, the size of the SDN-POA is not increased (Nordeen & Yahr, 1982), even though the implants affect the differentiation of sexual behavior. One possible explanation for this result, yet to be tested, is that estradiol may masculinize the SDN-POA by acting at distant site(s). Another is that steroid hormones are required to act at multiple sites, both local and distal, to permit sexual dimorphism of this nucleus.

The HPOA contributes to the regulation of many sex-dependent functions (reviewed: Hart, 1974; Hayward, 1977; Kelley & Pfaff, 1978; Gorski, 1984; Gerall A. A. & L. Givon, Chapter 10). To date, however, no simple, direct relationship has been demonstrated between a specific sex difference in HPOA morphology and a specific sex-dependent function. Brown-Grant and colleagues (Brown-Grant, Murray, Raisman, & Sood, 1977) attempted to test the hypothesis that the sex differences in nonamygdaloid synapses in the dorsal preoptic area of rats (Raisman and Field, 1971, 1973) were related to the ability to maintain a cyclic pattern of gonadotropin secretion. Discrete lesions in this sexually dimorphic area result in a high incidence of repetitive diestrous animals. These effects, however, are transient, and animals eventually recover normal cycles. Jacobson and colleagues (Jacobson, Terkel, Gorski, & Sawyer, 1980) attempted to test the hypothesis that the SDN-POA of the rat is involved in the ability to display maternal behavior. However, disruption of maternal behavior following discrete POA/AH lesions has not correlated with the degree of SDN-POA damage (Jacobson, Terkel, *et al.*, 1980).

A positive correlation has been noted by some authors between the presence (Gorski *et al.*, 1978) and size (Anderson, Fleming, Rhees, & Kinghorn, 1986) of the SDN-POA on the one hand and the ability to display masculine copulatory behavior on the other. Arendash and Gorski (1983) directly tested the hypothesis that the SDN-POA of the rat was important for the display of masculine copulatory behavior. Although lesions of the SDN-POA do not decrease the display of masculine coital behaviors significantly, lesions of a more dorsal POA/AH site do (Christensen, Nance, & Gorski, 1977; Arendash & Gorski, 1983). Yahr and colleagues tested the hypothesis that the SDA of the gerbil POA/AH is important for the maintenance of scent-marking behavior and/or male copulatory behavior (Yahr, Commins, Jackson, & Newman, 1982; Commins & Yahr, 1984c). The data indicate that lateral SDA lesions may be more effective than lesions in other parts of the POA/AH in disrupting masculine sexual behavior and ventral scent marking (Commins & Yahr, 1984c). In male ferrets, small lesions that include the MN-POA/AH do not disrupt masculine coital behaviors; however, feminine proceptive responsiveness is enhanced (Cherry & Baum, 1988).

Resolution of these questions relating structure and function may come from a more detailed temporal description of stages within the critical period for sexual differentiation. The ferret, having a longer gestation period and longer hormone-dependent "critical period" than rats or gerbils, is a useful animal for such an analysis. In the male ferret, behavioral data suggest that the differentiation of the capacity to display masculine sexual behaviors is a two-stage process (Tobet & Baum, 1987). Estrogen is important prenatally for the differentiation of behavioral capacities, and testosterone postnatally (Baum, 1976; Baum, Gallagher, Martin, & Damassa, 1982; Baum, Canick, Erskine, Gallagher, & Shim, 1983; Baum & Erskine, 1984). Estrogen derived from the neural aromatization of androgen might sensitize the developing brain for the subsequent action of testosterone (Tobet & Baum, 1987). We hypothesize that the formation of the MN-POA/AH may provide the anatomic substrate of sensitization and that testosterone may significantly modify the functional connectivity of cells in the MN-POA/AH. Increased intensity of either the first or second stage of the process might compensate for low intensity in the other step. This two-step morphological hypothesis remains to be tested.

Although less well defined with respect to hormonal dependence and development/life-span stages, many types of neural sexual dimorphism have been reported in other parts of the HPOA. The earliest sex differences reported in this region were in the size of perikaryal nuclei (Pfaff, 1966; Arai & Kusama, 1968; Dorner & Staudt, 1968, 1969a, 1969b), although not all reports agree on the direction of the difference. In rats, the volume of nuclei other than the SDN is also dependent on sex. In some experiments the volume of the suprachiasmatic nuclei (Gorski et al., 1978; Robinson et al., 1986) and that of the ventromedial nuclei (Matsumoto & Arai, 1983) are larger in males than females. In some reported sex differences the dependent measures are greater in females than in males. For example, the density of cells in the periventricular region of the preoptic area is greater in females than males (Bleier et al., 1982; Watson, Hoffman, & Weigand, 1986).

Sex differences in dendritic organization are found in the POA/AH of hamsters (Greenough et al., 1977) and primates (macaque) (Ayoub, Greenough, & Juraska, 1983). These differences in dendritic structure are of both orientation (hamster) and spine and bifurcation density (macaque). In a landmark study, Raisman and Field (1971) described a sex difference in synaptic organization within a dorsal region of the preoptic area that was detectable by electron microscopy after lesions were made in the stria terminalis. The number of nonamygdaloid synapses on dendritic spines was greater in females than in males (Raisman & Field, 1971, 1973). Sex differences at the ultrastructural level are also found in the arcuate (Matsumoto & Arai, 1980, 1981) and ventromedial (Matsumoto & Arai, 1986a, 1986b) nuclei of the hypothalamus. In the arcuate nucleus, female rats have approximately twice as many synapses on dendritic spines as males, but half as many synapses on perikarya. In the ventrolateral portion of the ventromedial nucleus, males have approximately one-third more axodendritic synapses (spines and shafts) than do females. Additional ultrastructural sex differences have been noted in the suprachiasmatic nucleus. More synapses (5–30%) were observed in male than in female rats (LeBlond, Morris, Karakiulakus, Powell, & Thomas, 1982; Guldner, 1984), and more synapses on dendritic spines in males than females (Guldner, 1982). Based on experiments involving neonatal hormone manipulations, synaptic sex differences have been attributed to the action of neonatal hormones in the strial portion of the POA (Raisman & Field, 1973), the arcuate (Matsumoto & Arai, 1976, 1981), and ventromedial nuclei of the hypothalamus (Matsumoto & Arai, 1986a).

AMYGDALA

Significant sex differences in various morphological characteristics of the medial nucleus of the amygdala (NMA) are apparent in squirrel monkeys, toads, and rats. In squirrel monkeys, the diameters of perikaryal nuclei in the NMA are approximately 10% larger in males than in females (Bubenik & Brown, 1973). In toads, the volume of the medial and lateral amygdala is more than 50% greater in males than females (Takami & Urano, 1984). In rats, the volume of the NMA is approximately 20% larger in males than females, whereas the volume of the lateral nucleus of the amygdala is not significantly different between the sexes (Mizukami, Nishizuka, & Arai, 1983). Furthermore, at the ultrastructural level, males have approximately

25% more synapses ending on dendritic shafts (as opposed to dendritic spines or cell bodies) in the NMA than females (Nishizuka & Arai, 1981a, 1983).

The NMA may be part of a sexually dimorphic neural circuit (Simerly & Swanson, 1986, 1987) that also includes the SDN-POA (discussed previously) and the bed nucleus of the stria terminalis (BST). The medial posterior portion of the BST is 22% larger in volume in male than female rats (del Abril, Segovia, & Guillamon, 1987), and the encapsulated portion of the BST is 36% larger in male than female guinea pigs (Hines *et al.*, 1985). Sex differences in peptide content in these interrelated nuclei are discussed further below.

Sex differences in synapse number may depend on the early action of gonadal steroids. In the NMA of rats, neonatal castration of males decreases the number of shaft synapses visible in adulthood to female levels, and neonatal administration of testosterone (day 5) to females increases the number of shaft synapses to male levels (Nishizuka & Arai, 1981a). However, the effects of adult steroid treatment on the morphology of NMA neurons or synapses have not been described. Since synaptic characteristics of some neurons can be affected by exposure to gonadal steroids in adulthood, these neonatal experiments need to be reexamined in the context of equivalent adult hormonal states.

As previously noted for the POA/AH, sex differences in the amygdala are manifested after the period considered critical for hormone action. In the rat NMA the sex difference in synapse number develops after a latency period following the initial exposure to hormone(s). A single injection of testosterone given to 5-day-old female rats increases the NMA synapse number to levels seen in males by 90–100 days of age but is not evident by day 11. This parallels the finding that sex differences in NMA volume are not observed until postnatal day 21 during normal development (Mizukami *et al.*, 1983).

It has been hypothesized that the sexual dimorphism in the rat NMA may be estrogen dependent (Nishizuka & Arai, 1981b; Mizukami *et al.*, 1983). Treatment of female rats with estradiol for the first 30 days of life results in a phenotypic male NMA with respect to both shaft synapses (Nishizuka & Arai, 1981b) and gross nuclear volume (Mizukami *et al.*, 1983). Further experiments are needed to confirm this hypothesis, since the estradiol was administered over a long period of time, following a different paradigm than that used with testosterone. Additionally, no experiments were carried out to test the effectiveness of nonestrogenic androgen metabolites or to block endogenous estrogen or androgen action. Therefore, it is premature to conclude that synaptic or volumetric changes in the NMA only require the action of estrogen. As for the time of manifestation of the effect of testosterone on synapse number, the synaptogenic effect of estradiol treatment is not evident until postnatal day 21 (Nishizuka & Arai, 1981b). When the NMA of a neonatal female rat is transplanted into an eye of an adult gonadectomized female, estradiol administration to the host results in a greater number of shaft synapses in the graft as compared to grafts in vehicle-treated hosts (Nishizuka & Arai, 1982). Thus, the effect of estrogen on shaft synapse formation may be independent of neural afferents and indicate a direct action of estrogen in the NMA.

Although many sex-specific functions have been ascribed to the medial amygdala, none have been correlated with these observed sexual dimorphisms. Possible functions include roles in regulating gonadotropin secretion, sexual behavior, play behavior, and the onset of puberty (Kawakami & Terasawa, 1972; Baum & Goldfoot, 1975; Masco & Carrer, 1980; Meaney, Dodge, & Beatty, 1981; Baum, Tobet, Starr, & Bradshaw, 1982). Additionally, it is noteworthy that the NMA contributes functionally identifiable fibers to the preoptic and anterior hypothalamic areas (Dyer *et al.,* 1976) that regulate many of the same sex-specific functions.

Spinal Cord Motor Neurons

In the fifth and sixth lumbar segments of the male rat spinal cord are two motoneuron cell groups that innervate muscles attached to the penis. The first group includes the bulbocavernosus and levator ani muscles, which are innervated by the spinal nucleus of the bulbocavernosus (SNB), and the second group includes the ischiocavernosus muscle, which is innervated by the dorsolateral motor nucleus (DLN) (Breedlove, 1984). Normally these muscles are absent in adult females. Nevertheless, a DLN is present in females, where it contains fewer neurons of smaller size than in males (Jordan, Breedlove, & Arnold, 1982). Consistent with the absence of known comparable muscles, the SNB is not discernible as a nucleus in females, although comparable neurons, fewer in number (approximately 60 versus 200) and smaller in size than those in the male SNB, are located within the same region (Breedlove & Arnold, 1980, 1981) (Figure 6). As demonstrated by autoradiography (Breedlove & Arnold, 1980, 1983c), a characteristic of these neurons is their ability to bind testosterone or its metabolites. Homologous sex-dependent spinal motor nuclei also are found in mice (Wee & Clemens, 1987), dogs, and humans (Forger & Breedlove, 1986).

The size of neurons in the SNB and DLN in adulthood is affected by androgen

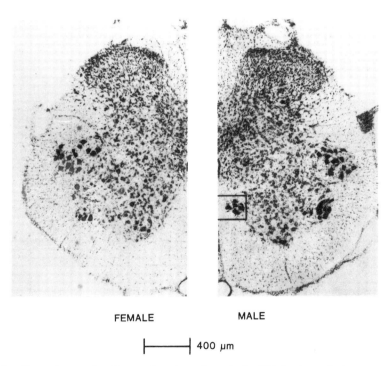

FEMALE   MALE

|——————| 400 μm

Figure 6. Photomicrographs of coronal sections through the fifth lumbar spinal segments of adult female and male rats. The spinal nucleus of the bulbocavernosus (SNB) is recognizable as a discrete nucleus only in the male. The arrows demarcate the region containing the SNB in the male and the homologous region in the female. (From Breedlove, 1984, with permission of Elsevier Science Publishers.)

levels in mature animals. The size of SNB neurons in males is decreased following adult castration and restored with testosterone replacement (Breedlove & Arnold, 1981). These changes in somal size are paralleled by changes in the extent of dendritic arborization (Kurz, Sengelaub, & Arnold, 1986), the amount of membrane with synaptic contacts (Leedy, Beattie, & Breshnahan, 1987; Matsumoto, Micevych, & Arnold, 1988), and the number of gap junctions between motor neurons (Matsumoto, Arnold, Zampighi, & Micevych, 1988). Testosterone administered to ovariectomized adult females results in larger neurons in the region comparable to the SNB (Breedlove & Arnold, 1981). In neither case, however, do adult hormonal manipulations reverse the significant sex difference in somal size, and the difference in neuronal number is not at all affected.

Hormone exposure during a critical period(s) early in development is essential for the normal SNB and DLN phenotype seen in adulthood. A single injection of testosterone propionate on day 2 of neonatal life increases the number of neurons in the SNB (Breedlove, Jacobson, Gorski, & Arnold, 1982) and DLN (Jordan *et al.*, 1982), although the somal size of these neurons is not affected. Treating females with testosterone or dihydrotestosterone for longer times during each of three different perinatal periods increases SNB somal size measured in adulthood when somal size is maximized by concurrent testosterone treatment (Breedlove & Arnold, 1983b). In contrast to somal size, prenatal and early postnatal (days 1, 3, and 5), but not late postnatal (days 7, 9, and 11), steroid administration increases neuronal number in adulthood (Breedlove & Arnold, 1983b).

Sex differences in SNB motor neuron number that appear over two sequential phases during perinatal life suggest a delineation of potential mechanisms of development. In the first phase, the prenatal phase, the number of SNB cells increases in both males and females, but with more cells being added in males than females (Nordeen, Nordeen, Sengelaub, & Arnold, 1985). In both sexes, many of these cells have been shown to send projections to the periphery (Sengelaub & Arnold, 1986). Since all potential SNB neurons are born before gestational days 14–15, testicular hormones do not affect neurogenesis (Breedlove, Jordan, & Arnold, 1983). However, motor neurons are migrating from the lateral motor column, medially, toward the SNB during the critical period and may contribute partially to the sex differences in neuron number (Nordeen *et al.*, 1985; Sengelaub & Arnold, 1986; Figure 7). The full magnitude of the sex difference, however, develops over the early postnatal period, when cell death decreases the number of motor neurons by 70% in females and 25% in males (Nordeen *et al.*, 1985). Thus, mechanisms that promote motor neuron survival or prevent cell death may be the major factors in establishing the male SNB phenotype. These mechanisms probably exist at the level of the SNB neurons and their target muscles, both of which are hormone sensitive, and not from supraspinal sources, since early spinal transection does not affect the differentiation of the SNB in females (Fishman & Breedlove, 1985).

Estrogen does not appear sufficient for SNB development. Neonatal administration of estradiol does not masculinize neuronal number, whereas testosterone does (Breedlove *et al.*, 1982). Thus, an androgen receptor mechanism is suggested for SNB development. In line with this hypothesis, the SNB is not discernible in rats in which a genetic defect in androgen receptor function prevents androgen but not estrogen action (Breedlove & Arnold, 1981). Additionally, blockade of the androgen receptor by administering flutamide to the mother during the last half of gestation significantly reduces both neuronal number and somal size in the SNB (Breedlove & Arnold, 1983a). All androgens, however, may not act in the same way,

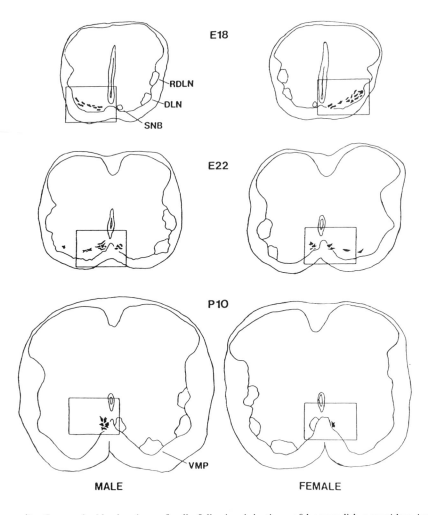

Figure 7. Camera lucida drawings of cells following injections of horseradish peroxidase into the bulbocavernosus muscle in male and female rats from embryonic day 18 (E18) to postnatal day 10 (P10). The drawings illustrate the distribution of cells that eventually migrate into the spinal nucleus of the bulbocavernosus (SNB). The changing number of neurons represented also illustrates the greater degree of cell death during the perinatal period in female rats. DLN, dorsolateral nucleus; RDLN, retrodorsolateral nucleus; VMP, ventral motor pool; calibration bar, 500 μm. (From Sengelaub & Arnold, 1986, with permission of the *Journal of Neuroscience*.)

because prenatal dihydrotestosterone does not influence neuronal number in the SNB, whereas prenatal testosterone does (Breedlove & Arnold, 1983b). Prenatal exposure to dihydrotestosterone, however, does influence the development of motor neurons outside the SNB (Breedlove, 1985). This suggests that the SNB is a neural site at which testosterone, rather than one of its metabolites, may be the active ligand.

The presence of a discernible SNB does not correlate with the ability to show postural aspects of masculine copulatory behavior in adult rats. Postnatal administration of dihydrotestosterone to females masculinizes the SNB, but it does not result in the differentiation of masculine copulatory behavior (Breedlove & Arnold, 1983b). Moreover, prenatal flutamide administration prevents the normal development of the SNB but does not influence such aspects of masculine sexual behavior as mounting (Breedlove & Arnold, 1983a). Indeed, the evidence suggests the importance of estrogenic metabolites in the differentiation of male copulatory behavior in the rat (reviewed: Baum, 1979; Goy & McEwen, 1980; Olsen, 1983). It is likely that the penile muscles that are innervated by the SNB and DLN are involved in penile erectile capacity and fertility (Sachs, 1982; Hart & Melese-D'Hospital, 1983; Wallach & Hart, 1983). The motoneuron contribution to these functions, however, has not been specified. Interestingly, spinal implants of testosterone restore penile reflexes in spinally transected, castrated rats without significant signs of peripheral leakage (Hart & Haugen, 1968), suggesting that hormonal activation of neurons in the lower cord such as the SNB, below the transection, is sufficient for the activation of the penile reflexes.

TELENCEPHALIC NUCLEI THAT CONTROL AVIAN VOCAL BEHAVIOR

Male canaries and zebra finches sing or vocalize in stereotypic manners not normally expressed in females (Kelley, 1978; Arnold, 1980a; Konishi & Gurney, 1982). Furthermore, the vocal behavior of male birds is androgen dependent (Kelley, 1978). The circuitry for the neural control of this vocal behavior in male birds has been described in great detail (Nottebohm, Stokes, & Leonard, 1976) (Figure 8). Significant sex differences in gross nuclear volumes exist in canaries and zebra finches in area X, hyperstriatum ventrale pars caudale (HVc), nucleus robustus archistriatalis (RA), magnocellular nucleus of the anterior neostriatum (MAN), and the tracheosyringeal portion of the hypoglossal motor nucleus (nXIIts) (Nottebohm & Arnold, 1976; Gurney & Konishi, 1980; Gurney, 1981, 1982) (Figure 9). These

CENTRAL CONTROL OF SONG IN THE CANARY

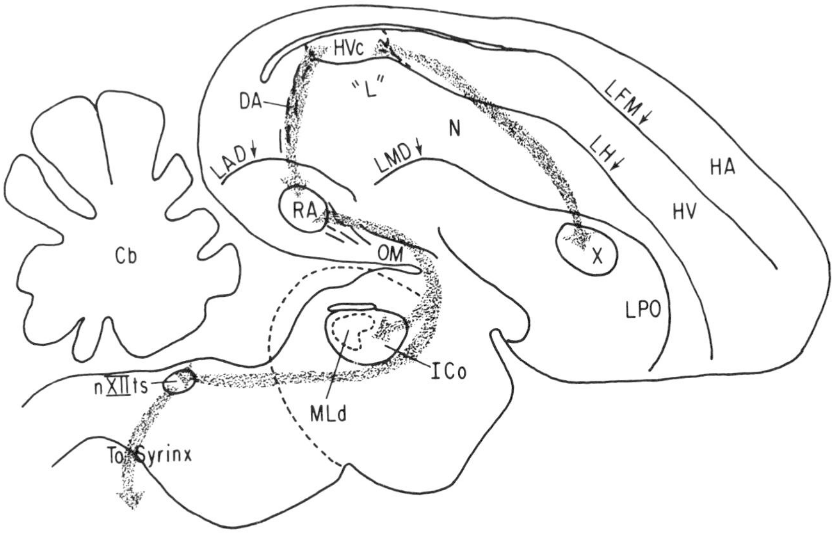

Figure 8. Schematic sagittal section of a canary brain illustrates the nuclei and pathways that are implicated in the control of singing behavior. (From Nottebohm *et al.,* 1976.) The major nuclei of interest are nucleus robustus archistriatalis (RA), hyperstriatum ventrale pars caudale (HVc), magnocellular nucleus of the anterior neostriatum (MAN), area X (X), and the tracheosyringeal portion of the hypoglossal nucleus (nXIIts), nucleus intercollicularis (ICo). (Copyright 1976 by Wiley-Liss; reprinted with permission of Wiley-Liss, a division of John Wiley & Sons, Inc.)

S. A. TOBET AND
T. O. FOX

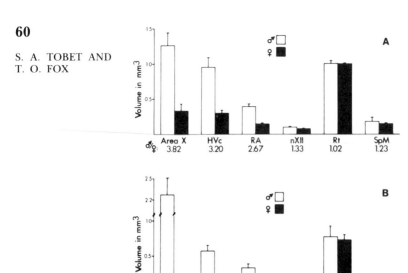

Figure 9. Volume measurements of neural regions involved in the control of song in adult male and female canaries (A) and zebra finches (B). Each bar indicates the total volume (left plus right) ± the standard deviation. The ratio of male to female volumes for each region are given below each region. Hyperstriatum ventrale pars caudale, HVc; robust nucleus of the archistriatum, RA; the hypoglossal nucleus of the medulla, nXII; Two thalamic nuclei are not associated with vocal control: nucleus rotundus, Rt; nucleus spiriformis medialis, SpM. (From Nottebohm & Arnold, 1976, copyright 1976 by the AAAS.)

nuclei are all larger in males (1.5- to 5.5-fold). Other types of morphological sex differences have also been reported in these systems. The size of perikarya and the extent of dendritic arborization of neurons in RA of canaries and zebra finches are greater in males than females (DeVoogd & Nottebohm, 1981a, 1981b; Gurney, 1981). Steroid binding capacity may also be sex specific by region. A greater percentage of autoradiographically detectable cells are found in the HVc and MAN of male than female zebra finches following a single injection of tritiated testosterone (Arnold & Saltiel, 1979; Nordeen, Nordeen, & Arnold, 1986, 1987). Similarly, more soluble tritiated dihydrotestosterone binding was detected in the male than female zebra finch telencephalon after DNA-cellulose chromatography (Siegel, Akutagawa, Fox, Konishi, & Politch, 1986). Thus, sex dependence appears to be the rule rather than the exception for numerous characteristics in nuclei in the songbird telencephalon.

Songbird brains provide a powerful example of the effect of hormones on the neural morphology of mature animals. In the canary song system, the gross nuclear boundaries of HVc and RA vary seasonally (Nottebohm, 1981) and are affected by adult gonadectomy and testosterone replacement (Nottebohm, 1980). The increase in RA volume is most likely related to an increase in dendritic arborization induced by androgen in adult birds (DeVoogd & Nottebohm, 1981b) and may be related to significant testosterone-induced increases in the number of synapses on RA neurons (DeVoogd, Nixdorf, & Nottebohm, 1985). These changes in the adult song system are species specific. For example, similar changes do not appear in the zebra finch song system (Arnold, 1980b). The effect of adult hormone treatment on dendritic arborization in the zebra finch song system has not been examined. However, in the zebra finch brainstem, adult treatment with either testosterone (Arnold, 1980b; Gurney, 1982) or dihydrotestosterone (Gurney, 1982) increased the volume of the hypoglossal motor nucleus (nXIIts).

Although there is significant plasticity in the vocal control system of some songbird species, as in mammalian systems, there is a significant degree of neural organization that is determined by exposure to gonadal hormones early in development (Konishi & Akutagawa, 1988). The temporal dependence on gonadal steroids is more easily described for the organization of singing capacity. Canaries and zebra finches learn their songs with different temporal dependence: male zebra finches learn one song early in development and can not change later in their life; in contrast, canaries learn new songs each season. Administering testosterone in sufficient quantities at hatching to genetic female zebra finches results in a phenotypic male vocal control system, but adult androgen has no effect on this behavior (Gurney & Konishi, 1980; Gurney, 1981, 1982). In canaries, adult androgen administration causes significant changes in neural morphology (e.g., the volume of telencephalic nuclei and characteristics of the neurons within them) and is required to produce new song.

Results from studies on telencephalic song nuclei support a dual role for testosterone. During the critical period of sexual differentiation in the zebra finch song system, testosterone may act as a prohormone, serving as a precursor for estradiol and dihydrotestosterone formation. These metabolites appear to masculinize the song control system in different ways. The androgenic metabolite dihydrotestosterone increases gross nuclear volume of RA by increasing cell number (Gurney, 1981). The androgenic metabolite estradiol enhances gross nuclear volume of RA by increasing somal size (Gurney, 1981; Pohl-Apel, 1985). In adult canaries testosterone acts directly by binding to androgen receptors and after metabolism by forming estradiol and dihydrotestosterone, although the combined administration of these two was not as effective as testosterone itself in inducing dendritic growth in RA neurons (DeVoogd & Nottebohm, 1981b).

In these neural song systems, effects of estradiol are puzzling, since estradiol uptake has not been seen in adult male zebra finch RA by autoradiographic methods (Arnold, 1979). In the zebra finch it was postulated (Gurney & Konishi, 1980) that estrogen receptors may be present for a brief critical period in development. Recently, estrogen receptors were characterized in the adult zebra finch telencephalon (Siegel *et al.*, 1986). In the adult canary it was suggested that an estrogen effect on RA neurons was not necessarily a direct action of estrogen on neurons within RA (DeVoogd & Nottebohm, 1981b). One hypothesis is that differentiation and appearance of sexual dimorphism result from a genomic action of specific steroids within the particular cells in which they are accumulated (Toran-Allerand *et al.*, 1980). Experiments will be needed in the songbird and other sexually dimorphic systems to fully test this hypothesis.

The vocal control system of birds has been a useful model for correlating structure (neural morphology) and function (ability to sing). The neural morphology is quantifiable and in some species shows a great deal of plasticity. Singing behavior is also quantifiable and is clearly hormone dependent. In both canaries and zebra finches there is a close correlation between the ability of various hormone treatments to induce singing and the ability of those same treatments to affect neural morphology in the telencephalic song control nuclei, RA and HVc (DeVoogd, 1984; Pohl-Apel, 1985). As a rule, gonadal hormones contribute to the organization of neural morphology in the song control nuclei when there is a critical period for the patterning of song. In canaries, where adult testosterone treatment mediates large changes in RA neurons, it can induce female birds to sing (Arnold, 1980b). Neonatal treatments with estradiol or testosterone, which affect the differentiation of somal size, produce females that sing as adults in response to concurrent steroid treatments that also affect neural morphology (testosterone or dihydrotestosterone) (Gurney, 1981, 1982; Pohl-Apel, 1985).

S. A. TOBET AND
T. O. FOX

Another type of "morphological" sex difference has become obvious with the application of immunocytochemical techniques for localizing many small molecules and macromolecules. Measurements of neurotransmitters in homogenates from selective brain regions have shown significant sex differences in the concentrations of various neurotransmitters, neuromodulators, and other molecules important for neural function (see Vaccari, 1980; McEwen, 1981). Localization of these same macromolecules by immunocytochemistry provides a morphological context in which to analyze sex-dependent gene expression. This biochemical morphology can reveal sex or hormone dependence in the same structure(s) that classically have been considered sites of morphological sex differences, such as the SDN-POA or periventricular preoptic area (Simerly, Swanson, & Gorski, 1984; Watson et al., 1986). The extra biochemical information gained in this context is aiding in determining the functional importance of many sexual dimorphisms already described. It is also indicating additional areas that are under hormonal control.

There is a growing list of neurotransmitter-associated (i.e., enzyme or transmitter) or neuropeptide systems that are partially characterized with respect to sex and hormone dependence. These include: *luteinizing hormone-releasing hormone* (LHRH) in the diffuse system in which it is distributed (King et al., 1978, 1987; King, Tobet, Snavely, & Arimura, 1980; Elkind-Hirsch, King, Gerall, & Arimura, 1981; Shivers, Harlan, Morrell, & Pfaff, 1983; Wray & Hoffman, 1986), *vasopressin* in the lateral septum and bed nucleus of the stria terminalis (De Vries et al., 1981; De Vries, Best, & Sluiter, 1983; De Vries, Buijs, & Sluiter, 1984; De Vries, Buijs, Van Leeuwen, Caffe, & Swaab, 1985; Van Leeuwen, Caffe, & De Vries, 1985; Mayes, Watts, McQueen, Fink, & Charlton, 1988), *substance P* in the amygdala and bed nucleus of the stria terminalis (Dees & Koslowski, 1984; Malsbury & McKay, 1987), *cholecystokinin* in the preoptic area, bed nucleus of the stria terminalis, and amygdala (Simerly, Gorski, & Swanson, 1986; Simerly & Swanson, 1987; Micevych, Park, Akesson, & Elde, 1987; Micevych, Akesson, & Elde, 1988), *opioid peptides* in the periventricular preoptic area (Watson et al., 1986; Watson, Wiegand, & Hoffman, 1988; Simerly, McCall, & Watson, 1988), *serotonin* in the preoptic area (Simerly et al., 1984; Simerly, Swanson, & Gorski, 1985b), and *tyrosine hydroxylase (dopamine)* in the anteroventral preoptic area (Simerly, Swanson, & Gorski, 1985a; Simerly, Swanson, Handa, & Gorski, 1985). This list includes only monoamine or peptide systems, but it should be noted that other sex- and hormone-dependent components or systems have been identified that involve neurotransmitter or drug binding sites: naloxone (Hammer, 1984; Ostrowski, Hill, Pert, & Pert, 1987), imipramine (Fischette, Biegnon, & McEwen, 1983), or α-bungarotoxin (Arimatsu, Seto, & Amano, 1981).

As for other sexual dimorphisms a primary question is, what effect does the concurrent hormonal milieu play in influencing the immunocytochemical results? In general, neuropeptide detectability is decreased following gonadectomy. This has been demonstrated for LHRH (King et al., 1987; Shivers et al., 1983), vasopressin (De Vries et al., 1985), substance P (Dees & Kozlowski, 1984), cholecytokinin (Simerly & Swanson, 1987), and Met-enkephalin (Watson et al., 1986). However, gonadectomy does not produce a steady-state condition, and peptide immunoreactivity may change with time after gonadectomy, indicating differences between acute and longer-term mechanisms. For example, although fewer immunopositive LHRH neurons are detected shortly after gonadectomy in male rats, with increasing time (2–3 weeks) there is no difference in the number of immunopositive cells, and there may even be an increase in the total number. Furthermore, the time course with which they become nondetectable and then detectable again is sex dependent (King *et al.*, 1987; Figure 10). For sexual dimorphisms of biochemical anatomy, the presence, absence, or time of absence of gonadal steroids may influence the regulation of individual genes (preprohormones) more dynamically than morphological changes that may involve many interrelated changes. Thus, for this type of sexual dimorphism, attention must be paid to the dynamic state of the hormonal milieu.

For monoaminergic systems the effect of the adult hormone environment is less dramatic than for neuropeptides. Although small immunocytochemical differences have been found with antibodies directed against the metabolic enzyme tyrosine hydroxylase (Simerly, Swanson, Handa, *et al.*, 1985), no differences are found with antibodies directed against the monoamine serotonin (Simerly *et al.*, 1985b); histofluorescence methods used to visualize catecholamines produce similar results (Hyppa, 1969). A major difference between the two types of systems is that monoamines are synthesized through cytoplasmic and vesicular enzymatic steps, whereas neuropeptides are synthesized by classical protein synthetic processes involving mRNA transcription and translation, rough endoplasmic reticulum, and Golgi apparati. The difference in hormone dependence and detectability of these compounds may relate directly to the level of regulation applied to molecules that are direct gene products versus products of enzyme activity and to the different mechanisms for their degradation. Since immunocytochemical experiments do not directly supply an index of turnover, the preceding data do not relate directly to changes in the effective capacity of the molecules in question.

Critical periods exist for most of the monoamines and neuropeptides examined, during which exposure to gonadal steroids has long-term consequences. Perinatal exposure of female rats to testosterone changes the immunocytochemical phenotype to that of normal males with respect to dopamine (Simerly, Swanson, Handa, *et al.*, 1985), serotonin (Simerly *et al.*, 1985b), Met-enkephalin (Watson *et al.*, 1986), and vasopressin (De Vries *et al.*, 1983). The effects of perinatal steroids on CCK have not yet been reported. There are more immunopositive LHRH cells in neonatal and adult males than females (King *et al.*, 1978, 1987). However, neonatal androgen treatment of females did not influence the number of LHRH-immunopositive cells when the animals were examined after long-term treatment with tonic amounts of estradiol (King *et al.*, 1980). This illustrates the importance of the adult hormonal environment for demonstrating sexual dimorphism. The hormonal state of an animal may obscure significant sex differences in the response characteristics of a population of peptidergic cells.

The case of LHRH also raises another issue concerning the immunocytochemical appearance of sexual dimorphism. Neonatal male rats had more detectable LHRH cells than females (King *et al.*, 1978), but this may be dependent on the antibody used (Wray & Hoffman, 1986). Furthermore, particular immunocytochemical procedures (peroxidase–antiperoxidase or avidin–biotin complex, Sternberger, 1986), by virtue of their sensitivity, may render quantitative differences more or less apparent. For sex differences in immunocytochemical patterns, the methodology employed remains a central concern.

The ontogeny of sex differences in periventricular preoptic Met-enkephalin fibers and lateral septal vasopressin fibers in rats suggests that hormone responsiveness and sex-dependent phenotype develop late in postnatal life (after day 10, De Vries *et al.*, 1981; Watson *et al.*, 1988). For LHRH, sex differences in cell number

S. A. TOBET AND
T. O. FOX

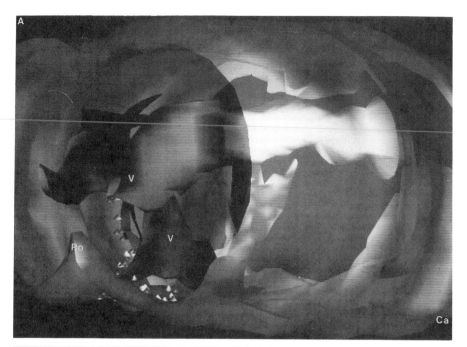

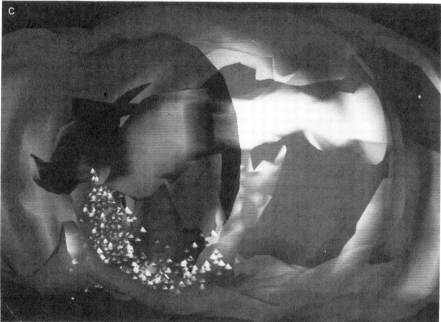

Figure 10. Three-dimensional computer reconstructions of the distribution of luteinizing hormone-releasing hormone-immunopositive cells in adult rats. Immunopositive cells are represented by pyramids. Animals represented in A and B are composites of intact females and males, respectively. Animals represented in C and D are composites of females and males, respectively, 1 day after gonadectomy. Labels in panel A apply to all panels; rostral, Ro; caudal, Ca; aspects of portions of the lateral and third ventricles contain a V. (Adapted from King *et al.*, 1987.)

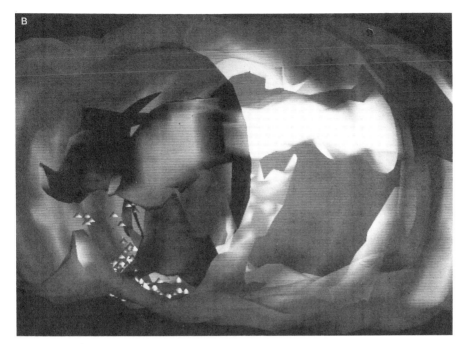

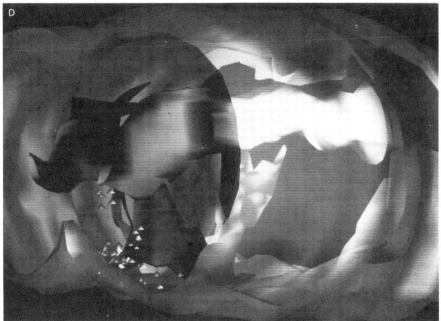

Figure 10. (*continued*)

were detected earlier in development, specifically within the first postnatal week (King *et al.*, 1978). For sexual dimorphisms in biochemical anatomy, the issue of sex-dependent results relating to a cascade of events such that the particular molecule examined is directly affected by the hormone, as opposed to a secondary effect, is difficult to determine. Resolution of the issue of primary versus secondary effects might occur by using the advancing techniques of molecular biology to look directly at steroid hormones and their receptors as regulators of specific genes. For example, the cloning and sequencing of individual peptide hormone genes should identify receptor-binding regulatory sequences.

The specificity of effective steroid metabolites is apparent in two systems. The sexually dimorphic vasopressin projection to the lateral septum appears dependent on both estrogen and androgen receptor action, since full expression of immunoreactivity is seen only after treatment in adulthood with estradiol and dihydrotestosterone together (De Vries *et al.*, 1983). The sexually dimorphic Met-enkephalin projection to the periventricular preoptic area appears estrogen dependent, since estradiol treatment of gonadectomized adult females restores full peptide immunoreactivity (Watson *et al.*, 1986).

Immunocytochemistry utilizing monoclonal antibodies generated against classes of previously unidentified antigens (McKay & Hockfield, 1982; Hawkes, Niday, & Matus, 1982; Sternberger, Harwell, & Sternberger, 1982; Levitt, 1984) has developed as an approach to determining factors that contribute to and/or result from the molecular bases of sexual differentiation. Antibodies have been generated that recognize antigens in insects that are present in one sex but not the other. For example, one antibody recognizes an antigen that is present only in cells that are found in the sensory system of the male moth, *Manduca sexta* (Hishinuma, Hockfield, McKay, & Hildebrand, 1988a, 1988b).

Critical periods reflect distinctive neural milieus. As determinants of these neural environments, some antigens may be expressed only transiently. If the time course of development or aging differs between males and females, transiently expressed antigens will exhibit sex dependence (Fox, 1987). The search for age- or sex-dependent antigens may be greatly aided by the development of techniques that allow the relatively selective generation of markers for molecules that differ between two groups. One approach that has been used to detect antigenic differences between two groups involves induced immunosuppression. With these procedures the production of hybridoma cell lines that make antibodies to antigens in common to the two groups is suppressed, increasing the probability of obtaining monoclonal antibodies to the unique antigens (Matthew & Sandrock, 1987; Hockfield, 1987). One antibody generated using such techniques recognizes an antigen detected in a selected population of monoaminergic cells in the brainstem (Tobet & Fox, 1989) and tanycytes in the hypothalamus (Tobet, Blank, & Fox, 1987) in an age-dependent pattern. In the brainstem, this antigen is found in more cells in females than males only at a particular postnatal age in one subgroup of noradrenergic cells located above the fourth ventricle. The elucidation of the structure and function of this antigen with age may aid in understanding the differentiation of sex differences in monoamine function (Vaccari, 1980).

GLIAL CELLS

The model systems above refer to apparent neuronal effects of hormone exposure. The importance, however, of different types of glial cells in the regulation of neuronal migration and function (reviewed: Varon & Somjen, 1979) makes consideration of glial cells important in discussing hormone action throughout the life span. There are relatively few observations on the interaction of gonadal steroid hormones and glial cells, but these few observations do suggest important avenues for future studies.

Beginning early in development, radial glial cells are believed to play an important role in cell migration in many regions of the neuraxis (Rakic, 1972; Dupouey, Benjelloun, & Gomes, 1985). Although the available experimental evidence for such a role is derived from experiments primarily on structures with laminar organization, these migration-guidance actions may also be effective in nonlaminar subcortical regions such as the hypothalamus (Levitt & Rakic, 1980). As a cell type, radial glial cells are present only during restricted periods of development over different time courses, depending on CNS region. Moreover, some radial glial cells may ultimately transform into astrocytes. In studies of transiently expressed molecules that depend on hormone exposure, one monoclonal antibody detects an antigen in radial glial cells during development (Tobet & Fox, 1989). In the medial basal hypothalamus more radial glia were detected on postnatal day 1 in females than males, whereas more radial glia were detected at embryonic day 19 in the male than female preoptic area. These findings address two issues. First, transient molecular sex differences are detectable. Second, with these methods an antibody-recognized event occurs in a cell type thought to regulate neuronal migration in regions of the CNS in a sex-dependent manner in regions of the brain known to develop sex-dependent morphology later in life. As has been suggested (Jacobson *et al.*, 1985), cytoarchitectonic sex differences in the hypothalamus may result in part from differential success in neuronal migration, with radial glia playing an important role in the process.

Another cell type that may be related to radial glia are tanycytes. Tanycytes are nonciliated ependymal cells that, depending on the author, may or may not be restricted to the lining of the ventral portion of the third ventricle. In contrast to radial glia in other parts of the neuraxis, tanycytes exist throughout life and may play a role in the transport of molecules between the cerebrospinal fluid in the ventricles and various cellular elements in brain or in brain capillaries (for review, see Flament-Durand & Brion, 1985). A role for a subset of tanycytes in sexual differentiation has been proposed by Akmayev and Fidelina (1976, 1981), who have noted sex differences by enzyme histochemistry in the early postnatal period that can be manipulated with hormonal treatment.

As noted earlier, radial glial cells may be transformed over time to astrocytes in parts of the neuraxis. Astrocytic activity, as measured by the accumulation of dense bodies or granules, increases with age in astrocytes of the arcuate nuclei of the hypothalamus, and this increase is dependent in part on long-term exposure to ovarian products. Thus, males do not exhibit this increase, nor do females ovariectomized at 2 months of age (Schipper, Brawer, Nelson, Felicio, & Finch, 1981). The increase in glial reactivity (microglia and astroglia) is also seen when younger female rats are injected with a large dose of estradiol valerate (Brawer, Naftolin, Martin, & Sonnenschein, 1978), and this effect is also dependent on ovarian product(s) (Brawer, Schipper, & Naftolin, 1980). Although the increase in microglial reactivity can be attributed to phagocytosis of neuronal debris, the increase in astrocytic activity can not (Brawer *et al.*, 1978).

In the hippocampus and globus pallidus, but not the arcuate nucleus, estrogen exposure increases the amount of glial fibrillary acidic protein detectable in glial processes by immunocytochemical methods (Tranque, Suarez, Olmos, Fernandez, & Garcia-Segura, 1987). In the songbird telencephalon the production of glial cells may be influenced by testosterone administration (Goldman & Nottebohm, 1983). No conclusions can yet be drawn as to whether gonadal steroids act within glial cells. Gonadal steroids have been localized within a limited number of ependymal and glial cells using autoradiographic techniques (Pfaff & Keiner, 1973; Stumpf, Sar, &

Keefer, 1975). However, autoradiographic procedures to detect estrogen uptake in brain have been optimized for localization in nuclei of neurons (Keefer, 1981). Since the kinetics of estrogen interactions with receptors differ for brain compared to pituitary and uterus (Fox, 1977; Linkie, 1977), these procedures may be suboptimal for detecting specific interactions in glial nuclei or in neural membranes. With regard to androgen metabolism, the steroidogenic enzyme 5α-reductase may be localized preferentially in glia, because enzymatic activity in an *in vitro* culture system does not decrease following kainic acid treatment, whereas in the same cultures the enzyme aromatase was greatly decreased (Canick *et al.*, 1986).

SUMMARY: A LIFE-SPAN APPROACH

Gonadal steroids differentially influence morphological characteristics depending on the stage of development at which they are present. During the period when the nervous system is developing, exposure or lack of exposure to gonadal steroids can have long-lasting consequences for observed morphology and function. In line with the results of behavioral and neuroendocrinological experiments, many researchers, starting with Phoenix, Goy, Gerall, and Young (1959), have referred to the early and long-lasting effects of gonadal steroids as organizational actions. Because the periods during which hormones exert their organizational effects are temporally limited, they have been referred to as critical periods. Transient steroid effects, usually thought of with respect to adulthood, that depend on an organized morphological substrate have been termed activational.

Although the likelihood of a critical period may be highest early in development, the numerous examples of adult effects of gonadal steroids on neural morphology suggest that sensitive periods may also occur during adult life. Varying levels of gonadal steroids circulate in the course of an individual's life span. At specific times for each sex, levels of particular steroids are elevated. Periods when steroid levels are elevated can also be critical if the prevailing environment is susceptible to the molecular action of the particular steroid(s).

As an example, suppose the actions of proteins A and B are required to stabilize the formation of a synapse. If protein B is present only at particular times and protein A is dependent on gonadal steroid action, then hormone exposure will result in synapse formation only when protein B is also present. Therefore, for synapse formation, the period when protein B is present and gonadal steroids are also present would constitute a critical period. For heuristic purposes, this formulation emphasizes the molecular environment in which the steroid acts as the determinant for critical periods and not simply the presence of any particular steroid or any particular age (Harlan, Gordon, & Gorski, 1979). Thus, a key area of investigation to be pursued with increasing frequency will be the determination of the molecular environment(s) within which gonadal steroids will have actions that are considered to be organizational (Phoenix *et al.*, 1959) or permanent (Arnold & Breedlove, 1985).

Acknowledgments

Supported by NIH Research Grant HD-20327, Training Grant HD-07251, MR Core Grant HD-04147, and in part by the Department of Mental Retardation of the Commonwealth of Massachusetts (Contract: 3403-8403-306). We thank Mr.

R. C. Whorf for assistance in the preparation of the figures. We thank Dr. M. J. Baum for helpful comments in the preparation of this manuscript.

69

NEURONAL
MORPHOLOGY

REFERENCES

Akmayev, I. G., & Fidelina, O. V. (1976). Morphological aspects of the hypothalamic–hypophyseal system. VI. The tanycytes: Their relation to the sexual differentiation of the hypothalamus. An enzyme histochemical study. *Cell Tissue Research, 173,* 407–416.

Akmayev, I. G., & Fidelina, O. V. (1981). Tanycytes and their relation to hypophyseal gonadotrophic function. *Brain Research, 210,* 253–260.

Allen, L. S., Hines, M., Shryne, J. E., & Gorski, R. A. (1989). Two sexually dimorphic cell groups in the human brain. *Journal of Neuroscience, 9,* 497–506.

Anderson, R. H., Fleming, D. E. Rhees, R. W., & Kinghorn, E. (1986). Relationships between sexual activity, plasma testosterone, and the volume of the sexually dimorphic nucleus of the preoptic area in prenatally stressed and non-stressed rats. *Brain Research, 370,* 1–10.

Arai, Y., & Kusama, T. (1968). Effect of neonatal treatment with estrone on hypothalamic neurons and regulation of gonadotrophin secretion. *Neuroendocrinology, 3,* 107–114.

Arendash, G. W., & Gorski, R. A. (1983). Effects of discrete lesions of the sexually dimorphic nucleus of the preoptic area or other medial preoptic regions on the sexual behavior of male rats. *Brain Research Bulletin, 10,* 147–154.

Arimatsu, Y., Seto, A., & Amano, T. (1981). Sexual dimorphism in alpha-bungarotoxin binding capacity in the mouse amygdala. *Brain Research, 213,* 432–437.

Arnold, A. P. (1979). Hormone accumulation in the brain of the zebra finch after injection of various steroids and steroid competitors. *Society for Neuroscience Abstracts, 5,* 437.

Arnold, A. P. (1980a). Sexual differences in the brain. *American Scientist, 68,* 165–173.

Arnold, A. P. (1980b). Effects of androgens on volumes of sexually dimorphic brain regions in the zebra finch. *Brain Research, 185,* 441–444.

Arnold, A. P., & Breedlove, S. M. (1985). Organizational and activational effects of sex steroids on brain and behavior: A reanalysis. *Hormones and Behavior, 19,* 469–498.

Arnold, A. P., & Gorski, R. A. (1984). Gonadal steroid induction of structural sex differences in the central nervous system. *Annual Reviews of Neuroscience, 7,* 413–442.

Arnold, A. P., & Saltiel, A. (1979). Sexual difference in the pattern of hormone accumulation in the brain of a songbird. *Science, 205,* 702–705.

Arnold, A. P., Bottjer, S. W., Brenowitz, E. A., Nordeen, E. J., & Nordeen, K. W. (1986). Sexual dimorphisms in the neural vocal control system in song birds: Ontogeny and phylogeny. *Brain Behavior and Evolution, 28,* 22–31.

Ayoub, D. M., Greenough, W. T., & Juraska, J. M. (1983). Sex differences in dendritic structure in the preoptic area of the juvenile macaque monkey brain. *Science, 219,* 197–198.

Baum, M. J. (1976). Effects of testosterone propionate administered perinatally on sexual behavior of female ferrets. *Journal of Comparative Physiology and Psychology, 90,* 399–410.

Baum, M. J. (1979). Differentiation of coital behavior in mammals: A comparative analysis. *Neuroscience Biobehavioral Review, 3,* 265–284.

Baum, M. J., & Erskine, M. S. (1984). Effect of neonatal gonadectomy and administration of testosterone on coital masculinization in the ferret. *Endocrinology, 115,* 2440–2444.

Baum, M. J., & Goldfoot, D. A. (1975). Effect of amygdaloid lesions on gonadal maturation in male and female ferrets. *American Journal of Physiology, 228,* 1646–1651.

Baum, M. J., Gallagher, C. A., Martin, J. T., & Damassa, D. A. (1982). Effects of testosterone, dihydrotestosterone, or estradiol administered neonatally on sexual behavior of female ferrets. *Endocrinology, 111,* 773–780.

Baum, M. J., Tobet, S. A., Starr, M. S., & Bradshaw, W. G. (1982). Implantation of dihydrotestosterone propionate into the lateral septum or medial amygdala facilitates copulation in castrated male rats given estradiol systemically. *Hormones and Behavior, 16,* 208–223.

Baum, M. J., Canick, J. A., Erskine, M. S., Gallagher, C. A., & Shim, J. H. (1983). Normal differentiation of masculine sexual behavior in male ferrets despite neonatal inhibition of brain aromatase or 5-alpha-reductase activity. *Neuroendocrinology, 36,* 277–284.

Bleier, R., Byne, W., & Siggelkow, I. (1982). Cytoarchitectonic sexual dimorphisms of the medial preoptic and anterior hypothalamic areas in guinea pig, rat, hamster and mouse. *Journal of Comparative Neurology, 212,* 118–130.

Bloch, G. J., & Gorski, R. A. (1988). Cytoarchitectonic analysis of the SDN-POA of the intact and gonadectomized rat. *Journal of Comparative Neurology, 275,* 604–612.

Brawer, J. R. (1971). The role of the arcuate nucleus in the brain–pituitary–gonad axis. *Journal of Comparative Neurology, 143,* 411–446.

Brawer, J. R., Naftolin, F., Martin, J., & Sonnenschein, C. (1978). Effects of a single injection of estradiol valerate on the hypothalamic arcuate nucleus and on reproductive function in the female rat. *Endocrinology, 103,* 501–512.

Brawer, J. R., Schipper, H., & Naftolin, F. (1980). Ovary-dependent degeneration in the hypothalamic arcuate nucleus. *Endocrinology, 107,* 274–279.

Breedlove, S. M. (1983). Regional sex differences in steroid accumulation in the nervous system. *Trends in Neuroscience, 6,* 403–406.

Breedlove, S. M. (1984). Steroid influences on the development and formation of a neuromuscular system. *Progress in Brain Research, 61,* 147–170.

Breedlove, S. M. (1985). Hormonal control of the anatomical specificity of motoneuron-to-muscle connections in rats. *Science, 227,* 1357–1359.

Breedlove, S. M., & Arnold, A. P. (1980). Hormone accumulation in a sexually dimorphic motor nucleus of the rat spinal cord. *Science, 210,* 564–566.

Breedlove, S. M., & Arnold, A. P. (1981). Sexually dimorphic motor nucleus in the rat lumbar spinal cord: Response to adult hormone manipulation, absence in androgen-insensitive rats. *Brain Research, 225,* 297–307.

Breedlove, S. M., & Arnold, A. P. (1983a). Hormonal control of a developing neuromuscular system I. Complete demasculinization of the male rat spinal nucleus of the bulbocavernosus using the anti-androgen flutamide. *Journal of Neuroscience, 3,* 417–423.

Breedlove, S. M., & Arnold, A. P. (1983b). Hormonal control of a developing neuromuscular system II. Sensitive periods for the androgen-induced masculinization of the rat spinal nucleus of the bulbocavernosus. *Journal of Neuroscience, 3,* 424–432.

Breedlove, S. M., & Arnold, A. P. (1983c). Sex differences in the pattern of steroid accumulation by motoneurons of the rat lumbar spinal cord. *Journal of Comparative Neurology, 215,* 211–216.

Breedlove, S. M., Jacobson, C. D., Gorski, R. A., & Arnold, A. P. (1982). Masculinization of the female rat spinal cord following a single neonatal injection of testosterone propionate but not estradiol benzoate. *Brain Research, 237,* 173–181.

Breedlove, S. M., Jordan, C. L., & Arnold, A. P. (1983). Neurogenesis of motoneurons in the sexually dimorphic spinal nucleus of the bulbocavernosus in rats. *Developmental Brain Research, 9,* 39–43.

Brown-Grant, K., Murray, M. A. F., Raisman, G., & Sood, M. C. (1977). Reproductive function in male and female rats following extra- and intra-hypothalamic lesions. *Proceedings of the Royal Society of London, B, 198,* 267–278.

Bubenik, G. A., & Brown, G. M. (1973). Morphologic sex differences in primate brain areas involved in regulation of reproductive activity. *Experientia, 29,* 619–621.

Byne, W., & Bleier, R. (1987). Medial preoptic sexual dimorphisms in the guinea pig. I. An investigation of their hormonal dependence. *Journal of Neuroscience, 7,* 2688–2696.

Byne, W., Warren, J. T., & Bleier, R. (1987). Medial preoptic sexual dimorphisms in the guinea pig. II. An investigation of medial preoptic neurogenesis. *Journal of Neuroscience, 7,* 2697–2702.

Calaresu, F. R., & Henry, J. L. (1971). Sex differences in the number of sympathetic neurons in the spinal cord of the cat. *Science, 173,* 343–344.

Callard, F. V., Petro, Z., & Ryan, K. J. (1978). Conversion of androgen to estrogen and other steroids in the vertebrate brain. *American Zoologist, 18,* 511–523.

Canick, J. A., Vaccaro, D. E., Livingston, E. M., Leeman, S. E., Ryan, K. J., & Fox, T. O. (1986). Localization of aromatase and 5-alpha-reductase to neuronal and non-neuronal cells in the fetal rat hypothalamus. *Brain Research, 372,* 277–282.

Cherry, J. A., & Baum, M. J. (1988). Effects of preoptic area lesions on the sexual behavior of male ferrets. *Society for Neuroscience Abstracts, 14,* 98.

Cherry, J. A., Basham, M. E., Weaver, C. E., Krohmer, R. W., & Baum, M. J. (1990). Ontogeny of the sexually dimorphic male nucleus in the preoptic/anterior hypothalamus of ferrets and its manipulation by gonadal steroids. *Journal of Neurobiology, 21,* 844–857.

Christensen, L. W., Nance, D. M., & Gorski, R. A. (1977). Effects of hypothalamic and preoptic lesions on reproductive behavior in male rats. *Brain Research Bulletin, 2,* 137–141.

Commins, D., & Yahr, P. (1984a). Acetylcholinesterase activity in the sexually dimorphic area of the gerbil brain: Sex differences and influences of adult gonadal steroids. *Journal of Comparative Neurology, 224,* 123–131.

Commins, D., & Yahr, P. (1984b). Adult testosterone levels influence the morphology of a sexually dimorphic area in the mongolian gerbil brain. *Journal of Comparative Neurology, 224,* 132–140.

Commins, D., & Yahr, P. (1984c). Lesions of the sexually dimorphic area disrupt mating and marking in male gerbils. *Brain Research Bulletin, 13*, 185–193.

Commins, D., & Yahr, P. (1985). Autoradiographic localization of estrogen and androgen receptors in the sexually dimorphic area and other regions of the gerbil brain. *Journal of Comparative Neurology, 231*, 473–489.

Dees, W. L., & Kozlowski, G. P. (1984). Effects of castration and ethanol on amygdaloid substance P immunoreactivity. *Neuroendocrinology, 39*, 231–235.

del Abril, A., Segovia, S., & Guillamon, A. (1987). The bed nucleus of the stria terminalis in the rat: Regional sex differences controlled by gonadal steroids early after birth. *Developmental Brain Research, 32*, 295–300.

DeVoogd, T. J. (1984). The avian song system: Relating sex differences in behavior to dimorphism in the central nervous system. *Progress in Brain Research, 61*, 171–184.

DeVoogd, T. J., & Nottebohm, F. (1981a). Sex differences in dendritic morphology of a song control nucleus in the canary: A quantitative Golgi study. *Journal of Comparative Neurology, 196*, 309–316.

DeVoogd, T. J., & Nottebohm, F. (1981b). Gonadal hormones induce dendritic growth in the adult avian brain. *Science, 214*, 202–204.

DeVoogd, T. J., Nixdorf, B., & Nottebohm, F. (1985). Synaptogenesis and changes in synaptic morphology related to acquisition of a new behavior. *Brain Research, 329*, 304–308.

De Vries, G. J., Buijs, R. M., & Swaab, D. F. (1981). Ontogeny of the vasopressinergic neurons of the suprachiasmatic nucleus and their extrahypothalamic projections in the rat brain—presence of a sex difference in the lateral septum. *Brain Research, 218*, 67–78.

De Vries, G. J., Best, W., & Sluiter, A. A. (1983). The influence of androgens on the development of a sex difference in the vasopressinergic innervation of the rat lateral septum. *Developmental Brain Research, 8*, 377–380.

De Vries, G. J., Buijs, R. M., & Sluiter, A. A. (1984). Gonadal hormone actions on the morphology of the vasopressinergic innervation of the adult rat brain. *Brain Research, 298*, 141–145.

De Vries, G. J., De Bruin, J. P. C., Uylings, H. B. M., & Corner, M. A. (1984). *Sex differences in the brain.* Amsterdam: Elsevier/North-Holland.

De Vries, G. J., Buijs, R. M., Van Leeuwen, F. W., Caffe, A. R., & Swaab, D. F. (1985). The vasopressinergic innervation of the brain in normal and castrated rats. *Journal of Comparative Neurology, 233*, 236–254.

Dohler, K.-D. (1978). Is female sexual differentiation hormone-mediated? *Trends in Neuroscience, 1*, 138–140.

Dohler, K.-D., Coquelin, A., Davis, F., Hines, M., Shryne, J. E., & Gorski, R. A. (1982). Differentiation of the sexually dimorphic nucleus in the preoptic area of the rat brain is determined by the perinatal hormone environment. *Neuroscience Letters, 33*, 295–298.

Dohler, K.-D., Hines, M., Coquelin, A., Davis, F., Shryne, J. E., & Gorski, R. A. (1982). Pre- and postnatal influence of diethylstilbestrol on differentiation of the sexually dimorphic nucleus of the female rat brain. *Neuroendocrinology Letters, 4*, 361–365.

Dohler, K.-D., Coquelin, A., Davis, F., Hines, M., Shryne, J. E., & Gorski, R. A. (1984). Pre- and postnatal influence of testosterone propionate and diethylstilbestrol on differentiation of the sexually dimorphic nucleus of the preoptic area in male and female rats. *Brain Research, 302*, 291–295.

Dohler, K.-D., Srivastava, S. S., Shryne, J. E., Jarzab, B., Sipos, A., & Gorski, R. A. (1984). Differentiation of the sexually dimorphic nucleus in the preoptic area of the rat brain is inhibited by postnatal treatment with an estrogen antagonist. *Neuroendocrinology, 38*, 297–301.

Dorner, G., & Staudt, J. (1968). Structural changes in the anterior hypothalamic area of the male rat, following neonatal castration and androgen substitution. *Neuroendocrinology, 3*, 136–140.

Dorner, G., & Staudt, J. (1969a). Structural changes in the hypothalamic ventromedial nucleus of the male rat, following neonatal castration and androgen treatment. *Neuroendocrinology, 4*, 278–281.

Dorner, G., & Staudt, J. (1969b). Perinatal structural sex differentiation of the hypothalamus in rats. *Neuroendocrinology, 5*, 103–106.

Dupouey, P., Benjelloun, S., & Gomes, D. (1985). Immunohistochemical demonstration of an organized cytoarchitecture of the radial glia in the CNS of the embryonic mouse. *Developmental Neuroscience, 7*, 81–93.

Dyer, R. G., MacLeod, N. K., & Ellendorf, F. (1976). Electrophysiological evidence for sexual dimorphism and synaptic convergence in the preoptic and anterior hypothalamic areas of the rat. *Proceedings of the Royal Society of London, B. 193*, 33–40.

Elkind-Hirsch, K., King, J. C., Gerall, A. A., & Arimura, A. A. (1981). The luteinizing hormone-releasing hormone (LHRH) system in normal and estrogenized neonatal rats. *Brain Research, 7*, 645–654.

Erskine, M. S., & Baum, M. J. (1982). Plasma concentrations of testosterone and dihydrotestosterone during perinatal development in male and female ferrets. *Endocrinology, 111*, 767–772.

Fischette, C. T., Biegnon, A., & McEwen, B. S. (1983). Sex differences in serotonin 1 receptor binding in rat brain. *Science, 222,* 333–335.

Fishman, R. B., & Breedlove, M. (1985). The androgenic induction of spinal sexual dimorphism is independent of supraspinal afferents. *Developmental Brain Research, 23,* 255–258.

Flament-Durand, J., & Brion, J. P. (1985). Tanycytes: Morphology and functions: A review. *International Reviews in Cytology, 96,* 121–155.

Forger, N. G., & Breedlove, S. M. (1986). Sexual dimorphism in human and canine spinal cord: Role of early androgen. *Proceedings of the National Academy of Sciences, U.S.A., 83,* 7527–7531.

Fox, T. O. (1977). Conversion of the hypothalamic estradiol receptor to the "nuclear" form. *Brain Research, 120,* 580–583.

Fox, T. O. (1987). Alpha-fetoprotein in sexual differentiation of mouse and rat brain. In G. J. Mizejewski & H. I. Jacobson (Eds.), *Biological activities of alpha₁-fetoprotein,* Vol. I (pp. 181–187). Boca Raton: CRC Press.

Fox, T. O., Olsen, K. L., Vito, C. C., & Wieland, S. J. (1982). Putative steroid receptors: Genetics and development. In F. O. Schmitt, S. J. Bird, & F. E. Bloom, (Eds.), *Molecular Genetic Neuroscience* (pp. 289–306). New York: Raven Press.

Goldman, S. A., & Nottebohm, F. (1983). Neuronal production, migration and differentiation in a vocal control nucleus of the adult female canary brain. *Proceedings of the National Academy of Sciences, U.S.A., 80,* 2390–2394.

Gorski, R. A. (1984). Critical role for the medial preoptic area in the sexual differentiation of the brain. *Progress in Brain Research, 61,* 129–146.

Gorski, R. A., Gordon, J. H., Shryne, J. E., & Southam, A. M. (1978). Evidence for a morphological sex difference within the medial preoptic area of the rat brain. *Brain Research, 148,* 333–346.

Gorski, R. A., Harlan, R. E., Jacobson, C. D., Shryne, J. E., & Southam, A. M. (1980). Evidence for the existence of a sexually dimorphic nucleus in the preoptic area of the rat. *Journal of Comparative Neurology, 193,* 529–539.

Gould, S. J. (1981). *The mismeasure of man.* New York: W. W. Norton.

Goy, R. W., & McEwen, B. S. (1980). *Sexual differentiation of the brain.* Cambridge: MIT Press.

Greenough, W. T., Carter, C. S., Steerman, C., & DeVoogd, T. J. (1977). Sex differences in dendritic patterns in hamster preoptic area. *Brain Research, 126,* 63–72.

Guldner, F. H. (1982). Sexual dimorphism of axo-spine synapses and postsynaptic density material in the suprachiasmatic nucleus of the rat. *Neuroscience Letters, 28,* 145–150.

Guldner, F. H. (1984). Suprachiasmatic nucleus: Numbers of synaptic appositions and various types of synapses, a morphogenetic study on male and female rats. *Cell and Tissue Research, 235,* 449–452.

Gurney, M. E. (1981). Hormonal control of cell form and number in the zebra finch song system. *Journal of Neuroscience, 1,* 658–673.

Gurney, M. E. (1982). Behavioral correlates of sexual differentiation in the zebra finch song system. *Brain Research, 231,* 153–172.

Gurney, M. E., & Konishi, M. (1980). Hormone-induced sexual differentiation of brain and behavior in zebra finches. *Science, 208,* 1380–1383.

Hammer, R. P. (1984). The sexually dimorphic region of the preoptic area in rats contains denser opiate receptor binding sites in females. *Brain Research, 308,* 172–176.

Hammer, R. P., & Jacobson, C. D. (1984). Sex difference in dendritic development of the sexually dimorphic nucleus of the preoptic area in the rat. *International Journal of Developmental Neuroscience, 2,* 77–85.

Handa, R. J., Corbier, P., Shryne, J. E., Schoonmaker, J. N., & Gorski, R. A. (1985). Differential effects of the perinatal steroid environment on three sexually dimorphic parameters of the rat brain. *Biology of Reproduction, 32,* 855–864.

Harlan, R. E., Gordon, J. H., & Gorski, R. A. (1979). Sexual differentiation of the brain: Implications for Neuroscience. *Reviews of Neuroscience, 4,* 31–71.

Hart, B. L. (1974). Medial preoptic–anterior hypothalamic area and sociosexual behavior of male dogs: A comparative neuropsychological analysis. *Journal of Comparative and Physiological Psychology, 386,* 328–349.

Hart, B. L., & Haugen, C. M. (1968). Activation of sexual reflexes in male rats by spinal implantation of testosterone. *Physiology and Behavior, 3,* 735–738.

Hart, B. L., & Melese-D'Hospital, P. Y. (1983). Penile mechanisms and the role of the striated penile muscles in penile reflexes. *Physiology and Behavior, 31,* 807–813.

Hawkes, R., Niday, E., & Matus, A. (1982). Monoclonal antibodies identify novel neural antigens. *Proceedings of the National Academy of Sciences, U.S.A., 79,* 2410–2414.

Hayward, J. N. (1977). Functional and morphological aspects of hypothalamic neurons. *Physiological Reviews, 57,* 574–658.

Hines, M., Davis, F., Coquelin, A., Goy, R. W., & Gorski, R. A. (1985). Sexually dimorphic regions in the medial preoptic area and the bed nucleus of the stria terminalis of the guinea pig brain: A description and an investigation of their relationship to gonadal steroids in adulthood. *Journal of Neuroscience, 5,* 40–47.

Hines, M., Alsum, P., Roy, M., Gorski, R. A., & Goy, R. W. (1987). Estrogenic contributions to sexual differentiation in the female guinea pig: Influences of diethylstilbestrol and tamoxifen on neural, behavioral, and ovarian development. *Hormones and Behavior, 21,* 402–417.

Hishinuma, A., Hockfield, S., McKay, R., & Hildebrand, J. G. (1988a). Monoclonal antibodies reveal cell-type-specific antigens in the sexually dimorphic olfactory system of *Manduca sexta:* I. Generation of monoclonal antibodies and partial characterization of the antigens. *Journal of Neuroscience, 8,* 296–307.

Hishinuma, A., Hockfield, S., McKay, R., & Hildebrand, J. G. (1988b). Monoclonal antibodies reveal cell-type-specific antigens in the sexually dimorphic olfactory system of *Manduca sexta:* II. Expression of antigens during postembryonic development. *Journal of Neuroscience, 8,* 308–315.

Hockfield, S. (1987). A MAb to a unique cerebellar neuron generated by immunosuppression and rapid immunization. *Science, 237,* 67–70.

Holbrook, P. G., & Baum, M. J. (1983). Characterization of estradiol receptors in brain cytosols from perinatal ferrets. *Developmental Brain Research, 7,* 1–8.

Hsu, H. K., Chen, F. N., & Peng, M. T. (1980). Some characteristics of the darkly stained area of the medial preoptic area of rats. *Neuroendocrinology, 31,* 327–330.

Hyyppa, M. (1969). A histochemical study of the primary catecholamines in the hypothalamic neurons of the rat in relation to the ontogenetic and sexual differentiation. *Zeitschrifr fur Zellforschung, 98,* 550–560.

Ifft, J. D. (1964). The effect of endocrine gland extirpations on the size of nucleoli in rat hypothalamic neurons. *Anatomical Record, 148,* 599–604.

Ito, S., Murakami, S., Yamanouchi, K., & Arai, Y. (1986). Prenatal androgen exposure, preoptic area and reproductive functions in the female rat. *Brain and Development, 4,* 463–468.

Jacobson, C. D., & Gorski, R. A. (1981). Neurogenesis of the sexually dimorphic nucleus of the preoptic area in the rat. *Journal of Comparative Neurology, 196,* 519–529.

Jacobson, C. D., Shryne, J. E., Shapiro, F., & Gorski, R. A. (1980). Ontogeny of the sexually dimorphic nucleus of the preoptic area. *Journal of Comparative Neurology, 193,* 541–548.

Jacobson, C. D., Terkel, J., Gorski, R. A., & Sawyer, C. H. (1980). Effects of small medial preoptic area lesions on maternal behavior: Retrieving and nest building in the rat. *Brain Research, 194,* 471–478.

Jacobson, C. D., Csernus, V. J., Shryne, J. E., & Gorski, R. A. (1981). The influence of gonadectomy, androgen exposure, or a gonadal graft in the neonatal rat on the volume of the sexually dimorphic nucleus of the preoptic area. *Journal of Neuroscience, 1,* 1142–1147.

Jacobson, C. D., Davis, F. C., & Gorski, R. A. (1985). Formation of the sexually dimorphic nucleus of the preoptic area: Neuronal growth, migration and changes in cell number. *Developmental Brain Research, 21,* 7–18.

Jacobson, C. D., Arnold, A. P., & Gorski, R. A. (1987). Steroid autoradiography of the sexually dimorphic nucleus of the preoptic area. *Brain Research, 414,* 349–356.

Jordan, C. L., Breedlove, S. M., & Arnold, A. P. (1982). Sexual dimorphism and the influence of neonatal androgen in the dorsolateral motor nucleus of the rat lumbar spinal cord. *Brain Research, 249,* 309–314.

Jouan, P., & Samperez, S. (1980). Metabolism of steroid hormones in the brain. In M. Matta (Ed.), *The endocrine functions of the brain* (pp. 95–115). New York: Raven Press.

Kawakami, M., & Terasawa, E. (1972). A possible role of the hippocampus and amygdala in the androgenized rat: Effect of electrical or electrochemical stimulation of the brain on gonadotropin secretion. *Endocrinologica Japonica, 19,* 349–358.

Keefer, D. A. (1981). Nuclear retention characteristics of [^{3}H]estrogen by cells in four estrogen target regions of the rat brain. *Brain Research, 229,* 224–229.

Kelley, D. B. (1978). Neuroanatomical correlates of hormone sensitive behaviors in frogs and birds. *American Zoologist, 18,* 477–488.

Kelley, D. B., & Pfaff, D. W. (1978). Generalizations from comparative studies on neuroanatomical and endocrine mechanisms of sexual behavior. In J. B. Hutchison (Ed.), *Biological determinants of sexual behavior* (pp. 225–254). Chichester: John Wiley & Sons.

King, J. C. (1972). *Ultrastructure analysis of arcuate neurons in normal and androgenized female rats.* Ph.D. Thesis, Tulane University.

King, J. C., Elkind, K. E., Gerall, A. A., & Millar, R. P. (1978). Investigation of the LH-RH system in the normal and neonatally steroid-treated male and female rat. In D. E. Scott, G. P. Kozlowski, & A. Weindl (Eds.), *Brain–Endocrine Interaction III* (pp. 97–107). Basel: S. Karger.

King, J. C., Tobet, S. A., Snavely, F. L., & Arimura, A. A. (1980). The LHRH system in normal and neonatally androgenized female rats. *Peptides, Supplement, 1*, 85–100.

King, J. C., Kugel, G., Zahniser, D., Wooledge, K., Damassa, D. A., & Alexsavich, G. (1987). Changes in populations of LHRH-immunopositive cell bodies following gonadectomy. *Peptides, 8*, 721–735.

Konishi, M., & Akutagawa, E. (1985). Neuronal growth, atrophy and death in a sexually dimorphic song nucleus in the zebra finch brain. *Nature, 315*, 145–147.

Konishi, M., & Akutagawa, E. (1988). A critical period for estrogen action on neurons of the song control system in the zebra finch. *Proceedings of the National Academy of Sciences, U.S.A., 85*, 7006–7007.

Konishi, M., & Gurney, M. E. (1982). Sexual differentiation of brain and behavior. *Trends in Neuroscience, 5*, 20–23.

Kurz, E. M., Sengelaub, D. R., & Arnold, A. P. (1986). Androgens regulate the dendritic length of mammalian motoneurons in adulthood. *Science, 232*, 395–398.

LeBlond, C. B., Morris, S., Karakiulakus, G., Powell, R., & Thomas, P. J. (1982). Development of sexual dimorphism in the suprachiasmatic nucleus of the rat. *Journal of Endocrinology, 95*, 137–145.

Leedy, M. G., Beattie, M. S., & Breshnahan, J. C. (1987). Testosterone-induced plasticity of synaptic inputs to adult mammalian motoneurons. *Brain Research, 424*, 386–390.

Levitt, P. (1984). A monoclonal antibody to limbic system neurons. *Science, 223*, 299–301.

Levitt, P., & Rakic, P. (1980). Immunoperoxidase localization of glial fibrillary acidic protein in radial glial cells and astrocytes of the developing rhesus monkey brain. *Journal of Comparative Neurology, 193*, 815–840.

Linkie, D. M. (1977). Estrogen receptors in different target tissues: Similarities of form—dissimilarities of transformation. *Endocrinology, 101*, 1862–1870.

Lisk, R. D., & Newlon, M. (1963). Estradiol: Evidence for its direct effect on hypothalamic neurons. *Science, 139*, 223–224.

Loy, R., & Milner, T. A. (1980). Sexual dimorphism in extent of axonal sprouting in rat hippocampus. *Science, 208*, 1282–1284.

Lydic, R., Albers, H. E., Tepper, B., & Moore-Ede, M. C. (1982). Three-dimensional structure of the mammalian suprachiasmatic nuclei: A comparative study of five species. *Journal of Comparative Neurology, 204*, 225–237.

MacLusky, N. J., & Naftolin, F. (1981). Sexual differentiation of the central nervous system. *Science, 211*, 1294–1303.

Malsbury, C. W., & McKay, K. (1987). A sex difference in the pattern of substance P-like immunoreactivity in the bed nucleus of the stria terminalis. *Brain Research, 420*, 365–370.

Martini, L. (1978). Role of the metabolism of steroid hormones in the brain in sex differentiation and sexual maturation. In G. Dorner & M. Kawakamo (Eds.), *Hormones and brain development* (pp. 3–12). New York: Elsevier/North-Holland Biomedical Press.

Masco, D. H., & Carrer, H. F. (1980). Sexual receptivity in female rats after lesion or stimulation in different amygdaloid nuclei. *Physiology and Behavior, 24*, 1073–1080.

Matsumoto, A., & Arai, Y. (1976). Effect of estrogen on early postnatal development of synaptic formation in the hypothalamic arcuate nucleus of female rats. *Neuroscience Letters, 2*, 79–82.

Matsumoto, A., & Arai, Y. (1980). Sexual dimorphism in wiring pattern in the hypothalamic arcuate nucleus and its modification by neonatal hormonal environment. *Brain Research, 190*, 238–242.

Matsumoto, A., & Arai, Y. (1981). Effect of androgen on sexual differentiation of synaptic organization in the hypothalamic arcuate nucleus: An ontogenetic study. *Neuroendocrinology, 33*, 166–169.

Matsumoto, A., & Arai, Y. (1983). Sex difference in volume of the ventromedial nucleus of the hypothalamus in the rat. *Endocrinologica Japonica, 30*, 277–280.

Matsumoto, A., & Arai, Y. (1986a). Development of sexual dimorphism in synaptic organization in the ventromedial nucleus of the hypothalamus in rats. *Neuroscience Letters, 68*, 165–168.

Matsumoto, A., & Arai, Y. (1986b). Male–female difference in synaptic organization of the ventromedial nucleus of the hypothalamus in the rat. *Neuroendocrinology, 42*, 232–236.

Matsumoto, A., Arnold, A. P., Zampighi, G. A., & Micevych, P. E. (1988). Androgenic regulation of gap junctions between motoneurons in the rat spinal cord. *Journal of Neuroscience, 8*, 4177–4183.

Matsumoto, A., Micevych, P. E., & Arnold, A. P. (1988). Androgen regulates synaptic input to motoneurons of the adult rat spinal cord. *Journal of Neuroscience, 8*, 4168–4176.

Matthew, W. D., & Sandrock, A. W. Jr. (1987). Cyclophosphamide treatment used to manipulate the immune response for the production of monoclonal antibodies. *Journal of Immunological Methods, 100*, 73–82.

Mayes, C. R., Watts, A. G., McQueen, J. K., Fink, G., & Charlton, H. M. (1988). Gonadal steroids influence neurophysin II distribution in the forebrain of normal and mutant mice. *Neuroscience, 25,* 1013–1022.

McEwen, B. S. (1981). Neural gonadal steroid actions. *Science, 211,* 1303–1311.

McKay, D. G., & Hockfield, S. J. (1982). Monoclonal antibodies distinguish antigenically discrete neuronal types in the vertebrate central nervous system. *Proceedings of the National Academy of Sciences, U.S.A., 79,* 6747–6751.

Meaney, M. J., Dodge, A. M., & Beatty, W. W. (1981). Sex-dependent effects of amygdaloid lesions on the social play of prepubertal rats. *Physiology and Behavior, 26,* 467–472.

Micevych, P. E., Park, S. S., Akesson, T. R., & Elde, R. (1987). Distribution of cholecystokinin-immunoreactive cell bodies in the male and female rat: I. Hypothalamus. *Journal of Comparative Neurology, 255,* 124–136.

Micevych, P. E., Akesson, T. R., & Elde, R. (1988). Distribution of cholecystokinin-immunoreactive cell bodies in the male and female rat: II. Bed nucleus of the stria terminalis and amygdala. *Journal of Comparative Neurology, 269,* 381–391.

Mizukami, S., Nishizuka, M., & Arai, Y. (1983). Sexual difference in nuclear volume and its ontogeny in the rat amygdala. *Experimental Neurology, 79,* 569–575.

Nance, D. M., Shryne, J., & Gorski, R. A. (1975). Facilitation of female sexual behavior in male rats by septal lesions: An interaction with estrogen. *Hormones and Behavior, 6,* 289–299.

Nishizuka, M., & Arai, Y. (1981a). Sexual dimorphism in synaptic organization in the amygdala and its dependence on neonatal hormone environment. *Brain Research, 212,* 31–38.

Nishizuka, M., & Arai, Y. (1981b). Organizational action of estrogen on synaptic pattern in the amygdala: Implications for sexual differentiation of the brain. *Brain Research, 213,* 422–426.

Nishizuka, M., & Arai, Y. (1982). Synapse formation in response to estrogen in the medial amygdala developing in the eye. *Proceedings of the National Academy of Sciences, U.S.A., 79,* 7024–7026.

Nishizuka, M., & Arai, Y. (1983). Regional difference in sexually dimorphic synaptic organization of the medial amygdala. *Experimental Brain Research, 49,* 462–465.

Nordeen, E. J., & Yahr, P. (1982). Hemispheric asymmetries in the behavioral and hormonal effects of sexually differentiating mammalian brain. *Science, 218,* 391–394.

Nordeen, E. J., Nordeen, K. W., Sengelaub, D. R., & Arnold, A. P. (1985). Androgens prevent normally occurring cell death in a sexually dimorphic spinal nucleus. *Science, 229,* 671–673.

Nordeen, K. W., Nordeen, E. J., & Arnold, A. P. (1986). Estrogen establishes sex differences in androgen accumulation in zebra finch brain. *Journal of Neuroscience, 6,* 734–738.

Nordeen, E. J., Nordeen, K. W., & Arnold, A. P. (1987). Sexual differentiation of androgen accumulation within the zebra finch brain through selective cell loss and addition. *Journal of Comparative Neurology, 259,* 393–399.

Nottebohm, F. (1980). Testosterone triggers growth of brain vocal control nuclei in adult female canaries. *Brain Research, 189,* 429–436.

Nottebohm, F. (1981). A brain for all seasons: Cyclical anatomical changes in song control nuclei of the canary brain. *Science, 214,* 1368–1370.

Nottebohm, F., & Arnold, A. P. (1976). Sexual dimorphism in vocal control areas of the songbird brain. *Science, 194,* 211–213.

Nottebohm, F., Stokes, T. M., & Leonard, C. M. (1976). Central control of song in the canary, *Serinus canarius. Journal of Comparative Neurology, 165,* 457–486.

Numan, M. (1974). Medial preoptic area and maternal behavior in the female rat. *Journal of Comparative and Physiological Psychology, 87,* 746–759.

Olsen, K. L. (1983). Genetic determinants of sexual differentiation. In J. Balthazart, E. Prove, & R. Gilles (Eds.), *Hormones and behavior in higher vertebrates* (pp. 138–158). Berlin: Springer-Verlag.

Ostrowski, N. L., Hill, J. M., Pert, C. B., & Pert, A. (1987). Autoradiographic visualization of sex differences in the pattern and density of opiate receptors in hamster hypothalamus. *Brain Research, 421,* 1–13.

Paxinos, G., & Watson, C. (1982). *The rat brain in stereotaxic coordinates.* New York: Academic Press.

Pfaff, D. W. (1966). Morphological changes in the brains of adult male rats after neonatal castration. *Journal of Endocrinology, 36,* 415–416.

Pfaff, D., & Keiner, M. (1973). Atlas of estradiol-concentrating cells in the central nervous system of the female rat. *Journal of Comparative Neurology, 151,* 121–157.

Phoenix, C. H., Goy, R. W., Gerall, A. A., & Young, W. C. (1959). Organizing action of prenatally administered testosterone propionate on the tissues mediating mating behavior in the female guinea pig. *Endocrinology, 65,* 369–382.

Pohl-Apel, G. (1985). The correlation between the degree of brain masculinization and song quality in estradiol treated female zebra finches. *Brain Research, 336,* 381–383.

Rainbow, T. C., Parsons, B., & McEwen, B. S. (1982). Sex differences in rat brain oestrogen and progestin receptors. *Nature, 300,* 648–649.

Raisman, G., & Field, P. M. (1971). Sexual dimorphism in the preoptic area of the rat. *Science, 173,* 731–733.

Raisman, G., & Field, P. M. (1973). Sexual dimorphism in the neuropil of the preoptic area of the rat and its dependence on neonatal androgen. *Brain Research, 54,* 1–29.

Rakic, P. (1972). Mode of cell migration to the superficial layers of fetal monkey neocortex. *Journal of Comparative Neurology, 145,* 61–84.

Ratner, A., & Adamo, N. J. (1971). Arcuate nucleus region in androgen-sterilized female rats: Ultrastructural observations. *Neuroendocrinology, 8,* 26–35.

Robinson, S. M., Fox, T. O., & Sidman, R. L. (1985). A genetic variant in the morphology of the medial preoptic area in mice. *Journal of Neurogenetics, 2,* 381–388.

Robinson, S. M., Fox, T. O., Dikkes, P., & Pearlstein, R. A. (1986). Sex differences in the shape of the sexually dimorphic nucleus of the preoptic area and suprachiasmatic nucleus of the rat: 3-D computer reconstructions and morphometrics. *Brain Research, 371,* 380–384.

Sachs, B. D. (1982). Role of striated penile muscles in penile reflexes, copulation, and induction of pregnancy in the rat. *Journal of Reproduction and Fertility, 66,* 433–443.

Schipper, H., Brawer, J. R., Nelson, J. F., Felicio, L. S., & Finch, C. E. (1981). Role of the gonads in the histologic aging of the hypothalamic arcuate nucleus. *Biology of Reproduction, 25,* 413–419.

Schoonmaker, J. N., Breedlove, S. M., Arnold, A. P., & Gorski, R. A. (1983). Accumulation of steroid in the sexually dimorphic nucleus of the preoptic area in the neonatal rat hypothalamus. *Society for Neuroscience Abstracts, 9(2),* 1094.

Sengelaub, D. R., & Arnold, A. P. (1984). Ontogeny of sexual dimorphism in a rat spinal nucleus: II. Formation and loss of early projections. *Society for Neuroscience Abstracts, 10,* 453.

Sengelaub, D. R., & Arnold, A. P. (1986). Development and loss of early projections in a sexually dimorphic rat spinal nucleus. *Journal of Neuroscience, 6,* 1613–1620.

Shivers, B. D., Harlan, R. E., Morrell, J. I., & Pfaff, D. W. (1983). Immunocytochemical localization of luteinizing hormone-releasing hormone in male and female rat brains. *Neuroendocrinology, 36,* 1–12.

Siegel, L. I., Akutagawa, E., Fox, T. O., Konishi, M., & Politch, J. A. (1986). Androgen and estrogen receptors in adult zebra finch brain. *Journal of Neuroscience Research, 16,* 617–628.

Simerly, R. B., & Swanson, L. W. (1986). The organization of neural imputs to the medial preoptic nucleus of the rat. *Journal of Comparative Neurology, 246,* 312–342.

Simerly, R. B., & Swanson, L. W. (1987). Castration reversibly alters levels of cholecystokinin immunoreactivity within cells of three interconnected sexually dimorphic forebrain nuclei in the rat. *Proceedings of the National Academy of Sciences, U.S.A., 84,* 2087–2091.

Simerly, R. B., Swanson, L. W., & Gorski, R. A. (1984). Demonstration of a sexual dimorphism in the distribution of serotonin-immunoreactive fibers in the medial preoptic nucleus of the rat. *Journal of Comparative Neurology, 225,* 151–166.

Simerly, R. B., Swanson, L. W., & Gorski, R. A. (1985a). The distribution of monoaminergic cells and fibers in a periventricular preoptic nucleus involved in the control of gonadotropin release: Immunohistochemical evidence for a dopaminergic sexual dimorphism. *Brain Research, 330,* 55–64.

Simerly, R. B., Swanson, L. W., & Gorski, R. A. (1985b). Reversal of the sexually dimorphic distribution of serotonin-immunoreactive fibers in the medial preoptic nucleus by treatment with perinatal androgen. *Brain Research, 340,* 91–98.

Simerly, R. B., Swanson, L. W., Handa, R. J., & Gorski, R. A. (1985). Influence of perinatal androgen on the sexually dimorphic distribution of tyrosine hydroxylase-immunoreactive cells and fibers in the anteroventral periventricular nucleus of the rat. *Neuroendocrinology, 40,* 501–510.

Simerly, R. B., Gorski, R. A., & Swanson, L. W. (1986). Neurotransmitter specificity of cells and fibers in the medial preoptic nucleus: An immunohistochemical study in the rat. *Journal of Comparative Neurology, 246,* 343–363.

Simerly, R. B., McCall, L. D., & Watson, S. J. (1988). Distribution of opioid peptides in the preoptic region: Immunohistochemical evidence for a steroid-sensitive enkephalin sexual dimorphism. *Journal of Comparative Neurology, 276,* 442–459.

Sternberger, L. A. (1986). *Immunocytochemistry,* (3rd ed.). New York: John Wiley & Sons.

Sternberger, L. A., Harwell, L. W., & Sternberger, N. H. (1982). Neurotypy: Regional individuality in rat brain detected by immunocytochemistry with monoclonal antibodies. *Proceedings of the National Academy of Sciences, U.S.A., 79,* 1326–1330.

Stumpf, W. E., Sar, M., & Keefer, D. A. (1975). Atlas of estrogen target cells in rat brain. In W. E. Stumpf & L. D. Grant (Eds.), *Anatomical neuroendocrinology* (pp. 104–119). Basel: S. Karger.

Swaab, D. F., & Fliers, E. (1985). A sexually dimorphic nucleus in the human brain. *Science, 228,* 1112–1115.

Swaab, D. F., & Hofman, M. A. (1984). Sexual differentiation of the human brain: A historical perspective. *Progress in Brain Research, 61,* 361–374.

Szentagothai, J., Flerko, B., Mess, B., & Halasz, B. (1962). *Hypothalamic control of the anterior pituitary* Budapest: Akademiai Kiado.

Takami, S., & Urano, A. (1984). The volume of the toad medial amygdala–anterior preoptic complex is sexually dimorphic and seasonally variable. *Neuroscience Letters, 44,* 253–258.

Tobet, S. A., & Baum, M. J. (1987). Role for prenatal estrogen in the development of masculine sexual behavior in the male ferret. *Hormones and Behavior, 21,* 419–429.

Tobet, S. A., & Fox, T. O. (1989). Sex- and hormone-dependent antigen immunoreactivity in developing rat brain. *Proceedings of the National Academy of Sciences, U.S.A., 86,* 382–386.

Tobet, S. A., & Fox, T. O. (1989). Androgen regulation of an antigen expressed in regions of developing brainstem monoaminergic cell groups. *Developmental Brain Research, 46,* 253–261.

Tobet, S. A., Baum, M. J., Tang, H. B., Shim, J. H., & Canick, J. A. (1985). Aromatase activity in the perinatal rat forebrain: Effects of age, sex and intrauterine position. *Developmental Brain Research, 23,* 171–178.

Tobet, S. A., Shim, J. H., Osiecki, S. T., Baum, M. J., & Canick, J. A. (1985). Androgen aromatization and 5alpha-reduction in ferret brain during perinatal development: Effects of sex and testosterone manipulation. *Endocrinology, 116,* 1869–1877.

Tobet, S. A., Zahniser, D. J., & Baum, M. J. (1986a). Sexual dimorphism in the preoptic/anterior hypothalamic area of ferrets: Effects of adult exposure to sex steroids. *Brain Research, 364,* 249–257.

Tobet, S. A., Zahniser, D. J., & Baum, M. J. (1986b). Differentiation in male ferrets of a sexually dimorphic nucleus of the preoptic/anterior hypothalamic area requires prenatal estrogen. *Neuroendocrinology, 44,* 229–308.

Tobet, S. A., Blank, D., & Fox, T. O. (1987). Hypothalamic tanycytes and brainstem neurons identified by monoclonal antibody 3D10. *Society for Neuroscience Abstracts, 13,* 1167.

Toran-Allerand, C. D. (1976). Sex steroids and the development of the newborn mouse hypothalamus and preoptic area *in vitro:* Implications for sexual differentiation. *Brain Research, 106,* 407–412.

Toran-Allerand, C. D. (1978). Gonadal hormones and brain development: Cellular aspects of sexual differentiation. *American Zoologist, 18,* 553–565.

Toran-Allerand, C. D. (1980). Sex steroids and the development of the newborn mouse hypothalamus and preoptic area *in vitro.* II. Morphological correlates and hormonal specificity. *Brain Research, 189,* 413–427.

Toran-Allerand, C. D. (1984). On the genesis of sexual differentiation of the central nervous system: Morphogenetic consequences of steroidal exposure and possible role of alpha-fetoprotein. *Progress in Brain Research, 61,* 63–98.

Toran-Allerand, C. D., Gerlach, J. L., & McEwen, B. S. (1980). Autoradiographic localization of [³H]estradiol related to steroid responsiveness in cultures of the newborn mouse hypothalamus and preoptic area. *Brain Research, 184,* 517–522.

Tranque, P. A., Suarez, I., Olmos, G., Fernandez, B., & Garcia-Segura, L. M. (1987). Estradiol-induced redistribution of glial fibrillary acidic protein immunoreactivity in the rat brain. *Brain Research, 406,* 348–351.

Ulibarri, C., & Yahr, P. (1988). Role of neonatal androgens in sexual differentiation of brain structure, scent marking, and gonadotropin secretion in gerbils. *Behavioral Neural Biology, 49,* 27–44.

Vaccari, A. (1980). Sexual differentiation of monoamine neurotransmitters. In H. Parvez & S. Parvez (Eds.), *Biogenic amines in development* (pp. 327–352). Amsterdam: Elsevier-North Holland.

van Leeuwen, F. W., Caffe, A. R., & De Vries, G. J. (1985). Vasopressin cells in the bed nucleus of the stria terminalis of the rat: Sex differences and the influence of androgens. *Brain Research, 325,* 391–394.

Varon, S. S., & Somjen, G. G. (1979). Neuron–glia interactions. *Neurosciences Research Program Bulletin, 17.*

Viglietti-Panzica, C., Panzica, G. C., Fiori, M. G., Calcagni, M., Anselmetti, G. C., & Balthazart, J. (1986). A sexually dimorphic nucleus in the quail preoptic area. *Neuroscience Letters, 64,* 129–134.

Vito, C. C., Baum, M. J., Bloom, C., & Fox, T. O. (1985). Androgen and estrogen receptors in perinatal ferret brain. *Journal of Neuroscience, 5,* 268–274.

Wallach, S. J. R., & Hart, B. L. (1983). The role of the striated penile muscles of the male rat in seminal plug dislodgement and deposition. *Physiology and Behavior, 31,* 815–821.

Watson, R. E., Hoffman, G. E., & Wiegand, S. J. (1986). Sexually dimorphic opioid distribution in the preoptic area: Manipulation by gonadal steroids. *Brain Research, 398,* 157–163.

Watson, R. E., Wiegand, S. J., & Hoffman, G. E. (1988). Ontogeny of a sexually dimorphic opioid system in the preoptic area of the rat. *Developmental Brain Research, 44,* 49–58.

Wee, B. E. F., & Clemens, L. G. (1987). Characteristics of the spinal nucleus of the bulbocavernnosus are influenced by genotype in the house mouse. *Brain Research, 424,* 305–310.

Weisz, J., & Ward, I. L. (1980). Plasma testosterone and progesterone titers of pregnant rats, their male and female fetuses, and neonatal offspring. *Endocrinology, 106,* 306–316.

Wray, S., & Hoffman, G. (1986). A developmental study of the quantitative distribution of LHRH neurons within the central nervous system of postnatal male and female rats. *Journal of Comparative Neurology, 252,* 522–531.

Yahr, P., Commins, D., Jackson, J. C., & Newman, A. (1982). Independent control of sexual and scent marking behaviors of male gerbils by cells in or near the medial preoptic area. *Hormones and Behavior, 16,* 304–322.

Yanai, J. (1979). Strain and sex differences in the rat brain. *Acta Anatomica, 103,* 150–158.

Zambrano, D., & de Robertis, E. (1968). The effect of castration upon the ultrastructure of the rat hypothalamus II. Arcuate neurons and the outer zone of the median eminence. *Zeitschrift fur Zellforschung, 87,* 409–421.

ADDENDUM

INTRODUCTION

In this chapter we discussed approaches to establishing the existence and understanding of sexual dimorphisms in brains of laboratory animals. Particular attention was devoted to examining the questions that can and cannot be answered experimentally concerning the effect of steroids and other factors during development and adulthood on morphology and behavior. One hallmark of neurobiological studies with laboratory animals is the ability of investigators to control a wide range of variables throughout life that can affect the brain's function, and hence behavior. Ultimately, we want to relate information obtained with laboratory animals to the origins of human behaviors and eventually comprehend the relationships between the dynamically changing structures of the human brain and social behavior.

Prior to 1991, investigators in the field of sexual dimorphisms could not agree that sexual dimorphisms in human brains had been unequivocally demonstrated. Accordingly, considerable interest was raised by the publication and attendant publicity of a study by LeVay (1991) in which he reported that men who had identified themselves as having a homosexual orientation had a grouping of neurons in the hypothalamus whose volume was more similar in average magnitude to that found in women than in presumed heterosexual men. In an earlier study, Allen *et al.* (1989) identified two groups of cells in the human preoptic area/anterior hypothalamus that differed in volume in males versus females. LeVay's (1991) report may prove to be especially significant in that it confirms one of the findings of Allen *et al.* (1989) and, further, suggests that men identified as homosexual and heterosexual tend to distribute at different ends of the wide size range for the area called Interstitial Nucleus of the Anterior Hypothalamus-3 (INAH-3).

In the broad public discussion following LeVay's report, an underlying commonly made assumption is that sex differences in neural morphology have been established for regions within the human brain. Given that assumption, the new issue was whether morphological differences could be correlated with self-reported sexual orientations. We must first ask, however, whether replicated evidence exists for differences in characteristics of brain morphology between male and female human brains. If differences between males and females can be demonstrated, based on reports in the existing literature, then we will have a better understanding of which variables are important (e.g., age, hormonal state, experience).

In behavioral, biochemical, and morphological studies in laboratory animals, gender characteristics vary within a single sex of a population. Measures of sexually dimorphic characteristics often overlap between the sexes, but are distinguished by significantly different mean values. For example, in one study the nuclear volume of the sexually dimorphic nucleus of the preoptic area (SDN-POA) of intact rats, one female was in the range of the group of males (Robinson *et al.*, 1986); a similar result was found in mice (Robinson *et al.*, 1985). A sexually dimorphic nucleus in the preoptic area of male ferrets apparent as a population of large cells is not apparent as a nucleus in females. Among female ferrets, however, a variable number of larger cells are found dispersed within this region (Tobet *et al.*, 1986; Tobet & Baum, 1991). Interpretations of studies on human brains should also focus on determining the variability within each sexual group and whether average values for each sex can be distinguished.

It is now believed that masculine traits and feminine traits in the same individual can range independently from low to high (Goy & McEwen, 1980). Actions via estrogen and androgen receptors occurring at different times (Fox *et al.*, 1982) may result in a mosaic of behavioral phenotypes (Baum, 1979). For example, animals can be defeminized or not, or masculinized or not, independently, resulting in animals with low or high levels of both behaviors at the extremes. This is a particularly useful perspective for approaching issues of sexual dimorphisms in humans. Hypotheses that assume either a strict dichotomy between male and female or a linear relationship for sexuality, with masculine at one end and feminine at the other, are inappropriate for many behaviors and probably for attitudes and preferences.

POSSIBLE SEXUAL DIMORPHISMS IN HUMAN PREOPTIC AREA/HYPOTHALAMUS

In 1985, Swaab and Fliers reported a sexual dimorphism in the preoptic area of humans. They indicated that the volume of, and number of cells in, a particular subdivision of the preoptic area were two to three times greater in males than in females. They suggested that this nucleus was analogous to sexually dimorphic nuclei in rats (SDN-POA) and other mammals. Swaab and Hofman (1988) found that the number of cells in this region reached a maximum in both boys and girls aged 2–4 years. With increasing age the number of cells in the sexes diverged, with females losing more cells than males, analogous to the pattern of neuronal development of the spinal nucleus of the bulbocavernosus (SNB) in rats (Nordeen *et al.*, 1985). In 1989, Allen, Hines, Shryne, and Gorski reported four groups of cells which they termed Interstitial Nuclei of the Anterior Hypothalamus (INAH-1 to 4). Two of these, INAH-2 and INAH-3, differed in average volume for males and females; INAH-3 was two to three times larger and INAH-2 two times larger in males than in females. However, for three women aged 20–32, the INAH-2 volumes were similar to the male mean volume. They suggested that INAH-2 might be larger in women of child-bearing age, stressing the importance of age and hormonal state.

Allen *et al.* (1989) suggest that the nucleus reported by Swaab and Fliers (1985) to be sexually dimorphic is the same as the nucleus Allen *et al.* refer to as INAH-1. However, the two groups sectioned tissue at different thicknesses and angles; used different embedding media, which could cause significant differential shrinkage; and used different sampling procedures for measuring volumes. Therefore a definitive comparison of these studies is difficult. Based on our examination of these and LeVay's (1991) reports, it may be that Swaab and Fliers (1985) examined a combination of regions including INAH-1 and INAH-2. In the absence of further data, the results of these two studies suggest a possible sex difference in average nuclear volume, but disagree on the volumes and locations of the respective nuclei.

LeVay's (1991) study examined the four nuclei identified by Allen *et al.* (1989) in three groups of specimens. Reasoning that these brain structures might correlate with sexual orientation, LeVay compared the INAHs in females and presumed heterosexual and homosexual males. LeVay (1991) found that the average INAH-3 volumes in females and homosexual men were smaller than that in heterosexual men. Several important aspects of the data are apparent. First, INAH-3 volumes of all three groups overlapped to a high degree. Three individuals (out of 16) in the homosexual group had large INAH-3 volumes that overlapped with the main cluster for heterosexual men, and the rest had smaller nuclei, overlapping with most of the females and with the other samples from heterosexual men. Whether the samples with larger values represent normal variation for this nucleus among all groups or limitations in the self-assignment of sexual orientation, a unique attribute in human studies, is an important question. Second, LeVay (1991) confirmed the sex difference reported by Allen *et al.* (1989) in INAH-3, but not in INAH-2. As LeVay (1991) notes, this might be a result of the ages of the available sample, since Allen *et al.* (1989) reported a large age dependence associated with their reported sex difference in INAH-2.

The following conclusion appears justifiable. One nucleus, INAH-3, has been shown in two laboratories (Allen *et al.*, 1989; LeVay, 1991) to distribute distinctively in men and women, and one laboratory (LeVay, 1991) reports a further distinction for men with a homosexual orientation. While awaiting additional human-studies replications, care must be exercised in considering the hypothetical causes and effects. In a separate region, putatively INAH-1 (and possibly INAH-1 and INAH-2), Swaab and Hofman found a sex difference but no difference between heterosexual and homosexual men (Hofman & Swaab, 1989). LeVay found neither a sex difference nor a difference between heterosexual men and homosexual men in INAH-1 and INAH-2. While the results for INAH-2 are controversial with regard to a male–female comparison, the findings represent agreement with regard to sexual orientation.

INTERPRETATIONS

Studies on laboratory animals have helped define limits for ascribing causative relationships between structures and behaviors. It would be a mistake, however, to assess studies of humans only using guidelines developed through research on laboratory animals. Some variables are clearly harder to control experimentally and ethically (e.g., castration and hormone replacement experiments cannot be performed in humans as they are with animals). However, human studies offer challenges and access to variables that may be unique to humans. Humans can report feelings and perceived motivations. Consequently, a range of psychological and clinical conditions can be considered that might not be apparent, and some of which might not occur, in laboratory animals. Furthermore, experimental laboratory research arises heuristically from historical, conceptual, and technical influences, and from the constraints of the scientific society. These factors intercalate with medical and political forces in determining the foci for research on human biology. Accordingly, while research on human biological material might not always follow the accepted experimental mores of the controlled laboratory, the results still must be judged rigorously and with scientific skepticism.

Factors that can contribute to gender differences include genetic background, developmental and later hormonal states, age, and experience. With laboratory animals it is possible to control, at least partially, for each of these factors. By judiciously manipulating these factors, investigators are often able to distinguish between cause and effect. In correlating developmental events and morphological parameters with sexual behavior, the ability to quantify specific behaviors is essential. Even though a great many factors can influence behavioral traits and their measurement, in laboratory animals it is possible to delimit testing conditions such that reproducible, statistically significant data are obtained. The data are "objective" to the extent that they are quantifiable and the evaluations do not depend on self reports.

With the exception of age, genetic, hormonal, or past experiential factors are very difficult to control in studies of humans. Patients whose gonads have been removed for clinical reasons have reported on the libidinal effects of therapeutically administered hormones. Data on brain morphology following hormone administration have been obtained from transsexual patients administered exogenous hormone (Hofman & Swaab, 1988). In addition, relevant natural experiments do occur. These include genetic defects (Fox *et al.,* 1982), such as changes in the gene for androgen receptors in testicular feminization, resulting in reduced responses of cells to circulating androgens, and other congenital defects such as excessive circulating steroids in congenital adrenal hyperplasia. Studies on human CNS morphology, however, still rely on available autopsy material, and as a result genetic backgrounds, hormone levels, conditions of health, and ages cannot be determined prior to obtaining brains for morphological or chemical analyses.

In the recent studies on human brain morphology, variations exist among laboratories in the ages of specimens, the technical choices for morphological methods, and the extent of heterogeneity in the populations from which samples are obtained. For studies of homosexual men in particular, the possible complication of infection and death from AIDS has been raised and remains a concern.

The LeVay (1991) report concludes that the difference seen in INAH-3 suggests that sexual orientation has a biological substrate. This statement has occasioned considerable response. It means either that behavior depends on brain structure or that the structural difference in this report contributed to sexual orientation. If it is the first, then that premise exists independently of the results in the report. It is a premise accepted by many as a tenet of neurobiology. If the latter was intended, the data in this study do not address the possibility of a causal relationship. The differences reported in INAH-3 might be a result rather than a cause of sexual orientation, reflecting the actions of neuronal activity, variable hormone levels, and environmental factors. It is possible that a combination of causative and resulting factors influence the size of nuclei involved in specific behavioral patterns.

In experimental systems, examples exist of brain morphology changing in response to circulating hormones (see chapter). Seasonal differences in the volume of the canary song nucleus HVc (Nottebohm, 1981) have been attributed to changes in the apparent size of cells in the periphery of the nucleus (Gahr, 1990). Consequently, the apparent volume of the nucleus is a result of variable hormonal levels. In the preoptic area of gerbils, sexually dimorphic acetylcholinesterase staining is dependent upon circulating levels of androgen (Commins and Yahr, 1984). Although it is possible to control hormone levels in animal studies, it has not been done in human studies. It is possible that the difference observed in INAH-3 by LeVay is a direct result of hormonal differences between the heterosexual and homosexual male groups. Moreover, it has been suggested that androgen levels may be decreased in individuals infected with the AIDS virus (cited in Hendricks *et al.*, 1989). In the study of LeVay (1991) all of the homosexual subjects and a third of the heterosexual subjects had been infected with the AIDS virus.

Both the Hofman–Swaab group and the Allen–Gorski group reported additional areas of possible sex differences in human brain. Allen and Gorski (1990) reported a 2- to 3-fold larger volume in males in the "darkly staining posteromedial" component of the bed nucleus of the stria terminalis (BNST-dspm). Swaab and Hofman (1990) examined the suprachiasmatic nucleus and found it to be almost twice as large in homosexual men compared to controls, even though a difference had not been found between males and females (Hofman and Swaab, 1989). These findings from both laboratories await replication.

Although beyond the scope of this review, an additional area of controversy for sex differences in brain morphology revolves around the shape, size, and content of the corpus callosum. Since the initial report of differences between human males and females in the shape and sub-regional sizes of the corpus callosum by de Lacoste-Utamsing and Holloway (1982), several other groups have attempted similar studies using various methods (e.g., autopsy material and magnetic resonance imaging). Separate groups have failed to replicate most of the sexually dimorphic results (see Byne *et al.*, 1988). As for our discussion of hypothalamic nuclear groups above, with more data and closer attention to factors such as age and common measureable parameters, some pattern of consistency may emerge (Allen *et al.*, 1991; Witelson, 1991). In conjunction with basic findings from animal models (reviewed in Denenberg, 1981), an understanding of the morphological actions of gonadal steroids is likely to emerge.

Acknowledgment

The authors thank M. J. Baum and J. C. King for their support and suggestions.

REFERENCES

Allen, L. S., Hines, M., Shryne, J. E., & Gorski, R. A. (1989). Two sexually dimorphic cell groups in the human brain. *Journal of Neuroscience, 9*, 497–506.

Allen, L. S., & Gorski, R. A. (1990). Sex difference in the bed nucleus of the stria terminalis of the human brain. *Journal of Comparative Neurology, 302*, 697–706.

Allen, L. S., Richey, M. F., Chai, Y. M., & Gorski, R. A. (1991). Sex differences in the corpus callosum of the living human being. *Journal of Neuroscience, 11*, 933–942.

Baum, M. J. (1979). Differentiation of coital behavior in mammals: A comparative analysis. *Neuroscience Biobehavior Review, 3*, 265–284.

Byne, W., Bleier, R., & Houston, L. (1988). Variations in human corpus callosum do not predict gender: a study using magnetic resonance imaging. *Behavioral Neuroscience, 102*, 222–227.

Commins, D., & Yahr, P. (1984). Acetylcholinesterase activity in the sexually dimorphic area of the gerbil brain: Sex differences and influences of adult gonadal steroids. *Journal of Comparative Neurology, 224*, 123–131.

de Lacoste-Utamsing, C., & Holloway, R. L. (1982). Sexual dimorphism in the human corpus callosum. *Science, 216*, 1431–1432.

Denenberg, V. H. (1981). Hemispheric laterality in animals and the effects of early experience. *Behavioral Brain Sciences, 4*, 1–49.

Fox, T. O., Olsen, K. L., Vito, C. C., & Wieland, S. J. (1982). Putative steroid receptors: Genetics and development. In F. O. Schmitt, S. J. Bird, and F. E. Bloom (Eds.), *Molecular Genetic Neuroscience*, (pp. 289–306). Raven Press: New York.

Gahr, M. (1990). Delineation of a brain nucleus: Comparisons of cytochemical, hodological, and cytoarchitectonic views of the song control nucleus HVC of the adult canary. *Journal of Comparative Neurology, 294,* 30–36.

Goy, R. W., & McEwen, B. S. (1980). Sexual differentiation of the brain. MIT Press: Cambridge.

Hendricks, S. E., Graber, B., & Rodriquez-Sierra, J. F. (1989). Neuroendocrine responses to exogenous estrogen: No differences between heterosexual and homosexual men. *Psychoneuroendocrinology, 14,* 177–185.

Hofman, M. A., & Swaab, D. F. (1989). The sexually dimorphic nucleus of the preoptic area in the human brain: a comparative morphometric study. *Journal of Anatomy, 164,* 55–72.

LeVay, S. (1991). A difference in hypothalamic structure between heterosexual and homosexual men. *Science, 253,* 1034–1037.

Nordeen, E. J., Nordeen, K. W., Sengelaub, D. R., & Arnold, A. P. (1985). Androgens prevent normally occurring cell death in a sexually dimorphic spinal nucleus. *Science, 229,* 671–673.

Nottebohm, F. (1981). A brain for all seasons: cyclical anatomical changes in song control nuclei of the canary brain. *Science, 214,* 1368–1370.

Robinson, S. M., Fox, T. O., & Sidman, R. L. (1985). A genetic variant in the morphology of the medial preoptic area in mice. *Journal of Neurogenetics, 2,* 381–388.

Robinson, S. M., Fox, T. O., Dikkes, P., & Pearlstein, R. A. (1986). Sex differences in the shape of the sexually dimorphic nucleus of the preoptic area and suprachiasmatic nucleus of the rat: 3-D computer reconstructions and morphometrics. *Brain Research, 371,* 380–384.

Swaab, D. F., & Fliers, E. (1985). A sexually dimorphic nucleus in the human brain. *Science, 228,* 1112–1115.

Swaab, D. F., & Hofman, M. A. (1988). Sexual differentiation of the human hypothalamus: ontogeny of the sexually dimorphic nucleus of the preoptic area. *Developmental Brain Research, 44,* 314–318.

Swaab, D. F., & Hofman, M. A. (1990). An enlarged suprachiasmatic nucleus in homosexual men. *Brain Research, 537,* 141–148.

Tobet, S. A., Zahniser, D. J., & Baum, M. J. (1986). Sexual dimorphism in the preoptic/anterior hypothalamic area of ferrets: Effects of adult exposure to sex steroids. *Brain Research, 364,* 249–257.

Tobet, S. A., & Baum, M. J. (1991). Estradiol binding neurons in the sexually dimorphic male nucleus of the preoptic/anterior hypothalamic area of adult ferrets. *Brain Research, 546,* 345–350.

Witelson, S. F. (1991). Neural sexual mosaicism: Sexual differentiation of the human temporo-parietal region for functional asymmetry. *Psychoneuroendocrinology, 16,* 131–153.

3

Gonadal Hormones and Sex Differences in Nonreproductive Behaviors

WILLIAM W. BEATTY

INTRODUCTION

HORMONE ORGANIZATION AND ACTIVATION OF BEHAVIOR

Experimental investigations of the influence of gonadal hormones on sex differences in nonreproductive behaviors have been guided by concepts and methods applied to the study of sexual behavior. In 1959, Phoenix, Goy, Gerall, and Young reported that administration of testosterone propionate to pregnant guinea pigs caused the female offspring to display high levels of male mounting behavior but very low levels of female receptive behavior when tested as adults. This original observation has been confirmed and extended in literally hundreds of published reports. In general the findings indicated that mammals are biased to develop feminine patterns of reproductive function and will only display male-like characteristics if exposed to testicular hormones early in life. The important concept that emerged from this work is that gonadal hormones exert two distinct actions, which are termed *organizational* and *activational* effects.

Activational effects refer to reversible changes in morphology, physiology, or behavior that depend on the continued presence of the hormone in the body and on the functional integrity of the receptors on which it acts. Within broad limits the age of the organism has only a minor influence on the activational effects of gonadal

WILLIAM W. BEATTY Department of Psychiatry and Behavioral Sciences, University of Oklahoma Health Sciences Center, Oklahoma City, Oklahoma 73190.

Sexual Differentiation, Volume 11 of *Handbook of Behavioral Neurobiology,* edited by Arnold A. Gerall, Howard Moltz, and Ingeborg L. Ward, Plenum Press, New York, 1992.

hormones, which are qualitatively similar from puberty to senescence. In contrast, the organizational effects of gonadal hormones produce enduring changes in morphology, physiology, and behavior, which arise from relatively brief exposure to the hormone during a limited period of perinatal development. In all mammals that have been studied, a relatively brief period exists during which the developing organism is responsive to the organizational actions of gonadal hormones. What varies is the timing of this "critical" (or sensitive) period in relation to birth. For species with relatively long gestation periods, such as the rhesus monkey, the critical period is completed before birth. For species with short gestation periods, such as the rat, the critical period begins prenatally but extends into the first several days of postnatal life (Goy & McEwen, 1980).

Mechanisms of Testicular Hormone Action: Aromatization and Reduction

In the original conception of hormonal organization it was assumed that androgens acted directly to accentuate the development of tissues essential for the display of masculine copulatory behavior and simultaneously to suppress the development of tissues responsible for female patterns of mating behavior and other aspects of neuroendocrine function (e.g., ovulation). Later in adulthood exposure to testosterone would activate the male behavioral capacities organized early in perinatal life. Subsequent studies in the rat and mouse demonstrated that neonatal treatment with estradiol, but not dihydrotestosterone (DHT), effectively suppressed the capacity to ovulate and display feminine sexual behavior and enhanced the potential to display male mating behavior (see Goy & McEwen, 1980, for references). Since earlier work (see Feder, 1978, for review) showed that either testosterone or estradiol could activate male reproductive behavior in adult male castrates or female mating patterns in ovariectomized females, the possibility arose that both the organizational and the activational effects of testosterone were actually mediated by its enzymatic conversion to estradiol.

Figure 1 summarizes the metabolism of testosterone in a highly simplified way. Testosterone and certain other androgens (e.g., androstenedione) are acceptable substrates for the aromatase enzyme, which catalyzes their conversion to estrogens. Androgens with this property are often called aromatizable androgens. Other androgens, such as DHT, are not acceptable substrates for the aromatase enzyme and hence are "nonaromatizable."

The discovery of aromatizing enzymes in fetal and neonatal brain tissue (Reddy, Naftolin, & Ryan, 1974) inspired a flurry of research to evaluate the possibility that the organizational effects of testosterone on mating and other sexually dimorphic functions were mediated by intraneuronal metabolism to estradiol (the so-called

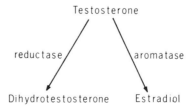

Figure 1. Two important metabolic "fates" of testosterone.

aromatization hypothesis). In principle, this idea is easily tested by performing the set of experiments outlined in Table 1 in a species such as the rat, in which the sensitive period for hormonal organization extends into early postnatal life. In such a species the predicted treatment effects are the same for females or males that are gonadectomized soon after birth. In addition, if the aromatization hypothesis is correct, treatments that block production of estrogens or interfere with estrogen action on intracellular estrogen receptors should attenuate the organizational effects of endogenous testosterone in gonadally intact males. Further, if aromatization is the only mechanism of testosterone action, then agents that prevent reduction of testosterone or block androgen binding to androgen receptors should be ineffective. Obviously, the same experiments can be (and have been) performed in males castrated as adults to test the idea that aromatization of testosterone to estradiol is essential for the activation of male mating behavior.

Despite problems of interpretation posed by the fact that none of the enzyme inhibitors or hormone antagonists is perfectly selective in its biochemical action, in general, the findings support the view that aromatization is an important mechanism mediating both the organizational and activational effects of testosterone on mating behavior in several species of rodents. The extent to which this model is broadly applicable to other species is as yet unclear. For example, considerable evidence exists that masculinization (i.e., enhanced potential to display male-like behavior) and defeminization (i.e., reduced capacity to exhibit female patterns of neuroendocrine function and behavior) are separable processes that are differentially affected by the organizational actions of testosterone and its metabolites. The extent to which hormones masculinize and defeminize behavior varies among species (see Baum, 1979, for review). Even in rodents such as the rat, in which exposure to testosterone early in development causes both masculinization and defeminization, the mechanisms responsible for these organizational effects are clearly not identical. Although the sensitive periods for defeminization and masculinization began prenatally and extend into the early postnatal period, brief neonatal exposure to testosterone or estradiol causes marked defeminization but only modest masculinization. Perhaps this is so because the sensitive periods for hormonal organization of masculine sexual behavior (i.e., mounting, intromission, ejaculation) occur at slightly different times in development (see Ward & Ward, 1985, for review). Moreover, although aromatization of testosterone to estradiol is sufficient to

TABLE 1. BEHAVIORAL EFFECTS AS PREDICTED BY THE AROMATIZATION HYPOTHESIS OF TESTOSTERONE ACTION

Treatment	Mimics (+) or does not mimic (−) effect of testosterone (T) treatment
Estradiol (E)	+
Dihydrotestosterone	−
T + aromatase inhibitor	−
T + reductase inhibitor	+
E + aromatase inhibitor	+
T or E + estrogen antagonist	−
T + androgen antagonist	+

cause defeminization, masculinization of mating behavior may involve direct effects on androgen receptor systems as well (e.g., Davis, Chaptal, & McEwen, 1979).

Sexually Dimorphic Behaviors: Varying Dependence on Organizational and Activational Effects of Hormones

It was once supposed that the expression of all sexually dimorphic behavior patterns depended on both the organizational and activational influences of gonadal hormones. Mating behavior, especially in rodents, exemplifies this pattern of hormonal control. As described above, the presence or absence of testosterone early in life fosters the development of capacities to display masculine or feminine behavior patterns, but in the absence of gonadal hormone stimulation in adulthood, little mating behavior of any kind occurs.

Examination of sex differences in nonreproductive behaviors reveals that the degree to which organizational and activational effects of hormones contribute to behavioral dimorphisms varies greatly. For some behaviors, such as the play fighting of juvenile rats and monkeys, only organizational actions of testosterone influence the display of the behavior; activational influences are not required, and, in fact, hormone manipulation after the end of the sensitive period has no effect (Meaney & Stewart, 1983). For other behaviors such as yawning, which is performed more frequently by male than by female rhesus monkeys, activational influences of hormones are principally, if not exclusively, responsible for the sex differences (Goy & Resko, 1972). Finally, the possibility must be considered that there may be sexually dimorphic behaviors that are not affected by gonadal hormones at all. No such behaviors can be identified at present, but there are obviously a great many genetic and experiential mechanisms that could influence sex differences in behavior without altering hormone function. Of greater importance is the likelihood that genetic, environmental, and hormonal factors interact, sometimes in complex ways, to determine behavioral output.

Activity

Sex differences in motor activity in rats have usually been studied in running wheels or in the open-field test. On both measures adult females are more active. The sex difference is generally not observed until around puberty (e.g., Beatty & Fessler, 1976; Blizard, Lippman, & Chen, 1975), although Stewart, Skavarenina, and Pottier (1975) found a significant sex difference in open-field activity as early as 30 days of age.

Wheel Running

The pattern of organizational and activational effects of gonadal hormones varies greatly for the two measures of activity. Wheel running is strongly dependent on activational effects of gonadal hormones in both sexes. Gonadectomy depresses activity in both sexes, abolishing the sex difference (Gentry & Wade, 1976a; Hitchcock, 1925). Hormone replacement stimulates wheel running in both males and females, and activity is positively correlated with estrogen dosage in both sexes. Testosterone activates wheel running via aromatization to estradiol (Roy & Wade, 1975), perhaps in the anterior hypothalamic–preoptic area, where estradiol benzoate (EB) implants stimulate running (Colvin & Sawyer, 1969; Wade & Zucker, 1970). Lesions in this area block the EB stimulation of wheel running (King, 1979).

Organizational effects on wheel-running activity have also been reported. Neonatal treatment of female rats with high doses of testosterone propionate (TP) (500–1,250 μg) reduces response to activating effects of estrogen later in life (Gentry & Wade, 1976a; Gerall, Stone, & Hitt, 1972). Gentry and Wade (1976a) suggested that the major difference between females and males or neonatally androgenized females was that the latter two groups required longer treatment with estrogen before activity was stimulated but that the level of activity after prolonged treatment was similar for all three groups. More recently, Blizard (1983) reported that males and neonatally androgenized females exhibited significantly smaller increases in activity than normal females and that the difference persisted for about 2 months, the duration of the experiment. The organizational effect of androgen on wheel-running activity apparently does not involve aromatization to estrogen, since neonatal treatment of females with EB in doses as high as 500 μg does not reduce subsequent responsiveness to activating doses of estrogen in adulthood (Gerall, 1967; Stern & Janowiak, 1973). The other obvious possibility, that androgens act directly through an androgen receptor mechanism, has not been empirically tested.

OPEN-FIELD ACTIVITY

Activational influences of gonadal steroids on open-field activity appear to be quite modest in both sexes. Gonadectomy in adulthood has no effect on open-field activity in males (Bengelloun, Nelson, Zent, & Beatty, 1976; Slob, Bogers, & van Stolk, 1981) and inconsistent effects in females. Some researchers have reported that ovariectomy reduces open-field activity (Blizard et al., 1975; Slob et al., 1981), but others have not observed any effect (Bengelloun et al., 1976; Bronstein & Hirsch, 1974). The consequences of estradiol treatment are equally unclear. Blizard et al. (1975) found that estradiol injections increased open-field activity in ovariectomized females, and Slater and Blizard (1976) reported a similar finding after estradiol treatment in intact females. On the other hand, Slob et al. (1981) found no influence of estradiol treatment in ovariectomized females, and Fahrbach, Meisel, and Pfaff (1985) observed that implants of estradiol into the preoptic area stimulated wheel running but not open-field activity. Since Quadagno, Shryne, Anderson, and Gorski (1972) reported that open-field activity was somewhat higher at estrus than at diestrus, failure to control for stage of the estrous cycle of the intact controls may account for some of the inconsistencies, but this factor cannot explain the ineffectiveness of estradiol treatment in ovariectomized females (Slob et al., 1981). At present the most reasonable conclusion is that ovarian hormones weakly stimulate open-field activity, but this influence cannot account for more than a small portion of the overall sex difference.

This suggests that gonadal hormones should exert prominent organizational effects on open-field behavior, and the experimental results demonstrate that this is the case. Neonatal treatment of females with TP consistently depresses open-field activity (e.g., Blizard & Denef, 1973; Blizard et al., 1975; Gray, Levine, & Broadhurst, 1965; Pfaff & Zigmond, 1971; Stewart et al., 1975). Males castrated neonatally or exposed to the androgen antagonist cyproterone acetate during the first 10 days of postnatal life are generally more active than controls or males castrated at 30 days of age (Bengelloun et al., 1976; Joseph, Hess, & Birecree, 1978). The combination of prenatal exposure to cyproterone acetate plus neonatal castration augments activity of males to levels that approximate those observed in females (Scouten, Grotelueschen, & Beatty, 1975).

These latter findings implicate an androgen receptor mechanism in the organizational actions of testosterone on open-field activity. Consistent with this interpretation, Stewart, Vallentyne, and Meaney (1979) observed that neonatal treatment of female rats with the nonaromatizable androgen DHT depressed activity. However, neonatal treatment of female rats (Stevens & Goldstein, 1981) and hamsters (Swanson, 1967a) with EB duplicates the effects of TP on open-field activity, suggesting that aromatization of androgens to estrogens is also important. The observation that combined treatment of female rats with EB and DHT during the neonatal period has a greater effect on open-field activity than either treatment alone (Stewart *et al.*, 1979) suggests that the organizational actions of testosterone on open-field behavior involve both aromatization of androgens to estrogens and an androgen receptor mechanism. Since aromatization does not appear to be involved in the organizational effect of androgens on wheel-running activity, it seems clear that the hormonal influences on the development of sex differences in wheel running and open-field activity involve somewhat different metabolic pathways.

Ovarian hormones also exert an organizational influence on open-field behavior. Stewart and Cygan (1980) demonstrated that ovariectomy at 1 or 8 days of age but not at 23 days of age or later reduced open-field activity. Exposing neonatally gonadectomized males or females to low doses of estradiol increased their open-field activity. Thus, low levels of estrogen seem to feminize open-field behavior, whereas higher doses have the opposite effect. It is interesting to note that a comparable dose-dependent effect of estradiol has been observed using various indices of feminine sexual behavior as the endpoint (e.g., Döhler *et al.*, 1984; Gerall, Dunlap, & Hendricks, 1973; S.E. Hendricks, Chapter 4, this volume).

Sex differences in open-field behavior are not as consistently observed in mice as in rats (Archer, 1975). Gray (1979) suggests that the frequent use of highly inbred strains of mice versus the general practice of using outbred strains of rats may account for this discrepancy. He contends that outbred strains of mice exhibit the same pattern of higher activity by females typical of rats. The general implication of Gray's proposal, namely, that sex differences and hormone effects depend on genotype, receives some support from the literature. For example, Kemble and Enger (1984) found that male and female northern grasshopper mice were equally active in the open field and, further, that none of the sex differences in nonreproductive behaviors usually found in the laboratory rat (e.g., play fighting, avoidance conditioning) could be observed in this species of mice.

HOME CAGE ACTIVITY

The important influence of the genotype on hormonal control of activity has been demonstrated directly by Broida and Svare (1983). Home cage activity was measured in male mice of three different strains over a 24-hour period using an automated photocell detector that is highly sensitive to ambulation, rearing, and sniffing. Castration in adulthood reduced activity in the three strains they studied, but the effect was larger in the inbred C57BL strain than in the outbred Rockland–Swiss (RS) or inbred DBA strains. However, during a 30-minute test in a novel cage, castration reduced activity in both inbred strains but not in the outbred RS strain. Under both testing conditions testosterone replacement restored activity to normal levels. In a subsequent study Broida and Svare (1984) found that female mice of the

RS strain were more active than males in the home cage situation. Gonadectomy in adulthood reduced home cage activity in both sexes but did not eliminate the sex difference. Males castrated on the day of birth were more active than animals castrated at 5 days or later. Females treated with TP on the day of birth were less active than controls or animals treated at 10 days of age. These findings suggest that both organizational and activational influences of gonadal hormones can influence activity in mice. Depending on the way in which activity is measured, hormonal influences may vary with the genotype.

Social Behaviors

Intraspecific Aggression

It is often claimed that males are more aggressive than females. Such a global statement is quite inaccurate, as the high level of aggression displayed by mothers when a strange intruder approaches their young clearly demonstrates. Males, on the other hand, are generally more aggressive than females in situations that involve territorial intrusion and threat to their social rank (Moyer, 1974). In such situations aggression by males is primarily directed against other males and is facilitated by the presence of females and by prior social isolation (Brain, Benton, Howell, & Jones, 1980). The present review is focused on the influence of gonadal hormones on intraspecific aggression by pairs of nonmaternal adults. For a recent review of maternal aggression, see Svare (1983).

MICE.   The most detailed information on hormonal control of aggression is available for the mouse. Adult male mice behave quite aggressively toward other males, at least until dominance relationships are established. Females, by contrast, rarely fight. A common procedure for studying aggression is to introduce two individuals that were previously housed in social isolation either into a neutral arena or into the home cage of one member of the test pair.

With this approach, early studies (Beeman, 1947; Edwards, 1969) demonstrated the activational effects of gonadal steroids on aggressive behavior in male mice. Castration in adulthood abolished fighting, and hormone replacement with testosterone restored fighting in a dose-dependent manner. Developmental studies (e.g., Barkley & Goldman, 1977a) indicated a close correspondence between the onset of aggressive behavior and the rapid increase of androgen secretion at puberty and attested to the importance of hormones for behavioral activation.

Initial attempts to determine the mechanism by which testosterone exerted its activational effect led to great confusion about the relative potency of testosterone, estradiol, and other gonadal steroids. In an earlier review (Beatty, 1979) I suggested that some of the confusion resulted from differences in the characteristics of the opponents used in the assessment of agonistic behavior in various experiments. Since then the important role of chemosensory information in the control of aggression in mice has been thoroughly documented.

Early work (Mugford & Nowell, 1970) demonstrated the existence of both an aggression-promoting pheromone (produced by males) and an aggression-inhibiting pheromone (produced by females). The aggression-promoting pheromone is androgen dependent. Castration reduces the capacity of a male to elicit aggression, and testosterone but not estrogen treatment restores it (Mugford, 1974; Simon & Gandelman, 1978a).

Since the pheromones produced by the opponent, as well as the opponent's behavior, will obviously influence the aggressive behavior of the target animal, most recent studies have attempted to standardize the opponent as much as possible by using males rendered anosmic by olfactory bulbectomy. These animals elicit attacks from other animals, but they do not initiate attacks, nor do they fight back if attacked (Denenberg, Gaulin-Kremer, Gandelman, & Zarrow, 1973).

Activation of Aggression by Testosterone. When anosmic opponents are used, testosterone and estradiol are nearly equipotent in activating aggression in castrated male mice (e.g., Bowden & Brain, 1978; Simon & Gandelman, 1978b). This might suggest that aromatization is critical if the activating effect of testosterone is to occur. Several studies support this view. Administration of the aromatase inhibitor ATD (4-androsten-3,6,17-trione) inhibited the arousal of aggression in testosterone-treated castrates in a dose-dependent manner (Bowden & Brain, 1978), but the same agent had no effect on aggression stimulated by estradiol. Further, Clark and Nowell (1979) found that treatment with the antiestrogen CI-628 blocked testosterone-maintained aggression in castrates, whereas treatment with the antiandrogen flutamide was ineffective (Clark & Nowell, 1980). On the other hand, Simon, Gandelman, and Howard (1981) found that the antiestrogen MER-25 completely blocked activation of fighting by estradiol in castrated males but had only a slight effect on aggression elicited by testosterone. This suggests that in males testosterone may activate fighting by a direct action on an androgen receptor system or following conversion to estradiol.

In females, activation of fighting by testosterone appears to depend on an androgen receptor mechanism. A number of studies demonstrate that females or neonatally castrated males will respond to testosterone treatment in adulthood by fighting. Given sufficiently long exposure to adult TP treatment, levels of fighting observed in females are as high as in males (Barkley & Goldman, 1977b; Gandelman, 1980; Svare, Davis, & Gandelman, 1974; Vom Saal, Gandelman, & Svare, 1976) and sometimes higher (e.g., Simon, Gandelman, & Gray, 1984). However, a longer duration of treatment is required to induce fighting in females than in males. Aggression can be induced in adult females by treatment with the nonaromatizable androgens DHT or methyltrienolone (R1881) (Schechter, Howard, & Gandelman, 1981; Simon, Whalen, & Tate, 1985) but not by treatment with estradiol or diethylstilbestrol.

Neonatal Exposure to Hormones. Exposure to gonadal hormones during the perinatal period also influences adult fighting. Early work (Bronson & Desjardins, 1968, 1970; Edwards, 1969) suggested that exposure of mice to androgens or their estrogenic metabolites during the early postnatal period might be required for the expression of activational effects seen later in life. It was generally presumed that early exposure to testosterone organized the development of brain tissue, rendering it responsive to the activating properties of testosterone on aggression. Since early exposure to androgen is not essential for activation of fighting by adult testosterone treatment, Gandelman (1980) has argued that early exposure to androgen simply increases later sensitivity to the aggression-promoting properties of testosterone treatment in adulthood without necessarily organizing neural development. Although the distinction between an organizational effect and sensitization is somewhat semantic, the label "organizational" has usually been restricted to effects of gonadal hormones that only occur during a brief period early in development. Edwards' (1970) finding that a prolonged testosterone treatment regimen of female mice beginning at 30 days of age enhanced their display of aggression when they were later given TP as adults would not be considered an organizational effect by the usual definition. Gandelman's concept of sensitization can encompass Edwards' results as well as the more general finding that early exposure to hormones shortens the period of later exposure to testosterone required to produce fighting (Svare *et al.*, 1974; vom Saal *et al.*, 1976).

A variety of hormone treatments during the neonatal period can facilitate the subsequent response to testosterone. In addition to testosterone and estradiol, dihydrotestosterone (Schechter *et al.*, 1981) and the androgenic progestin norethindrone (Gandelman, Howard, & Reinisch, 1981) are effective.

Prenatal Influences. Hormonal influences on the differentiation of aggressive behaviors in mice are not limited to the neonatal period. Vom Saal *et al.* (1976) observed that males that were castrated on the day of birth began to fight after fewer days of testosterone treatment in adulthood than females that were ovariectomized at birth. This observation suggested that hormonal influences on aggression begin prior to birth. Subsequent studies confirmed this hypothesis. Prenatal exposure to testosterone potentiates later activation of fighting by testosterone in females (Gandelman, Simon, & McDermott, 1979; vom Saal, 1979). Exposure to testosterone any time between fetal days 13 and 18 is sufficient to produce this effect (Gandelman, Rosenthal, & Howard, 1980). By contrast, prenatal exposure to estrogen does not potentiate the activation of fighting by adult testosterone treatment (Gandelman, Peterson, & Hauser, 1982).

Studies of the influence of uterine position provide additional evidence that the prenatal hormone environment exerts important influences on the development of aggressive behavior. Female mice that develop *in utero* between two males (2M females) behave more aggressively than females that were not contiguous to males in the uterus (0M females) (Gandelman, vom Saal, & Reinisch, 1977; vom Saal & Bronson, 1978). Other data demonstrate that, relative to 0M females, 2M females are masculinized and defeminized on a number of anatomic and behavioral measures, presumably because of exposure to the high levels of testosterone that have been demonstrated in the amniotic fluid and plasma of 2M animals (vom Saal, 1983; vom Saal & Bronson, 1980).

Uterine position also influences the behavior of males. Male mice that developed between two males (2M males) are more aggressive when given testosterone as adults than males that developed between two females (0M males) (vom Saal, Grant, McMullen, & Laves, 1983). The mechanism responsible for this effect is unknown, but it is unlikely that estradiol is the critical metabolite, since 0M males had higher estradiol levels in amniotic fluid than 2M males. Further, prenatal exposure to estradiol does not enhance aggression in female mice (Gandelman *et al.*, 1982).

In summary, perinatal exposure to androgens appears to have two distinct influences on aggression in mice. First, early hormone exposure increases sensitivity to the activational effects of adult treatment with testosterone. This sensitization effect can occur during a broad time period beginning prior to birth and extending to puberty. Aromatization of androgens to estrogens clearly is not essential for the development of increased sensitivity to the aggression-promoting properties of testosterone in adulthood. Second, since female mice do not fight if given estradiol in adulthood, early exposure to androgen or estrogen is required to induce responsiveness to the aggression-promoting properties of estrogen. This effect can occur during the early postnatal period, but the available data do not establish whether or not the period of sensitivity is restricted to early postnatal life. Neonatal treatments with either testosterone or estradiol are equally effective in inducing responsiveness to the activational effects of estrogen exposure in adulthood (Simon & Gandelman, 1978b; Simon *et al.*, 1984). Hence, aromatization of testicular androgens to estrogens may be essential to confer responsiveness to later estrogen treatment. The available data are consistent with this possibility but do not rule out other mechanisms.

RATS. Aggression has not been studied as extensively in the laboratory rat as in the mouse. Although sex differences have been demonstrated in rats, the magnitude of the difference is not great. This is true, regardless of whether aggression is elicited by electric shock (i.e., shock-elicited fighting) or studied in other settings. Regardless of the technique used to elicit aggression, activational influences of testosterone can be demonstrated consistently. Thus, castration reduces shock-elicited fighting (Bernard & Paolino, 1975; Hutchinson, Ulrich, & Azrin, 1965) as well as aggression directed toward an intruder placed in the resident's home cage (Barfield, Busch, & Wallen, 1972; Christie & Barfield, 1979) or intermale aggression stimulated by the presence of an estrous female (Taylor, Haller, Rupich, & Weiss, 1984). Castration also increases the threshold current necessary to elicit attack by electrical stimulation of the hypothalamus (Bermond, Mos, Meelis, van der Poel, & Kruk, 1982). Replacement with testosterone restores aggressive behavior to precastration levels if the dose is sufficient (Bermond *et al.*, 1982; Christie & Barfield, 1979; Conner, Levine, Wertheim, & Cummer, 1969; Taylor *et al.*, 1984).

The mechanism by which testosterone activates aggression in male rats has received less attention. Christie and Barfield (1979) found that testosterone and estradiol were equally potent in restoring aggression toward an intruder by castrated resident males. Other aromatizable androgens (e.g., androstenedione) were also effective, whereas nonaromatizable androgens (e.g., dihydrotestosterone) were less potent but more effective than control injections. van de Poll, Swanson, and van Oyen (1981) also found that estradiol stimulated fighting in male rats about as effectively as testosterone.

Very little work has been done with hormone antagonists. Although Taylor *et al.* (1984) reported that the antiandrogen flutamide reduced intermale aggression, Prasad and Sheard (1981) found no effect with the androgen receptor blocker cyproterone acetate. Since aggression was induced by the presence of an estrous female in the former study and by electric shock in the latter study, the possible explanations for the discrepancy are numerous.

Little work has been done on aggression in female rats. DeBold and Miczek (1984) introduced an intruder into the cage occupied by two residents (one intact male, one intact female). Male residents attacked intact male intruders more frequently than castrated male intruders. Female intruders, regardless of their endocrine state, were rarely attacked by male residents. By contrast, female residents rarely attacked males unless they were castrated and received estrogen and progesterone. Instead, the resident females attacked female intruders, and the frequency of attack varied with the intruders' endocrine condition. Ovariectomized females, regardless of whether or not they received estrogen and progesterone replacement, were attacked more often than ovariectomized females given testosterone or lactating females. Although male residents attacked male intruders more frequently than female residents attacked female intruders, DeBold and Miczek emphasized the relatively high levels of aggression displayed by the resident females, given an opponent of the proper gender and hormone state. Furthermore, aggression by female residents was not altered by gonadectomy, which caused marked reduction in the aggressive behavior of resident males.

An unresolved question is whether or not there are organizational effects of gonadal hormones on aggressive behavior in rats. Barr, Gibbons, and Moyer (1976) observed that males were more likely to initiate attacks than females and that neonatal castration greatly reduced attack even when testosterone replacement was given in adulthood. The responsiveness of neonatally gonadectomized males to adult testosterone could be restored by neonatal testosterone treatment. Comparable findings for shock-elicited aggression have been reported by Conner *et al.* (1969).

By contrast, van de Poll, de Jonge, van Oyen, van Pelt, and de Bruin (1981) found no sex difference in the overall amount of fighting induced by adult testosterone treatment. Males and females did differ in their response to estradiol, which stimulated fighting in males but not in females (van de Poll, de Jonge, *et al.*, 1981). In these experiments aggression was studied in relatively inexperienced animals. A marked sex difference was observed when the outcome of aggressive encounters was manipulated (van de Poll, Smeets, van Oyen, & van der Zwan, 1982). In testosterone-treated males that consistently experienced victory, aggression increased on subsequent test fights, whereas in testosterone-treated males that were consistently defeated, aggression was inhibited. Testosterone-treated females and oil-treated animals of both sexes exhibited little differential response to victory or defeat.

The sex difference in response to the activational effect of estrogen on aggression is similar in rats and mice. In mice this difference appears to depend on exposure to hormones during the neonatal period. Whether or not this is true of rats is unknown. The possibility that male and female mice exhibit androgen-dependent differences in response to victory and defeat has not been studied, but the influence of adrenocortical hormones on these responses in mice is well known (see Leshner, Merkle, & Mixon, 1981, for review).

Another approach to studying hormonal influences on aggressive behavior capitalizes on the fact that electrical stimulation of the hypothalamus elicits aggressive behavior in both male and female rats. Although there is some evidence that males display more intense forms of attack than females, the similarities in the behavioral patterns of aggression, the current intensities required to elicit aggression, as well as the location of effective sites within the hypothalamus are more striking than the modest differences between the sexes (Kruk *et al.*, 1984). Neither ovariectomy nor estradiol treatment affected hypothalamically elicited aggression in female rats, but castration elevated the threshold current required to elicit aggression in males (Bermond *et al.*, 1982).

Kruk *et al.* (1984) suggested that the apparent similarity in the basic organization of hypothalamic mechanisms controlling aggression in males and females was consistent with the idea that organizational influences of gonadal hormones are less important in the development of aggressive behavior in the rat than in the mouse. In support of this conclusion, they noted the quantitatively similar response of male and female rats to the activational effects of testosterone reported by van de Poll, de Jonge, *et al.* (1981). Although this analysis may prove to be correct, for two reasons it is surprising that Kruk *et al.* (1984) failed to observe a sex difference in the threshold current necessary to elicit aggressive behavior. First, based on their earlier finding (Bermond *et al.*, 1982) that castration elevated the threshold required to elicit aggression in males, it would have been expected that less intense stimulation would be required to induce aggression in gonadally intact males than in females. Second, Kruk *et al.* (1984) used male opponents for all of their tests. Based on DeBold and Miczek's (1984) finding that females rarely attack males, a sex difference would have been predicted. The failure to observe this difference raises questions about the sensitivity of the electrical stimulation technique and the extent to which it models aggression exhibited under less artificial conditions.

Play Fighting

Play fighting (also called rough-and-tumble play) is the most commonly reported and most extensively studied form of juvenile social play. Although the behavioral components of play fighting vary from species to species, they generally resemble the motor acts associated with intraspecific aggression in adults. For example, "mounting" or attempting to get on top of an opponent is a frequent component of play fighting in juveniles of many species. However, during a play fight, biting and other responses that might seriously injure an opponent are at least partially inhibited, and the roles of attacker and target frequently and rapidly reverse (Aldis, 1975). Unlike the agonistic encounters of adults, play fighting lacks those components that are related to the communication of threat or submission (Bekoff, 1974; Symons, 1974). In some species there are explicit "play signals" that appear to be unique to play fighting (Symons, 1978).

Play fighting is sexually dimorphic in a number of species of primates including humans and great apes (Blurton-Jones, 1976; Braggio, Nadler, Lance, & Miseyko, 1978) as well as both New-World and Old-World monkeys (e.g., Baldwin & Baldwin, 1974; Harlow, 1965; Kummer, 1968; Raleigh, Flannery, & Ervin, 1979). In these primates, males engage in more play fighting than females, and it is claimed that male play fighting is more vigorous. Among primates less than 1 year of age, play groups are typically composed of both males and females. As the animals grow older and the intensity of male play fighting increases, females leave the play groups (e.g., DeVore, 1963; Kummer, 1968). It might be supposed that the sex difference in play fighting arises because of the tendency of females to avoid the intense play fights typical of males. This may be a factor, but it cannot be the entire explanation for the sex difference. Goy (1978) has shown that the sex difference in play fighting in rhesus monkeys appears prior to the time when males and females form separate play groups.

Sex differences in play fighting have also been reported in domestic sheep (Sachs & Harris, 1978), stellar sea lions (Gentry, 1974), golden hamsters (Goldman & Swanson, 1975), and Norway rats (e.g., Meaney & Stewart, 1981a). Again, males play fight more frequently than females. By contrast, there appear to be no gender differences in play fighting by juvenile timber wolves, coyotes, domestic dogs and cats, or northern grasshopper mice, a predatory rodent (Barrett & Bateson, 1978; Bekoff, 1974; Davies & Kemble, 1983; Traylor, 1982). The significance of sex differences in play patterns and of play in general for the development of adult behavior remains unclear (see Fagen, 1981; Smith, 1982), but it does appear that sex differences in play fighting are more common in group-living polygynous species that exhibit marked sexual dimorphism in body size and social roles (Meaney, Stewart, & Beatty, 1985).

INFLUENCE OF GONADAL HORMONES.   Information about the role of steroid hormones on the development of sex differences in play fighting is available only for rhesus monkeys and Norway rats. In both species sexual differentiation of play fighting is controlled by organizational influences of androgens during a sensitive period in early development. Activational effects of steroid hormones appear to be of little importance.

The sex difference in play fighting by juvenile rhesus monkeys is large in magnitude (males play fight 5–20 times as frequently as females) and appears across widely varying conditions of social rearing early in life (see Mitchell, 1979). Prenatal exposure to TP or DHTP causes equivalent increases in play fighting by female rhesus monkeys, although animals so treated do not play fight as frequently as males (Goy, 1978). Since neonatal castration has no effect on play fighting by juvenile males (Goy, 1978), and postnatal androgen treatment does not stimulate play fighting by juvenile females (Joslyn, 1973), the prenatal period constitutes the most sensitive period for the organizational action of androgens on play fighting in this species. Neonatal ovariectomy does not affect play fighting (Goy, 1970), so hormones of ovarian origin apparently have no role in the control of sex differences in play fighting in rhesus monkeys, at least during the postnatal period.

A similar pattern of hormonal control of sexually dimorphic patterns of play fighting exists in the rat, although the most sensitive period for hormone action occurs during the early postnatal period. If males are castrated shortly after birth, the frequency of their play fighting resembles that of females. If castration is delayed until 10 days of age or later, play fighting is already masculinized, and the castrated animals play fight as frequently as intact males (Beatty, Dodge, Traylor, & Meaney, 1981; Meaney & Stewart, 1981b) (see Table 2). These findings suggest that activational effects of androgens are unimportant in the control of play fighting, a conclusion that is supported by the failure to find an effect of TP injections during the juvenile period (Meaney & Stewart, 1983). Hormones of ovarian origin seem to be relatively unimportant as well, since neonatal ovariectomy has no effect on play fighting by females (Meaney & Stewart, 1981b).

Recent findings suggest that the critical period for hormonal organization of sex differences in play fighting in rats may begin during the prenatal period. Stehm (1985) has found that exposure to prenatal stress, a treatment known to reduce testosterone levels in fetal males, also reduces play fighting in juvenile males (see I. L. Ward, Chapter 5, this volume, for review).

TABLE 2.   INFLUENCE OF AGE OF
CASTRATION ON PLAY FIGHTING
IN MALE RATS[a]

Treatment	Frequency of play fighting[b]
Intact males	132.6
Day 20 castrates	122.5
Day 10 castrates	121.1
Day 6 castrates	100.0
Day 1 castrates	90.5
Intact females	99.5

[a] Data are from Beatty *et al.*
[b] Total play fighting responses over 15 day period.

MECHANISM OF HORMONE ORGANIZATION.   Several lines of evidence support the view that androgens exert their organizational effect on play fighting by a direct action on androgen-sensitive target tissues. Neonatal treatment with testosterone or DHT produces comparable stimulation of play fighting by females, but estradiol has no effect (Meaney & Stewart, 1981b). This latter observation suggests that the aromatization of androgen to estrogen, an important mechanism in the sexual differentiation of some reproductive behaviors in the rat (Baum, 1979; Goy & McEwen, 1980), is not crucial for the hormonal organization of sexually dimorphic patterns of play fighting. Consistent with this conclusion, males that were exposed to the aromatase inhibitor ATD from birth to 10 days of age (i.e., during the sensitive period for hormone organization) exhibited as much play fighting as intact untreated males (Meaney & Stewart, 1981b), whereas treatment with the androgen antagonist flutamide during the same period reduced play fighting by males to levels typical of normal females (Meaney, Stewart, Poulin, & McEwen, 1983). Further, male rats of the King–Holtzman strain with the *tfm* mutation played about as frequently as normal females and significantly less than normal males (Meaney *et al.*, 1983). Animals with the *tfm* mutation have normal plasma levels of gonadal hormones, a normal complement of estrogen receptors, but greatly reduced numbers of androgen receptors (Attardi, Geller, & Ohno, 1976; Naess *et al.*, 1976). Considered together, these findings suggest that hormonal stimulation of an androgen receptor system (presumably in the brain) during the neonatal period is responsible for the development of sex differences in play fighting that emerge during the juvenile period. The only evidence that challenges this conclusion is the report of Hines, Döhler, and Gorski (1982) that exposure to the synthetic estrogen diethylstilbestrol (DES) from day 16 of gestation to day 10 of postnatal life increased play fighting by female rats. Since the untreated males and females in this study did not differ in play fighting, the effect of DES is difficult to interpret. The findings reviewed above indicate that estrogens play no significant role in sexual differentiation of play fighting during postnatal life. An influence of estrogen during prenatal life cannot be ruled out.

Whether testosterone exerts its organizational effect by acting directly on androgen receptor systems or after conversion to DHT is not yet clear. The comparable effects of T and DHT injections on play fighting in female rats and rhesus monkeys support the latter mechanism, but Meaney and Stewart (1981b) found that male rats exposed to the reductase inhibitor testosterone 17β-carboxylic acid during the first 10 days of postnatal life engaged in as much play fighting as male controls and significantly more than females. These findings may indicate that either T or DHT can stimulate the androgen receptor systems that control sexual differentiation of play fighting. Alternatively, it might be argued that DHT is the critical hormone, and the degree of reductase inhibition achieved by Meaney and Stewart (1981b) was inadequate to demonstrate this effect.

Progesterone may also play a role in the sexual differentiation of play fighting. Birke and Sadler (1983, 1984) found that neonatal exposure to either progesterone or medroxyprogesterone acetate reduced play fighting in general, and play initiation in particular, by juvenile rats of both sexes. Since neonatal ovariectomy did not affect play (Meaney & Stewart, 1981a), this finding may seem surprising, but it is important to remember that the adrenal cortex is a significant source of progesterone. In fact, Shapiro, Goldman, Bongiovanne, and Marino (1976) found that during the neonatal period both serum and adrenal progesterone levels were higher in females than in males. Thus, progesterone might influence sexual differentiation of play fighting by acting as an endogenous antiandrogen, as suggested by Meaney *et al.* (1985). However, it should be noted that Weisz and Ward (1980) found no sex differences in serum progesterone levels in fetal or neonatal rats, so this hypothesis is highly speculative.

ADRENOCORTICAL HORMONES.   Glucocorticoids also influence the development of play fighting in rats, and the effect is sex-dependent. Meaney, Stewart, and Beatty (1982) administered corticosterone or the synthetic glucocorticoid dexamethasone at various times during the first 10 days of postnatal life. Treatment on days 1 and 2 or on days 3 and 4 reduced play fighting by males, but if treatment was delayed until days 9 and 10, there was no effect. Glucocorticoid treatment did not influence play fighting by females. In a follow-up study, Meaney and Stewart (1983) found that concurrent treatment with corticosterone had no effect on play fighting by juvenile males.

The glucocorticoid treatments that reduced play did not influence plasma testosterone levels in neonates or disrupt any aspect of masculine sexual behavior when it was measured in adulthood (Meaney *et al.*, 1982). Since M. J. Meaney (personal communication) has found that corticosterone does not act as an androgen antagonist, the most plausible hypothesis is that early exposure to glucocorticoids alters development of the neural circuitry that controls play fighting in a way that is independent of androgen receptor systems.

INFLUENCE OF TESTING METHOD.   Although the hormonal influences on play fighting are quite similar in rhesus monkeys and rats, the magnitude of the sex difference is much smaller in rats. In the rat, sex differences in play fighting have most frequently been observed when the animals are housed and studied in resident groups using a focal observation technique. With this procedure the observer scans several groups of animals, observing each in turn for a brief period (e.g., 20 seconds). The occurrence or nonoccurrence of play fighting and play initiation by each member of the group is recorded during the observation period. Usually the data from 70–100 observation periods made per day are pooled over 10–15 daily sessions. Most of the studies in which the early or adult hormone environment has been manipulated have used this approach.

The other widely used method for studying play fighting in rats is called the paired encounter technique. With this procedure pairs of animals are studied during brief, typically 5- to 10-minute-long sessions, usually conducted in a neutral arena. To stimulate play, animals are usually caged singly (i.e., socially isolated) except during testing. The frequency of pinning (i.e., one rat rolls the other onto its back and stands over it) is the most frequently employed dependent variable. This measure is positively correlated with other indices of play (Panksepp & Beatty, 1980). Using this method, several investigators (e.g., Birke & Sadler, 1983; Olioff & Stewart, 1978; Thor & Holloway, 1982) have reported sex differences, but more commonly, sex differences are found to be absent or so small in magnitude that they can be detected only with very large samples (Panksepp, Siviy, & Normansell, 1984; Thor & Holloway, 1984a, 1984b). The reasons for the apparent influences of testing procedure on the reliability of sex differences in play are unknown, but it is worth noting that studies with primates, in which the largest sex differences in play fighting are found, have generally observed behavior in intact groups.

Panksepp *et al.* (1984) have argued that the focal observation procedure does not adequately control for differences in social reinforcement history. Further, they suggest that the slight sex difference that is found by the paired encounter technique emerges slowly in parallel to the development of increasingly greater differences in body weight between males and females. For example, in one experiment they found that 66-day-old females that were 12% heavier than the 46-day-old males they were paired with pinned the males twice as frequently as the males pinned the heavier females. Since a careful analysis of the influence of body weight on play fighting of rats tested with the focal observation procedure has not yet been performed, the importance of this factor is unknown. However, the observation of Meaney and Stewart (1981a) that males tend to play fight more with males than with females indicates that the sex difference observed with the focal observation technique cannot simply arise because heavier animals seek out lighter adversaries. Further, Thor and Holloway (1986) found that both male and neonatally androgenized female rats exhibited more frequent play initiation than control females. Differences in play were apparent before reliable body weight differences emerged between the control and testosterone-treated females.

In a broader sense I would agree with Panksepp *et al.* (1984) that the social reinforcement history of the animals is critical to the emergence of the sex difference. As Meaney and Stewart (1981a) demonstrated, play fights initiated by males are likely to progress to wrestling and/or boxing and continue until one rat pins (dominates) the other. By contrast, females are more likely to withdraw before the play fight sequence is completed (i.e., before one rat pins the other). It is probable that these behavioral differences require social experience of the sort that is most likely to occur under conditions of relatively undisturbed group living. Perhaps, as Panksepp *et al.* (1984) suggest, females play fight less and withdraw from play more frequently because they learn that they cannot compete effectively with males. However, these experiential factors cannot explain all of the sex difference in play fighting, since the frequency of play initiation by male rats is increased by rearing them in isosexual groups. Isosexual rearing had no effect on the frequency of play initiation by females (Thor & Holloway, 1986). Further, the analysis of Panksepp *et al.* cannot explain the sex difference in play in rhesus monkeys, which is similar in magnitude for males and females raised in heterosexual or isosexual groups (Goldfoot & Wallen, 1979).

PLAY FIGHTING, AGGRESSION, AND SOCIAL DEVELOPMENT

Despite the superficial resemblance of play fighting and adult aggression, it is now quite clear that play fighting and aggression are sexually dimorphic behavior patterns under separate neurobiological controls. First, play fighting is influenced by organizational but not by activational effects of gonadal hormones, whereas adult aggression is affected by both organizational and activational effects of these hormones, with the activational effects being the more prominent. This is not surprising since play fighting is most frequent during the prepubertal period, when gonadal hormone levels are low relative to the levels observed during perinatal life or after puberty. Second, in male rodents organizational influences of androgens on aggression depend importantly on aromatization to estrogens, whereas the organizational actions of androgens on play fighting involve a direct effect on androgen receptor systems. Third, adult aggression in rodents is critically dependent on olfactory and other chemosensory information (see above). By contrast, play fighting is little affected by olfactory bulbectomy (Beatty & Costello, 1983) or intranasal infusion of zinc sulfate (Thor & Holloway, 1982). Finally, play fighting and aggression can be dissociated pharmacologically. For example, morphine increases play fighting (Panksepp, 1981; Panksepp, Jalowiec, DeEskinazi, & Bishop, 1985) but inhibits many forms of aggression (Miczek & Barry, 1976). Opiate antagonists decrease play (Beatty & Costello, 1982; Panksepp, Herman, Vilberg, Bishop, & DeEskinazi, 1980) but have little effect on intermale aggression (Rodgers & Hendrie, 1982). The effects of amphetamine on aggression are mixed, with some forms being increased after drug treatment (Miczek & Barry, 1976), but amphetamine consistently reduces play fighting, even in very low doses (Beatty, Costello, & Berry, 1984; Beatty, Dodge, Dodge, White, & Panksepp, 1982).

Despite the unambiguous conclusion that play fighting is not simply the juvenile form of adult aggression, play fighting might still predict adult aggression. The evidence on this point is mixed. Taylor (1980) studied play fighting and aggression in groups of male rats from 14 to 109 days of age. Individual differences were stable over time; that is, the animals that engaged in the most play fighting as juveniles were the most aggressive as adults. In a second experiment, play fighting between 40 and 49 days of age predicted aggressive behavior at 100 to 109 days of age. Furthermore, animals that behaved most aggressively on the adult fighting tests were most dominant on a resource competition test.

Adams and Boice (1983) did not replicate this result in an extensive study of a mixed-sex colony of rats maintained under seminaturalistic conditions. Before 150 days of age, social interactions among males consisted mostly of play fighting. The dominance hierarchies based on these interactions were unstable, unrelated to copulatory performance, and not predictive of adult dominance. Among older males, stable dominance relationships emerged. Dominant males engaged in most of the social interactions including copulation, won most of the encounters with other males with chasing–biting attacks, and were most active in attacking intruders. Adams and Boice suggested that the aggressive behavior observed by Taylor (1980) was really play fighting.

During the last decade the amount of research on animal play behavior in general and play fighting in particular has increased greatly. Although the description of the neuroanatomy and pharmacology of play fighting has progressed (see Panksepp *et al.*, 1984; Thor & Holloway, 1984b; for reviews), the long-term consequences, if any, of play fighting for adult behavior remain a mystery. Logical arguments have been made in support of hypotheses that play fighting facilitates the development of specific motor skills required for success in adult agonistic or predatory encounters (e.g., Aldis, 1975) or fosters the development of skills essential to integration within the social group (e.g., Meaney *et al.*, 1985) or simply provides exercise to stimulate cardiovascular and muscular development (e.g., Fagen, 1976). None of these ideas is very convincingly supported by evidence from longitudinal or experimental studies.

Since play is very costly, as it consumes energy and exposes the young animal to considerable risk, it has been assumed that important benefits must be associated with it. In a recent review Martin and Caro (1985) concluded that the empirical evidence indicates that the costs of play are, in fact, quite minimal, and when play becomes too costly, as in times when food is scarce, it is drastically reduced. From this perspective, play need not have any major long-term consequence for adult behavior.

Despite the fact that play is easily suspended when resources are scarce, some minimal amount of play may be essential for the development of some aspects of adult social behavior. For example, the impairments in male copulatory behavior resulting from social isolation (e.g., Gerall, Ward, & Gerall, 1967) may arise, in part, because social isolation deprives animals of the opportunity for social play (see I. L. Ward, Chapter 5, this volume, for further discussion).

FEEDING AND BODY WEIGHT REGULATION

With a few exceptions (e.g., the hamster), male mammals are larger than females; they eat more and weigh more. In the rat, sex differences in body size are small in magnitude until around puberty. Thereafter, a marked divergence in body size and weight begins and increases throughout life. Both organizational and activational effects of gonadal hormones contribute to this sexual dimorphism, but the activational influences appear to be predominant.

ACTIVATIONAL EFFECTS: ESTROGEN

Ovarian hormones, specifically estrogen, reduce food intake and body weight. In adult cycling females this influence is manifested by relatively higher intake during the postovulatory phase of the reproductive cycle, a relationship that has been reported in a number of species including rats (e.g., Ter Harr, 1972), guinea pigs (Czaja & Goy, 1975), rhesus monkeys (Czaja, 1975), baboons (Bielert & Busse, 1983), and humans (Pliner & Fleming, 1983). Following ovariectomy food intake and body weight increase. These changes can be prevented or reversed by administering estrogen but not progesterone (see Wade, 1976). Experimental studies (e.g., Gray & Wade, 1981; Wade, 1976) consistently indicate that progestins do not regulate feeding and body weight in the absence of estrogen, but progestins can antagonize the effects of estrogens. Hence, in intact females cyclic variation in feeding reflects the modulatory influence of progesterone on the inhibitory action of estrogen. Although estrogen lowers feeding in the majority of mammalian species studied, this effect is not universal. In the hamster, the effect of estrogen on feeding and body weight has been reported to be quite modest (Morin & Fleming, 1978; Swanson, 1967b; Zucker, Wade, & Zigler, 1972), but Gerall and Theil (1975) found that ovariectomy caused substantial increases in body weight that were reversed by estrogen treatment. In gerbils, estrogen actually stimulates feeding (Roy, Maass, & Wade, 1977).

A major influence of estrogen on feeding does not occur until around puberty and may depend on attainment of a minimum accumulation of body fat (Zucker, 1972). Prior to puberty growth hormone stimulates skeletal growth and lean body mass but suppresses the accumulation of fat. Hypophysectomy, which removes growth hormone, renders juvenile females responsive to estrogen's suppression of feeding, and growth hormone injections reverse this effect (Wade, 1974). Although considerably less responsive than mature females, prepuberal female rats are not completely refractory to estrogenic suppression of feeding. As much as 12 days prior to vaginal opening, cyclic changes in meal duration that resemble those observed in adults can be demonstrated (Sieck, Nance, Ramaley, Taylor, & Gorski, 1977). These changes are eliminated by ovariectomy. Nance (1983) suggests that these findings indicate that the fundamental control mechanisms through which estrogen acts to control feeding are mature long before puberty but are simply masked by pituitary growth hormone.

The foregoing description implies that estrogen inhibits feeding, thereby lowering body weight, principally by reducing adiposity. Although this influence is surely important, estrogen affects body weight in other ways as well. Following ovariectomy, not only does body fat accumulate, but linear growth increases as well (Clark & Tarttelin, 1982; Mueller & Hsiao, 1980). Further, the amount of weight gain after ovariectomy is greater than would be expected from the increase in food intake (Kanarek & Beck, 1980), and ovariectomized rats gain weight even if intake is restricted to control levels (Roy & Wade, 1977). This is not surprising, since estrogen increases activity (see above) and heat production (Laudenslager, Wilkinson, Carlisle, & Hammel, 1980), perhaps by increasing lipogenesis in brown adipose tissue (Kemnitz, Glick, & Bray, 1983).

The mechanism by which estrogen regulates feeding and body weight has been studied extensively, but no consensus has emerged. It is clear that estrogenic control of feeding is not mediated by the vagus nerve or the pituitary or by insulin or insulin-dependent mechanisms (Dudley, Gentry, Silverman, & Wade, 1979; Eng, Gold, & Wade, 1979; Nance, 1983). Nance (1983) has proposed that estrogen acts on cells in the ventromedial hypothalamus principally concerned with regulating changes in feeding in response to changes in long-term energy balance. Carbohydrates are seen as playing a central role in the operation of the regulatory system. One argument for the view that estrogen affects long-term regulation is the relationship between body weight and estrogen-induced anorexia described earlier. Second, estrogen does not potentiate the short-term inhibitory effects of intragastric glucose loads but does enhance the delayed effects of glucose on feeding. In fact, after 48 hours of deprivation, the latency between the first and second meals is actually shorter in estrogen-treated ovariectomized females than in oil-treated ovariectomized controls (Young, Nance, & Gorski, 1978b). This observation indicates that, although estrogen leads to a chronic reduction in meal size (Blaustein & Wade, 1976), this shift does not imply that estrogen simply enhances short-term satiety. Finally, the anorexic response to estradiol is highly correlated with body weight among rats fed a high-glucose diet, suggesting that when simple carbohydrates constitute the principal energy source, body weight directs food intake. When fats constitute a major energy source, body weight is not correlated with responsiveness to estrogen. However, high-fat diets induce marked glucose intolerance, and the degree of intolerance is highly correlated with the inhibition of feeding induced by estrogen (Young, Schlussman, & Dixit, 1979). These findings also indicate that glucose metabolism exerts important influences on the response to estrogen.

The recent finding that estrogen reduces the proportion of calories consumed from carbohydrates more than from other macronutrient sources (Bartness & Waldbillig, 1984) can be interpreted to support the view that carbohydrates play a particularly important role in the influences of estrogen on long-term regulation, although other studies indicate that ovarian hormones do not always influence dietary self-selection (Kanarek & Beck, 1980).

CENTRAL VERSUS PERIPHERAL ACTION OF ESTROGEN

The ventromedial hypothalamus (VMH) is clearly an important site for estrogenic regulation of feeding. Implants of crystalline estradiol into the VMH consistently have been found to suppress eating in gonadectomized males and females (Beatty, O'Briant, & Vilberg, 1974; Donohoe & Stevens, 1982; Janowiak & Stern, 1974; Wade & Zucker, 1970), although similar implants in other hypothalamic areas are generally ineffective. Lesions of the VMH reduce the anorexic effects of estrogen, but only very large lesions that involve damage to adjacent areas of the medial hypothalamus are effective (Beatty, O'Briant, & Vilberg, 1975; King & Cox, 1973; Nance, 1976). This suggests that the VMH is not the only neural target involved in the estrogenic regulation of feeding. Consistent with this view, Donohoe and Stevens (1981) found that estradiol implants in the corticomedial nuclei of the amygdala suppress feeding.

The neural pathways involved in mediating estrogenic influences on feeding are not known. Knife cuts that sever connections between the VMH and more rostral and lateral structures do not abolish the anorexic effect of estradiol but severely impair compensation for dietary dilution (i.e., the increase in food intake usually observed in response to reduction in the caloric density of the diet) (Nance, Saporta, & Phelps, 1981). Compensation for dietary dilution is enhanced by estrogen (Nance & Gorski, 1977).

An alternative hypothesis regarding the mechanism of estrogenic regulation of feeding and body weight was proposed by Wade and Gray (1979). According to this view estradiol reduces feeding by inhibiting the enzyme lipoprotein lipase (LPL), which is necessary for the storage of fat. As a consequence of high estrogen, plasma triglyceride levels would rise, directly reducing intake. Consistent with this hypothesis, Ramirez (1981) showed that reductions in LPL activity preceded reductions in feeding caused by estrogen. However, changes in feeding induced by estradiol are not closely correlated with plasma triglyceride levels (Gray & Wade, 1981; Ramirez, 1980), and reductions in food intake following medial hypothalamic implants of estradiol occur before changes in LPL activity can be demonstrated in adipose tissue (Nunez, Gray, & Wade, 1980). As a unitary explanation of the mechanism by which estrogen influences feeding, the hypothesis of Wade and Gray (1979) is clearly inadequate, as their own research indicates. However, by demonstrating the important influences of estrogen on peripheral metabolic events, Wade and Gray compelled consideration of extrahypothalamic actions of estrogen on feeding and body weight regulation. Although the critical peripheral processes have not yet been identified, Young (1986) interprets his finding that thyroxine reduces the suppression of feeding by estrogen in terms of generalized influences on metabolism rather than specific interactions of estrogen on putative hypothalamic targets.

ACTIVATIONAL EFFECTS: ANDROGENS

Androgens exert clear activational effects on feeding and body weight gain. Following castration in adulthood, growth is reduced almost immediately; after some delay, food intake declines as well. Both food intake and body weight gain can be restored by replacement therapy with moderate testosterone doses (less than 1 mg/day) (Gentry & Wade, 1976b). Higher doses of testosterone can actually depress feeding and body weight below the level of oil-treated castrates or castrates given low doses of testosterone (Gentry & Wade, 1976b; Gray, Nunez, Siegel, & Wade, 1979; Siegel, Nunez, & Wade, 1981). Such treatments also reduce carcass fat. Depressed feeding, growth, and reduced body fat content following higher doses of testosterone result from aromatization to estradiol and can be antagonized by treatment with progesterone or an aromatase inhibitor (Gentry & Wade, 1976b; Gray *et al.*, 1979). Sensitivity to the growth-promoting effects of testosterone can be demonstrated in weanling male rats (Nunez & Grundman, 1982), but the physiological significance of this finding is unclear, since castration does not affect feeding and growth until puberty (Slob & van der Werff ten Bosch, 1975).

Testosterone also alters dietary selection and carcass composition. Following testosterone, protein intake is preferentially increased, and carcass protein content is increased. Treatment with DHT also stimulates food intake and body weight gain, but DHT is much less effective than testosterone (Gentry & Wade, 1976b; Siegel *et al.*, 1981). Dihydrotestosterone slightly increases protein consumption but does not alter carcass composition (Siegel *et al.*, 1981). Further, high doses of DHT do not inhibit food intake or weight gain. Taken together these observations indicate that the growth-promoting effects of low doses of testosterone are probably not mediated by reduction to DHT, whereas the growth-inhibiting effects of high doses of testosterone depend on aromatization to estradiol. The ability of central implants of testosterone into the ventromedial hypothalamus to suppress intake presumably depends on aromatization within the brain (Nunez, Siegel, & Wade, 1980). The growth-promoting effects of androgens are not observed in all species. In hamsters, unlike rats, castration increases growth and body weight in males (Gerall & Theil, 1975; Swanson, 1967b).

POSSIBLE ORGANIZATIONAL EFFECTS

During the perinatal period gonadal hormones have important influences on linear and ponderal growth that persist throughout life. In the female rat, neonatal treatment with testosterone increases growth and the level at which body weight will be regulated in an irreversible fashion. Neonatal exposure to estradiol has a qualitatively similar effect, although the magnitude of the effect is not as great (e.g., Bell & Zucker, 1971; Donohoe & Stevens, 1983). The effectiveness of early hormone treatment is limited to a brief period during the first few days of neonatal life (Tarttellin, Shryne, & Gorski, 1975), and even 20 days of high dosages of testosterone does not modify weight regulation when given to prepuberal females (Beatty, 1973).

Sex differences in response to the weight-limiting effects of estradiol have been reported in adult gonadectomized rats (Young, Nance, & Gorski, 1979; Zucker, 1969). Interpretation of the findings with estradiol treatments is complicated because males and females were given the same fixed amount of hormones, resulting in a lower effective dose for males if dosage is expressed on a body-weight basis. Rowland, Perrings, and Thommes (1980), who matched males and females in terms of body weight (necessarily confounding age with sex), found that males and females showed equivalent gains in weight after testosterone treatment. Unfortunately, they did not examine response to estradiol.

Czaja (1984) studied the problem in guinea pigs, employing a statistical correction for baseline differences in feeding and body weight between males and females. Regardless of whether gonadectomy was performed neonatally or in adulthood, adult treatment with estradiol suppressed food intake and body weight loss less in males than in females. Conversely, when gonadectomized and given testosterone in adulthood, males increased their body weight more than females, but their feeding was comparable. These findings indicate that organizational effects of androgens during the prenatal period reduce the subsequent response of male guinea pigs to the weight-limiting effects of estradiol. Whether or not such exposure also increases sensitivity to the growth-promoting effects of adult testosterone treatment is unclear. Since testosterone stimulated food intake to an equivalent degree in males and females, the differential effect on body weight gain might simply reflect a sex difference in response to the anabolic actions of testosterone on peripheral structures.

Still unresolved is the issue of whether or not neonatal exposure to hormones alters sensitivity of the female rat to the activational influences of gonadal hormones in adult life. Bell and Zucker (1971) suggested that neonatal exposure to testosterone or estradiol increased sensitivity to the growth-promoting actions of androgens and reduced sensitivity to the growth-limiting actions of estrogens (i.e., sensitivity to adult hormone was "masculinized"). Donohoe and Stevens (1983) reported similar results with respect to changes in sensitivity to estradiol among neonatally androgenized or estrogenized females, but their animals responded normally to the antiestrogen CI-628 (which is estrogenic in its effect on feeding, Roy & Wade, 1976). Other workers (Beatty, Powley, & Keesey, 1970; Dubuc, 1976; Slob & van der Werff ten Bosch, 1975; Tarttellin *et al.*, 1975) have found that neonatally androgenized females, although larger and heavier than oil-treated controls, exhibit equivalent suppression of feeding and comparable amounts of body weight loss when given estradiol in adulthood.

It might be reasoned that the organizational effects of testosterone on feeding and weight regulation in rats begin during the prenatal period and extend into early neonatal life. If this were the case, then, perhaps a combination of pre- and postnatal exposure to androgens would produce a more complete shift in sensitivity to hormone stimulation in adulthood. Somewhat surprisingly, the limited evidence available (Slob & van der Werff ten Bosch, 1975; Ward, 1969) indicates that females exposed to testosterone prenatally are significantly smaller than control females, and this persists in adulthood. In an earlier review (Beatty, 1979), I suggested that the effect of prenatal exposure to testosterone on growth and body weight in females probably represented a nonspecific pharmacological effect of no physiological significance. More recent evidence suggests that this conclusion may be incorrect. Meisel and Ward (1981) reported that female rats whose mounting behavior as adults was masculinized by prenatal exposure to hormones secreted by their male littermates *in utero* were lighter at birth than females that were not exposed to such hormonal stimulation prenatally. I. L. Ward (personal communication) reported that body weight differences apparent at birth tended to persist into adulthood, but the difference was not statistically significant.

In summary, organizational effects of androgens during the neonatal period clearly promote body growth in rats but not in hamsters and may reduce the weight-limiting effects of estrogen. The magnitude of the latter effect is small in rats and guinea pigs and apparently absent in hamsters. Clear evidence that organizational effects of neonatal hormone exposure modify the adult response to the growth-promoting effects of androgen is even more limited. The importance of prenatal exposure to androgens is unclear.

EVOLUTIONARY ASPECTS

Hoyenga and Hoyenga (1982) recently proposed that as a result of evolutionary pressures female mammals have become more resistant to starvation in times of famine than males. As a consequence of this adaptation, females are also more likely to develop obesity in times of relative plenty. The theory was formulated to account for the higher incidence of obesity, especially "superobesity," among women. In addition, the generally poorer survival of males under conditions of starvation (Widdowson, 1976) supports the hypothesis. Other findings are more equivocal. For example, the theory predicts that females should be more likely to develop dietary obesity, which results when rats are fed a variety of highly palatable foods.

selected from a constantly changing menu. Such differences have sometimes been reported in rats (e.g., Sclafani & Gorman, 1977; Young, Nance, & Gorski, 1978a), but not always (e.g., Jen, Greenwood, & Brasel, 1981). Genotypic factors appear to be particularly important, as the sex difference in dietary obesity varies with the strain in rats (Schemmel, Mickelsen, & Gill, 1970).

Sex differences in response to nutritional challenges including forced or voluntary exercise (Applegate, Upton, & Stern, 1982; Hirsch, Ball, & Goodkin, 1982; Nance, Bromiley, Bernard, & Gorski, 1977; Nikoletseas, 1980) and dietary dilution (Nance & Gorski, 1977) have also been reported. In these studies females are better able to maintain body weight in response to the challenge, supporting the basic tenet of the hypothesis. In part such regulation is achieved by compensatory increases in food intake, but metabolic changes are probably also involved. Finally, the intriguing observation that females are much more likely than males to hoard food when they are not deprived (Coling & Herberg, 1982) is consistent with the overall hypothesis and indicates that nonregulatory factors are involved in gender differences in adaptation for feeding. The role of gonadal hormones in this sexually dimorphic response to conditions of relatively high and low states of energy balance is as yet unclear but should be a fruitful area for future research since it has obvious practical as well as theoretical implications.

SENSORY FACTORS

The scattered literature on sex differences in response to sensory stimuli in humans and other animals has been reviewed elsewhere (Gandelman, 1983; Parlee, 1983). The present section reviews hormonal influences on sex differences in response to taste stimuli.

TASTE PREFERENCE

Beyond puberty female rats exhibit a greater preference than males for sweet (Valenstein, Kakolewski, & Cox, 1967; Wade & Zucker, 1969a) and salt (Krecek, Novakova, & Stibral, 1972) solutions. Although the ontogeny of the sex difference in preference for sweet or salty solutions is similar, activational influences of gonadal hormones are different. Acquisition of the typical feminine pattern of preference for saccharin depends on the ovarian hormones, estrogen and progesterone, acting in combination (Zucker, 1969). Once established, saccharin preference persists after ovariectomy. Ovariectomy after 10 days of age has no effect on salt preference, and adult castration of males has little effect on their preference for either salt or sweet solutions (Krecek et al., 1972; Zucker, 1969).

Organizational actions of gonadal hormones also influence the development of sex differences in taste preferences. Neonatal testosterone treatment of female rats reduces their subsequent preference for sweet or salty solutions (Krecek, 1973; Wade & Zucker, 1969b), whereas neonatal castration increases salt preference in males (Krecek, 1976). Suppression of the development of a feminine pattern of preference by neonatal exposure to testosterone appears to depend on an androgen receptor mechanism, since neonatal treatment of females with estradiol increases saccharin and salt preference (Krecek, 1978; Wade & Zucker, 1969b). Further, androgen-insensitive male rats exhibit a female pattern of saccharin preference (Shapiro & Goldman, 1973). This presumably reflects the failure of androgen-dependent organization of the usual male preference, because eliminating activational influences of gonadal hormones has no effect on saccharin preference of adult males.

Ovarian hormones may also play an organizational role. Krecek (1978) found that neonatal ovariectomy reduced salt preference in female rats. This observation suggests that the increased preference for salt observed after neonatal estradiol injection probably reflects a physiological influence rather than a pharmacological effect.

Taste Aversion

Male rats extinguish a conditioned taste aversion more slowly than females (Chambers, 1976). Activational effects of testicular hormones are principally responsible for this dimorphism. Gonadectomy reduces the persistence of the aversion in males but has no effect in females, whereas testosterone injections increase resistance to extinction in both sexes (Chambers, 1976). The presence or absence of testosterone during extinction controls the persistence of the aversion, but the hormonal status during acquisition is irrelevant (Chambers & Sengstake, 1979). The influence of testosterone clearly involves an androgenic mechanism: DHT is almost as effective as testosterone in maintaining an aversion (Chambers, 1980; Earley & Leonard, 1978), but estradiol does not maintain the aversion. If given alone to castrated males, estradiol speeds extinction. In combination with dihydrotestosterone, estradiol partially antagonizes the effectiveness of dihydrotestosterone in retarding extinction (Earley & Leonard, 1979). Progesterone has no effect on extinction rate (Chambers, 1980).

Weinberg, Gunnar, Brett, Gonzalez, and Levine (1982) observed that when females were first reexposed to the testing conditions after acquisition of a taste aversion, their plasma corticosterone levels were depressed relative to base line, whereas males maintained elevated corticosterone levels. Gonadectomy eliminated this differential pattern of corticosterone activity and the sex difference in resistance to extinction of the taste aversion. These findings raise the possibility that differences in ACTH levels at the time of reexposure during extinction tests were responsible for the sex differences and the influence of testosterone on taste aversion learning, since ACTH is known to prolong extinction of taste aversions as well as other avoidance behaviors (Kendler, Hennessy, Smotherman, & Levine, 1976). Chambers (1982) evaluated this possibility by comparing the effects of ACTH on extinction of aversion in intact and castrated males. ACTH retarded extinction only in intact males. In these animals ACTH also increased testosterone levels, suggesting that androgens may mediate the effects of ACTH on retention of taste aversions. This implies that androgens may have a direct influence on mechanisms that control retention of this task.

Learning

Active Avoidance

Among adult rats, females generally acquire active avoidance responses more rapidly than males, although the magnitude of effect varies from strain to strain (Levine & Broadhurst, 1963; Ray & Barrett, 1975). In other rodent species such as the gerbil and the hamster there is no sex difference (see Beatty, 1979), whereas in northern grasshopper mice the sex difference is reversed (Kemble & Enger, 1984).

Bauer (1978) studied the ontogeny of two-way avoidance learning (i.e., shuttle-box avoidance) in male and female rats ranging in age from 17 to 200 days. Performance improved until about 50 days of age, but no sex differences were apparent until 90 days of age.

ACTIVATIONAL EFFECTS. Since the sex difference in avoidance behavior in the rat does not emerge until well after puberty, it might be suspected that activational influences of gonadal hormones exert important regulatory effects. However, the experimental data do not support this deduction. In our own studies (Beatty & Beatty, 1970; Scouten et al., 1975) gonadectomy in adulthood failed to alter acquisition of shuttlebox avoidance in either sex. Comparable findings have been reported concerning the effects of gonadectomy on lever-press avoidance (van Oyen, Walg, & van de Poll, 1981). However, Sfikakis, Spyraki, Sitaras, and Varonos (1978) found that shuttle avoidance acquisition varied with the stage of the estrous cycle. Performance was better if training occurred during proestrus, coincident with the peak plasma estradiol concentration, than at diestrus. The performance of ovariectomized females was intermediate between that of the proestrus and diestrus groups. Although the difference between the diestrus and ovariectomized groups is puzzling, the findings of Sfikakis et al. (1978) can perhaps be reconciled with our own negative findings regarding the effects of ovariectomy in adulthood on avoidance acquisition. Since we (Beatty & Beatty, 1970) did not monitor the stages of estrus of our intact females, better performance by rats trained at proestrus might have been canceled by inferior performance by rats trained during diestrus, resulting in our finding of no difference between intact and ovariectomized groups.

More recent data indicate that testosterone exerts an activational influence in males. Mora, Nasello, Mandelli-Lopes, and Diaz-Véliz (1983) found that adult castration improved avoidance acquisition in males, and replacement with testosterone reversed the effect of castration. Overall, activational effects of gonadal hormones appear to have modest effects on avoidance acquisition in both sexes.

ORGANIZATIONAL EFFECTS. On the whole, organizational actions of androgens appear to be primarily responsible for sex differences in active avoidance acquisition in rats, but the developmental period for these influences is not temporally limited. Although neonatal castration of males did not affect their performance (Denti & Negroni, 1975; Scouten et al., 1975), a combined treatment consisting of prenatal exposure to the antiandrogen cyproterone acetate plus neonatal castration improved performance by males to levels typical of untreated females (Scouten et al., 1975) (see Table 3). More recently, using outbred RS mice and a conditioned lever-press avoidance task, Hauser and Gandelman (1983) found that females located in utero between two females acquired avoidance responding more rapidly than females that had been located between two males. Performance by the latter groups of females was indistinguishable from that of males, for whom uterine position was of no consequence.

These findings indicate that exposure to androgens during prenatal life may be sufficient to produce the lower levels of avoidance responding typical of males, but other data indicate that postnatal exposure to androgens can have a similar effect. Denti and Negroni (1975) reported that a single injection of TP given neonatally depressed avoidance acquisition by female rats. However, Beatty and Beatty (1970)

TABLE 3. INFLUENCE OF PERINATAL
HORMONE MANIPULATIONS ON
AVOIDANCE BEHAVIOR: EFFECTS OF
GONADECTOMY (Gx) AND PRENATAL
CYPROTERONE ACETATE (CA)[a]

Group	Mean number of avoidance responses
Effect of gonadectomy	
Normal females	115.5
Neonatally Gx males	97.8
Intact males	97.5
Effect of CA	
CA females	123.1
Control females	126.1
CA + Gx males	127.5
Control males	105.7

[a] Data are from Scouten *et al.* (1975).

found that neither neonatal nor adult treatment with TP affected avoidance by females if given alone, but the combination of neonatal and adult TP treatment impaired performance. Working in our laboratory, Scouten (1972) found that a series of postnatal testosterone injections beginning at birth and extending until 75 days of age tended to depress avoidance behavior by females tested 2 months after the end of injections.

The mechanism by which androgens influence development of avoidance behavior is unknown. Although cyproterone acetate is antiandrogenic, it is also a weak inhibitor of aromatase and a strong progestin (McEwen *et al.*, 1979). The possible importance of the progestational action is suggested by the finding (Snyder & Hull, 1980) that perinatal exposure to progesterone improved avoidance behavior of rats of both sexes.

Prenatal stress, which increases lordosis and depresses ejaculatory behavior in male rats (Ward, 1972), does not affect their avoidance behavior (Meisel, Dohanich, & Ward, 1979). Prenatal stress also reduces testosterone secretion during the prenatal period (Ward & Weisz, 1980), which presumably accounts for its effect on reproductive behavior. However, if the organizational effects of androgens on avoidance behavior occur over a broad time period during development, prenatal stress or any other treatment that depressed androgen secretion only during the prenatal period or only during the postnatal period would not be expected to alter avoidance behavior.

PASSIVE AVOIDANCE

Adult female rats generally exhibit inferior performance on tests of passive avoidance behavior (e.g., Beatty, Gregoire, & Parmiter, 1973; Denti & Epstein, 1972). The ontogeny of the sex difference has not been studied extensively, but van Oyen, van de Poll, and De Bruin (1979) found no difference between 60-day-old males and females, whereas among prepubertal animals, females exhibited stronger response suppression (i.e., better passive avoidance conditioning). The sex difference that ultimately develops among adults is independent of differences in body weight and is observed across a range of shock intensities (Drago, Bohus, Scapagnini, & De Wied, 1980; van Oyen *et al.*, 1979), including intensities well above the pain threshold. Hence, sex differences in sensitivity to shock are not likely to be responsible for the differences in passive avoidance behavior.

Although sex differences are sometimes found if passive avoidance is tested immediately after shock presentation (van Oyen, van de Poll, & de Bruin, 1980), the influence of sex is generally clearer if a delay is imposed between acquisition and retention. This raises the possibility that females perform more poorly because they do not remember the punishing experience as well as males. van Haaren and van de Poll (1984) tested this hypothesis by exposing males and females to two distinctive compartments for three trials each. Both compartments were connected to the same platform. Shock was then administered in one of the compartments, and the animals were allowed to escape to the platform. On the retention test females were more likely than males to leave the "safe" platform. However, the females that did not "passively avoid" did not return to the compartment in which they had been shocked, indicating that they remembered the experience. At present, the most likely explanation of the sex differences in avoidance behavior is that male rats are more likely than females to freeze in situations in which they have been exposed to electric shock.

At the present time the influence of gonadal hormones on passive avoidance behavior is quite unclear. Van Oyen *et al.* (1980) found that gonadectomy in adulthood impaired passive avoidance in males but not in females. In contrast, Bengelloun *et al.* (1976) found no consistent effects of gonadectomy in males or females. Sex differences in pituitary adrenocortical function apparently do not contribute to differences in avoidance conditioning. Treatment with high doses of dexamethasone that were sufficient to suppress endogenous adrenocortical activity does not affect acquisition or extinction of active avoidance responses (Heinsbroek, van Oyen, van de Poll, & Boer, 1983; Scouten & Beatty, 1971) or retention of passive avoidance behavior (Heinsbroek, van Oyen, & van de Poll, 1984).

Maze Learning

With very few exceptions adult male rats make fewer errors than females while learning complex mazes such as the Lashley III maze or variations of the Hebb–Williams maze (see Beatty, 1979, for references). The sex difference does not appear until after puberty (Krasnoff & Weston, 1976), but activational influences of gonadal hormones do not seem to be very important, since gonadectomy in adulthood has little effect on performance by either sex (Joseph *et al.*, 1978).

Manipulations of the neonatal hormone environment have consistent effects on the maze behavior of both sexes, suggesting that organizational influences of gonadal hormones are mainly responsible for the sex differences evident in adulthood. Neonatal treatment with TP improves maze learning by females (Dawson, Cheung, & Lau, 1975; Joseph *et al.*, 1978; Stewart *et al.*, 1975), whereas neonatal castration increases the number of errors by males (Dawson *et al.*, 1975; Joseph *et al.*, 1978).

The mechanism by which neonatal exposure to androgen exerts an organizational effect on maze learning is unclear. Joseph *et al.* (1978) reported that neonatal exposure to cyproterone acetate (CA) impaired (i.e., feminized) maze performance by gonadally intact males, but this result is difficult to interpret because in the same study neonatal treatment with CA improved performance of females. As described above, CA is antiandrogenic, progestational, and a weak aromatase inhibitor. Its effect on maze performance in males could reflect any or all of these actions. The progestational action of CA alone could account for its effect in males, since Hull, Franz, Snyder, and Nishita (1980) showed that perinatal exposure to progesterone impaired performance of males in a Lashley III maze. At present, aromatization does not appear to be an important pathway for the organizational effects of androgens on maze behavior, since Dawson *et al.* (1975) found that neonatal estradiol treatment impaired performance of intact or neonatally castrated males. Cyproterone is also weakly androgenic (Bloch & Davidson, 1971), which could explain its effect on females. Unfortunately, there is no published experiment in which the effects of neonatal treatment with T, DHT, and E were compared under conditions in which a sex difference between intact males and females could be demonstrated. Until this basic information is available, the mechanism by which gonadal hormones affect maze learning will remain unclear.

An equally important issue is whether the sex difference in learning complex mazes represents a difference in spatial ability or is simply another manifestation of the sex difference in activity. Since the complex mazes that yield clear sex differences resemble open fields with additional walls, it is easy to imagine that the higher error scores of females might be secondary to their higher levels of activity and exploration. The pattern of prominent organizational effects and weak activational influences of gonadal hormones is similar for open-field behavior and complex maze learning in rats, supporting the interpretation that the two tasks are measuring the same capacity.

In an earlier review (Beatty, 1979) I suggested that the radial maze task devised by Olton and Samuelson (1976) might provide a test of spatial ability in rats that was relatively uncontaminated by sex differences in activity. The apparatus resembles a rimless wagon wheel set on a pedestal. In the standard version of the task food is placed into each arm of the maze at the outset of the daily test. Testing continues until the rat visits every arm and finds all of the food. Since food is never replenished during the run, optimal performance entails visiting each arm only once. Reentries into previously visited arms are considered working memory errors (see Olton, 1978).

Using this procedure Einon (1980) observed a small but significant sex difference favoring males, but only among animals raised in an enriched environment from weaning until testing. Tees, Midgley, and Nesbit (1981) also reported that males performed more accurately than females, but the sex difference was larger among animals reared in the dark than in controls raised in standard laboratory cages. More recently Juraska, Henderson, and Muller (1984) found no evidence of sex differences in rats that were raised in restricted or enriched environments. Our own experience parallels that of Juraska *et al.* (1984). In a series of unpublished studies we have failed to find consistent sex differences using the standard testing procedure described above.

Beatty (1984) reported preliminary findings using a modified testing procedure that permitted simultaneous measurement of both working and reference memory. Instead of baiting all of the arms at the outset of the session, only six arms were supplied with food. The other two arms (the same two arms each day) were never baited. Hence, the rat had to learn to avoid the two unbaited arms and avoid reentering baited arms. Entries into arms that were never baited were considered reference memory errors, whereas reentries into baited arms were considered working memory errors. In the initial experiment males performed much better than females on both the working and reference memory components of the task. However, in subsequent attempts to study the influence of gonadal hormones on radial maze performance we have been unable to replicate the original findings.

In summary, in rats sex differences in the radial maze are small in magnitude and not readily reproduced. At present there is no reason to conclude that the sex differences observed in other complex mazes reflect differences in spatial ability. Rather, such differences are likely to be secondary to sex differences in activity.

Brain Laterality

Although asymmetries of brain–behavioral function were once thought to be exclusive to humans, recent research suggests that both behavioral and brain asymmetries occur in subhuman mammals and even in nonmammalian species (see reviews by Denenberg, 1981; Robinson, Becker, Camp, & Mansour, 1985). On a number of measures of postural and motor function modest sex differences in the degree of asymmetry of function have been reported. Interestingly, where sex differences exist, females are often more strongly lateralized.

Postural Asymmetry

For example, Collins (1977) reported that significantly more female than male mice exhibited strong paw preferences when reaching for food. As with paw preferences in other species, the difference was in the extent to which the preferred paw was used; the proportion of "left-pawed" and "right-pawed" males and females was about the same.

In the rat, sex differences have been reported in the direction of turning bias when leaving the corner in an open field and in the postural asymmetry produced by a sustained but mild tail pinch. Among rats that were not handled during development, females exhibit a left bias when leaving the corner of an open field, whereas males exhibit no directional bias (Camp, Robinson, & Becker, 1984; Sherman, Garbanati, Rosen, Yutzey, & Denenberg, 1980; Sherman et al., 1983). Handling from birth to weaning induced a leftward bias in males but eliminated the bias in females (Camp et al., 1984; Sherman et al., 1980, 1983). Nonhandled females, but not males, also exhibited a postural bias, in this case a rightward bias, in response to tail pinch. Nonhandled females also spent significantly more time in a lateralized posture than males. Among males and females that were handled there were no postural asymmetries (Camp et al., 1984).

Neonatal rats have been reported to display asymmetries in the position in which their tails are held. In two studies (Denenberg et al., 1982; Ross, Glick, & Meibach, 1981) females have been observed to have significantly stronger leftward tail bias.

Rotational Behavior

The most thoroughly studied rodent behavioral asymmetry is rotational behavior. Ungerstedt (1971) first demonstrated that after unilateral destruction of the dopamine neurons of the substantia nigra, rats display spontaneous rotational behavior toward the side of the lesion. Although rotational behavior declines with time after the lesion, it can be reinstated with dopamine agonists such as amphetamine or apomorphine. Rotational behavior can also be induced by unilateral electrical stimulation of the nigrostriatal pathway as it courses through the lateral hypothalamus (Arbuthnott & Ungerstedt, 1975). Such stimulation produces turning away from the stimulated side. Thus, rotation occurs away from the nigrostriatal pathway with the greater activity.

Subsequent work demonstrated that rotational behavior could be induced by amphetamine treatment in normal animals (Jerussi & Glick, 1974) and that females exhibit more vigorous rotational behavior than males (Brass & Glick, 1981; Robinson, Becker, & Ramirez, 1980). A similar sex difference in the magnitude of rotational behavior can be induced by treatment with cocaine (Glick, Hinds, & Shapiro, 1983). There is some evidence that females are more likely to exhibit a right-turning bias, whereas males tend to display a left-turning bias, but the difference is very slight (see Robinson *et al.*, 1985). Thus, the major difference between males and females is the magnitude of the difference between left and right turning on the rotational task.

Repeated treatment with psychomotor stimulants such as amphetamine often results in sensitization (also called reverse tolerance). That is, prior exposure to the drug results in a more vigorous behavioral response to subsequent treatment. Rotational behavior can be sensitized by a single injection of amphetamine. Although sensitization occurs in both sexes, its magnitude and duration are greater in female rats (Robinson, 1984; Robinson, Becker, & Presty, 1982).

Although sex differences in striatal dopamine content have sometimes been reported (Crowley, O'Donohue, & Jacobowitz, 1978), sex differences in the magnitude of amphetamine-induced rotational behavior are more closely correlated with differences in the magnitude of asymmetries in the dopamine content of the nigrostriatal pathway (Robinson *et al.*, 1980). Females also display more vigorous rotational behavior when asymmetries of nigrostriatal dopamine function are experimentally induced by unilateral dopamine-depleting lesions of the substantia nigra (Robinson & Becker, 1983; Robinson *et al.*, 1985) or unilateral electrical stimulation of the lateral hypothalamus (Robinson, Camp, Jacknow, & Becker, 1982). Again, the sex difference appears as an increase in the vigor of rotational behavior without a change in the direction of rotation.

Gonadal hormones appear to influence the magnitude of sex differences in rotational behavior. Among female rats, rotational behavior elicited by amphetamine or electrical stimulation of the nigrostriatal pathway varies in magnitude with the estrous cycle, peaking on the day of estrus (Becker, Robinson, & Lorenz, 1982; Robinson *et al.*, 1982). Gonadectomy attenuates rotation in females but has no effect in males. Since estradiol increases amphetamine-induced rotational behavior in males (Hruska & Silbergeld, 1980a), varying levels of this hormone may explain both the sex difference in rotational behavior and the variation of rotational behavior with the estrous cycle.

Many possible mechanisms might explain the apparent influence of estradiol on rotational behavior. For example, the amphetamine-stimulated release of dopamine from striatal tissue varies with the estrous cycle in the rat. Ovariectomy reduces dopamine release from striatal tissue, and replacement with estrogen and progesterone increases dopamine release in tissue from ovariectomized females (Becker & Ramirez, 1981). Estrogen also modulates dopamine receptor sensitivity (e.g., Gordon & Perry, 1983; Hruska & Silbergeld, 1980b) and turnover (Jori, Colturani, Dolfini, & Rutczynski, 1976). Although chronic estrogen treatment often potentiates the effects of dopamine agonists (e.g., Chiodo, Caggiula, & Saller, 1981;

Nausieda, Koller, Weiner, & Klawans, 1979), acute estrogen treatment can have the opposite effect (Gordon, Gorski, Borison, & Diamond, 1980). In fact, some behavioral responses to dopamine agonists are initially suppressed but later enhanced by exposure to estrogen (e.g., Gordon, 1980; Joyce, Smith, & Van Hartesveldt, 1982), and different behavioral responses to dopamine agonists can vary in intensity independently during the estrous cycle (Joyce & Van Hartesveldt, 1984). Because the interactions of estradiol with the brain dopamine systems are exceedingly complex (see Van Hartesveldt & Joyce, 1986, for review), no conclusion can be drawn regarding the neurochemical mechanism by which estrogen influences rotational behavior.

Since the major influence of gonadal hormones, particularly estrogen, seems to be on the magnitude of rotational behavior, it might be argued that the hormone influence on rotational behavior is secondary to an increase in general activity. The most persuasive evidence against this position is the failure of ovariectomy to affect rotation elicited by stimulation of the pontine reticular formation (Robinson, Camp, et al., 1982). If ovarian hormones increased rotation simply by increasing activity, a comparable influence of ovariectomy on rotational behavior elicited by different manipulations would be predicted.

Limited evidence supports the view that there may be sex differences in the effects of unilateral brain lesions on some behaviors. Sherman et al. (1980, 1983) studied the effect of left or right neocortical ablation on the directional bias displayed by rats when released from the corner of an open field. Following lesions all rats displayed an increased tendency to turn toward the side of the lesion. For males that were not given handling during early development, left hemisphere ablations produced a greater turning preference toward the lesioned side than right hemisphere lesions. In males that were handled, or in handled or nonhandled females, there was no difference in the effect of left or right hemisphere lesions.

Robinson et al. (1985) examined the effects of unilateral 6-hydroxydopamine lesions to the "dominant" or "nondominant" nigrostriatal pathway. For each rat the dominant pathway was defined as the one on the side of the brain contralateral to the direction of rotation produced by an amphetamine challenge. For rats that sustained lesions resulting in 90% depletion of dopamine, there was no difference in the direction of rotation postoperatively. All animals rotated toward the side of the lesion. Among animals with less extensive depletion of dopamine (20–90%), nearly all of the rats with lesions of the nondominant side rotated toward the lesioned side when given amphetamine, but the majority of rats with lesions of the dominant side displayed the opposite pattern. They turned away from the lesioned side. There was no sex difference in the effect of dominant versus nondominant lesions on the direction of rotational behavior, but females exhibited more vigorous rotational behavior than males.

Robinson et al. (1985) also studied the effects of dominant or nondominant lesions of the nigrostriatal pathway on long-term body weight regulation. In both males and females the extent to which body weight was chronically reduced was greater with lesions that caused more extensive dopamine depletion. For males, the magnitude of this effect was significantly greater if the lesion was to the dominant side (defined as described above). For females the effects of dominant and nondominant lesions were equivalent, and overall, lesions caused a greater reduction in the level of body weight regulation in males than in females.

In earlier work Lenard and colleagues (Lenard, 1977; Lenard, Sarkasian, & Szabo, 1975) reported that bilateral destruction of the nigrostriatal pathway at the level of the globus pallidus produced far more severe and longer-lasting disturbances of feeding in males than in females. The possible importance of neonatal exposure to gonadal hormones was suggested by the observation that neonatal gonadectomy of males eliminated the sex difference in response to pallidal lesions (Hahn & Lenard, 1977). At present there are no data regarding the influence of gonadal hormones on the effects of unilateral nigrostriatal pathway lesions on body weight regulation, but it is obvious that such information would be most interesting.

SEX AND LATERALITY: OVERVIEW

The available data suggest that there may be sex differences in the extent of behavioral and brain asymmetries in rodents, but no simple statement can be made as to which sex is more strongly lateralized. If one considers various postural and motor asymmetries, then females appear to be more lateralized than males, and only females exhibit consistent hemispheric differences in the pattern of glucose utilization in various regions of the brain (Ross *et al.*, 1981). However, the results for studies of unilateral brain injury suggest the opposite conclusion, namely, that males may be more strongly lateralized.

It has been argued that in humans males are more strongly lateralized than females for both verbal and visuospatial abilities (see McGlone, 1980, for review). Although hotly disputed, this idea has led to the expectation that men should be more strongly lateralized than women for all asymmetric functions. In fact, several reports (Annett, 1980; Porac & Coren, 1981) indicate that women are more strongly right-sided than men on a number of measures of manual skill. More recent studies of humans with unilateral cerebral injury (see Kimura & Harshman, 1984) make it clear that the simple generalization that men are more strongly lateralized than women for cognitive functions is also incorrect. For some verbal tasks, such as oral fluency, disruption of performance in females was associated with damage to a more restricted region of the left hemisphere than was true for males. For other verbal tasks, such as vocabulary, both left and right hemisphere damage affected performance by females, but only left hemisphere lesions influenced performance by males.

These findings make it clear that although there may be sex differences in the pattern of cerebral organization, neither sex is more strongly lateralized in any simple, general sense. Further, sex differences in laterality are likely to depend on a complex interplay of genotypic, hormonal, and experiential influences. The relative importance and pattern of interaction of these factors are likely to vary from one asymmetric function to another.

CONCLUSIONS

Sex differences have been described in a number of behavioral tasks that have no direct or obvious relationship to reproduction. These differences are typically small in magnitude and vary with the genotype and prior behavioral history of the individual. Both organizational and activational actions of gonadal hormones influence performance on sexually dimorphic tasks. The relative influence of organizational and activational effects varies with the behavioral measure and interacts with genotypic and experiential variables.

Perhaps some day it will be possible to relate sex differences in laboratory tests of nonreproductive behaviors to the evolutionary history of individual species. Such an objective would focus attention on the differences in the behavioral demands placed on males and females of different species (i.e., the "roles" that males and females assume). If successful, such a venture might provide a theoretical basis for predicting the existence of sex differences in species and tasks that have not yet been studied as well as a scheme for comprehending the data already available. Developing such a model poses formidable obstacles, not the least of which is the fact that the relationship between performance on laboratory tests and behavior in the field is generally unknown, although not, in principle, unknowable. Nevertheless, without some attempt to comprehend the functional significance of sex differences in nonreproductive behavior, it is unlikely that our understanding will advance much beyond the current descriptive level.

REFERENCES

Adams, N., & Boice, R. (1983). A longitudinal study of dominance in an outdoor colony of domestic rats. *Journal of Comparative Psychology, 97,* 24–33.

Aldis, O. (1975). *Play fighting.* New York: Academic Press.

Annett, M. (1980). Sex differences in laterality—meaningfulness versus reliability. *The Brain and Behavioral Sciences, 3,* 227–228.

Applegate, E. A., Upton, D. E., & Stern, J. S. (1982). Food intake, body composition and blood lipids following treadmill exercise in male and female rats. *Physiology and Behavior, 28,* 917–920.

Arbuthnott, G. W., & Ungerstedt, V. (1975). Turning behavior induced by electrical stimulation of the nigro-neostriatal system of the rat. *Experimental Neurology, 47,* 162–172.

Archer, J. (1975). Rodent sex differences in emotional and related behavior. *Behavioral Biology, 14,* 451–479.

Attardi, B., Geller, L. N., & Ohno, S. (1976). Androgen and estrogen receptors in brain cytosol from male, female and *Tfm* mice. *Endocrinology, 98,* 864–874.

Baldwin, J. D., & Baldwin, J. J. (1974). Exploration and social play in squirrel monkey (*Samiri*). *American Zoologist, 14,* 303–314.

Barfield, R. S., Busch, D. E., & Wallen, K. (1972). Gonadal influence on agonistic behavior in the male domestic rat. *Hormones and Behavior, 3,* 247–259.

Barkley, M. S., & Goldman, B. D. (1977a). A quantitative study of serum testosterone, sex accessory organ growth, and the development of intermale aggression in the mouse. *Hormones and Behavior, 8,* 208–218.

Barkley, M. S., & Goldman, B. D. (1977b). Testosterone-induced aggression in adult female mice. *Hormones and Behavior, 9,* 76–84.

Barr, G. A., Gibbons, J. L., & Moyer, K. E. (1976). Male–female differences and the influence of neonatal and adult testosterone on intraspecific aggression in rats. *Journal of Comparative and Physiological Psychology, 90,* 1169–1183.

Barrett, P. T., & Bateson, P. P. G. (1978). Development of play in cats. *Behaviour, 66,* 106–120.

Bartness, T. J., & Waldbillig, R. J. (1984). Dietary self-selection in intact, ovariectomized and estradiol-treated female rats. *Behavioral Neuroscience, 98,* 125–137.

Bauer, R. H. (1978). Ontogeny of two-way avoidance behavior in male and female rats. *Developmental Psychobiology, 11,* 103–116.

Baum, M. J. (1979). Differentiation of coital behavior in mammals: A comparative analysis. *Neuroscience and Biobehavioral Reviews, 3,* 265–284.

Beatty, W. W. (1973). Postneonatal testosterone treatment fails to alter hormonal regulation of body weight and food intake of female rats. *Physiology and Behavior, 10,* 627–628.

Beatty, W. W. (1979). Gonadal hormones and sex differences in nonreproductive behaviors in rodents: Organizational and activational influences. *Hormones and Behavior, 12,* 112–163.

Beatty, W. W. (1984). Hormonal organization of sex differences in play fighting and spatial behavior. *Progress in Brain Research, 61,* 315–330.

Beatty, W. W., & Beatty, P. A. (1970). Hormonal determinants of sex differences in avoidance behavior and reactivity to electric shock in the rat. *Journal of Comparative and Physiological Psychology, 73,* 446–455.

Beatty, W. W., & Costello, K. B. (1982). Naloxone and play fighting in juvenile rats. *Pharmacology, Biochemistry and Behavior, 17,* 905–907.

Beatty, W. W., & Costello, K. B. (1983). Olfactory bulbectomy and play fighting in juvenile rats. *Physiology and Behavior, 30,* 525–528.

Beatty, W. W., & Fessler, R. G. (1976). Ontogeny of sex differences in open field behavior and sensitivity to electric shock. *Physiology and Behavior, 16,* 413–417.

Beatty, W. W., Powley, T. L., & Keesey, R. E. (1970). Effects of neonatal testosterone injections and hormone replacement in adulthood on body weight and body fat in female rats. *Physiology and Behavior, 5,* 1093–1098.

Beatty, W. W., Gregoire, K. C., & Parmiter, L. P. (1973). Sex differences in retention of passive avoidance behavior in rats. *Bulletin of the Psychonomic Society, 2,* 99–100.

Beatty, W. W., O'Briant, D. A., & Vilberg, T. R. (1974). Suppression of feeding by intrahypothalamic implants of estradiol in male and female rats. *Bulletin of the Psychonomic Society, 3,* 273–274.

Beatty, W. W., O'Briant, D. A., & Vilberg, T. R. (1975). Effects of ovariectomy and estradiol injections on food intake and body weight in rats with ventromedial hypothalamic lesions. *Pharmacology, Biochemistry and Behavior, 3,* 539–544.

Beatty, W. W., Dodge, A. M., Traylor, K. L., & Meaney, M. J. (1981). Temporal boundary of the sensitive period for hormonal organization of social play in juvenile rats. *Physiology and Behavior, 26,* 241–243.

Beatty, W. W., Dodge, A. M., Dodge, L. J., White, K., & Panksepp, J. (1982). Psychomotor stimulants, social deprivation and play in juvenile rats. *Pharmacology, Biochemistry and Behavior, 16,* 417–422.

Beatty, W. W., Costello, K. B., & Berry, S. L. (1984). Suppression of play fighting by amphetamine: Effects of catecholamine antagonists, agonists and synthesis inhibitors. *Pharmacology, Biochemistry and Behavior, 20,* 747–755.

Becker, J. B., & Ramirez, V. D. (1981). Sex differences in the amphetamine-stimulated release of catecholamines from rat striatal tissue *in vitro. Brain Research, 204,* 361–372.

Becker, J. B., Robinson, T. E., & Lorenz, K. A. (1982). Sex differences and estrous cycle variations in amphetamine-elicited rotational behavior. *European Journal of Pharmacology, 80,* 65–72.

Beeman, E. (1947). The effect of male hormone on aggressive behavior in mice. *Physiological Zoology, 20,* 373–405.

Bekoff, M. (1974). Social play and play-soliciting by infant canids. *American Zoologist, 14,* 323–341.

Bell, D. D., & Zucker, I. (1971). Sex differences in body weight: Organization and activation by gonadal hormones in the rat. *Physiology and Behavior, 7,* 27–34.

Bengelloun, W. A., Nelson, D. J., Zent, H. M., & Beatty, W. W. (1976). Behavior of male and female rats with septal lesions: Influence of prior gonadectomy. *Physiology and Behavior, 16,* 317–330.

Bermond, B., Mos, J., Meelis, W., van der Poel, A. M., & Kruk, M. R. (1982). Aggression induced by stimulation of the hypothalamus: Effects of androgens. *Pharmacology, Biochemistry and Behavior, 16,* 41–45.

Bernard, B. K., & Paolino, R. M. (1975). Temporal effects of castration on emotionality and shock-induced aggression in adult male rats. *Physiology and Behavior, 14,* 201–206.

Bielert, C., & Busse, C. (1983). Influences of ovarian hormones on the food intake and feeding of captive and wild chacma baboons (*Papio ursinus*). *Physiology and Behavior, 30,* 103–111.

Birke, L. I. A., & Sadler, D. (1983). Progestin-induced changes in play behavior of the prepubertal rat. *Physiology and Behavior, 30,* 341–347.

Birke, L. I. A., & Sadler, D. (1984). Modification of juvenile play and other social behavior in the rat by neonatal progestins: Further studies. *Physiology and Behavior, 33,* 217–219.

Blaustein, J. D., & Wade, G. N. (1976). Ovarian influence on the meal patterns of rats. *Physiology and Behavior, 17,* 201–208.

Blizard, D. A. (1983). Sex differences in running-wheel behaviour in the rat: The inductive and activational effects of gonadal hormones. *Animal Behaviour, 31,* 378–384.

Blizard, D. A., & Denef, C. (1973). Neonatal androgen effects on open field activity and sexual behavior in the female rat: The modifying influence of ovarian secretions during development. *Physiology and Behavior, 14,* 601–608.

Blizard, D. A., Lippman, H. R., & Chen, J. J. (1975). Sex differences in open field behavior in the rat: The inductive and activational role of gonadal hormones. *Physiology and Behavior, 14,* 601–608.

Bloch, G. J., & Davidson, J. M. (1971). Behavioral and somatic responses to the antiandrogen cyproterone. *Hormones and Behavior, 2,* 11–26.

Blurton-Jones, N. (1976). Rough-and-tumble play among nursery school children. In J. S. Bruner, A. Jolly, & K. Sylva (Eds.), *Play* (pp. 352–363). New York: Penguin.

Bowden, N. J., & Brain, P. F. (1978). Blockade of testosterone-maintained intermale fighting in albino laboratory mice by an aromatization inhibitor. *Physiology and Behavior, 20,* 543–546.

Braggio, J. T., Nadler, R. D., Lance, J., & Miseyko, D. (1978). Sex differences in apes and children. *Recent Advances in Primatology, 1,* 529–532.

Brain, P. F., Benton, D., Howell, P. A., & Jones, S. E. (1980). Resident rats' aggression toward intruders. *Animal Learning and Behavior, 8,* 331–335.

Brass, C. A., & Glick, S. D. (1981). Sex differences in drug-induced rotation in two strains of rats. *Brain Research, 223,* 229–234.

Broida, J., & Svare, B. (1983). Genotype modulates testosterone dependent activity and reactivity in male mice. *Hormones and Behavior, 17,* 76–85.

Broida, J., & Svare, B. (1984). Sex differences in the activity of mice: Modulation by postnatal gonadal hormones. *Hormones and Behavior, 18,* 65–78.

Bronson, F. H., & Desjardins, C. H. (1968). Aggression in adult mice: Modification by neonatal injections of gonadal hormones. *Science, 161,* 705–706.

Bronson, F. H., & Desjardins, C. H. (1970). Neonatal androgen administration and adult aggressiveness in female mice. *General and Comparative Endocrinology, 15,* 320–325.

Bronstein, P. M., & Hirsch, S. M. (1974). Reactivity in the rat: Ovariectomy fails to affect open field behaviors. *Bulletin of the Psychonomic Society, 3,* 257–260.

Camp, D. M., Robinson, T. E., & Becker, J. B. (1984). Sex differences in the effects of early experience on the development of behavioral and brain asymmetries in rats. *Physiology and Behavior, 33,* 433–439.

Chambers, K. C. (1976). Hormonal influences on sexual dimorphism in rate of extinction of a conditioned taste aversion. *Journal of Comparative and Physiological Psychology, 90,* 851–856.

Chambers, K. C. (1980). Progesterone, estradiol, testosterone and dihydrotestosterones: Effects on rate of extinction of a conditioned taste aversion in rats. *Physiology and Behavior, 24,* 1061–1065.

Chambers, K. C. (1982). Failure of ACTH to prolong extinction of a conditioned taste aversion in the absence of the testes. *Physiology and Behavior, 29,* 915–920.

Chambers, K. C., & Sengstake, C. B. (1979). Temporal aspects of the dependency of a dimorphic rate of extinction on testosterone. *Physiology and Behavior, 22,* 53–56.

Chiodo, L. A., Caggiula, A. R., & Saller, C. F. (1981). Estrogen potentiates the stereotypy induced by dopamine agonists in the rat. *Life Sciences, 28,* 827–835.

Christie, M. H., & Barfield, R. J. (1979). Effect of aromatizable androgens on aggressive behavior among rats (*Rattus norvegicus*). *Journal of Endocrinology, 83,* 17–26.

Clark, C. R., & Nowell, N. W. (1979). The effect of the antiestrogen CI-628 on androgen-induced aggressive behavior in castrated male mice. *Hormones and Behavior, 12,* 205–210.

Clark, C. R., & Nowell, N. W. (1980). The effect of the non-steroidal antiandrogen flutamide on neural receptor binding of testosterone and intermale aggressive behavior in mice. *Psychoneuroendocrinology, 5,* 39–45.

Clark, R. G., & Tarttelin, M. F. (1982). Some effects of ovariectomy and estrogen replacement on body composition in the rat. *Physiology and Behavior, 28,* 963–969.

Coling, J. G., & Herberg, L. J. (1982). Effect of ovarian and exogenous hormones on defended body weight, actual body weight, and the paradoxical hoarding of food by female rats. *Physiology and Behavior, 29,* 687–691.

Collins, R. L. (1977). Toward an admissible genetic model for the inheritance of the degree and direction of asymmetry. In S. Harnad, R. W. Doty, L. Goldstein, J. Jaynes, & G. Krauthamer (Eds.), *Lateralization in the nervous system* (pp. 137–150). New York: Academic Press.

Colvin, G. B., & Sawyer, C. H. (1969). Induction of running activity by intracerebral implants of estrogen in ovariectomized rats. *Neuroendocrinology, 4,* 309–320.

Conner, R. L., Levine, S., Wertheim, G. A., & Cummer, J. F. (1969). Hormonal determinants of aggressive behavior. *Annals of the New York Academy of Science, 159,* 760–776.

Crowley, W. R., O'Donohue, T. L., & Jacobowitz, D. A. (1978). Sex differences in catecholamine content in discrete brain nuclei of the rat: Effects of neonatal castration or testosterone treatment. *Acta Endocrinologica, 89,* 20–28.

Czaja, J. A. (1975). Food rejection by rhesus monkeys during the menstrual cycle and early pregnancy. *Physiology and Behavior, 14,* 579–588.

Czaja, J. A. (1984). Sex differences in the activational effects of gonadal hormones on food intake and body weight. *Physiology and Behavior, 33,* 553–558.

Czaja, J. A., & Goy, R. W. (1975). Ovarian hormones and food intake in guinea pigs and rhesus monkeys. *Hormones and Behavior, 6,* 329–349.

Davies, J. A., & Kemble, E. D. (1983). Social play behaviors and insect predation in Northern grasshopper mice (*Onochomys leucogaster*). *Behavioral Processes, 8,* 197–204.

Davis, P. G., Chaptal, C. V., & McEwen, B. S. (1979). Independence of the differentiation of masculine and feminine sexual behavior in rats. *Hormones and Behavior, 12,* 12–19.

Dawson, J. L. M., Cheung, Y. M., & Lau, R. T. S. (1975). Developmental effects of neonatal sex hormones on spatial and activity skills in the white rat. *Biological Psychology, 3,* 213–229.

DeBold, J. F., & Miczek, K. A. (1984). Aggression persists after ovariectomy in female rats. *Hormones and Behavior, 18,* 177–190.

Denenberg, V. H. (1981). Hemispheric laterality in animals and effects of early experience. *The Behavioral and Brain Sciences, 4,* 1–50.

Denenberg, V. H., Gaulin-Kremer, M., Gandelman, R., & Zarrow, M. X. (1973). The development of standard stimulus animals for mouse (*Mus musculus*) aggression testing by means of olfactory bulbectomy. *Animal Behaviour, 21,* 590–598.

Denenberg, V. H., Rosen, G. D., Hofman, M., Gall, J., Stockler, J., & Yutzey, D. A. (1982). Neonatal postural asymmetry and sex differences in the rat. *Developmental Brain Research, 2,* 417–419.

Denti, A., & Epstein, A. (1972). Sex differences in the acquisition of two kinds of avoidance behavior in rats. *Physiology and Behavior, 8,* 611–615.

Denti, A., & Negroni, J. A. (1975). Activity and learning in neonatally hormone treated rats. *Acta Physiologica Latino America, 25,* 99–106.

DeVore, I. T. (1963). Mother–infant relations in baboons. In H. Rheingold (Ed.), *Maternal behavior in mammals* (pp. 305–335). New York: John Wiley & Sons.

Döhler, K. D., Hancke, J. L., Srivastava, S. S., Hofman, C., Shryne, J. E., & Gorski, R. A. (1984). Participation of estrogens in female sexual differentiation of the brain: Neuroanatomical, neuroendocrine and behavioral evidence. *Progress in Brain Research, 61,* 99–117.

Donohoe, T. P., & Stevens, R. (1981). Modulation of food intake by amygdaloid estradiol implants in female rats. *Physiology and Behavior, 27,* 105–114.

Donohoe, T. P., & Stevens, R. (1982). Modulation of food intake by hypothalamic implants of estradiol benzoate, estrone, estriol, and CI-628 in female rats. *Pharmacology, Biochemistry and Behavior, 16,* 93–99.

Donohoe, T. P., & Stevens, R. (1983). Effects of ovariectomy, estrogen treatment and CI-628 on food intake and body weight in female rats treated neonatally with gonadal hormones. *Physiology and Behavior, 31,* 325–329.

Drago, F., Bohus, B., Scapagnini, U., & De Wied, D. (1980). Sexual dimorphism in passive avoidance behavior of rats. Relation to body weight, age, shock intensity and retention interval. *Physiology and Behavior, 24,* 1161–1165.

Dubuc, P. (1976). Bodyweight regulation in female rats following neonatal testosterone. *Acta Endocrinologica, 81,* 215–224.

Dudley, S. D., Gentry, R. T., Silverman, B. S., & Wade, G. N. (1979). Estradiol and insulin: Independent effects on eating and body weight in rats. *Physiology and Behavior, 22,* 63–68.

Earley, C. J., & Leonard, B. E. (1978). Androgenic involvement in conditioned taste aversion. *Hormones and Behavior, 11,* 1–11.

Earley, C. J., & Leonard, B. E. (1979). Effects of prior exposure on conditioned taste aversion in the rat: Androgen- and estrogen-dependent events. *Journal of Comparative and Physiological Psychology, 93,* 793–805.

Edwards, D. A. (1969). Early androgen stimulation and aggressive behavior in male and female mice. *Physiology and Behavior, 4,* 333–338.

Edwards, D. A. (1970). Post-neonatal androgenization and adult aggressive behavior in female mice. *Physiology and Behavior, 5,* 465–467.

Einon, D. (1980). Spatial memory and response strategies in rats: Age, sex and rearing differences in performance. *Quarterly Journal of Experimental Psychology, 32,* 473–489.

Eng, R., Gold, R. M., & Wade, G. N. (1979). Ovariectomy-induced obesity is not prevented by subdiaphragmatic vagotomy in rats. *Physiology and Behavior, 22,* 353–356.

Fagen, R. (1976). Exercise, play and physical training in animals. In P. P. G. Bateson & P. H. Klopfer (Eds.), *Perspectives in ethology* (Vol. 2, pp. 189–219). New York: Plenum Press.

Fagen, R. (1981). *Animal play behavior.* New York: Oxford University Press.

Fahrbach, S. E., Meisel, R. L., & Pfaff, D. W. (1985). Preoptic area implants of estradiol increase wheel running but not the open field activity of female rats. *Physiology and Behavior, 35,* 985–992.

Feder, H. H. (1978). Specificity of steroid hormone activation of sexual behaviour in rodents. In J. B. Hutchinson (Ed.), *Biological determinants of sexual behaviour.* New York: John Wiley & Sons.

Gandelman, R. (1980). Gonadal hormones and the induction of intraspecific fighting in mice. *Neuroscience and Biobehavioral Reviews, 4,* 133–140.

Gandelman, R. (1983). Gonadal hormones and sensory function. *Neuroscience and Biobehavioral Reviews, 7,* 1–18.

Gandelman, R., vom Saal, F. S., & Reinisch, J. M. (1977). Contiguity to male foetuses affects morphology and behavior of female mice. *Nature, 266,* 722–724.

Gandelman, R., Simon, N. G., & McDermott, N. J. (1979). Prenatal exposure to testosterone and its precursors influences morphology and later behavioral responsiveness to testosterone of female mice. *Physiology and Behavior, 23,* 23–26.

Gandelman, R., Rosenthal, C., & Howard, S. M. (1980). Exposure of mouse fetuses of various ages to testosterone and later activation of intraspecific fighting. *Physiology and Behavior, 25,* 333–335.

Gandelman, R., Howard, S. M., & Reinisch, J. M. (1981). Perinatal exposure to 19-nor-17α-ethynyltestosterone (norethindrone) influences morphology and aggressive behavior of female mice. *Hormones and Behavior, 15,* 404–415.

Gandelman, R., Peterson, C., & Hauser, A. M. (1982). Mice: Fetal estrogen exposure does not facilitate later activation of fighting by testosterone. *Physiology and Behavior, 29,* 397–399.

Gentry, R. L. (1974). The development of social behavior through play in the stellar sea lion. *American Zoologist, 14,* 391–403.

Gentry, R. T., & Wade, G. N. (1976a). Sex differences in sensitivity of food intake, body weight, and running-wheel activity to ovarian steroids in rats. *Journal of Comparative and Physiological Psychology, 90,* 747–754.

Gentry, R. T., & Wade, G. N. (1976b). Androgenic control of food intake and body weight in male rats. *Journal of Comparative and Physiological Psychology, 90,* 18–25.

Gerall, A. A. (1967). Effects of early postnatal androgen and estrogen injections on estrous activity cycles and mating behavior of rats. *Anatomical Record, 157,* 97–104.

Gerall, A. A., & Theil, A. R. (1975). Effects of perinatal gonadal secretions on parameters of receptivity and weight gain in hamsters. *Journal of Comparative and Physiological Psychology, 89,* 580–589.

Gerall, H. D., Ward, I. L., & Gerall, A. A. (1967). Disruption of the male rat's sexual behaviour induced by social isolation. *Animal Behaviour, 15,* 54–58.

Gerall, A. A., Stone, L. S., & Hitt, J. C. (1972). Neonatal androgen depresses female responsiveness to estrogen. *Physiology and Behavior, 8,* 17–20.

Gerall, A. A., Dunlap, J. L., & Hendricks, S. E. (1973). Effects of ovarian secretions on female behavioral potentiality in the rat. *Journal of Comparative and Physiological Psychology, 82,* 449–465.

Glick, S. D., Hinds, P. A., & Shapiro, R. M. (1983). Cocaine induced rotation: Sex-dependent differences between left- and right-sided rats. *Science, 221,* 775–777.

Goldfoot, D. A., & Wallen, K. (1979). Development of gender role behaviors in heterosexual and isosexual groups of infant rhesus monkeys. *Recent Advances in Primatology, 1,* 155–159.

Goldman, L., & Swanson, H. H. (1975). Developmental changes in pre-adult behavior in confined colonies of golden hamsters. *Developmental Psychobiology, 8,* 137–150.

Gordon, J. H. (1980). Modulation of apomorphine-induced stereotypy by estrogen: Time course and dose response. *Brain Research Bulletin, 5,* 679–682.

Gordon, J. H., & Perry, K. O. (1983). Pre- and postsynaptic neurochemical alterations following estrogen-induced striatal dopamine hypo- and hypersensitivity. *Brain Research Bulletin, 10,* 425–428.

Gordon, J. H., Gorski, R. A., Borison, R. L., & Diamond, B. I. (1980). Postsynaptic efficacy of dopamine: Possible suppression by estrogen. *Pharmacology, Biochemistry and Behavior, 12,* 515–518.

Goy, R. W. (1970). Early hormone influences on the development of sexual and sex-related behavior. In F. O. Schmitt (Ed.), *The neurosciences: Second study program* (pp. 196–206). New York: Rockefeller University Press.

Goy, R. W. (1978). Development of play and mounting behavior in female rhesus virilized prenatally with esters of testosterone or dihydrotestosterone. *Recent Advances in Primatology, 1,* 155–160.

Goy, R. W., & McEwen, B. S. (1980). *Sexual differentiation of the brain.* Cambridge, MA: MIT Press.

Goy, R. W., & Resko, J. A. (1972). Gonadal hormones and behavior of normal and pseudohermaphroditic nonhuman female primates. *Recent Progress in Hormone Research, 28,* 707–733.

Gray, J. A. (1979). Sex differences in the emotional behaviour of laboratory rodents: Comment. *British Journal of Psychology, 70,* 35–36.

Gray, J. A., Levine, S., & Broadhurst, P. L. (1965). Gonadal hormone injections in infancy and adult emotional behaviour. *Animal Behaviour, 13,* 33–45.

Gray, J. M., & Wade, G. N. (1981). Food intake, body weight, and adiposity in female rats: Actions and interactions of progestins and antiestrogens. *American Journal of Physiology, 240,* E474–E481.

Gray, J. M., Nunez, A. A., Siegel, L. I., & Wade, G. N. (1979). Effects of testosterone on body weight and adipose tissue: Role of aromatization. *Physiology and Behavior, 23,* 465–469.

Hahn, Z., & Lenard, L. (1977). Elimination of the sex-dependence of short term body weight changes after bilateral pallidal lesion by neonatal castration. *Acta Physiologica Academicae Scientia Hungarica, 50,* 229–231.

Harlow, H. F. (1965). Sexual behavior in rhesus monkeys. In F. A. Beach (Ed.), *Sex and behavior* (pp. 234–265). New York: John Wiley & Sons.

Hauser, H., & Gandelman, R. (1983). Contiguity to males *in utero* affects avoidance responding in adult female mice. *Science, 220,* 437–438.

Heinsbroek, R. P. W., van Oyen, H. G., van de Poll, N. E., & Boer, G. J. (1983). Failure of dexamethasone to influence sex differences in acquisition of discriminated lever press avoidance. *Pharmacology, Biochemistry and Behavior, 19,* 599–604.

Heinsbroek, R. P. W., van Oyen, H. G., & van de Poll, N. E. (1984). The pituitary–adrenocortical system is not involved in the sex difference in passive avoidance. *Pharmacology, Biochemistry and Behavior, 20,* 663–668.

Hines, M., Döhler, K. D., & Gorski, R. A. (1982). *Rough play in female rats following pre- and postnatal treatment with diethylstilbestrol or testosterone.* Paper presented at the Conference on Reproductive Behavior, East Lansing, MI.

Hirsch, E., Ball, E., & Goodkin, L. (1982). Sex differences in the effects of voluntary activity on sucrose-induced obesity, *Physiology and Behavior, 29,* 253–262.

Hitchcock, F. A. (1925). Studies in vigor: V. The comparative activity of male and female albino rats. *American Journal of Physiology, 75,* 205–210.

Hoyenga, K. B., & Hoyenga, K. T. (1982). Gender and energy balance: Sex difference in adaptations for feast and famine. *Physiology and Behavior, 28,* 545–563.

Hruska, R. E., & Silbergeld, E. K. (1980a). Increased dopamine receptor sensitivity after estrogen treatment using the rat rotation model. *Science, 208,* 1466–1468.

Hruska, R. E., & Silbergeld, E. K. (1980b). Estrogen treatment enhances dopamine receptor sensitivity in the rat striatum. *European Journal of Pharmacology, 61,* 397–400.

Hull, E. M., Franz, J. R., Snyder, A. M., & Nishita, J. K. (1980). Perinatal progesterone and learning, social and reproductive behavior in rats. *Physiology and Behavior, 24,* 251–256.

Hutchinson, R., Ulrich, R., & Azrin, N. (1965). Effects of age and related factors on the pain aggression reaction. *Journal of Comparative and Physiological Psychology, 59,* 365–369.

Janowiak, R., & Stern, J. J. (1974). Food intake and body weight modifications following medial hypothalamic hormone implants in female rats. *Physiology and Behavior, 12,* 875–879.

Jen, C. K. L., Greenwood, M. R. C., & Brasel, J. A. (1981). Sex differences in the effect of high-fat feeding on behavior and carcass composition. *Physiology and Behavior, 29,* 161–166.

Jerussi, T. P., & Glick, S. D. (1974). Amphetamine-induced rotation in rats without lesions. *Neuropharmacology, 13,* 283–286.

Jori, A., Colturani, F., Dolfini, E., & Rutczynski, M. (1976). Modifications of striatal dopamine metabolism during the estrous cycle in mice. *Neuroendocrinology, 21,* 262–266.

Joseph, R., Hess, S., & Birecree, E. (1978). Effect of hormone manipulations and exploration on sex differences in maze learning. *Behavioral Biology, 24,* 364–377.

Joslyn, W. D. (1973). Androgen-induced social dominance in infant female rhesus monkeys. *Journal of Child Psychology and Psychiatry, 14,* 137–145.

Joyce, J. N., & Van Hartesveldt, C. (1984). Behaviors induced by intrastriatal dopamine vary across the estrous cycle. *Pharmacology, Biochemistry and Behavior, 20,* 551–557.

Joyce, J. N., Smith, R. L., & Van Hartesveldt, C. (1982). Estradiol suppresses then enhances intracaudate dopamine-induced contralateral deviation. *European Journal of Pharmacology, 81,* 117–122.

Juraska, J. M., Henderson, C., & Muller, J. (1984). Differential rearing experience, gender and radial maze performance. *Developmental Psychobiology, 17,* 209–216.

Kanarek, R. B., & Beck, J. M. (1980). Role of gonadal hormones in diet selection and food utilization in female rats. *Physiology and Behavior, 24,* 381–386.

Kemble, E. D., & Enger, J. M. (1984). Sex differences in shock motivated behaviors, activity, and discrimination learning of northern grasshopper mice (*Onychomys leucogaster*). *Physiology and Behavior, 32,* 375–380.

Kemnitz, J. W., Glick, Z., & Bray, G. A. (1983). Ovarian hormones influence brown adipose tissue. *Pharmacology, Biochemistry and Behavior, 18,* 563–566.

Kendler, K., Hennessy, J. W., Smotherman, W. P., & Levine, S. (1976). An ACTH effect on recovery from conditioned taste aversion. *Behavioral Biology, 17,* 225–229.

Kimura, D., & Harshman, R. A. (1984). Sex differences in brain organization for verbal and non-verbal functions. *Progress in Brain Research, 61,* 423–441.

King, J. M. (1979). Effects of lesions of the amygdala, preoptic area and hypothalamus on estradiol-induced activity in the female rat. *Journal of Comparative and Physiological Psychology, 93,* 360–367.

King, J. M., & Cox, V. C. (1973). The effects of estrogens on food intake and body weight following ventromedial hypothalamic lesions. *Physiological Psychology, 1,* 261–264.

Krasnoff, A., & Weston, L. M. (1976). Puberal status and sex differences: Activity and maze behavior in rats. *Developmental Psychobiology, 9,* 261–269.

Krecek, J. (1973). Sex differences in salt taste: The effect of testosterone. *Physiology and Behavior, 10,* 683–688.

Krecek, J. (1976). The pineal gland and the development of the salt intake pattern in male rats. *Developmental Psychobiology, 8,* 181–188.

Krecek, J. (1978). The effects of ovariectomy of females and oestrogen administration to males during the neonatal critical period on the salt intake in adulthood in rats. *Physiologica Bohemoslovaca, 27,* 1–5.

Krecek, J., Novakova, V., & Stibral, K. (1972). Sex differences in the taste preference for a salt solution in the rat. *Physiology and Behavior, 8,* 183–188.

Kruk, M. R., van der Laan, C. E., Mos, J., van der Poel, A. M., Meelis, W., & Olivier, B. (1984). Comparison of aggressive behavior induced by electrical stimulation in the hypothalamus of male and female rats. *Progress in Brain Research, 61,* 303–314.

Kummer, H. (1968). *Social organization of hamadryas baboons: A field study.* Chicago: University of Chicago Press.

Laudenslager, M. L., Wilkinson, C. W., Carlisle, H. J., & Hammel, H. T. (1980). Energy balance in ovariectomized rats with and without estrogen replacement. *American Journal of Physiology, 238,* R400–R405.

Lenard, L. (1977). Sex-dependent body weight loss after bilateral 6-hydroxydopamine injection into globus pallidus. *Brain Research, 128,* 559–568.

Lenard, L., Sarkasian, J., & Szabo, I. (1975). Sex-dependent survival of rats after bilateral pallidal lesions. *Physiology and Behavior, 15,* 389–397.

Leshner, A. I., Merkle, D. A., & Mixon, J. F. (1981). Pituitary–adrenocortical effects on learning and memory in social situations. In J. L. Martinez, Jr., R. A. Jensen, R. B. Messing, H. Rigter, & J. L. McGaugh (Eds.), *Endogenous peptides and learning and memory processes* (pp. 159–180). New York: Academic Press.

Levine, S., & Broadhurst, P. L. (1963). Genetic and ontogenetic determinants of adult behavior in the rat. *Journal of Comparative and Physiological Psychology, 56,* 423–428.

Martin, P., & Caro, T. M. (1985). On the functions of play and its role in behavioral development. *Advances in the Study of Behavior, 15,* 59–103.

McEwen, B. S., Lieberburg, I., Chaptal, C., Davis, P. G., Krey, L. C., MacLusky, N. J., & Roy, E. J. (1979). Attenuating the defeminization of the neonatal rat brain: Mechanisms of action of cyproterone acetate, 1,4,6-androstatriene-3,17,-dione and a synthetic progestin, R5020. *Hormones and Behavior, 13,* 269–281.

McGlone, J. (1980). Sex differences in human brain asymmetry: A critical survey. *The Brain and Behavioral Sciences, 3,* 215–263.

Meaney, M. J., & Stewart, J. (1981a). A descriptive study of social development in the rat (*Rattus norvegicus*). *Animal Behaviour, 29,* 34–45.

Meaney, M. J., & Stewart, J. (1981b). Neonatal androgens influence the social play of prepubescent male and female rats. *Hormones and Behavior, 15,* 197–213.

Meaney, M. J., & Stewart, J. (1983). The influence of exogenous testosterone and corticosterone on the social behavior of prepubertal male rats. *Bulletin of the Psychonomic Society, 21,* 232–234.

Meaney, M. J., Stewart, J., & Beatty, W. W. (1982). The effects of neonatal glucocorticoids on the social play of juvenile rats. *Hormones and Behavior, 16,* 475–491.

Meaney, M. J., Stewart, J., Poulin, P., & McEwen, B. S. (1983). Sexual differentiation of social play in rat pups is mediated by the neonatal androgen-receptor system. *Neuroendocrinology, 37,* 85–90.

Meaney, M. J., Stewart, J., & Beatty, W. W. (1985). Sex differences in social play: The socialization of sex roles. *Advances in the Study of Behavior, 15,* 1–58.

Meisel, R. L., & Ward, I. L. (1981). Fetal female rats are masculinized by male littermates located caudally in the uterus. *Science, 213,* 239–242.

Meisel, R. L., Dohanich, G. P., & Ward, I. L. (1979). Effects of prenatal stress on avoidance acquisition, open field performance and lordotic behavior in male rats. *Physiology and Behavior, 22,* 527–530.

Miczek, K. A., & Barry, H. A. III. (1976). Pharmacology of sex and aggression. In S. D. Glick & J. Goldfarb (Eds.), *Behavioral pharmacology* (pp. 176–257). St. Louis: C. V. Mosby.

Mitchell, G. D. (1979). *Behavioral sex differences in nonhuman primates.* Princeton, NJ: Van Nostrand-Rheinhold.

Mora, S., Nasello, A. G., Mandelli-Lopes, M., & Diaz-Véliz, G. (1983). LHRH and rat avoidance behavior: Influence of castration and testosterone. *Physiology and Behavior, 30,* 19–22.

Morin, L. P., & Fleming, A. S. (1978). Variation of food intake and body weight with estrous cycle, ovariectomy, and estradiol benzoate treatment in hamsters (*Mesocricetus auratas*). *Journal of Comparative and Physiological Psychology, 92,* 1–6.

Moyer, K. E. (1974). Sex differences in aggression. In R. C. Friedman, R. M. Richart, & R. L. Vande Wiele (Eds.), *Sex differences in behavior* (pp. 335–372). New York: John Wiley & Sons.

Mueller, K., & Hsiao, S. (1980). Estrus- and ovariectomy-induced body weight changes: Evidence for two estrogenic mechanisms. *Journal of Comparative and Physiological Psychology, 94,* 1126–1134.

Mugford, R. A. (1974). Androgenic stimulation of aggression eliciting cues in adult opponent mice castrated at birth, weaning or maturity. *Hormones and Behavior, 5,* 93–102.

Mugford, R. A., & Nowell, N. W. (1970). Pheromones and their effect on aggression in mice. *Nature, 226,* 967–968.

Naess, D., Haug, E., Attrramadal, A., Aakvaag, A., Hansson, V., & French, F. (1976). Androgen receptors in the anterior pituitary and the central nervous system of the "androgen insensitive" (*Tfm*) rat: Correlation between receptor binding and effects of androgens on gonadotropin secretion. *Endocrinology, 49,* 1295–1303.

Nance, D. M. (1976). Sex differences in the hypothalamic regulation of feeding behavior in the rat. In A. H. Riesen & R. F. Thompson (Eds.), *Advances in psychobiology* (Vol. 3, pp. 75–123). New York: John Wiley & Sons.

Nance, D. M. (1983). The developmental and neural determinants of the effects of estrogen on feeding behavior in the rat: A theoretical perspective. *Neuroscience and Biobehavioral Reviews, 7,* 189–211.

Nance, D. M., & Gorski, R. A. (1977). Sex and hormone dependent alterations in responsiveness to caloric dilution. *Physiology and Behavior, 19,* 679–683.

Nance, D. M., Bromiley, B., Bernard, R. J., & Gorski, R. A. (1977). Sexually dimorphic effects of forced exercise on food intake and body weight in the rat. *Physiology and Behavior, 19,* 155–158.

Nance, D. M., Saporta, S., & Phelps, C. P. (1981). Effects of hypothalamic knife cuts on estrogenic control of feeding behavior. *Brain Research Bulletin, 6,* 259–266.

Nausieda, P. A., Koller, W. C., Weiner, W. J., & Klawans, H. L. (1979). Modification of postsynaptic dopaminergic sensitivity by female sex hormones. *Life Sciences, 25,* 521–526.

Nikoletseas, M. M. (1980). Food intake in the exercising rat: A brief review. *Neuroscience and Biobehavioral Reviews, 4,* 265–267.

Nunez, A. A., & Grundman, M. (1982). Testosterone affects food intake and body weight of weanling male rats. *Pharmacology, Biochemistry and Behavior, 16,* 933–936.

Nunez, A. A., Gray, J. M., & Wade, G. N. (1980). Food intake and adipose tissue lipoprotein lipase activity after hypothalamic estradiol benzoate implants in rats. *Physiology and Behavior, 25,* 595–598.

Nunez, A. A., Siegel, L. I., & Wade, G. N. (1980). Central effects of testosterone on food intake in male rats. *Physiology and Behavior, 24,* 469–472.

Olioff, M., & Stewart, J. (1978). Sex differences in play behavior of prepubescent rats. *Physiology and Behavior, 20,* 113–115.

Olton, D. S. (1978). Characteristics of spatial memory. In S. H. Hulse, H. Fowler, & W. K. Honig (Eds.), *Cognitive processes in animal behavior* (pp. 342–373.) Hillsdale, NJ: Lawrence Erlbaum.

Olton, D. S., & Samuelson, R. J. (1976). Remembrance of places passed: Spatial memory in rats. *Journal of Experimental Psychology: Animal Behavior Processes, 2,* 97–116.

Panksepp, J. (1981). Brain opioids—a neurochemical substrate. In S. Cooper (Ed.), *Theory in psychopharmacology* (pp. 149–175). London: Academic Press.

Panksepp, J., & Beatty, W. W. (1980). Social deprivation and play in rats. *Behavioral and Neural Biology, 30,* 197–206.

Panksepp, J., Herman, B. H., Vilberg, T., Bishop, P., & DeEskinazi, F. G. (1980). Endogenous opioids and social behavior, *Neuroscience and Biobehavioral Reviews, 4,* 473–487.

Panksepp, J., Siviy, S., & Normansell, L. (1984). The psychobiology of play: Theoretical and methodological perspectives. *Neuroscience and Biobehavioral Reviews, 8,* 465–492.

Panksepp, J., Jalowiec, J., DeEskinazi, F. G., & Bishop, P. (1985). Opiates and play dominance in juvenile rats. *Behavioral Neuroscience, 99,* 441–453.

Parlee, M. B. (1983). Menstrual rhythms in sensory processes: A review of fluctuations in vision, olfaction, audition, taste, and touch. *Psychological Bulletin, 93,* 539–548.

Pfaff, D. W., & Zigmond, R. E. (1971). Neonatal androgen effects on sexual and nonsexual behaviors of adult rats tested under various hormone regimes. *Neuroendocrinology, 7,* 129–145.

Phoenix, C. H., Goy, R. W., Gerall, A. A., & Young, W. C. (1959). Organizing action of prenatally administered testosterone propionate on the tissues mediating mating behavior in the female guinea pig. *Endocrinology, 65,* 369–382.

Pliner, P., & Fleming, A. S. (1983). Food intake, body weight, and sweetness preferences over the menstrual cycle in humans. *Physiology and Behavior, 30,* 663–666.

Porac, C., & Coren, S. (1981). *Lateral preferences and human behavior.* Berlin: Springer-Verlag.

Prasad, V., & Sheard, M. H. (1981). The effect of cyproterone acetate on shock elicited aggression. *Pharmacology, Biochemistry and Behavior, 15,* 691–694.

Quadagno, D. M., Shryne, J., Anderson, C., & Gorski, R. A. (1972). Influence of gonadal hormones on social, sexual, emergence, and open field behavior in the rat (*Rattus norvegicus*). *Animal Behaviour, 20,* 732–740.

Raleigh, M. H., Flannery, J. W., & Ervin, F. R. (1979). Sex differences in behavior among juvenile vervet monkeys (*Cercopithecus aethiops sabacus*). *Behavioral and Neural Biology, 26,* 455–465.

Ramirez, I. (1980). Relation between estrogen-induced hyperlipemia and food intake and body weight in rats. *Physiology and Behavior, 25,* 511–518.

Ramirez, I. (1981). Estradiol-induced changes in lipoprotein lipase, eating, and body weight in rats. *American Journal of Physiology, 240,* E533–E538.

Ray, O. S., & Barrett, R. S. (1975). Behavioral, pharmacological, and biochemical analysis of genetic differences in rats. *Behavioral Biology, 15,* 391–417.

Reddy, V. V. R., Naftolin, F., & Ryan, K. J. (1974). Conversion of androstenedione to estrone by neural tissues from fetal and neonatal rats. *Endocrinology, 94,* 117–121.

Robinson, T. E. (1984). Behavioral sensitization: Characterization of enduring changes in rotational behavior produced by intermittent injections of amphetamine in male and female rats. *Psychopharmacology, 84,* 466–475.

Robinson, T. E., & Becker, J. B. (1983). The rotational behavior model: Asymmetry in the effects of unilateral 6-OHDA lesions of the substantia nigra in rats. *Brain Research, 264,* 127–131.

Robinson, T. E., Becker, J. B., & Ramirez, V. D. (1980). Sex differences in amphetamine-elicited rotational behavior and the lateralization of striatal dopamine in rats. *Brain Research Bulletin, 5,* 539–545.

Robinson, T. E., Becker, J. B., & Presty, S. K. (1982). Long-term facilitation of amphetamine-induced rotational behavior and striatal dopamine release produced by a single exposure to amphetamine: Sex differences. *Brain Research, 253,* 231–241.

Robinson, T. E., Camp, D. E., Jacknow, D. S., & Becker, J. B. (1982). Sex differences and estrous cycle dependent variation in rotational behavior elicited by electrical stimulation of the mesostriatal dopamine system. *Behavioural Brain Research, 6,* 273–287.

Robinson, T. E., Becker, J. B., Camp, D. M., & Mansour, A. (1985). Variation in the pattern of behavioral and brain asymmetries due to sex differences. In S. Glick (Ed.), *Cerebral lateralization in nonhuman species* (pp. 185–231). New York: Academic Press.

Rodgers, R. J., & Hendrie, C. A. (1982). Agonistic behavior in rats: Evidence for non-involvement of opioid mechanisms. *Physiology and Behavior, 29,* 85–90.

Ross, D. A., Glick, S. D., & Meibach, R. C. (1981). Sexually dimorphic brain and behavioral asymmetries in the neonatal rat. *Proceedings of the National Academy of Sciences, U.S.A., 78,* 1958–1961.

Rowland, D. L., Perrings, T. S., & Thommes, J. A. (1980). Comparison of androgenic effects on food intake and body weight in adult rats. *Physiology and Behavior, 24,* 205–209.

Roy, E. J., & Wade, G. N. (1975). Role of estrogens in androgen-induced spontaneous activity in male rats. *Journal of Comparative and Physiological Psychology, 89,* 573–579.

Roy, E. J., & Wade, G. N. (1976). Estrogenic effects of an antiestrogen, MER-25, on eating and body weight in rats. *Journal of Comparative and Physiological Psychology, 89,* 573–579.

Roy, E. J., & Wade, G. N. (1977). Role of food intake in estradiol-induced body weight changes in female rats. *Hormones and Behavior, 8,* 255–274.

Roy, E. J., Maass, C. A., & Wade, G. N. (1977). Central action and a species comparison of the estrogenic effects of an antiestrogen on eating and body weight. *Physiology and Behavior, 18,* 137–140.

Sachs, B. D., & Harris, V. S. (1978). Sex differences and developmental changes in selected juvenile activities (play) of domestic lambs. *Animal Behaviour, 26,* 678–684.

Schechter, D., Howard, S. M., & Gandelman, R. (1981). Dihydrotestosterone promotes fighting behavior of female mice. *Hormones and Behavior, 15,* 233–237.

Schemmel, R., Mickelsen, R., & Gill, J. L. (1970). Dietary obesity in rats: Body weight and body fat accretion in seven strains of rats. *Journal of Nutrition, 100,* 1040–1048.

Sclafani, A., & Gorman, A. N. (1977). Effects of age, sex and prior body weight on the development of dietary obesity in adult rats. *Physiology and Behavior, 18,* 1021–1026.

Scouten, C. W. (1972). *Hormonal effects on avoidance in rats.* Unpublished M.S. thesis, North Dakota State University.

Scouten, C. W., & Beatty, W. W. (1971). Adrenocortical function and sex differences in acquisition and extinction of active avoidance behavior in the rat. *Psychological Reports, 29,* 1011–1018.

Scouten, C. W., Grotelueschen, L. K., & Beatty, W. W. (1975). Androgens and the organization of sex differences in active avoidance behavior in the rat. *Journal of Comparative and Physiological Psychology, 88,* 264–270.

Sfikakis, A., Spyraki, C., Sitaras, N., & Varonos, D. (1978). Implication of estrous cycle on conditioned avoidance behavior in the rat. *Physiology and Behavior, 21,* 441–446.

Shapiro, B. H., & Goldman, A. S. (1973). Feminine saccharin preference in the genetically androgen insensitive male rat pseudohermaphrodite. *Hormones and Behavior, 4,* 371–375.

Shapiro, B. H., Goldman, A. S., Bongiovanni, A. M., & Marino, J. M. (1976). Neonatal progesterone and feminine sexual development. *Nature, 264,* 795–796.

Sherman, G. F., Garbanati, J. A., Rosen, G. D., Yutzey, D. A., & Denenberg, V. H. (1980). Brain and behavioral asymmetries for spatial preference in rats. *Brain Research, 192,* 61–67.

Sherman, G. F., Garbanati, J. A., Rosen, G. D., Hofman, M., Yutzey, D. A., & Denenberg, V. H. (1983). Lateralization of spatial preference in the female rat. *Life Sciences, 33,* 189–194.

Sieck, G. C., Nance, D. M., Ramaley, J. A., Taylor, A. N., & Gorski, R. A. (1977). Pubertal cyclicity in feeding behavior and body weight regulation in the female rat. *Physiology and Behavior, 18,* 299–305.

Siegel, L. I., Nunez, A. A., & Wade, G. N. (1981). Effects of androgens on dietary self-selection and carcass composition in male rats. *Journal of Comparative and Physiological Psychology, 95,* 529–539.

Simon, N. G., & Gandelman, R. (1978a). Aggression-promoting and aggression-eliciting properties of estrogen in male mice. *Physiology and Behavior, 21,* 161–164.

Simon, N. G., & Gandelman, R. (1978b). The estrogenic arousal of aggressive behavior in female mice. *Hormones and Behavior, 10,* 118–127.

Simon, N. G., Gandelman, R., & Howard, S. M. (1981). MER-25 does not inhibit the activation of aggression by testosterone in adult Rockland–Swiss mice. *Psychoneuroendocrinology, 6,* 131–137.

Simon, N. G., Gandelman, R., & Gray, J. L. (1984). Endocrine induction of intermale aggression in mice: A comparison of hormonal regimens and their relationship to naturally occurring behavior. *Physiology and Behavior, 33,* 379–383.

Simon, N. G., Whalen, R. E., & Tate, M. P. (1985). Induction of male-typical aggression by androgens but not by estrogens in adult female mice. *Hormones and Behavior, 19,* 204–212.

Slater, J., & Blizard, D. A. (1976). A reevaluation of the relation between estrogen and emotionality in female rats. *Journal of Comparative and Physiological Psychology, 90,* 755–764.

Slob, A. K., & van der Werff ten Bosch, J. J. (1975). Sex differences in body growth in the rat. *Physiology and Behavior, 14,* 353–361.

Slob, A. K., Bogers, H., & van Stolk, M. A. (1981). Effects of gonadectomy and exogenous gonadal steroids on sex differences in open-field behaviour of adult rats. *Behavioural Brain Research, 2,* 347–362.

Smith, P. K. (1982). Does play matter? Functional and evolutionary aspects of animal and human play. *The Behavioral and Brain Sciences, 5,* 139–184.

Snyder, A. M., & Hull, E. M. (1980). Perinatal progesterone affects learning in rats. *Psychoneuroendocrinology, 5,* 113–119.

Stehm, K. E. (1985). *The effects of prenatal stress on the play behavior of juvenile rats.* Masters thesis, Villanova University.

Stern, J. J., & Janowiak, P. (1973). No effect of neonatal estrogen stimulation or hypophysectomy on spontaneous activity in female rats. *Journal of Comparative and Physiological Psychology, 85,* 409–412.

Stevens, R., & Goldstein, R. (1981). Effects of neonatal testosterone and estrogen on open-field behaviour in rats. *Physiology and Behavior, 26,* 551–553.

Stewart, J., & Cygan, D. (1980). Ovarian hormones act early in development to feminize adult open-field behavior in the rat. *Hormones and Behavior, 14,* 20–32.

Stewart, J., Skavarenina, A., & Pottier, J. (1975). Effects of neonatal androgens on open field behavior and maze learning in the prepubescent and adult rat. *Physiology and Behavior, 14,* 291–295.

Stewart, J., Vallentyne, S., & Meaney, M. J. (1979). Differential effects of testosterone metabolities in the neonatal period on open-field behavior and lordosis in the rat. *Hormones and Behavior, 13,* 282–292.

Svare, B. (1983). Psychobiological determinants of maternal aggressive behavior. In M. Hahn, E. Simmel, & C. Walters (Eds.), *Genetic and neural aspects of aggression: Synthesis and new directions* (pp. 129–146). Hillsdale, NJ: Lawrence Erlbaum.

Svare, B., Davis, P. G., & Gandelman, R. (1974). Fighting behavior in female mice following chronic androgen treatment during adulthood. *Physiology and Behavior, 12,* 399–403.

Swanson, H. H. (1967a). Alteration of sex-typical behaviour of hamsters in open field and emergence tests by neonatal administration of androgen or estrogen, *Animal Behaviour, 15,* 209–216.

Swanson, H. H. (1967b). Effects of pre- and post-puberal gonadectomy on sex differences in growth, adrenal, and pituitary weights, *Journal of Endocrinology, 39,* 555–564.

Symons, D. (1974). Aggressive play and communication in rhesus monkeys (*Macaca mulatta*). *American Zoologist, 14,* 317–322.

Symons, D. (1978). *Play and aggression: A study of rhesus monkeys.* New York: Columbia University Press.

Tarttellin, M. F., Shryne, J. E., & Gorski, R. A. (1975). Patterns of body weight change following neonatal hormone manipulation: A "critical" period for androgen-induced growth increases. *Acta Endocrinologica, 79,* 177–191.

Taylor, G. T. (1980). Fighting in juvenile rats and the ontogeny of agonistic behavior. *Journal of Comparative and Physiological Psychology, 94,* 953–961.

Taylor, G. T., Haller, J., Rupich, R., & Weiss, J. (1984). Testicular hormones and intermale aggressive behavior in the presence of a female rat. *Journal of Endocrinology, 100,* 315–321.

Tees, R. C., Midgley, G., & Nesbit, J. C. (1981). The effect of early visual experience on spatial maze learning in rats. *Developmental Psychobiology, 14,* 425–438.

Ter Harr, M. B. (1972). Circadian and estrous rhythms in food intake in the rat. *Hormones and Behavior, 3,* 213–219.

Thor, D. H., & Holloway, W. R., Jr. (1982). Anosmia and play fighting behavior in prepubescent rats, *Physiology and Behavior, 29,* 281–285.

Thor, D. H., & Holloway, W. R., Jr. (1984a). Sex and social play in juvenile rats (*R. norvegicus*). *Journal of Comparative Psychology, 98,* 276–284.

Thor, D. H., & Holloway, W. R., Jr. (1984b). Social play in juvenile rats: A decade of methodological and experimental research. *Neuroscience and Biobehavioral Reviews, 8,* 455–464.

Thor, D. H., & Holloway, W. R., Jr. (1986). Social play soliciting by male and female juvenile rats: Effects of neonatal androgenization and sex of cage mates. *Behavioral Neuroscience, 100,* 275–279.

Traylor, K. L. (1982). *Social interactions among captive wolves (Canis lupus).* Unpublished M.A. thesis, North Dakota State University.

Ungerstedt, U. (1971). Striatal dopamine release after amphetamine or nerve degeneration revealed by rotational behavior. *Acta Physiologica Scandanavica, 82*(Supplement 367), 49–68.

Valenstein, E. S., Kakolewski, J. W., & Cox, V. C. (1967). Sex differences in taste preferences for glucose and saccharin solutions. *Science, 156,* 942–943.

van de Poll, N. E., de Jonge, F., van Oyen, H. G., van Pelt, J., & de Bruin, J. P. C. (1981). Failure to find sex differences in testosterone activated aggression in two strains of rats. *Hormones and Behavior, 15,* 94–105.

van de Poll, N. E., Swanson, H. H., & van Oyen, H. G. (1981). Gonadal hormones and sex differences in aggression in rats. In P. F. Brain & D. Benton (Eds.), *The Biology of Aggression* (pp. 243–252). Alphen aan den Rijn, The Netherlands: Sijthoff & Noordhoff.

van de Poll, N. E., Smeets, J., van Oyen, H. G., & van der Zwan, S. M. (1982). Behavioral consequences of agonistic experience in rats: Sex differences and the effects of testosterone. *Journal of Comparative and Physiological Psychology, 96,* 893–903.

van Haaren, F., & van de Poll, N. E. (1984). Effect of a choice alternative on sex differences in passive avoidance behavior. *Physiology and Behavior, 32,* 211–216.

Van Hartesveldt, C., & Joyce, J. N. (1986). Effects of estrogen on the basal ganglia. *Neuroscience and Biobehavioral Reviews, 10,* 1–14.

van Oyen, H. G., van de Poll, N. E., & de Bruin, J. P. C. (1979). Sex, age and shock-intensity as factors in passive avoidance. *Physiology and Behavior, 23,* 915–918.

van Oyen, H. G., van de Poll, N. E., & de Bruin, J. P. C. (1980). Effects of retention interval and gonadectomy on sex differences in passive avoidance behavior. *Physiology and Behavior, 25,* 854–862.

van Oyen, H. G., Walg, H., & van de Poll, N. E. (1981). Discriminated lever press avoidance conditioning in male and female rats. *Physiology and Behavior, 26,* 313–317.

vom Saal, F. S. (1979). Prenatal exposure to androgen influences morphology and aggressive behavior of male and female mice. *Hormones and Behavior, 12,* 1–11.

vom Saal, F. S. (1983). Models of early hormonal effects on intrasex aggression in mice. In B. Svare (Ed.), *Hormones and aggressive behavior* (pp. 197–222). New York: Plenum Press.

vom Saal, F. S., & Bronson, F. H. (1978). *In utero* proximity of female mouse fetuses to males: Effect on reproductive performance during later life. *Biology of Reproduction, 19,* 842–853.

vom Saal, F. S., & Bronson, F. H. (1980). Sexual characteristics of adult female mice are correlated with their blood testosterone levels during prenatal development. *Science, 208,* 597–599.

vom Saal, F. S., Gandelman, R., & Svare, B. (1976). Aggression in male and female mice: Evidence for changed neural sensitivity in response to neonatal but not adult androgen exposure. *Physiology and Behavior, 17,* 53–57.

vom Saal, F. S., Grant, W. M., McMullen, C. W., & Laves, K. S. (1983). High fetal estrogen concentrations: Correlations with increased adult sexual activity and decreased aggression in male mice. In B. Svare (Ed.), *Hormones and aggressive behavior* (pp. 197–222). New York: Plenum Press.

Wade, G. N. (1974). Interaction between estradiol-17β and growth hormone in control of food intake in weanling rats. *Journal of Comparative and Physiological Psychology, 86,* 359–362.

Wade, G. N. (1976). Sex hormones, regulatory behaviors, and body weight. *Advances in the Study of Behavior, 6,* 201–279.

Wade, G. N., & Gray, J. M. (1979). Gonadal effects on food intake and adiposity: A metabolic hypothesis, *Physiology and Behavior, 22,* 583–594.

Wade, G. N., & Zucker, I. (1969a). Hormonal and developmental influences in rat saccharin preferences. *Journal of Comparative and Physiological Psychology, 69,* 291–300.

Wade, G. N., & Zucker, I. (1969b). Taste preferences of female rats: Modification by neonatal hormones, food deprivation and prior experience. *Physiology and Behavior, 4,* 935–943.

Wade, G. N., & Zucker, I. (1970). Modulation of food intake and locomotor activity in female rats by diencephalic hormone implants. *Journal of Comparative and Physiological Psychology, 72,* 328–336.

Ward, I. L. (1969). Differential effect of pre- and postnatal androgen on sexual behavior of intact and spayed rats. *Hormones and Behavior, 1,* 25–36.

Ward, I. L. (1972). Prenatal stress feminizes and demasculinizes the behavior of males. *Science, 175,* 82–84.

Ward, I. L., & Ward, O. B. (1985). Sexual behavior differentiation: Effects of prenatal manipulations in rats. In N. Adler, D. Pfaff, & R. W. Goy (Eds.), *Handbook of behavioral neurobiology* (Vol. 7, pp. 77–98). New York: Plenum Press.

Ward, I. L., & Weisz, J. (1980). Maternal stress alters plasma testosterone in fetal males. *Science, 207,* 328–329.

Weinberg, J., Gunnar, M. R., Brett, L. P., Gonzales, C. A., & Levine, S. (1982). Sex differences in biobehavioral responses to conflict in a taste aversion paradigm. *Physiology and Behavior, 29,* 201–210.

Weisz, J., & Ward, I. L. (1980). Plasma testosterone and progesterone titers of pregnant rats, their male and female fetuses, and neonatal offspring. *Endocrinology, 106,* 306–315.

Widdowson, E. M. (1976). The response of the sexes to nutritional stress. *Proceedings of the Nutrition Society, 35,* 175–180.

Young, J. K. (1986). Thyroxine treatment reduces the anorectic effect of estradiol in rats. *Behavioral Neuroscience, 100,* 284–287.

Young, J. K., Nance, D. M., & Gorski, R. A. (1978a). Dietary effects upon food and water intake and responsiveness to estrogen, 2-deoxy-glucose and glucose in female rats. *Physiology and Behavior, 21,* 395–403.

Young, J. K., Nance, D. M., & Gorski, R. A. (1978b). Effects of estrogen upon feeding inhibition produced by intragastric glucose loads. *Physiology and Behavior, 21,* 423–430.

Young, J. K., Nance, D. M., & Gorski, R. A. (1979). Sexual dimorphism in the regulation of caloric intake and body weight of rats fed different diets. *Physiology and Behavior, 23,* 577–582.

Young, J. K., Schlussman, S., & Dixit, P. K. (1979). Relationship between glucose tolerance and effects of estrogen upon feeding. *Physiology and Behavior, 23,* 679–684.

Zucker, I. (1969). Hormonal determinants of sex differences in saccharin preference, food intake, and body weight. *Physiology and Behavior, 4,* 595–602.

Zucker, I. (1972). Body weight and age as factors determining estrogen responsiveness in the rat feeding system. *Behavioral Biology, 7,* 527–542.

Zucker, I., Wade, G. N., & Ziegler, R. (1972). Sexual and hormonal influences on eating, taste preferences, and body weight of hamsters. *Physiology and Behavior, 8,* 101–111.

4

Role of Estrogens and Progestins in the Development of Female Sexual Behavior Potential

SHELTON E. HENDRICKS

INTRODUCTION

The demonstration by Phoenix, Goy, Gerall, and Young (1959) that injection of pregnant guinea pigs with testosterone modified the sexual behavior of female offspring has been followed by a multitude of studies attesting to the fact that mammalian sexual behavior observed at maturity is strongly influenced by gonadal hormones present during particular periods of development. Following the models proposed for sexual differentiation of the genital tracts (Jost, 1947, 1953), Phoenix *et al.* (1959) postulated that gonadal hormones act on developing tissues, "organizing" them in a manner such that they mediate reproductive behaviors typical of the male of the species. Tissues mediating female-typical behaviors were hypothesized to develop in the absence of significant gonadal steroid titers. Over the past 25 years much evidence has accumulated that is consistent with this organizational hypothesis. Researchers have repeatedly demonstrated that the presence of gonadal steroids during periods of fetal and/or neonatal life masculinizes and/or defeminizes adult sexual behavior potential, whereas their absence feminizes and/or demasculinizes behavior. These same manipulations modify characteristics of the nervous system that appear to constitute sexually dimorphic substrates that may mediate sex differences in behavior (Arnold & Breedlove, 1985; Gorski, 1985).

Phoenix *et al.* (1959) distinguished organizational effects of hormones, which

SHELTON E. HENDRICKS   Department of Psychology, University of Nebraska at Omaha, Omaha, Nebraska 68182.

Sexual Differentiation, Volume 11 of *Handbook of Behavioral Neurobiology*, edited by Arnold A. Gerall, Howard Moltz, and Ingeborg L. Ward, Plenum Press, New York, 1992.

ascribe lifelong consequences to hormones present at rather specific times, usually prior to maturity, from activational effects of hormones, which are more transient, lasting from a few hours to perhaps months and occurring throughout the organism's life cycle. Reviews of the further development of this hypothesis and some of the accumulated data supporting it have been provided by a number of authors (Gorski, 1973; Money & Ehrhardt, 1971; Ward, 1974; Whalen, 1968; Young, Goy, & Phoenix, 1964).

Challenges to the organizational hypothesis will serve to highlight issues discussed in this chapter. One of these challenges questions the uniqueness of testicular secretions in the process of sexual differentiation. In particular, the possibility has been raised that ovarian hormones play an active and significant role in the feminization of behavioral potential. Early evidence indicating that ovarian hormones feminize and/or demasculinize sexual behavior potential has been provided by some investigations (Lisk, 1969; Gerall, Dunlap, & Hendricks, 1973). Ovaries present during prepuberal development have been shown to, at least temporarily, facilitate the female sexual behavior of adult rats. Other investigators have presented evidence that complete development of the female sexual behavior potential requires the action of estrogenic steroids (Döhler, 1978; Resko, 1974; Shapiro, Goldman, Steinbeck, & Neuman, 1976). Döhler and Hancke (1978) and Resko (1974) have proposed theoretical models that specify the necessity for steroidal stimulation, ovarian, maternal, or otherwise, for complete development of female sexual behavior.

A second challenge is to the basic distinction between organizational and activational effects of hormones. The possibility of permanent effects on behavioral potential of exposure to hormones after maturity questions the clarity of the distinction between the two modes of hormonal action. Arnold and Breedlove (1985) have suggested that it would be more productive to look on organization of behavioral potential as a process that can occur throughout the life cycle. They give particular emphasis to the failure of research to date to identify cellular processes uniquely associated with either organizational or activational effects of hormones.

This chapter reviews data and theory relevant to the significance of gonadal steroids to the feminization of tissues mediating sexual behavior. Further, the organization–activation distinction is examined with a particular focus on the value of viewing ovarian hormones as having organizational effects throughout life rather than only during early development. Emphasis is placed on relevant data collected on laboratory rodents, particularly rats. Findings from other species that address the general organizational theory are also included.

THE ROLE OF THE OVARY IN SEXUAL DIFFERENTIATION

THE OVARY AND THE DEVELOPING ENDOCRINE SYSTEM

Ojeda, Andrews, Advis, and White (1980) have suggested that the time between birth and puberty in the rat can be viewed as comprised of four phases: a neonatal period (birth to 7 days of age), an infantile period (7–21 days of age), a juvenile period (21–32 days of age), and a peripubertal phase, the days just before and after the first ovulation (34–40 days of age). These phases roughly correspond to changing roles and functional status of the developing ovary in the female rats and thus provide a useful framework for the present discussion.

In the rat, the critical period for sexual differentiation of behavioral potentiality is generally thought to be late prenatal and early postnatal life (Ward, 1974). Evidence that the mammalian ovary is an active endocrine organ during the perinatal period has been difficult to obtain. The paucity of positive evidence that the ovaries function during early development has been interpreted as support for the traditional organizational hypothesis, which asserts a unique endocrine role for the testis in sexual differentiation. However, the present state of knowledge does not justify rejection of the hypothesis of endocrine activity in the perinatal rodent ovary.

Evidence for endocrine activity in the fetal ovary has been provided by findings of sex differences in estradiol levels in mice. Female fetuses have higher blood concentrations of estradiol than their male litter mates (vom Saal, Grant, McMullen, & Laves, 1983). Further, males that develop *in utero* surrounded by females have higher estradiol concentrations than males that develop surrounded by male fetuses (vom Saal *et al.*, 1983). Although the source of the estradiol measured in these fetuses cannot be definitely ascribed to the fetal ovary, because of the sex difference in estradiol concentration and the effects in males of proximity to females, it is likely that some or most of this estradiol is of ovarian origin. Alternatively, there may be nonovarian mechanisms of steroid metabolism in female but not male mouse fetuses.

Another technique employed to evaluate the functional status of the developing ovary has been to measure compensatory ovarian hypertrophy. When hemiovariectomized adult female rats exhibit a compensatory increase in the size and weight of the remaining ovary, this response can be taken as suggestive evidence, but not proof, of a functional hypothalamic–pituitary–ovarian axis and, more specifically, of endocrine activity by the ovary. Gerall and Dunlap (1971) and Dunlap, Preis, and Gerall (1972) observed compensatory ovarian hypertrophy within 10 days for female rats hemiovariectomized at birth. Although other investigators failed to observe compensatory ovarian hypertrophy in similarly aged and similarly treated rats (Baker & Kragt, 1969; Ojeda & Ramirez, 1972), it appears that the detection of compensatory hypertrophy in neonatal rats simply requires delicate and consistent dissecting and weighing procedures (Dunlap & Gerall, 1973).

Assaying hormonal concentrations in neonatal rats has not yielded evidence of perinatal ovarian function. Plasma estrogen levels are elevated during the first 2 days of life, perhaps because of the residual consequences of fetal and/or maternal secretions, and then decline and remain low until about 10 days of age (Csernus, 1986; Döhler & Wuttke, 1975; Meijs-Roelofs, Uilienbroek, de Jong, & Welschen, 1973; Pang, Caggiula, Gay, Goodman, & Pang, 1979; Presl, Herzman, & Horsky, 1969). Plasma progesterone levels also are low during the first 10 days of life (Döhler & Wuttke, 1975; Meijs-Roelofs, de Greef, & Uilienbroek, 1975; Weisz & Ward, 1980), and the progesterone present is likely of adrenal origin (Csernus, 1986).

The neonatal ovary contains follicles composed of no more than 60–100 cells. These follicles develop from a nonproliferating to a proliferating pool and then become atretic. Interstitial tissue, assumed to be steroidogenic, is recognizable in the rat ovary within the first week of life, and *in vitro* conversion of progesterone to estrogens is observed in ovaries as young as 5 days (Quattropani & Weisz, 1973). Early evidence indicated that during the first week of life the ovary is unresponsive to gonadotropins (Divila, Fresl, Krabec, & Horsky, 1971). However, in a later study, Funkenstein, Nimrod, and Lindner (1980) found that ovaries taken from 4-day-old rats, when cultured, produced small and decreasing amounts of estrogen and progesterone. The addition of LH to the culture medium enhanced progesterone synthesis only after 4 days of incubation. Neither FSH nor testosterone affected progesterone release, but they enhanced the effects of LH when all three hormones were added to the culture. The appearance of endocrine responsivity to LH stimulation occurred at 4–8 days of culture, regardless of presence or absence of previous hormonal exposures. Funkenstein *et al.* (1980) concluded that the ability of the neonatal ovary to respond to LH is evidenced by the eighth day of life and apparently involves an autonomous developmental process, i.e., does not depend on hormonal stimulation.

Extensive evidence documents that the endocrine function of the rat ovary begins after 10 days of age. During this infantile period the ovary responds to both endogenous and exogenous gonadotropins. Between 10 and 25 days of age high estrogen levels are observed in female rats (Döhler & Wuttke, 1975; Ojeda & Ramirez, 1972). However, much of this estrogen may be of adrenal origin (Weisz & Gunsalus, 1973), and physiological consequences of these hormones may be attenuated by the concurrent presence of high circulating levels of α-fetoprotein, which binds estrogen (Andrews & Ojeda, 1977). The negative feedback on pituitary gonadotropin release exerted by the ovary at this age appears to be a consequence of ovarian secretion of testosterone (Ojeda *et al.*, 1980).

Beginning after approximately 12 days of age, a series of events occurs that initiates the pubertal ovulatory cycle (Ojeda, Aguado, & Smith, 1983). By the end of the infantile period the hypothalamic–pituitary axis is able to respond to increases in circulating estrogens with enhanced gonadotropin release (Ojeda *et al.*, 1980). Since the reestablishment of a negative-feedback gonadotropin control mechanism apparently occurs only after the pubertal gonadotropin surge (Andrews, Advis, & Ojeda, 1981), the most significant factor for determining the onset of puberty is the development of the ovary to the point where it can release estrogens in a magnitude sufficient to induce a preovulatory surge of gonadotropins (Ojeda *et al.*, 1980). Follicular development in the rat requires about 19 days. Thus, those follicles destined to ovulate at puberty must have begun their development at about 14 days of age.

The ovarian response to gonadotropins increases during the juvenile period (Advis, Andrews, & Ojeda, 1979). This increased responsivity reflects, among other factors, an increase in LH receptors (Ojeda *et al.*, 1980). During this period the ovary plays an important role in the development of patterns of LH minisurges. When neonatal or juvenile rats are ovariectomized, these surges are not detectable but can be induced by estrogen infusion (Urbanski & Ojeda, 1986).

Possible permanent effects of the ovary on the development of endocrine function during prepuberal life have been the subject of a number of studies. In the pioneering work of Pfeiffer (1936) neonatal ovariectomy did not alter the capacity of adult female rats to exhibit estrous cycles when ovaries were implanted at adulthood. A more recent study have confirmed this finding (Gogan, Beattie, Hery, LaPlante, & Kordon, 1980). Similarly, the presence of the ovary was not necessary for the development of the feminine pattern of thyrotropin-releasing hormone induction of thyrotropin release from the pituitary in female rats (Watanabe & Takebe, 1987a). However, ovarian actions during neonatal life apparently influence the development of the feminine pattern of dopaminergic regulation of prolactin secretion in rats. Watanabe and Takebe (1987b) report permanent effects of the neonatal ovary on this function. In contrast to intact female rats, those ovariectomized within 24 hours of birth did not exhibit significant release of prolactin in response to injection of sulpiride (a putative D_2-receptor blocker). However, significant sulpiride release of prolactin was exhibited by female rats ovariectomized at 1 week of age or later. Also, the anterior pituitary content of prolactin increased with increasing age of ovariectomy. Female rats ovariectomized within 24 hours of birth and injected daily with estradiol and/or progesterone between 3 and 9 days of age exhibited, at adulthood, prolactin secretion to sulpiride injection. Thus, the feminization of the pattern of prolactin release inducible by dopamine receptor blockers appears to be dependent on ovarian secretion during the first 10 days of life. In the absence of ovarian secretions, injected ovarian hormones produce comparable effects.

The Ovary and Changes in the Mature Endocrine System

The ovary is obviously a necessary actor in the transition from the anovulatory prepubertal period to the fertile mature organism. In response to both endocrine and neural influences, ovulation occurs, and mature sexual physiology is established (Ojeda *et al.*, 1983). A multitude of somatic and endocrine changes occur in response to the pubertal elevation in ovarian steroids. Also, the roles of ovarian secretions in reproductive cycles of estrus, pregnancy, and lactation are well known. The ovary may also have a role in effecting relatively permanent changes in neuroendocrine function throughout an organism's adult life.

Although there are age-related changes in the rat ovary (Aschheim, 1964/65; Mandl & Shelton, 1959; Wolfe, 1943), the eventual failure of reproductive function in rats appears not to result from failure in ovarian function. Young ovaries implanted into ovariectomized aged rats that were either persistently estrus or pseudopregnant prior to ovariectomy fail to reinstate estrous cycles, whereas young ovariectomized hosts grafted with ovaries from old rats exhibit regular estrous cycles (Aschheim 1964/65; Peng & Huang, 1972). Current evidence points to hypothalamic–pituitary dysfunction as the primary cause of reproductive failure in the aged female rat (Finch, 1979). However, ovarian secretions may be partially responsible for age-related changes in hypothalamic and/or pituitary function.

In rats (Blake, Elias, & Huffman, 1983; Gosden & Bancroft, 1976; Gray, Tennent, Smith, & Davidson, 1980; Lu *et al.*, 1977) and mice (Nelson, Felicio, & Finch, 1980) early ovariectomy prolongs the capacity of injected estrogen and progesterone to release an LH surge. However, old rats, unlike younger females, will not exhibit an elevation in serum concentrations of LH in response to injection of estrogen alone. These findings are consistent with the hypothesis proposed by Aschheim (1964/65) that aging of the neuroendocrine mechanisms controlling the rat estrous cycle is suspended in the absence of ovarian steroids. The sparing effect may, at least in part, result from the continued ability of the hypothalamus to secrete luteinizing hormone-releasing hormone. Injection of this peptide elevates serum LH regardless of ovarian status, although some deficit in this regard appears in 2-year-old rats (Blake *et al.*, 1983). Early ovariectomy of rats also spares them from the effects of aging structural characteristics of the hypothalamic arcuate nucleus (Brawer, Schipper, & Naftolin, 1980; Schipper, Brawer, Nelson, Felicio, & Finch, 1981), an area critically important in the modulation of pituitary function.

Studies of the effects of exogenous ovarian hormones on reproductive aging also support the hypothesis of ovarian influences on the hypothalamic–pituitary axis. However, these findings suggest that it is not simply the presence or absence of the ovary but its secretory pattern that is critical. Treatment of adult female rats with a long-acting estrogen, estradiol valerate, accelerates the appearance of persistent vaginal estrus and induces degenerative changes in the arcuate nucleus comparable to those observed in older rats (Brawer, Naftolin, Martin, & Sonnenschein, 1978). Similar effects were more recently demonstrated in mice (Mobbs *et al.*, 1984). However, treatment of mature female rats with progesterone has effects opposite those of treatment with estradiol. Repeated treatment with progesterone delays the onset of persistent vaginal estrus and the loss of reproductive capacity in female rats (Lu, LaPolt, Nass, Matt, & Judd, 1985; DePaolo & Rowlands, 1986; LaPolt, Matt, Judd, & Lu, 1986).

The complex interactions of these hormones in affecting reproductive life span were recently reviewed by vom Saal and Finch (1988). It appears that timing, dose, and species must be considered when predicting effects of exposure to ovarian hormones. Physiological effects on reproductive life span of patterns of ovarian hormone secretion are suggested by demonstrations of the effects varying numbers of and timing of pregnancies (LaPolt *et al.*, 1986; Nelson, 1981; Gerall, Dunlap, & Sonntag, 1980). The fact that ovarian hormones acting postpubertally alter neuroendocrine function over the life span has led to the conclusion that there are permanent effects of ovarian steroids on the substrates underlying age-related changes in neuroendocrine function (vom Saal & Finch, 1988). However, it has also been argued that the absence of positive feedback of the hypothalamic–pituitary–ovarian axis of older female rats may be caused by an acute effect of the high concentrations of circulating estrogens in the persistently estrous female. Lu, Gilman, Meldrum, Judd, and Sawyer (1981) reported that whereas injection of estrogen and progesterone fails to elicit an LH surge in formerly persistent estrus, recently ovariectomized, old rats; such treatment does produce an LH surge, albeit smaller than in young females, when repeated 5 weeks after ovariectomy. Considered in the light of the transplant studies cited above, this evidence of some positive feedback in the aged hypothalamic–pituitary axis cannot be taken as evidence that it is capable of maintaining ovarian cycles. However, reproductive competence is not completely compromised in the old, persistently estrous female rat. If these rats are permitted access to males, they mate and frequently exhibit evoked ovulations (Hendricks & Blake, 1981; Hendricks, Lehman, & Oswalt, 1979; Meites, Huang, & Simpkins, 1978).

A model that has proved instructive to our understanding of long-term effects of ovarian secretion on neuroendocrine function is the delayed anovulation syndrome (DAS). The DAS describes the effects observed in female rats and other female laboratory rodents given small doses of steroid hormones during neonatal life. Normal estrous cycles may be observed immediately after puberty but cease within the first few months (Mobbs, Kannegieter, & Finch, 1985; Gorski, 1968; Swanson & van der Werff ten Bosch, 1964).

The development of the DAS is apparently a function of age-related changes in hypothalamic–pituitary function, as the responsivity of the ovaries of DAS females to gonodotropin is normal (Mobbs *et al.*, 1985; Harlan & Gorski, 1977). It has been suggested that the DAS is a useful model for studying neuroendocrine factors in aging, including the role of the ovary in determining reproductive life span in rodents (Finch *et al.*, 1980; Gerall & Dunlap, 1978; Lehman, McArthur, & Hendricks, 1978). Arai (1971) and Kikuyama and Kawashima (1966) found that rats that were lightly neonatally androgenized and ovariectomized prior to puberty were able to maintain ovulatory cycles in grafted ovaries at ages at which similarly androgenized intact rats could not. The LH surge in response to injected estrogen and progesterone is present in neonatally androgenized rats ovariectomized prior to puberty at ages at which that response is absent in similarly androgenized rats exhibiting persistent estrus (Harlan & Gorski, 1978). Further, whereas control rats exhibit LH surges to estrogen alone or to combined estrogen–progesterone administration, rats androgenized lightly during neonatal life respond only to the estrogen–progesterone combination (Handa & Gorski, 1985).

The capacity of the ovary to influence reproductive life span in neonatally androgenized rats appears to be sufficiently complex to suggest that a simple conclusion that ovarian activity acts to limit the reproductive life span of the rat is not warranted. Hendricks, McArthur, and Pickett (1977) found that neither the presence of the ovary nor the administration of estrogen or androgen beginning at 25 days of age altered the time course of the DAS. Handa and Gorski (1985) found that estrogen injected between 26 and 30 days of age slowed the development of the DAS, whereas estrogen injections between 33 and 37 days of age hastened its development. Apparently the timing and dose of ovarian steroids interact in a complex manner with the hypothalamic–pituitary axis to determine the span of reproductive competence in lightly androgenized rats.

OVARIAN INFLUENCES ON BEHAVIORAL POTENTIAL

EFFECTS OF NEONATAL OVARIECTOMY. Most descriptions of the organizational effects of gonadal steroids on mammalian sexual differentiation have suggested that the ovary is not a significant participant in this process (Valenstein, 1968; Young et al., 1964). Several studies have suggested that the presence of the ovary during early life can facilitate the later expression of adult female sexual behavior in the rat (Gerall et al., 1973; Lisk, 1969). One consequence of neonatal ovariectomy apparently is to diminish the initial effectiveness of injected steroids in inducing female sexual behavior. An example of this finding is presented in Figure 1. The differences between neonatally and postpubertally ovariectomized rats are observed only for initial tests of female sexual behavior following injections of ovarian steroids. With subsequent hormone injection and behavioral testing, these differences disappear.

In male rats castration has permanent and irreversible behavioral consequences only if the gonadectomy occurs during the neonatal period. Castration later than 5 days of age is often without noticeable behavioral effects when gonadal hormones are injected at adulthood (Gerall, Hendricks, Johnson, & Bounds, 1967; Grady, Phoenix, & Young, 1965). The effects of prepubertal ovariectomy are less discrete. Presently available data do not permit specification of a significant or critical period, if there is one, when ovariectomy can relatively permanently alter the potential to display reproductive behavior. However, the observed behavioral effects of neonatal ovariectomy appear to be a consequence of an absence of ovarian influences during the prepubertal rather than the neonatal period (Gerall et al., 1973). As noted above, it is during these later infantile and juvenile periods that the developing ovary is more clearly demonstrated to be an active endocrine organ (Ojeda et al., 1980).

ANTIANDROGEN FUNCTION OF THE DEVELOPING OVARY. The ovary present in the rat during prepuberal life may counter the concurrent or previous masculinizing influences of exposure to endogenous or exogenous androgen. The effects attributable to ovarian influences in the perinatally androgenized rat are somewhat

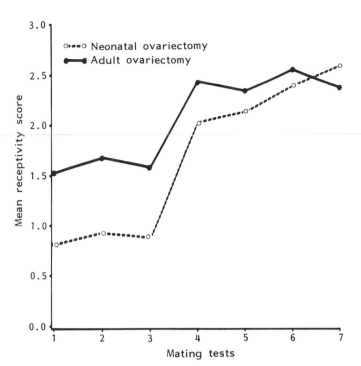

Figure 1. Mean sexual receptivity scores obtained in sequential weekly mating tests by female rats ovariectomized at 1, 5, or 10 days of age (neonatal ovariectomy) or at 35 or 60 days of age (adult ovariectomy). (From Gerall *et al.*, 1973. Copyright 1973 by the American Psychological Association.)

different from those previously described for normally developing females. In the nonandrogenized female the ovary appears to facilitate sexual receptivity after initial injections of ovarian hormones, but the presence or absence of the ovary during prepubertal life is without apparent effect on behavior observed after the first several weeks of hormone injections and behavioral tests. In addition, neonatally castrated males implanted with ovaries during early development have been found to be more behaviorally responsive to ovarian hormones than males neonatally castrated but not given ovarian implants, and this difference in responsiveness is not limited to the initial few behavioral tests following injection of ovarian hormones (Dunlap, Gerall, & McLean, 1973; Gerall *et al.*, 1973). It has also been demonstrated that the presence of the ovary during development has a significant and permanent facilitating influence on the development of female sexual behavior potential in female rats injected neonatally with testosterone (Hendricks & Duffy, 1974; Nikels, 1976) or exposed to testosterone *in utero* (Dunlap, Gerall, & Carlton, 1978). Similar antiandrogenic effects of ovarian influences in neonatally androgenized female rats have been reported for sexually dimorphic characteristics of open-field behavior (Blizard & Denef, 1973), steroid metabolism of the liver (Denef & DeMoor, 1972), and pituitary content of growth hormone (Jansson & Frohman, 1987).

The mechanism(s) by which the ovary might counteract the defeminizing influences of testosterone are not known. Since these effects on behavioral development are not dependent on the ovary being present at the time of androgen exposure, modification of androgen metabolism is not a likely possibility. Apparently, the ovary is either capable of augmenting the development of tissues important for female sexual behavior, thus masking defeminizing effects of androgen, or it directly reconstructs androgenized tissue to a more feminine character. As noted above, the steroidogenic activity of the juvenile and peripubertal ovary modifies functional characteristics of the developing neuroendocrine system (Urbanski & Ojeda, 1986). It is thus likely that prepubertal ovarian secretions act on developing neural tissues to sensitize them to subsequent exposure to ovarian hormones.

Cheng and Johnson (1973/74) found that at 10 days of age the serum concentration of estrogens in neonatally androgenized female rats was three times that observed in control females. A significant proportion of these estrogens appears to be secreted by the ovary, because ovariectomy at 5 days of age drastically reduced the serum concentrations of estrogens measured at 10 days of age. Estrogen levels declined in both groups between 10 and 15 days of age but were increased at 20 days of age, when they were 15 times higher in androgenized than in control females. After 40 days of age androgenized and control females exhibited comparable levels of serum estrogens when they were assessed on the morning of proestrus for the control females. Gonadotropin levels in neonatally androgenized females were approximately half of those of controls at 10 and 15 days of age but comparable after 20 days of age.

Cheng and Johnson (1973/74) suggest that these findings support the view that estrogens rather than androgens masculinize and defeminize the developing brain (see below for discussion of effects of exogenous estrogen). This hypothesis is supported by the recent report of vom Saal (1989) that male mice that develop *in utero* surrounded by females exhibit higher blood concentrations of estradiol and are masculinized in some respects (e.g., male sexual behavior, prostate weight, and volume of the sexual dimorphic nucleus of the preoptic area) although feminized or demasculinized in other respects (e.g., body weight and seminal vesicle weight) compared to males that develop *in utero* surrounded by other male fetuses. vom Saal (1989) hypothesizes that these differences reflect the degree to which circulating estradiol interacts with testosterone in various target tissues of the developing mouse. In view of the evidence presented above that neonatal ovariectomy increases the developing rat's sensitivity to the effects of neonatally administered androgens on development of female sexual behavior. It appears reasonable to suggest that ovarian hormones attenuate the effects of neonatally administered androgen on tissues important for the expression of this behavior.

Evidence of the capacity of prepuberal ovarian secretions to alter the effects of androgens present during perinatal development is also provided by studies of the long-term influence on behavioral potential of ovaries implanted into neonatally castrated male rats. Unlike female rats, which exhibit fairly constant levels of female sexual behavior in response to ovarian hormones over a several-month period, neonatally castrated males exhibit a rapid postpubertal decline in that potential (Dunlap, Gerall, & Hendricks, 1972; Dunlap *et al.*, 1978). Also, in neonatally castrated males the peak level of receptivity attained declines with increasing age of initial steroid injection and behavioral testing (Dunlap *et al.*, 1972). However, neonatally castrated males implanted with ovaries prepubertally exhibit the female pattern of a rather stable behavioral response to ovarian steroid throughout adult life (Dunlap *et al.*, 1973; Gerall *et al.*, 1973). These data suggest that the ovary present prepubertally may alter the age-related decline in feminine sexual behavior potential. The rapid decline in behavioral responsivity to ovarian hormones is not observed in neonatally androgenized female rats (Hendricks *et al.*, 1977), and neonatal ovariectomy does not alter age-related changes in the behavior of neonatally androgenized female rats (Dunlap *et al.*, 1978). Thus, the presence of the ovary can modify the course of age-related declines in feminine sexual behavior potential in the male but not in the normal or neonatally androgenized female rat. Perhaps events occurring prenatally account for this sex difference.

OVARIAN EFFECTS AFTER PUBERTY.   In rats the continued presence of a functioning ovary is necessary for the regular display of estrus. With respect to the acute behavioral effects, both the estrogen and progesterone secreted by the ovary are necessary for the induction of sexual receptivity. Ovariectomy prior to the release of progesterone that follows the surge of LH on the afternoon of proestrus completely prevents estrous behavior from occurring (Powers, 1970). Estrous behavior is observed in rats in which the proestrus LH surge was blocked by barbiturate injection (Everett, 1967). However, serum concentrations of progesterone during the afternoon and evening of proestrus are only partially suppressed by barbiturates (Butcher, Collins, & Fugo, 1974).

In addition to their acute effects on behavior at each estrous period, ovarian hormones also prime the system to be more responsive to subsequent exposure to the same hormones. In the absence of relatively recent exposure to ovarian hormones, behavioral responsivity to even superphysiological amounts of estrogen or estrogen and progesterone declines over time (Beach & Orndoff, 1974; Clark, MacLusky, Parsons, & Naftolin, 1981; Czaja & Butera, 1985; Damassa & Davidson, 1973; Parsons, MacLusky, Krieger, McEwen, & Pfaff, 1979). Possibly accounting for this behavioral refractoriness is an overall reduction in sensitivity and/or number of steroid receptors in the brains of ovariectomized females not recently exposed to ovarian steroids (Clark *et al.*, 1981; Parsons *et al.*, 1979). As would be anticipated, exposure of ovariectomized female rats to estrogen or to estrogen plus progesterone not only induces estrous behavior acutely but facilitates the sexual receptivity induced by hormones administered on subsequent tests (Beach & Orndoff, 1974; Gerall & Dunlap, 1973; Parsons *et al.*, 1979; Whalen & Nakayama, 1965). The cumulative facilitating effects of ovarian steroids on behavioral responsivity to subsequent steroid administration are reversible, and it is difficult to identify any permanent effects of such hormones on female sexual behavior potential (see next section).

ARE OVARIAN EFFECTS ORGANIZATIONAL?   As noted previously, organizational effects are permanent, occur early in life, and are limited to critical periods of development. Activational effects are temporary, usually occur in adulthood, and are not restricted to any tightly defined period. Arnold and Breedlove (1985) suggest that two additional criteria (perhaps more appropriately described as assumptions) for distinguishing organizational from activational effects have arisen *de facto* in the practices and writings of researchers in the field. The first of these is that organizational effects involve structural changes in the brain, at least at the cellular level, whereas activational effects are viewed as mediated by more subtle neurochemical events, the occurrence of which leave no significant trace. Secondly, organizational effects are asymmetric relative to sex, whereas activational effects are symmetrical. This latter assumption refers to the frequently observed phenomenon that in a given species hormonal action appears to be required for the development of sex-typical behavior in one sex but not in the other, whereas hormonal action is required to activate reproductive behaviors in both sexes. It could be argued that except for the first, permanence, all of these distinctions are superfluous to a concept of organization. Reason and biology strongly argue that if an event or experience permanently alters the behavior or behavioral potential of an organism, then its tissues have in some way been reorganized and structurally modified. In the absence of a permanent behavioral effect, demonstration of the remaining distinctions is of questionable significance. Further, if an effect is permanent, no matter when it occurs, whether or not it is symmetrical, and whether or not we can identify underlying structural changes, we would likely ascribe its occurrence to some kind of organizational process.

The fact that the differences between rats that develop with and without ovaries disappear with repeated exposure to ovarian hormones argues against the notion that the ovary present during development has permanent or organizational effects on systems mediating sexual behavior. However, if one takes initial responsivity to ovarian hormones after a period of deprivation of such hormonal stimulation as the behavioral trait influenced by the ovary during development, the possibility of an ovarian organizing effect still exists. As noted above, removal of ovarian influences for a period of time will decrease the subsequent behavioral responsivity to ovarian steroids. The behavior will recover with repeated exposures to ovarian steroids and behavioral testing (Beach & Orndoff, 1974; Clark et al., 1981; Czaja & Butera, 1985; Damassa & Davidson, 1973; Gerall & Dunlap, 1973). The differences between prepubertally and postpubertally ovariectomized female rats could simply reflect a longer period of hormonal deprivation in the former compared to the latter. Such a conclusion is supported by the report of Dunlap et al. (1978) that the initial receptivity of neonatally ovariectomized female rats evaluated at 40, 80, or 120 days of age was diminished with increasing age of first exposure to ovarian hormones and first test for female sexual behavior. A similar pattern was noted for females ovariectomized at 35 days of age except that the behavioral deficits at initial testing were not as great, especially in rats tested at 40 days of age.

The existing literature provides no definitive answer as to whether the presence of the ovary during any period of life can modify the effects of ovarian hormonal deprivation on subsequent behavioral responsivity to these same hormones.

Can the ovary or exogenous steroids permanently sensitize the organism to future stimulation by estrogen and progesterone? A recently completed study in my laboratory attempts to provide an answer to this question. In this study female rats were ovariectomized at 2, 25, 60, 180, or 360 days of age and tested for female sexual behavior in response to injected estrogen and progesterone beginning at approximately 60, 180, or 360 days of age. Representative results of this study are presented in Figure 2. This figure presents receptivity scores of females ovariectomized at 2, 180, or 360 days of age and tested for sexual behavior for the first time at 360 days of age. Although the group ovariectomized at 360 days exhibited significantly higher receptivity scores on the initial behavioral test, there were no significant between-group differences by the third behavioral test. Similar results were observed with all other age-of-ovariectomy and age-of-testing group combinations. Essentially, behavioral refractoriness to exogenous ovarian hormones was observed to vary as a function of prolonged deprivation of ovarian hormones with no significant effect attributable to the stage of life during which such deprivation occurred. These data, along with those of Dunlap et al. (1978), suggest that the behavioral effects of prepuberal ovariectomy, like those of ovariectomy anytime in life, are reversible. Thus, a critical-period phenomenon for ovarian influences on behavioral feminization has yet to be demonstrated.

It has been shown that the presence of an ovary during early development in neonatally castrated males and neonatally androgenized females has small but per-

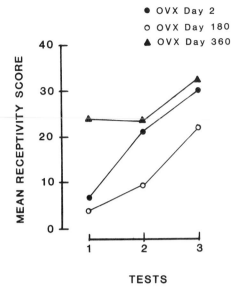

Figure 2. Mean female sexual receptivity scores obtained from three weekly behavior tests conducted when all animals were approximately 360 days of age. Female rats had been ovariectomized at 2, 180, or 360 days of age. These were the first behavior tests for all of these rats. Prior to each test rats were injected with 2 μg/kg estradiol benzoate and 500 μg progesterone.

manent effect on female behavioral potential. However, until some role for the ovary can be demonstrated in the behavioral development of normal females, the physiological significance of these effects remains obscure. One possibility is that the ovary offers some protection to developing female fetuses from the effects of hormones secreted by their male littermates *in utero*. Uterine position with respect to the sex of littermates has been shown to be a variable in the sexual development of rodents. Females that had males on either side of them *in utero* have been shown to be defeminized and masculinized compared to females that *in utero* had females to either side of them (Clemens, Gladue, & Coniglio, 1978; Gandelman, vom Saal, & Reinisch, 1977; vom Saal & Bronson, 1978, 1980). It has been suggested that the developing female fetus can be affected by testicular androgens diffused from nearby male littermates (Clemens, 1974). Meisel and Ward (1981) suggested that androgens in the venous flow from male fetuses diffuse to arterial flow and primarily affect females downstream from the males. It has also been found that female rats that develop *in utero* between two males and are given low dosages of androgen neonatally develop the delayed anovulation syndrome earlier than similarly treated females that developed *in utero* between two females (Tobet, Dunlap, & Gerall, 1982). If the ovaries acting perinatally or prepubertally alter the effects of androgen secreted by male intrauterine neighbors, it should be possible to demonstrate that some uterine position effects are amplified in neonatally ovariectomized rats.

MATERNAL AND PLACENTAL HORMONES AND SEXUAL DIFFERENTIATION

THE PROGRESSIVE HYPOTHESIS

In their "progressive hypothesis" of sexual differentiation Döhler and Hancke (1978) state that hormones of maternal and/or placental origin passed to the fetus might be critical for feminine sexual differentiation. Döhler *et al.* (1984) summarize evidence that fetal or neonatal ovariectomy does not eliminate all estrogenic influences in the developing female mammal. Adrenal, maternal, and placental sources of estrogenic influences remain. Further, it is well established that hormones of maternal origin pass easily to the fetus and can be found in the offspring up to 7 days later, postnatally (Friend, 1977). Therefore, the roles of estrogens or other ovarian hormones in female sexual differentiation cannot be evaluated conclusively based on the consequences of early gonadectomy. Ovarian hormones of maternal or placental origin are potential contributors to development in general and sexual differentiation in particular.

The progressive hypothesis, summarized diagrammatically in Figure 3, proposes that sexual differentiation of neural tissue and behavior is controlled by the amount of estrogen present during development. In the presence of moderate amounts of estrogen, the brain differentiates to yield neural tissue that, in adulthood, will respond in a female-typical manner to gonadal hormones. When "sufficiently higher" levels of estrogens are present, the brain differentiates such that neural tissues respond, in adulthood, in a male-typical fashion. The moderate levels of estrogen are provided to the developing fetus through the placenta. Döhler (1978) suggests that α-fetoprotein, rather than protecting the developing female rat from the effects of circulating hormones, actually attracts estrogens to brain sites undergoing sexual differentiation. This position is supported by the demonstration by Toran-Allerand (1980) of intracellular localization of α-fetoproteins in neural tissue. Evidence from both *in vivo* and *in vitro* research indicates that the major function of α-fetoprotein is to prevent the multiplication of estrogen-sensitive cells and that the α-fetoprotein's bonding of estrogen is incidental to this role (Sonnenschein & Soto, 1980). During early life α-fetoprotein may act to suppress the hyperplasia of many estrogen-sensitive cells. Such hyperplasia occurs at puberty under titers of estrogen comparable to those observed prepubertally. Thus, during perinatal development estrogenic influences could stimulate growth and otherwise influence the development of individual target cells, particularly neurons, without inducing extensive hyperplasia.

The progressive hypothesis proposes that the higher levels of estrogens needed for masculinization of developing neural tissues are derived via metabolism of tes-

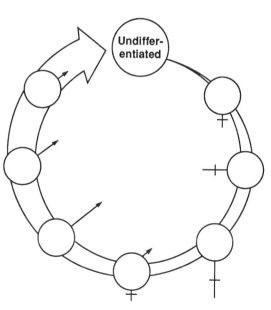

Figure 3. A diagrammatic representation of the effects of increasing degrees of exposure to estrogen on sexual differentiation as proposed by one version of the progressive hypothesis (after Döhler & Hanke, 1978). Moving clockwise around the circle indicates greater degrees of exposure to estrogen during sexual differentiation. The symbols indicate increasing feminization with small degrees of estrogen exposure that, with increasing estrogen, changes to defeminization and masculinization. Finally, more estrogen exposure leads to demasculinization, returning full circle.

tosterone. An extension of the progressive hypothesis is proposed to account for the effects of pharmacological doses of estrogen (Döhler & Hancke, 1978). Such doses are found to disrupt masculine sexual behavior and often result in an organism that is asexual with respect to neuroendocrine physiology and behavior (see discussion of antiandrogenic effects of injected estrogen below). In this extension of the progressive hypothesis, sexual differentiation is assumed to proceed in a circular manner, beginning with a sexually undifferentiated state and proceeding, as a function of increasing estrogen stimulation, through female, bisexual, male, and back to sexually undifferentiated (Döhler & Hancke, 1978).

The Effects of Antiestrogens

Much of the research attempting to evaluate the progressive hypothesis has involved the blockage of estrogenic effects through pharmacological means. It is reasoned that if ovarian hormones are not important for female sexual differentiation, then pharmacologically blocking the action of estrogen during early development should have little or no effect on female development. The progressive hypothesis, however, predicts that such disruption of estrogenic function should interfere with feminine sexual differentiation.

Administration of estrogen antagonists during the neonatal period produces acyclic, sterile adult female rats (Döhler, Von zur Muhlen, & Döhler, 1976; Gellert, Bakke, & Lawrence, 1971; McEwen, Lieberburg, Chaptal, & Krey, 1977). Döhler *et al.* (1984) found that unlike estrogen administration during the neonatal period, which both masculinizes and defeminizes sexual behavior potential in females, the antiestrogen tamoxifen given to female rats during neonatal life defeminized but did not masculinize behavior potential. Tamoxifen also inhibited the masculinization of the sexually dimorphic nucleus of the preoptic area. These results are interpreted as supporting the progressive hypothesis of sexual differentiation in that perinatal antagonism of estrogen leads to the development of neither male nor female sexual behavior.

Critics of this interpretation of the effects of neonatal administration of tamoxifen and other antiestrogen compounds point out that most antiestrogenic drugs have mild estrogenic as well as estrogen-antagonizing effects. It could be argued that these compounds defeminize by virtue of their estrogenic rather than antiestrogenic activity. However, concomitant administration of estradiol along with tomoxifen was found to attenuate the defeminizing effects of the antiestrogen (Döhler *et al.*, 1984). One would expect estrogenic effects of tamoxifen, if there are any, to add to those of estradiol. Instead, these data provide support for the progressive hypothesis and suggest a functional role for perinatal estrogens present during perinatal life in the process of female sexual differentiation.

Effects of Maternal Castration

Removal of the maternal ovary also offers another strategy for assessing the effects of maternal estrogens on sexual differentiation. Recently Witcher and Clemens (1987) evaluated male and female sexual behavior in rats delivered by females ovariectomized on day 10 of pregnancy. Pregnancies were maintained by exogenous progesterone. Additionally, some pregnant females were given injections of estrogen in addition to progesterone, and others were given an aromatase inhibitor in addition to progesterone. The latter treatment should have effectively blocked the biosynthesis of estrogens. Masculine behavior of offspring was unaffected by these treatments. However, feminine sexual behavior was enhanced in offspring from ovariectomized mothers that received only progesterone replacement compared to offspring from mothers receiving estrogen and progesterone or control animals. Offspring from mothers receiving the aromatase inhibitor exhibited the greatest facilitation of female sexual behavior.

These findings are strongly contradictory to the progressive hypothesis or any other suggestion that estrogens of maternal, or indeed any, origin have feminizing influences during sexual differentiation. To the contrary, when estrogenic influences were removed during fetal development, feminine sexual potential was enhanced. Witcher and Clemens (1987) note that the feminization observed could have been the result of the exogenous progesterone, which, as is discussed below, has been shown, under some circumstances, to facilitate feminine development. Nonetheless, no evidence for any feminizing influence of estrogens is observed. It would be difficult to develop a reasonable modification of the progressive hypothesis or any other suggestion that maternal estrogens have feminizing influence that would account for these data.

EXOGENOUS STEROIDS DURING DEVELOPMENT

ESTROGENS

THE "PARADOXICAL" EFFECTS OF EXOGENOUS ESTROGEN.   Early studies of the somatic effects of exogenous estrogen noted what was described as a paradoxical effect on the developing rat (Greene, Burrill, & Ivy, 1939a, 1940). These paradoxical effects included stimulation by injected estrogen of the development of male accessory glands in females. Normal development of feminine sexual tissues was also disrupted. Indeed, the female mouse injected neonatally with estrogen and progesterone has become a research model for the study of the iatrogenic syndrome observed in men and women exposed to steroids during early development (Warner, Yau, & Rosen, 1980). Subsequent research has clearly demonstrated that estrogen given to female rats during the neonatal period has potent defeminizing and masculinizing effects on sexual behavior development (Dorner, Docke, & Hinz, 1971; Edwards & Thompson, 1970; Hendricks, 1969; Hendricks & Gerall, 1970; Koster, 1943; Levine & Mullins, 1964; Whalen & Nadler, 1963; Wilson, 1943). Further, concomitant treatment with antiestrogens or aromatase inhibitors can limit the masculinizing effects of giving androgen to neonatal female rats (Clemens & Gladue, 1978; Fadem & Barfield, 1981; Gladue & Clemens, 1980; McDonald & Doughty, 1973/1974), and such dosages disrupt normal sexual differentiation in male rats (Booth, 1977; Södersten, 1978).

The initial confusion produced by the apparent paradoxical influences of estrogen on somatic and behavioral development was largely caused by the misconception that this class of steroids is female by nature. Subsequent research has demonstrated important roles for estrogens in male physiology. Estradiol, derived from aromatization of testosterone within neural tissue, has been shown to contribute significantly to the sexual differentiation of the brain and behavioral potential in the rat (Martini, 1978; McDonald et al., 1970). This view of the role of estradiol in rat sexual differentiation is referred to as the conversion or aromatization hypothesis. According to this theory, in females, the feminine potential of the developing neural tissue is protected from the effects of circulating estrogens by extracellular binding of estrogens to α-fetoproteins. Circulating estradiol is also found in males but masculinization of the brain is accomplished, at least in part, by the intracellular conversion of testosterone to estradiol, which then acts to masculinize developing tissues. Exogenously administered pharmacological amounts of estrogens presumably overwhelm the capacities of circulating α-fetoprotein to bind estrogen, resulting in masculinization and defeminization of the developing neural tissue.

FEMINIZING EFFECTS OF EXOGENOUS ESTROGEN. Attempts to identify doses of estrogen that would facilitate rather than disrupt the development of female sexual behavior in rats have been without much success. The "progressive hypothesis" of sexual differentiation assumes that there should be a dose range within which estrogen would feminize behavior potential because some level of estrogen is important for full development of feminine behavioral potential. Doses of estrogen or androgen that exceed this range result in masculine behavioral development (Döhler & Hancke, 1978). In support of this hypothesis, Döhler (1978) reported that administration of high levels of estrogen to female rats during late pregnancy feminized male offspring both behaviorally and physiologically. Males were also feminized if estrogen was injected during the neonatal period. However, Gerall *et al.* (1973) found that if neonatally castrated males were injected with estrogen at 3 or 5 days of age in doses ranging as low as 0.01 μg, either they had no behavioral effects from the injection or their female sexual behavior potential was suppressed. The different effects of neonatally injected estrogen in intact as opposed to neonatally castrated rats are addressed in the next section. Estrogen injections in neonatal female rats, as described above, consistently result in masculinization and defeminization of sexual behavior.

Other attempts to demonstrate a feminizing effect for exogenously administered estrogens on the development of female behavioral potential have involved administering the hormones later in the prepubertal period. Napoli and Gerall (1970) found that estrogen injections begun at 26 days of age had no effect on behavioral development in untreated or neonatally androgenized female rats. Hendricks and Weltin (1976) injected neonatally ovariectomized rats with estradiol for various periods during prepubertal development. In otherwise untreated or in neonatally androgenized females, estrogen treatment between 1–10 and 10–20 days of age suppressed female behavioral potential. Absence of adult estrous cycles has also been reported for female rats receiving estradiol implants during this period (Nass, Matt, Judd, & Lu, 1984). Consistent with the findings of Napoli and Gerall (1970), Hendricks and Weltin (1976) found that estradiol injections given to female rats 20–30 or 30–40 days of age were without effect on reproductive behavior development. A similar lack of effect for estrogen injections was observed in neonatally castrated males, except that injection of estrogen between 20 and 30 days of age did result in greater sexual receptivity in the initial tests conducted at adulthood. This finding of a facilitatory effect on the development of feminine behavior potential by injected estradiol in neonatally castrated male rats is similar to the previously described effects of ovarian influences on sexual behavior development in neonatally castrated males. Such parallel influences are consistent with an interpretation that these ovarian effects are a consequence of estrogenic activity on the part of the implanted ovaries in neonatally castrated males.

DEMASCULINIZING EFFECTS OF INJECTED ESTROGEN. As was first demonstrated for somatic development (Greene *et al.*, 1940), injected estrogens can interfere with the normal development of male behavior (Diamond, Llacuna, & Wong, 1973; Feder, 1967; Harris & Levine, 1965; Hendricks & Gerall, 1970; Mullins & Levine, 1968; Soulairac & Soulairac, 1974). One likely mechanism for these effects is the disruptive effect injected estrogen has on the developing testis (Maqueo & Kincl, 1964). However, androgen supplementation or replacement after castration in adulthood does not abolish the behavioral impairment. There are also suggestions that high dosages of injected estrogens have disruptive effects directly on the developing target tissues of testicular hormones (Harris & Levine, 1965; Simmons, 1971; Whalen, 1964; Zadina, Dunlap, & Gerall, 1979). It is likely that these target sites include developing neural tissue, since the effects of neonatally injected estrogen on peripheral tissues do not appear sufficient to account for the observed behavioral effects (Zadina et al., 1979).

In light of the aromatization hypothesis, which stipulates that estrogen is an important hormone in the development of male sexual behavior potential in rat, there should be a low dosage at which injected estrogen would facilitate rather than impede masculine development in males (Zadina et al., 1979). However, no such effect has been observed with any dose of estrogen injected neonatally in gonadally intact male rats. The same estrogen doses that, when injected neonatally, masculinize and defeminize female or neonatally castrated male rats demasculinize intact males. Any complete understanding of the role of estrogens in sexual differentiation will have to account for this apparent inconsistency. The effects of injecting intact neonatal males with estrogen may reflect pathogenic consequences of high doses rather than revealing a physiologically relevant role for hormones during normal sexual differentiation of the female.

PROGESTINS

ANTIANDROGENIC INFLUENCES OF PROGESTINS.   Although not as extensively studied as the effects of perinatally administered androgens and estrogens, exogenous progesterone given during early development has been demonstrated to affect sexual differentiation in rats. Citing evidence of antiandrogenic and antiestrogenic activity of progesterone, Dorfman (1967) proposed that one function of progesterone during pregnancy is to protect the fetus from high concentrations of androgens and/or estrogens. Similar suggestions for a protective role of progesterone have been made by Diamond (1967) and Diamond, Rust, and Westphal (1969). In adult animals progesterone can antagonize the effects of androgen on sexual behavior (Diamond, 1966; Erpino & Chappelle, 1971) and interacts with estrogen in a complex, often biphasic, manner to facilitate or inhibit female sexual behavior (Morin, 1977).

Evidence that progesterone can counter the effects of other steroids on behavioral development can be derived from a number of sources. Hull (1981) found that progesterone administered to lactating females resulted in the impairment of masculine sexual behavior in their suckled males. Progesterone injected neonatally to female rats has been reported to block the masculinizing effects of neonatally administered androgen (Cagnoni, Fantini, Morace, & Ghetti, 1965; Kincl & Maqueo, 1965) and estrogen (Kincl & Maqueo, 1965). Also, injection of progesterone during late gestation prevented the effects that neonatal androgenization has on the age of vaginal opening (Butterstein & Freis, 1978). Turkelson, Dunlap, MacPhee, and Gerall (1977) found that administration of an antiserum to progesterone together with a low dose of testosterone on the day of birth resulted in a more rapid development of sterility than the administration of testosterone alone. Nikels (1982) demon-

strated that pre- and/or postnatal progesterone administration enhanced the female sexual behavior potential shown by intact or neonatally castrated male rats.
These findings are summarized in Figure 4. Similar findings are reported by Griffo
(1974). Also, the synthetic progestin R5020, given neonatally to male rats, has been
shown to facilitate female sexual behavior in adulthood in a manner comparable to
the antiandrogen cyproterone acetate (McEwen *et al.*, 1979).

ANDROGENIC INFLUENCES OF PROGESTINS.   When given in high doses during
neonatal life, progesterone can have defeminizing effects. Diamond and Wong
(1969) report that 5 mg of progesterone injected to female rats at 3 days of age
suppressed female sexual behavior observed at adulthood. Strecke, Oettel, Tiroke,
and Ohme (1978) found that 2 mg/day of progesterone administered from birth to
14 days of age resulted in an inhibition of vaginal opening and infertility. When
given in such large doses progesterone has obvious androgenic effects, as indicated
by increased prostate and seminal vesicle weights in castrated male rats and clitoral
hypertrophy in female rats (Greene, Burrill, & Ivy, 1939b). The results of studies
utilizing pharmacological dosages provide information of practical significance to
the understanding of the risks of administering high doses of progestins during
pregnancy, but they do not permit us to conclude whether potential androgenic

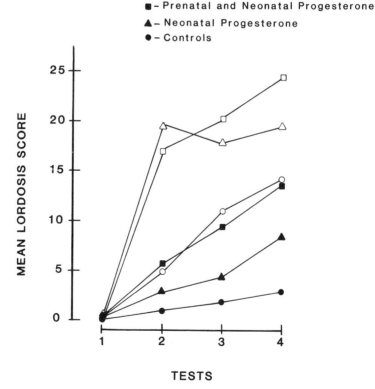

■ – Prenatal and Neonatal Progesterone
▲ – Neonatal Progesterone
● – Controls

Figure 4.   The effects of perinatal progesterone on lordosis scores of male rats. Open symbols represent
mean lordosis scores for rats castrated on the day of birth; closed symbols similarly represent rats
castrated at 30 days of age. Rats in the prenatal and neonatal progesterone conditions were from litters
born to females that received 2 mg progesterone on days 15, 16, and 17 of gestation and the pups
injected with 500 mg progesterone at 1 and 3 days of age. Rats in the neonatal progesterone condition
received progesterone at 1 and 3 days of age but were born to untreated females. (From Nikels, 1982.)

actions of progestins play a physiological role in sexual differentiation under normal circumstances.

147

FEMALE SEXUAL
BEHAVIOR

MECHANISMS OF PROGESTIN INFLUENCES.   Progesterone might influence the course of sexual differentiation by more than one mechanism. First, it might alter the production and/or availability of testosterone. Pointis, Latreille, Richard, Athis, and Cedard (1984) found that administration of progesterone to pregnant mice resulted in lower than normal testosterone levels in male fetuses and abnormally high testosterone levels in female fetuses. The reduced androgen levels are likely to be accounted for by decreased testicular activity, since the testes from fetuses exposed to progesterone produced less testosterone *in vitro* than testes from control males. However, similar effects were not observed in rats. The elevated androgen levels found in females treated with progesterone may be a consequence of progesterone augmentation of testosterone by competing with androgens for available metabolizing enzymes that break down both steroids (Pointis *et al.*, 1984). Another possibility is that exposure to high levels of progesterone neonatally alters the subsequent responsivity of target tissues to other gonadal hormones. For example, progesterone given neonatally to mice inhibits the later development of the mammary gland and blocks the facilitating effects of neonatally injected estrogen on mammary gland development (Warner *et al.*, 1980). The effects of progesterone treatment were detectable by the hormonal sensitivity of mammary explants in culture (Warner *et al.*, 1980). Thus, the alteration in mammary gland development reflects changes in the mammary gland itself and may not be dependent on progesterone effects on hormone secretion. However, this alteration of target tissue sensitivity may not apply to tissues mediating sexual behavior.

The occupation of estrogen receptors in the brain does not appear to be affected by progesterone (Bullock, Bardin, & Sherman, 1978). Data obtained for *Tfm*/Y mice with defective androgen receptors suggest that the androgenic and antiandrogenic effects of progesterone are mediated through androgen receptors. Progestins, whether they mimic, enhance, inhibit, or have no detectable effect on androgen action, all compete with triated testosterone for cytoplasmic binding sites. These findings are accounted for in a proposed "steric–allosteric" model of the androgen receptor. In this model the androgen receptor is hypothesized to exhibit different bonding characteristics in its active as compared to its inactive state (Bullock *et al.*, 1978). Such a model could be useful in accounting for the effects of progestins on sexual differentiation in that it would predict varying consequences of progesterone exposure, depending on the then-existing condition of androgen receptors.

The feminizing effects of neonatally administered progesterone must be considered in light of the fact that they have only been demonstrated in animals in which some androgenization has occurred. Thus, the feminizing effects of progesterone are similar to those observed for ovarian influences on the development of sexual behavior potential. It is possible that physiological amounts of progesterone present during fetal and/or neonatal life could serve to protect the developing female against the masculinizing and defeminizing effects of circulating androgens and estrogens. In rats, apparently no sex differences have been found in serum progesterone concentrations during perinatal life (Döhler & Wuttke, 1974, 1975; Weisz & Ward, 1980). However, based on studies in primates, it has been suggested that a critical factor in the hormonal milieu affecting sexual differentiation is the testosterone-to-progesterone ratio (Resko, 1974): high testosterone-to-progesterone ratios lead to the development of male-typical neural tissues, whereas low testosterone-to-progesterone ratios lead to the development of female-typical neural tissue. At times in fetal and neonatal life male rats do have, on the average, higher serum concentrations of testosterone than females (Weisz & Ward, 1980). However, these differences are small, and the values from the two sexes show extensive overlap. These overlapping values with small mean differences in testosterone concentration are troublesome to the organizational hypothesis, which attributes the discrete and impressive sex differences in behavior to differences in circulating hormones.

The physiological amounts of progesterone present may have a limited antiandrogenic influence, essentially buffering the testosterone present and amplifying the differential effects on sexual differentiation produced by the relatively small differences in testosterone titers. Or perhaps, as Resko (1974) suggests, variations in the ratios of progesterone to testosterone would yield a more distinct difference between sexes than testosterone alone. Techniques presently available for assaying steroids in fetal and neonatal rodent blood require that samples taken from individuals be pooled with others in order to obtain adequate volumes, thus precluding estimates of different hormone concentrations in the same individual. It would be expected that an elevation of progesterone concentrations could, to a point, inhibit androgenic action and promote behavioral feminization and demasculinization. However, pharmacological levels of progesterone would be expected to exert the opposite effect by virtue of progesterone's potential androgenic actions.

Conclusions

This review has summarized existing evidence that estrogens and progestins can feminize sexual behavior potential when present at a variety of times and under a number of circumstances. In the case of the role of ovarian hormones in the development of female sexual behavior potential, the step between demonstrating an experimental effect and concluding a physiological function has proved to be particularly difficult. Data provided by Döhler and his colleagues supporting the progressive hypothesis are impressive. However, questions about this hypothesis are raised by the possible estrogenic action of the antiestrogens used to generate these data and the evidence provided by Witcher and Clemens (1987) that estrogen deprivation in utero facilitates feminine development. Resko's hypothesis (Resko, 1974) that ratios of steroid concentrations, particularly the testosterone-to-progesterone ratio, are a critical consideration to the understanding of sexual differentiation also receives implicit support from the data reviewed.

Theories of a "protective" function of progesterone in female sexual differentiation such as that proposed by Dorfman (1967) also receive support from the available data. The role of the prepubertal ovary in the development of the potential of the female rat to exhibit its normal pubertal estrus is well established. However, these effects of the ovary on the sexual behavior potential of normal female rats appear to be reversible and not dependent on exposure to hormones during any as yet identified critical period. It is clear that any organizational actions of the ovary on the development of sexual behavior are likely to be much subtler than those so far established for the testis.

With respect to the value of the distinction between organizational and activational effects of hormones, the data reviewed in this chapter provide no compelling impetus for the abandonment of the distinction. The ovary does appear to play a ignificant role in setting limits on the female rat's reproductive life span. Most or all of these effects appear to occur after puberty and may involve relatively permanent changes in the function of the hypothalamic–pituitary–ovarian axis. However, there is no compelling evidence to support the concept of organizational actions of ovarian hormones during the postpubertal period on tissues mediating female sexual behavior in rodents. In this regard the hypothesis of Phoenix *et al.* (1959) still accounts for the available data. Sexual behavior potential is largely determined by events occurring during early development and is little, if at all, modified by events in adulthood that can even be approximately described as physiological. The fact that we have not identified a hormonal–cellular interaction that distinguishes organization from activational actions of hormones does not justify rejecting the hypothesis of a biological mechanism to account for the observed fundamental difference between the role of hormones on the development of sexual behavior potential and the role of hormones in proximate elicitation of sexual behavior.

REFERENCES

Advis, J. P., Andrews, W. W., & Ojeda, S. R. (1979). Changes in ovarian steroidal and prostaglandin E responsiveness to gonadotropins during the onset of puberty in the female rat. *Endocrinology, 104,* 653–658.

Andrews, W. W., & Ojeda, S. R. (1977). On the feedback actions of estrogen on gonadotropin and prolactin release in infantile female rats. *Endocrinology, 101,* 1517–1523.

Andrews, W. W., Advis, J. P., & Ojeda, S. R. (1981). The maturation of estradiol-negative feedback in female rats: Evidence that the resetting of the hypothalamic "gonadostat" does not preceed the first preovulatory surge of gonadotropins. *Endocrinology, 109,* 2022–2039.

Arai, Y. (1971). A possible process of the secondary sterilization: Delayed anovulation syndrome. *Experientia, 15,* 463–464.

Arnold, A. P., & Breedlove, S. M. (1985). Organizational and activational effects of sex steroids on brain and behavior: A reanalysis. *Hormones and Behavior, 19,* 469–498.

Aschheim, P. (1964/65). Résultats fournis par la greffe hétérochrone des ovaries dans l'étude de la régulation hypothalamo–hypophyso–ovarieene de la ratte sénile. *Gerontologia, 10,* 65–75.

Baker, F. D., & Kragt, C. L. (1969). Maturation of the hypothalamic–pituitary–gonadal negative feedback system. *Endocrinology, 85,* 522–528.

Beach, F. A., & Orndoff, R. K. (1974). Variation in the responsiveness of female rats to ovarian hormones as a function of preceeding hormonal deprivation. *Hormones and Behavior, 5,* 201–295.

Blake, C. A., Elias, K. A., & Huffman, L. J. (1983). Ovariectomy of young adult rats has a sparing effect on the ability of aged rats to release luteinizing hormone. *Biology of Reproduction, 28,* 575–585.

Blizard, D., & Denef, C. (1973). Neonatal androgen effects on open-field activity and sexual behavior in the female rat: The modifying influence of ovarian secretions during development. *Physiology and Behavior, 11,* 65–69.

Booth, J. E. (1977). Sexual behaviour of male rats injected with the anti oestrogen MER-25 during infancy. *Physiology and Behavior, 19,* 35–39.

Brawer, J. R., Naftolin, F., Martin, J., & Sonnenschein, C. (1978). Effects of a single injection of estradiol valerate on the hypothalamic arcuate nucleus and on reproductive function in the female rat. *Endocrinology, 103,* 501–512.

Brawer, J. R., Schipper, H., & Naftolin, F. (1980). Ovary-dependent degeneration in the hypothalamic arcuate nucleus. *Endocrinology, 107,* 274–279.

Bullock, L. P., Bardin, C. W., & Sherman, M. R. (1978). Androgenic, anti-androgenic, and synandrogenic actions of progestins: Role of steric and allosteric interactions with androgen receptors. *Endocrinology, 103,* 1768–1782.

Butcher, R. L., Collins, W. E., & Fugo, N. W. (1974). Altered secretion of gonadotropins and steroids resulting from delayed ovulation in the rat. *Endocrinology, 96,* 576–586.

Butterstein, G. M., & Freis, E. S. (1978). Effect of parturition time on the response to neonatal androgen in female rats. *Proceedings of the Society for Experimental Biology and Medicine, 158,* 179–182.

Cagnoni, M., Fantini, F., Morace, G., & Ghetti, A. (1965). Failure of testosterone propionate to induce "early androgen" syndrome in rats previously injected with progesterone. *Journal of Endocrinology, 33*, 527–528.

Cheng, H. C., & Johnson, D. C. (1973/74). Serum estrogens and gonadotropins in developing androgenized and normal female rats. *Neuroendocrinology, 13*, 357–365.

Clark, C. R., MacLusky, N. J., Parsons, B., & Naftolin, F. (1981). Effects of estrogen deprivation on brain estrogen and progesterone receptor levels and the activation of female sexual behavior. *Hormones and Behavior, 15*, 289–298.

Clemens, L. G. (1974). Neurohormonal control of male sexual behavior. In W. Montagna & W. A. Sadler (Eds.), *Advances in behavioral biology, Vol. 11: Reproductive behavior* (pp. 23–52). New York: Plenum Press.

Clemens, L. G., & Gladue, B. A. (1978). Feminine sexual behavior in rats enhanced by prenatal inhibition of androgen aromatization. *Hormones and Behavior, 11*, 190–201.

Clemens, L. G., Gladue, B. A., & Coniglio, L. P. (1978). Prenatal endogenous androgenic influences on masculine sexual behavior and genital morphology in male and female rats. *Hormones and Behavior, 10*, 40–53.

Csernus, V. (1986). Production of sexual steroids in rats during pre- and early postnatal life. *Experimental and Clinical Endocrinology, 88*, 1–5.

Czaja, J. A., & Butera, P. C. (1985). Behavioral consequences of hormonal deprivation on the responsiveness of female rats to estradiol. *Physiology and Behavior, 35*, 873–877.

Damassa, D., & Davidson, J. M. (1973). Effects of ovariectomy and constant light on responsiveness to estrogen in the rat. *Hormones and Behavior, 4*, 269–279.

Denef, C., & DeMoor, P. (1972). Sexual differentiation of steroid metabolizing enzymes in the rat liver. Further studies on predetermination of testosterone at birth. *Endocrinology, 91*, 374–384.

DePaolo, L. V., & Rowlands, K. L. (1986). Deceleration of age-associated changes in the preovulatory but not seconday follicle-stimulating hormone surge by progesterone. *Biology of Reproduction, 35*, 320–326.

Diamond, M. (1966). Progesterone inhibition of normal sexual behavior in the male guinea pig. *Nature, 209*, 1322–1324.

Diamond, M. (1967). Androgen-induced masculinization in the ovariectomized and hysterectomized guinea pig. *Anatomical Record, 157*, 47–52.

Diamond, M., & Wong, C. L. (1969). Neonatal progesterone: Effect on reproductive functions in the female rat. *Anatomical Record, 163*, 178.

Diamond, M., Rust, N., & Westphal, U. (1969). High-affinity binding of progesterone, testosterone, and cortisol in normal and androgen-treated guinea pigs during various reproductive stages: Relationship to masculinization. *Endocrinology, 84*, 1143–1151.

Diamond, M., Llacuna, A., & Wong, C. L. (1973). Sex behavior after neonatal progesterone, testosterone, estrogen, or antiandrogens. *Hormones and Behavior, 4*, 73–88.

Divila, F., Fresl, J., Krabec, Z., & Horsky, J. (1971). Developmental changes in the ovulatory and secretory response to a single dose of HCG in the rat. *Physiologia Bohemoslovia, 20*, 364–365.

Döhler, K.-D. (1978). Is female sexual differentiation hormone-mediated? *Trends in Neuroscience, 1*, 138–140.

Döhler, K.-D., & Hancke, J. L. (1978). Thoughts on the mechanism of sexual brain differentiation. In G. Dörner and M. Kawakami (Eds.), *Hormones and brain development, Developments in endocrinology, Vol. 3* (pp. 153–158). Amsterdam: Elsevier/North-Holland Biomedical Press.

Döhler, K.-D., & Wuttke, W. (1974). Serum LH, FSH, prolactin and progesterone from birth to puberty in female and male rats. *Endocrinology, 94*, 1003–1008.

Döhler, K.-D., & Wuttke, W. (1975). Changes with age in levels of serum gonadotropins, prolactin and gonadal steroids in prepubertal male and female rats. *Endocrinology, 97*, 898–907.

Döhler, K.-D., Von zur Muhlen, A., & Döhler, U. (1976). Estrogen–gonadotropins interaction in postnatal female rats, and the induction of anovulatory sterility by treatment with an estrogen antagonist. *Annales de Biologie Animale Biochimie, Biophysique, 16*, 363–372.

Döhler, K.-D., Hancke, J. L., Srivastava, S. S., Hofmann, C., Shryne, J. E., & Gorski, R. A. (1984). Participation of estrogens in female sexual differentiation of the brain: Neuroanatomical neuroendocrine and behavioral evidence. In G. E. DeVries, J. P. C. DeBruin, H. B. M. Uylings, and M. A. Corner (Eds.), *Progress in Brain Research, Vol. 61* (pp. 99–117). Amsterdam: Elsevier.

Dorfman, R. I. (1967). The anti-estrogenic and anti-androgenic activity of progesterone in defense of the normal fetus. *Anatomical Record, 157*, 547–558.

Dörner, G., Döcke, F., & Hinz, G. (1971). Paradoxical effects of estrogen on brain differentiation. *Neuroendocrinology, 7*, 146–155.

Dunlap, J. L., & Gerall, A. A. (1973). Compensatory ovarian hypertrophy can be obtained in neonatal rats. *Journal of Reproduction and Fertility, 32*, 517–519.

Dunlap. J. L., Gerall, A. A., & Hendricks, S. E. (1972). Female receptivity in neonatally castrated males as a function of age and experience. *Physiology and Behavior, 8,* 21–23.

Dunlap, J. L., Preis, L. K., & Gerall, A. A. (1972). Compensatory ovarian hypertrophy as a function of age and neonatal androgenization. *Endocrinology, 90,* 1309–1314.

Dunlap, J. L., Gerall, A. A., & McLean, L. D. (1973). Enhancement of female receptivity in neonatally castrated males by prepuberal ovarian transplants. *Physiology and Behavior, 10,* 701–705.

Dunlap, J. L., Gerall, A. A., & Carlton, S. F. (1978). Evaluation of prenatal androgen and ovarian secretions on receptivity in female and male rats. *Journal of Comparative and Physiological Psychology, 92,* 280–288.

Edwards, D. A., & Thompson, M. L. (1970). Neonatal androgenization and estrogenization and the hormonal induction of sexual receptivity in rats. *Physiology and Behavior, 5,* 1115–1119.

Erpino, M. J., & Chappelle, T. C. (1971). Interactions between androgens and progesterone in mediation of aggression in the mouse. *Hormones and Behavior, 2,* 265–272.

Everett, J. W. (1967). Provoked ovulation or long-delayed pseudopregnancy from coital stimuli in barbiturate-blocked rats. *Endocrinology, 78,* 145–154.

Fadem, B. H., & Barfield, R. J. (1981). Neonatal hormonal influences on the development of proceptive and receptive feminine sexual behavior in female rats. *Hormones and Behavior, 15,* 282–288.

Feder, H. H. (1967). Specificity of testosterone and estradiol in the differentiating neonatal rat. *Anatomical Record, 157,* 79–86.

Finch, C. E. (1979). Neuroendocrine mechanisms of aging. *Federation Proceedings, 38,* 178–183.

Finch, C. E., Felicio, L. S., Flurkey, K., Gee, D. M., Mobbs, C., Nelson, S. F., & Osterburg, H. H. (1980). Studies of ovarian–hypothalamic–pituitary interactions during reproductive aging in C57BL/6J mice. *Peptides, 1(Supplement 1),* 161–175.

Friend, J. P. (1977). Persistence of maternally derived ^{3}H-estradiol in fetal and neonatal rats. *Experientia, 33,* 1235–1236.

Funkenstein, B., Nimrod, A., & Lindner, H. R. (1980). The development of steroidogenic capability and responsiveness to gonadotropins in cultured neonatal rat ovaries. *Endocrinology, 106,* 98–106.

Gandelman, R., vom Saal, F. S., & Reinisch, J. M. (1977). Contiguity to male fetuses affects morphology and behavior of female mice. *Nature, 266,* 722–724.

Gellert, R. J., Bakke, J. L., & Lawrence, N. L. (1971). Persistent estrus and altered estrogen sensitivity in rats treated neonatally with clomiphene citrate. *Fertility and Sterility, 22,* 244–250.

Gerall, A. A., & Dunlap, J. L. (1971). Evidence that the ovaries of the neonatal rat secrete active substances. *Journal of Endocrinology, 50,* 529–530.

Gerall, A. A., & Dunlap, J. L. (1973). The effects of experience and hormones on the initial receptivity in female and male rats. *Physiology and Behavior, 10,* 851–854.

Gerall, A. A., & Dunlap, J. L. (1978). Fertility in aging normal and neonatally androgenized female rats. *Age, 1,* 35.

Gerall, A. A., Hendricks, S. E., Johnson, L. L., & Bounds, T. W. (1967). Effects of early castration in male rats on adult sexual behavior. *Journal of Comparative and Physiological Psychology, 62,* 370–375.

Gerall, A. A., Dunlap, J. L., & Hendricks, S. E. (1973). Effect of ovarian secretions on female behavior potentiality in the rat. *Journal of Comparative and Physiological Psychology, 82,* 449–465.

Gerall, A. A., Dunlap, J. L., & Sonntag, W. E. (1980). Reproduction in aging normal and neonatally androgenized female rats. *Journal of Comparative and Physiological Psychology, 94,* 556–563.

Gladue, B. A., & Clemens, L. G. (1980). Masculinization diminished by disruption of prenatal estrogen biosynthesis in male rats. *Physiology and Behavior, 25,* 589–593.

Gogan, F., Beattie, I. A., Hery, M., Laplante, E., & Kordon, C. (1980). Effect of neonatal administration of steroids or gonadectomy upon oestradiol-induced luteinizing hormone release in rats of both sexes. *Journal of Endocrinology, 85,* 69–74.

Gorski, R. A. (1968). Influence of age on the response to paranatal administration of a low dose of androgen. *Endocrinology, 82,* 1001–1004.

Gorski, R. A. (1973). Perinatal effects of sex steroids on brain development and function. In E. Zimmerman, W. H. Gispen, & B. H. Marks (Eds.), *Progress in brain research, Vol. 39: Drug effects on neuroendocrine regulation* (pp. 149–163). Amsterdam: Elsevier/North-Holland.

Gorski, R. A. (1985). The 13th J. A. F. Stevenson Memorial Lecture: Sexual differentiation of the brain: Possible mechanisms and implications. *Canadian Journal of Physiology and Pharmacology, 63,* 577–594.

Gosden, R. G., & Bancroft, L. (1976). Pituitary function in reproductivity senescent female rats. *Experimental Gerontology, 11,* 157–160.

Grady, K. L., Phoenix, C. H., & Young, W. C. (1965). Role of the developing testis in differentiation of the neural tissues mediating mating behavior. *Journal of Comparative and Physiological Psychology, 59,* 176–182.

Gray, G. G., Tennent, B., Smith, E. R., & Davidson, J. M. (1980). Luteinizing hormone regulation and sexual behavior in middle-aged female rats. *Endocrinology, 107,* 187–194.

Greene, R. R., Burrill, M. W., & Ivy, A. C. (1939a). Experimental intersexuality: The paradoxical effects of estrogens on sexual development of the female rat. *Anatomical Record, 74,* 429–438.

Greene, R. R., Burrill, M. W., & Ivy, A. C. (1939b). Progesterone is androgenic. *Endocrinology, 24,* 351–357.

Greene, R. R., Burrill, M. W., & Ivy, A. C. (1940). Experimental intersexuality: The effects of estrogens on the antenatal sexual development of the rat. *American Journal of Anatomy, 67,* 305–345.

Griffo, W. R. (1974). The disruption of the process of sex differentiation by progestational steroids (Doctoral dissertation, City University of New York, 1976). *Dissertation Abstracts International, 37(3-B),* 1476 (University Microfilms No. 76-21, 169).

Handa, R. J., & Gorski, R. A. (1985). Alterations in the onset of ovulatory failure and gonadotropin secretion following steroid administration to lightly androgenized female rats. *Biology of Reproduction, 32,* 248–256.

Harlan, R. E., & Gorski, R. A. (1977). Correlations between ovarian sensitivity, vaginal cyclicity and luteinizing hormone and prolactin secretion in lightly androgenized rats. *Endocrinology, 101,* 750–759.

Harlan, R. E., & Gorski, R. A. (1978). Effects of postpubertal ovarian steroids on reproductive function and sexual differentiation of lightly androgenized rats. *Endocrinology, 102,* 1716–1724.

Harris, G. W., & Levine, S. (1965). Sexual differentiation of the brain and its experimental control. *Journal of Physiology, 181,* 379–400.

Hendricks, S. E. (1969). Influence of neonatally administered hormones and early gonadectomy on rats' sexual behavior. *Journal of Comparative and Physiological Psychology, 69,* 408–413.

Hendricks, S. E., & Blake, C. A. (1981). Plasma prolactin and progesterone responses to mating are altered in aged rats. *Journal of Endocrinology, 90,* 179–191.

Hendricks, S. E., & Duffy, J. A. (1974). Ovarian influences on the development of sexual behavior in neonatally androgenized rats. *Developmental Psychobiology, 1,* 297–303.

Hendricks, S. E., & Gerall, A. A. (1970). Effect of neonatally administered estrogen on development of male and female rats. *Endocrinology, 87,* 435–439.

Hendricks, S. E., & Weltin, M. (1976). Effect of estrogen given during various periods of prepuberal life on the sexual behavior of rats. *Physiological Psychology, 4,* 105–110.

Hendricks, S. E., McArthur, D. A., & Pickett, S. (1977). The delayed anovulation syndrome: Influence of hormones and correlation with behavior. *Journal of Endocrinology, 75,* 15–22.

Hendricks, S. E., Lehman, J. R., & Oswalt, G. L. (1979). Effects of copulation on reproductive function in aged female rats. *Physiology and Behavior, 23,* 267–272.

Hull, E. M. (1981). Effects of neonatal exposure to progesterone on sexual behavior of male and female rats. *Physiology and Behavior, 26,* 401–405.

Jansson, J. O., & Frohman, L. A. (1987). Inhibitory effect of the ovaries on neonatal androgen imprinting of growth hormone secretion in female rats. *Endocrinology, 121,* 1417–1423.

Jost, A. (1947). Recherches sur la différenciation sexuelle de l'embryon de lapin. III. Rôle des gonades foetales dans la différenciation sexuelle somatique. *Archives Anatomic Microscopish Experimentalles, 36,* 271–316.

Jost, A. (1953). Problems of fetal endocrinology. *Recent Progress in Hormone Research, 8,* 379–418.

Kikuyama, S., & Kawashima, S. (1966). Formation of corpora lutea in ovarian grafts in ovariectomized adult rats subjected to early postnatal treatment with androgen. *Scientific Papers of the College of General Education, University of Tokyo, 16,* 69–74.

Kincl, F. A., & Maqueo, M. (1965). Prevention by progesterone of steroid-induced sterility in neonatal male and female rats. *Endocrinology, 77,* 859–862.

Koster, R. (1943). Hormone factors in male behavior of female rats. *Endocrinology, 33,* 337–348.

LaPolt, P. S., Matt, D. W., Judd, H. L., & Lu, J. K. H. (1986). The relation of ovarian steroid levels in young female rats to subsequent estrous cyclicity and reproductive function during aging. *Biology of Reproduction, 35,* 1131–1139.

Lehman, J. R., McArthur, D. A., & Hendricks, S. E. (1978). Pharmacological induction of ovulation in old and neonatally androgenized rats. *Experimental Gerontology, 13,* 107–114.

Levine, S., & Mullins, F., Jr. (1964). Estrogen administered neonatally affects adult sexual behavior in male and female rats. *Science, 144,* 185–187.

Lisk, R. D. (1969). Progesterone: Biphasic effects on the lordosis response in adult or neonatally gonadectomized rats. *Neuroendocrinology, 5,* 149–160.

Lu, K. H., Huang, H. H., Chen, H. T., Kurcz, M., Nioduszewski, R., & Meites, J. (1977). Positive feedback by estrogen and progesterone on LH release in old and young rats. *Proceedings of the Society for Experimental Biology and Medicine, 154,* 82–85.

Lu, J. K. H., Gilman, D. P., Meldrum, D. R., Judd, H. L., & Sawyer, C. H. (1981). Relationship between circulating estrogens and the central mechanisms by which ovarian steroids stimulate luteinizing hormone secretion in aged and young female rats. *Endocrinology, 108,* 836–841.

Lu, J. K. H., LaPolt, P. S., Nass, T. E., Matt, D. W., and Judd, H. L. (1985). Relation of circulating estradiol and progesterone to gonadotropin secretion and estrous cyclicity in aging female rats. *Endocrinology, 116,* 1953–1959.

Mandl, A. M., & Shelton, M. (1959). A quantitative study of oocytes in young and old nulliparous laboratory rats. *Journal of Endocrinology, 18,* 444–450.

Maqueo, M., & Kincl, F. A. (1964). Testicular histo-morphology of young rats treated with oestradiol-17β benzoate. *Acta Endocrinologica, 46,* 25–30.

Martini, L. (1978). Role of the metabolism of steroid hormones in the brain in sex differentiation and sexual maturation. In G. Dörner & M. Kawakami (Eds.), *Hormones and brain development, Development in endocrinology, Vol. 3* (pp. 3–25). Amsterdam: Elsevier/North-Holland Biomedical Press.

McDonald, P. G., & Doughty, C. (1973/1974). Androgen sterilization in the neonatal female rat and its inhibition by an estrogen antagonist. *Neuroendocrinology, 13,* 182–188.

McDonald, P., Beyer, C., Newton, F., Brien, B., Baker, R., Tan, H. S., Sampson, C., Kitching, P., Greenhill, R., & Pritchard, D. (1970). Failure of 5-α-dihydrotestosterone to initiate sexual behavior in the castrated male rat. *Nature, 227,* 964–965.

McEwen, B. S., Lieberburg, I., Chaptal, C., & Krey, L. C. (1977). Aromatization: Important for sexual differentiation of the neonatal rat brain. *Hormones and Behavior, 9,* 249–263.

McEwen, B. S., Lieberburg, I., Chaptal, C., Davis, P. G., Krey, L. C., MacLusky, N. J., & Roy, E. J. (1979). Attenuating the defeminization of the neonatal rat brain: Mechanisms of action of cyproterone acetate, 1,4,6-adrostatriene-3,17,-dione and a synthetic progestin, R5020. *Hormones and Behavior, 13,* 269–281.

Meijs-Roelofs, H. M. A., Uilenbroek, J. T. J., de Jong, F. H., & Welschen, R. (1973). Plasma estradiol-17 beta and its relationship to serum follicle-stimulating hormone in immature female rats. *Journal of Endocrinology, 59,* 295–304.

Meijs-Roelofs, H. M. A., de Greef, W. J., & Uilenbroek, J. T. J. (1975). Plasma progesterone and its relationship to serum gonadotrophins in immature female rats. *Journal of Endocrinology, 64,* 329–336.

Meisel, R. L., & Ward, I. L. (1981). Fetal female rats are masculinized by male littermates located caudally in the uterus. *Science, 213,* 239–242.

Meites, J., Huang, H. H., & Simpkins, J. W. (1978). Recent studies on neuroendocrine control of reproductive senescence in rats. In E. L. Schneider (Ed.), *The aging reproductive system, Vol. 4: Aging* (pp. 213–235). New York: Raven Press.

Mobbs, C. V., Flurkey, K., Gee, D., Yamamoto, K., Sinha, Y., & Finch, C. E. (1984). Estradiol-induced anovulatory syndrome in female C57BL/6J mice: Age-like neuroendocrine, but not ovarian, impairments. *Biology of Reproduction, 30,* 556–563.

Mobbs, C. V., Kannegieter, L. S., & Finch, C. E. (1985). Delayed anovulatory syndrome induced by estradiol in female C57BL/6J mice: Age-like neuroendocrine, but not ovarian, impairments. *Biology of Reproduction, 32,* 1010–1017.

Money, J., & Ehrhardt, A. A. (1971). Fetal hormones and the brain: Effect on sexual dimorphism of behavior—a review. *Archives of Sexual Behavior, 1,* 241–262.

Morin, L. P. (1977). Progesterone: Inhibition of rodent sexual behavior. *Physiology and Behavior, 18,* 701–715.

Mullins, R. R., & Levine, S. (1968). Hormonal determinants during infancy of adult sexual behavior in the male rat. *Physiology and Behavior, 3,* 339–343.

Napoli, A. M., & Gerall, A. A. (1970). Effect of estrogen and anti-estrogen on reproductive function in neonatally androgenized rats. *Endocrinology, 94,* 117–121.

Nass, T. E., Matt, D. W., Judd, H. L., & Lu, J. H. K. (1984). Prepubertal treatment with estrogen or testosterone precipitates the loss of regular estrous cyclicity and normal gonadotropin secretion in adult female rats. *Biology of Reproduction, 31,* 723–731.

Nelson, J. F. (1981). *Patterns of reproductive aging in C57Bl/6J mice: Estrous cycles, gonadal steroids, and uterine estradiol receptors.* Ph.D. Thesis, Department of Biological Sciences, University of Southern California, Los Angeles.

Nelson, J. F., Felicio, L. S., & Finch, C. E. (1980). Ovarian hormones and the etiology of reproductive aging in mice. In A. Dietz (Ed.), *Aging: Its chemistry* (pp. 64–81). New York: American Association of Clinical Chemistry.

Nikels, K. W. (1976). Ovarian modification of sexual behavior in neonatally androgenized female rats. *Bulletin of the Psychonomic Society, 7,* 59–60.

Nikels, K. W. (1982). *Effects of perinatal progesterone manipulations and maternal age on female sexual behav-*

ior in the rat. Doctoral dissertation, University of Nebraska. *Dissertation Abstracts International, 43,* 3067B.

Ojeda, S. R., & Ramirez, V. D. (1972). Plasma level of LH and FSH in maturing rats: Response to hemigonadectomy. *Endocrinology, 90,* 466–472.

Ojeda, S. R., Andrews, W. W., Advis, J. P., & White, S. S. (1980). Recent advances in the endocrinology of puberty. *Endocrine Reviews, 1,* 228–257.

Ojeda, S. R., Aguado, L. I., & Smith (White), S. (1983). Neuroendocrine mechanisms controlling the onset of female puberty: The rat as a model. *Neuroendocrinology, 37,* 306–313.

Pang, S. F., Caggiula, A. R., Gay, V. L., Goodman, R. L., & Pang, C. S. F. (1979). Serum concentrations of testosterone, oestrogens, luteinizing hormone and follicle-stimulating hormone in male and female rats during the critical period of neural sexual differentiation. *Journal of Endocrinology, 80,* 103–110.

Parsons, B., MacLusky, M. S., Krieger, B. S., McEwen, B. S., & Pfaff, D. W. (1979). The effects of long-term estrogen exposure on the induction of sexual behavior and measurements of brain estrogen and progesterone receptors in the female rat. *Hormones and Behavior, 13,* 301–313.

Peng, M. T., & Huang, H. H. (1972). Aging of hypothalamic–pituitary–ovarian function in the rat. *Fertility and Sterility, 23,* 575–442.

Pfeiffer, C. A. (1936). Sexual differences of the hypophyses and their determination by the gonads. *American Journal of Anatomy, 58,* 195–223.

Phoenix, C., Goy, R., Gerall, A., & Young, W. (1959). Organizing action of prenatally administered testosterone propionate on the tissues mediating mating behavior in the female guinea pig. *Endocrinology, 65,* 369–382.

Pointis, G., Latreille, M. T., Richard, M. O., Athis, P. D., & Cedard, L. (1984). Effect of maternal progesterone exposure on fetal testosterone in mice. *Biology of the Neonate, 45,* 203–208.

Powers, J. B. (1970). Hormonal control of sexual receptivity during the estrous cycle of the rat. *Physiology and Behavior, 5,* 831–835.

Presl, J., Herzman, J., & Horsky, J. (1969). Oestrogen concentration in blood of developing rats. *Journal of Endocrinology, 45,* 611–612.

Quattropani, S. L., & Weisz, J. (1973). Conversion of progesterone to estrone and estradiol *in vitro* by the ovary of the infantile rat in relation to the development of its interstitial tissue. *Endocrinology, 53,* 1269–1276.

Resko, J. A. (1974). The relationship between fetal hormones and the differentiation of the central nervous system in primates. In W. Montagna & W. A. Sadler (Eds.), *Advances in behavioral biology, Vol. 11: Reproductive behavior* (pp. 211–222). New York: Plenum Press.

Schipper, H., Brawer, J. R., Nelson, J. F., Felicio, L. S., & Finch, C. E. (1981). Role of the gonads in the histologic aging of the hypothalamic arcuate nucleus. *Biology of Reproduction, 25,* 413–419.

Shapiro, B. H., Goldman, A. S., Steinbeck, H. F., & Neuman, F. (1976). Is feminine differentiation of the brain hormonally determined? *Experientia, 32,* 650–651.

Simmons, J. E. (1971). Uptake of (1,2-^{3}H) testosterone in oestrogenized male rats. *Acta Endocrinologica, 67,* 535–543.

Södersten, P. (1978). Effects of anti-oestrogen treatment of neonatal male rats on lordosis behaviour and mounting behaviour in the adult. *Journal of Endocrinology, 76,* 241–249.

Sonnenschein, C., & Soto, A. (1980). The mechanism of estrogen action: The old and new paradigm. In J. A. McLachlan (Ed.), *Estrogens in the environment* (pp. 169–195). Amsterdam: Elsevier/North-Holland.

Soulairac, M.-L., & Soulairac, A. (1974). Compartement sexuel et tractus genital du rat male adulte apres oestrogenisation post-natale precoce. *Annales d'Endocrinologie, 35,* 577–578.

Strecke, J., Oettel, M., Tiroke, I., & Ohme, E. (1978). Influence of neonatally administered gestagens on fertility in rats. In G. Dörner & M. Kawakami (Eds.), *Hormones and brain development, Developments in endocrinology, Vol. 3* (pp. 205–213). Amsterdam: Elsevier/North-Holland.

Swanson, H. E., & van der Werff ten Bosch, J. J. (1964). The "early androgen" syndrome, its development and the response to hemi-spaying. *Acta Endocrinologica, 45,* 1–12.

Tobet, S. A., Dunlap, J. L., & Gerall, A. A. (1982). Influence of fetal position on neonatal androgen-induced sterility and sexual behavior in female rats. *Hormones and Behavior, 16,* 251–258.

Toran-Allerand, C. D. (1980). Coexistence of α-fetoprotein, albumin and transferrin immunoreactivity in neurons of the developing mouse brain. *Nature, 286,* 733–735.

Turkelson, C. M., Dunlap, J. L., MacPhee, A. A., & Gerall, A. A. (1977). Assay of perinatal testosterone and influence of anti-progesterone sterility. *Life Sciences, 21,* 1149–1158.

Urbanski, H. F., & Ojeda, S. R. (1986). The development of afternoon minisurges of luteinizing hormone secretion in prepubertal female rats is ovary dependent. *Endocrinology, 118,* 1187–1193.

Valenstein, E. S. (1968). Steroid hormones and the neuropsychology of development. In R. L. Isaacson (Ed.), *The neuropsychology of development: A symposium* (pp. 1–39). New York: John Wiley & Sons.

vom Saal, F. S. (1989). Sexual differentiation in litter-bearing mammals: Influence of sex of adjacent fetuses *in utero. Journal of Animal Science, 67*(7), 1824–1840.

vom Saal, F. S., & Bronson, F. H. (1978). *In utero* proximity of female mouse fetuses to males: Effect on reproductive performance during later life. *Biology of Reproduction, 19,* 842–853.

vom Saal, F. S., & Bronson, F. H. (1980). Variation in the length of the estrous cycle in mice due to former intrauterine proximity to male fetuses. *Biology of Reproduction, 22,* 777–780.

vom Saal, F. S., & Finch, C. E. (1988). Reproductive senescence: Phenomena and mechanisms in mammals and selected vertebrates. In E. Knobel & J. Neill, (Eds.), L. Ewing, G. Greenwald, C. Markert, & D. Pfaff, (Assoc. Eds.), *The physiology of reproduction* (pp. 2351–2413). New York: Raven Press.

vom Saal, F. S., Grant, W. M., McMullen, C. W., & Laves, K. S. (1983). High fetal estrogen concentrations: Correlations with increased adult sexual activity and decreased aggression in male mice. *Science, 220*(4603), 1306–1309.

Ward, I. L. (1974). Sexual behavior differentiation: Prenatal hormonal and environmental control. In R. C. Friedman, R. M. Richart, & R. L. Vande Wiele (Eds.), *Sex differences in behavior* (pp. 3–17). New York: John Wiley & Sons.

Warner, M. R., Yau, L., & Rosen, J. M. (1980). Long term effects of perinatal injection of estrogen and progesterone on the morphological and biochemical development of the mammary gland. *Endocrinology, 106,* 823–832.

Watanabe, H., & Takebe, K. (1987a). Role of postnatal gonadal function in the determination of thyrotropin (TSH) releasing hormone-induced TSH response in adult male and female rats. *Endocrinology, 120,* 1711–1718.

Watanabe, H., & Takebe, K. (1987b). Involvement of postnatal gonads in the maturation of dopaminergic regulation of prolactin secretion in female rats. *Endocrinology, 120,* 2212–2219.

Weisz, J., & Gunsalus, P. (1973). Estrogen levels in immature female rats: True or spurious—ovarian or adrenal? *Endocrinology, 93,* 1057–1065.

Weisz, J., & Ward, I. L. (1980). Plasma testosterone and progesterone titers of pregnant rats, their male and female fetuses, and neonatal offspring. *Endocrinology, 106,* 306–316.

Whalen, R. E. (1964). Hormone-induced changes in the organization of sexual behavior in the male rat. *Journal of Comparative and Physiological Psychology, 57,* 175–182.

Whalen, R. E. (1968). Differentiation of the neural mechanisms which control gonadotropin secretion and sexual behavior. In M. Diamond (Ed.), *Perspectives in reproduction and sexual behavior* (pp. 303–340). Bloomington: Indiana University Press.

Whalen, R. E., & Nadler, R. D. (1963). Suppression of the development of female mating behavior by estrogen administered in infancy. *Science, 141,* 273–274.

Whalen, R. E., & Nakayama, K. (1965). Induction of estrous behavior: Facilitation by repeated hormone treatments. *Journal of Endocrinology, 33,* 255–264.

Wilson, J. G. (1943). Reproductive capacity of adult female rats treated prepubertally with estrogenic hormone. *Anatomical Record, 86,* 341–363.

Witcher, J. A., & Clemens, L. G. (1987). A prenatal source for defeminization of female rats is the maternal ovary. *Hormones and Behavior, 21,* 36–43.

Wolfe, J. M. (1943). The effects of advancing age on the structure of the anterior hypophysis and ovaries of female rats. *American Journal of Anatomy, 72,* 361–383.

Young, W. C., Goy, R. W., & Phoenix, C. H. (1964). Hormones and sexual behavior. *Science, 143,* 212–218.

Zadina, J. E., Dunlap, J. L., & Gerall, A. A. (1979). Modifications induced by neonatal steroids in reproductive organs and behavior of male rats. *Journal of Comparative and Physiological Psychology, 93,* 314–322.

5

Sexual Behavior
The Product of Perinatal Hormonal and Prepubertal Social Factors

INGEBORG L. WARD

INTRODUCTION

The diversity of sexual patterns that characterize adult members of all mammalian species is the result of influences experienced during earlier stages of development. Although there is general agreement on this assumption, the scientific community has split as to the nature of these early influences and the specific life stages during which formative effects are exerted. Basically, the issue revolves around nature versus nurture, the old battlefield destined to be revisited by every generation of scientists struggling to uncover the mysteries underlying behavioral mechanisms.

With regard to sexual behavior, the two camps divide along clearly definable lines. Although there are exceptions, the perspective of clinicians is heavily influenced by work with patient populations. The training of psychiatrists and clinical psychologists traditionally has included liberal exposure to Freudian theory and has emphasized the role of social influences acting during infancy and early childhood as being most critical to the formation of gender identity and sexual behaviors expressed in adulthood. Genetic and physiological factors are vaguely acknowledged as somehow being involved, but social influences acting through learning, modeling, or the relationship between parent and child are viewed as sufficiently powerful to override any physiological predispositions with which an individual may have been born. The social primacy position stands in sharp contrast to the more recently emerging theories and findings of the behavioral endocrinologists. This class of investigators has accumulated an astonishing body of information docu-

INGEBORG L. WARD   Department of Psychology, Villanova University, Villanova, Pennsylvania 19085.

Sexual Differentiation, Volume 11 of *Handbook of Behavioral Neurobiology*, edited by Arnold A. Gerall, Howard Moltz, and Ingeborg L. Ward, Plenum Press, New York, 1992.

menting the profound and irreversible effects that exposure to androgenic steroids during specific stages of fetal life has on adult sexual behavior.

So which position is correct? Historically, many of the controversies that pitted nature against nurture were resolved by data demonstrating that both contribute. The same may be the case with regard to the expression of sexually dimorphic behaviors. A small but growing body of literature suggests that the relative exposure to androgenic steroids incurred during perinatal development establishes behavioral potentials. However, the extent to which these potentials can be activated by gonadal hormones beginning at the time of puberty is modulated by the particular social milieu the individual experienced during prepuberal life. In other words, the diversity of sexual behavior patterns found in adult organisms is the product of an interaction between fetal hormonal events and prepubertal social factors. The objective of the present chapter is to summarize briefly the literature dealing with perinatal hormonal and prepubertal social influences on sexual behavior development and then to describe the available data that suggest a functional link between these two seemingly unrelated determinative categories.

Perinatal Hormonal Influences

The critical influence on adult sexual patterns exerted by exposure to gonadal hormones during fetal development was not widely appreciated until 1959. In that year a paper by Phoenix, Goy, Gerall, and Young appeared that demonstrated that female guinea pigs exposed to testosterone during gestation showed severe deficiencies in their ability to display lordosis, the female receptivity pattern, but performed high levels of male-like mounting and intromission behavior. Hundreds of studies probing the hormonal mechanism(s) mediating the process of sexual behavior differentiation have appeared since the publication of this benchmark paper. Although much remains to be uncovered, the accumulated body of information suggests that there seem to be some general principles that hold across species as well as idiosyncratic variations that apply to particular categories of species.

Types of Organizing Hormones

In mammals, exposure of the developing organism to specific testicular hormones at discrete stages of early development leads to a masculinization of reproductive morphology, physiology, and behavior. If these hormones are not present in adequate amounts or over the correct temporal intervals, the natural tendency to retain female characteristics prevails. Under normal conditions, only the fetal male's gonads release the primary masculinizing steroid, testosterone. Once testosterone becomes available to target tissues, it can be utilized in a variety of forms. Some target cells respond directly to testosterone, whereas others contain enzymes that convert testosterone into a metabolite that, in turn, exerts the masculinizing action. For example, development of the male's epididymides, vas deferens, and seminal vesicles is accomplished when testosterone reaches the embryonic Wolffian duct system either through local diffusion or the vasculature. On the other hand, the prostate and the external genitalia are derived from different embryological anlagen, namely, the urogenital sinus and genital tubercle. These target tissues contain the enzyme 5α-reductase, which acts intracellularly to convert testosterone obtained from the circulation into dihydrotestosterone (DHT). Masculinization of this category of androgen-sensitive tissue is accomplished by the action of DHT, not testosterone. Males with a deficiency in the 5α-reductase system develop a normally masculinized duct system, but the prostate does not differentiate, and the external genitalia are those of a female (Imperato-McGinley, Guerrero, Gautier, & Peterson, 1974; Imperato-McGinley *et al.*, 1985; Imperato-McGinley, Binienda, Gedney, & Vaughan, 1986; Jost, 1985; Wilson, 1978; Wilson, Griffin, & George, 1980; Wilson, Griffin, George, & Leshin, 1981).

Similarly, there is a great deal of evidence to suggest that in some species the masculinization and defeminization of sexual behavior potentials involve not androgens but, rather, their aromatization products, estrogens. The aromatizing enzyme needed to accomplish this conversion is present during fetal development in brain areas known to be part of the central circuit that mediates adult sexual behavior (George & Ojeda, 1982; MacLusky, Philip, Hurlburt, & Naftolin, 1985; Reddy, Naftolin, & Ryan, 1974; Tobet, Baum, Tang, Shim, & Canick, 1985; Weisz, Brown, & Ward, 1982). Interestingly, although estrogens probably play a major role in the differentiation of sexual behavior in the rat (Goy & McEwen, 1980) and, perhaps, in the ferret (Baum & Tobet, 1986), they do not have this action in the rhesus monkey (Goy, 1981; Goy & Robinson, 1982; Thornton & Goy 1986). The relative degree to which estrogen or testosterone exerts a primary organizing action on the process of sexual behavior differentiation is a striking example of species differences and serves as a sharp reminder that phenomena discovered in one animal species cannot be assumed to generalize to all others.

TIMING OF PERINATAL HORMONE EXPOSURE

The timing of hormonal fluctuations during perinatal ontogeny is another major factor that determines the extent to which sexually dimorphic behaviors differentiate. Since the critical period of development is slightly different for various behaviors or, in some instances, for components making up complex behavioral patterns, the onset and duration of androgenic influences are important. Thus, if aberrant patterns of androgen exposure exist for only a limited period of perinatal development, the individual may show some behaviors that are congruent with genetic sex and some that are heterotypical.

For example, the normal copulatory sequence in the male rat consists of multiple mounts and intromissions displayed over a 10- to 20-minute period, ending in ejaculation. If the male is castrated on the day of birth, ejaculation is eliminated, and the number of intromissions is sharply reduced. Mounting, however, continues at a high rate (Gerall, Hendricks, Johnson, & Bounds, 1967; Grady, Phoenix, & Young, 1965). Similarly, female rats exposed to androgen only during the last week of fetal development show high levels of mounting and some intromission responses, but the ejaculatory pattern never occurs (Gerall & Ward, 1966; Ward, 1969; Ward & Renz, 1972). On the other hand, females injected with androgen neonatally exhibit high levels of intromission behavior, and a few are able to ejaculate, but their mount rate is no higher than that of untreated females (Harris & Levine, 1965; Thomas, McIntosh, & Barfield, 1980; Ward, 1969; Whalen & Edwards, 1967). These studies suggest that the three components of the male copulatory pattern have overlapping but not identical critical periods of differentiation. Mounting emerges during fetal development, the intromission pattern spans late fetal and early neonatal ontogeny, and ejaculation can only be achieved if androgen is present for at least a short period during neonatal life.

Similarly, giving perinatal female rats exogenous testosterone propionate (TP) impairs the development of normal estrous behavior patterns to varying degrees, depending on when the exposure occurred. Brief exposure limited to day 18 (Nadler, 1969), day 19 (Huffman & Hendricks, 1981), or days 17.5 and 18.5 (Hoepfner & Ward, 1988) of gestation is sufficient to alter the receptivity pattern shown by females when they reach adulthood. Exposure on day 16 (Huffman & Hendricks, 1981), days 15.5 and 16.5, or days 19.5 and 20.5 (Hoepfner & Ward, 1988) of gestation has no detectable consequences on lordosis. However, if elevated androgen levels during either of the latter two ineffective periods are combined with a small dosage of TP (5 μg) administered within the first few days after birth, the proportion of females capable of showing lordosis is markedly reduced (Hoepfner & Ward, 1988). A single injection of 5 μg of TP given to a normal neonatal female is insufficient by itself to impair development of estrous behavior.

The diverse pattern of sexual behavior shown by male and female rats whose androgenic milieu was manipulated during limited intervals of perinatal ontogeny has been instructive. It seems clear that the complex of intersexed behavioral patterns shown by animals at various life stages reflects the exact perinatal period(s) during which the androgenic milieu was heterotypical and the degree and duration of the abnormality.

It also is important to keep in mind that a variety of behaviors other than those involved in copulation differentiate in a sexually dimorphic manner as a function of relative exposure to androgen during perinatal development. These include patterns of juvenile play, the rate at which certain learning tasks are acquired, parental behaviors, and levels of aggression (for a review of this literature see Beatty, 1979; W. W. Beatty, Chapter 3). There are few data delineating the specific critical period during which sexually dimorphic behaviors other than copulation patterns develop. However, it is assumed that once the critical perinatal periods of differentiation are completed, it is not possible to reverse the effects of androgen exposure. One objective of the present chapter is to question the finality with which perinatal hormone exposure is viewed as determining sexually dimorphic behavioral potentials. Modulation through social influences may be possible.

PREPUBERTAL SOCIAL INFLUENCES

SOCIAL ISOLATION

Although it is well established that the expression of various sexually dimorphic behaviors reflects the perinatal hormonal milieu, adult copulatory patterns also are influenced by the social environment in which a young animal is reared. Particularly devastating effects occur in males reared in the absence of conspecifics. Young males raised in complete social isolation from an early age to the time of puberty appear sexually aroused when placed with receptive females in adulthood, but they do not show appropriate copulatory patterns. Instead, they appear hyperexcitable and engage in such stereotyped behaviors as head shaking, leaping, aberrant mounting patterns, and climbing over and under the female. This phenomenon has been demonstrated in a variety of species, including the guinea pig (Gerall, 1963; Valenstein, Riss, & Young, 1955), rat (Folman & Drori, 1965; Gerall, Ward, & Gerall, 1967; Gruendel & Arnold, 1974), dog (Beach, 1968), rhesus monkey (Mason, 1960; Missakian, 1969), and chimpanzee (Turner, Davenport, & Rogers, 1969). Complete social isolation is unlikely to occur with any frequency under natural conditions. However, similar disruptions in sexual behavior have been found in males reared in socially impoverished environments. For example, most male rats raised from 14 days of age onward in a compartmentalized cage that separated them from other males and/or females by only a wire mesh screen failed to mate when they became adults (Gerall, Ward, *et al.*, 1967; Spevak, Quadagno, & Knoeppel, 1973). Access during prepubertal development to the sight, sound, and smell of conspecifics in the absence of direct tactile contact was insufficient to normalize adult sexual behavioral potentials.

Precisely which aspects of early social stimulation critically affect the development of sexual behavior is not clear, but in rodents, at least, the importance of play has been suggested (Gerall, 1963; Gerall, Ward, *et al.*, 1967). According to this hypothesis, opportunities for play are essential if young males are to mature into adults that respond appropriately to an estrous female. Males unable to find an outlet for play at appropriate early stages of life may fail to replace play behavior with copulatory patterns when puberty is attained. Thus, an estrous female will sexually arouse an adult male rat that had been socially isolated during prepubertal life, but his retarded social development leads him to engage in high levels of juvenile play rather than the appropriate adult patterns of mounting, intromission, and ejaculation.

The extent to which a general tendency to display competing behaviors incompatible with mating can explain the sexual inadequacies of social isolates has been questioned. Gruendel and Arnold (1974) reared isolated male rats in cages containing mirrors, toys, and other devices that promoted juvenile play in a nonsocial environment. This strategy effectively reduced inappropriate play behavior in adulthood; nonetheless, the isolates reared in this "enriched" environment continued to show deficient sexual behavior. Gruendel and Arnold concluded that the tactile stimulation arising from the social interactions in which juveniles normally engage, particularly the stimulation derived from social grooming and sniffing, is critical for the formation of normal sexual patterns. The only treatment that has been reported to induce isolated male rats to begin to copulate is prolonged cohabitation with a female in adulthood (Dunlap, Zadina, & Gougis, 1978; Gerall, Ward, *et al.*, 1967; Hård & Larsson, 1968). This type of "therapy," however, is effective in only a limited proportion of isolated males. Interestingly, although the estrous behavior patterns of isolated female rats are normal, the low levels of male-like mounting that adult female rats often display are reduced by prepubertal social isolation (Duffy & Hendricks, 1973; Hansen, 1977). The effects of isolation on female guinea pigs are not yet clear. One study reported lordosis behavior in female guinea pigs reared in total social isolation to be normal (Harper, 1968), but another found it to be severely impaired (Goy & Young, 1957).

MATERNAL INFLUENCES IN RODENTS

Although opportunities to interact with peers during prepubertal life clearly are important if males are to be sexually proficient, at an even earlier age the mother also exerts an influence. Male rats reared apart from the mother beginning on the day of birth are unable to copulate in adulthood, even if raised together with their siblings (Gruendel & Arnold, 1969). If all the siblings are removed, leaving the male alone with the mother, sexual behavior develops normally in at least some of the animals (Gruendel & Arnold, 1969; Hård & Larsson, 1968).

Two lines of evidence suggest what the nature of some of the maternal influences might be. Among the many behavioral components that make up the parental behavior pattern in the rat is anogenital (A-G) licking. Rat mothers lick the genital area of their neonates, thereby stimulating urination and defecation. This licking behavior benefits the mother in that she ingests the urine, thereby recovering up to two-thirds of the fluid volume she previously had transferred to the pups in the form of milk (Friedman & Bruno, 1976). Anogenital licking begins at birth and continues at a relatively high rate until about 17 days of age (Gubernick & Alberts, 1983). Furthermore, the mother exhibits A-G licking toward male pups at a greater frequency than toward female pups (Moore & Morelli, 1979). Moore (1984) has shown that if males are deprived of high levels of A-G licking while neonates, adult sexual behavior is impaired. Specifically, the ability of rat mothers to perceive olfactory cues was eliminated by lining the nasal passages with polyethylene tubing, a manipulation that reduced the rate at which they licked their pups. Male offspring that received low levels of licking from their mother showed longer interintromission and ejaculation latencies and longer postejaculatory intervals when tested with estrous females in adulthood. Thus, the tactile stimulation of the genital area, which is an important component of mother–infant interactions in rats, may contribute to the development of normal patterns of male sexual behavior. The stimulus characteristics that cause the mothers to treat male pups differently from female pups derive from perinatal exposure to androgen, and female pups injected with TP on the day of birth are licked at the same high rate as their male siblings (Moore, 1982).

A second source of maternal stimulation shown to affect copulation in male rats derives from maternal odors associated with suckling (Fillion & Blass, 1986). Adult male rats achieved ejaculation rapidly when paired with receptive females that had a unique odorant applied to the perivaginal area, an odor that the male previously had encountered while still a neonate suckling the nipple of the mother. Ejaculation was delayed when the female sexual partner did not bear the maternal suckling odor.

Peer and Maternal Influences in Primates

The data accumulated on rodents demonstrate that in the absence of proper prepubertal socialization a normal prenatal hormonal milieu is insufficient to ensure the normal display of sexual behavior in adulthood. Even more instructive is a series of reports from the Wisconsin Regional Primate Research Center conducted on rhesus monkeys. The detrimental consequences of complete social isolation in monkeys are well known, and these more recent studies have concentrated on the impact of less severe social manipulations. A variety of behaviors can be altered by modifying the social environment of young monkeys, but the pattern most pertinent to the present discussion is the double foot-clasp mount. This species-specific behavior involves mounting the target animal from behind and clasping each hind leg with one foot. The pattern is displayed typically during copulation by adult males, although juvenile males also normally mount females or one another in this unique manner. The double foot-clasp mount in rhesus monkeys is another sexually dimorphic behavior whose incidence is proportional to the degree of exposure to androgen during prenatal but not postnatal development (Goy & Goldfoot, 1974). Further, two lines of evidence suggest that the activation of the double foot-clasp mount is not dependent on circulating androgen levels. First, the pattern appears in juvenile males many years before the onset of puberty with its attendant increase in circulating gonadal hormones. Second, the ontogeny of the pattern is the same in neonatally castrated juveniles as in intact males (Goy & Goldfoot, 1974; Wallen, Bielert, & Slimp, 1977).

The frequency with which the double foot clasp is displayed by juveniles normally increases with age. Young males that do not increase the rate at which this pattern is exhibited as they mature fail to copulate normally as adults (Goy & Goldfoot, 1974). Females occasionally perform the double foot-clasp mount but do so infrequently (Goldfoot & Wallen, 1979; Goldfoot, Wallen, Neff, McBriar, & Goy, 1984; Goy & Goldfoot, 1974). Because of the robust relationship between the ontogeny of the juvenile mounting pattern and normal adult copulatory behavior in males, the double foot-clasp mount has been used as a reliable predictor of the long-term consequences of various early manipulations. A heavy reliance on the predictive characteristics of this pattern is justified by the slow rate at which this primate species reaches sexual maturity, i.e., around 4–5 years of age.

A potential for high levels of double foot-clasp mounting requires more than exposure to adequate amounts of androgen during prenatal development. The frequency with which the pattern is exhibited by either male or female rhesus juveniles varies as a function of the amount of time the animal is allowed access to peers, the sex of peer playmates, and the age at which the animal is weaned from its mother. The earliest and highest incidence of foot-clasp mounting is seen among young males raised with continuous access to both their mother and other juveniles. This form of socialization also promotes the emergence of normal patterns of sexual behavior in adulthood (Goy & Goldfoot, 1974). Males allowed only $\frac{1}{2}$ hour of daily access to peers during the first 2 years of life rarely show double foot-clasp mounting as juveniles and do not copulate when they reach adulthood (Goy & Goldfoot, 1974; Goy & Wallen, 1979; Goy, Wallen, & Goldfoot, 1974; Wallen et al., 1977). This is true even if the animals are left continuously with the mother for the first year of life (Wallen, Goldfoot, & Goy, 1981). More foot-clasp mounting is seen among males reared in heterosexual peer groups than in isosexual groups (Goldfoot & Wallen, 1979; Goldfoot et al., 1984). The reverse is true for juvenile females: those living in groups consisting entirely of females show much higher levels of foot-clasp mounting than do those reared in mixed-sex groups (Goldfoot & Wallen, 1979).

INTERACTION BETWEEN FETAL HORMONAL AND PREPUBERTAL SOCIAL EVENTS

The wealth of accumulated data leaves little doubt that both perinatal hormonal events and prepubertal socialization influence the expression of sexually dimorphic behaviors at all stages of the life cycle. It is likely that these two categories of factors do not exert mutually exclusive control over the organization of sexual behavior potentials. However, little scientific attention has been directed to determining the extent to which perinatal hormones and early socialization interact to attenuate or exaggerate the effects that either would have acting independently. The remainder of this chapter summarizes data that do exist.

CHARACTERISTICS OF THE PRENATAL STRESS SYNDROME

A variety of animal models and experimental strategies could be used to assess whether the expression of normal sexual behavior involves an interaction between perinatal hormone exposure and later socialization. The preparation from which the greatest amount of data has been derived thus far is the prenatally stressed male rat. The so-called prenatal stress syndrome was first characterized in the male offspring of pregnant rats stressed three times daily by exposure to 200 footcandles of light coupled with restraint in a Plexiglas tube (Ward, 1972, 1977, 1984; Ward & Ward, 1985). The treatment was given during days 14–21 of the 23-day-long gestation period and caused elevated plasma levels of corticosterone in both the mothers and the fetuses they were carrying (Ward & Weisz, 1984). When tested in adulthood, many of the male offspring of the stressed mothers not only failed to mate with estrous females but showed the female lordosis pattern when mounted by a vigorous stud male.

A similar feminization and demasculinization of sexual behavior has been reported with other types of stressors, including social overcrowding (Dahlöf, Hård, & Larsson, 1977; Harvey & Chevins, 1984) and malnutrition (Rhees & Fleming, 1981). More recent data have made it possible to begin to generalize the syndrome to species other than the laboratory rat. For example, components of the prenatal stress syndrome have been characterized in mice (Harvey & Chevins, 1984, 1985; Kinsley & Svare, 1986; Politch & Herrenkohl, 1984; vom Saal, 1983) and perhaps in humans, although the data base supporting the claim that prenatal stress plays a role in the etiology of homosexual behavior in man is insecure (Dörner et al., 1980; Dörner, Schenk, Schmiedel, & Ahrens, 1983).

The effects of prenatal stress on developing female fetuses are less certain than those on males. In rats, abnormalities in estrous cycle length, in various measures of reproductive capability, and in maternal behavior have been reported (Herrenkohl, 1979; Herrenkohl & Politch, 1978; Herrenkohl & Scott, 1984; Kinsley & Bridges, 1988; Politch & Herrenkohl, 1984). Other investigators were unable to detect deficiencies in the reproductive behavior or competence of the female littermates of males manifesting the prenatal stress syndrome (Beckhardt & Ward, 1983; Dahlöf et al., 1977). The long-term consequences of prenatal stress on sexually dimorphic behaviors and on reproductive physiology in females remain to be clarified.

EFFECTS OF PRENATAL STRESS ON REPRODUCTIVE MORPHOLOGY AND ON THE CENTRAL NERVOUS SYSTEM.   The alterations in sexual behavior characteristically found in prenatally stressed males are particularly interesting, since they occur without concomitant changes in reproductive morphology. Although prenatally stressed male rats at the time of birth have low body weights, and the anogenital distance and weight of the testes are reduced (Beckhardt & Ward, 1983; Dahlöf, Hård, & Larsson, 1978; Herrenkohl & Whitney, 1976; Rhees & Fleming, 1981; Wilke, Tseu, Rhees, & Fleming, 1982), these androgen-dependent effects on morphology may or may not persist into adulthood. Adult males showing the behavioral manifestations of the prenatal stress syndrome have been found to be heavier (Dahlöf et al., 1977; Kinsley & Svare, 1986), lighter (Meisel, Dohanich, & Ward, 1979), or no different (McLeod & Brown, 1988; Ward, 1977) in body weight as compared with control animals. Other measures of morphology such as the weight of the testes or of the epididymides also have yielded equivocal results (Meisel et al., 1979; McLeod & Brown, 1988; Politch & Herrenkohl, 1984; Ward, 1977). In brief, there is no component of the peripheral reproductive morphology by which a prenatally stressed male can be distinguished reliably from a nonstressed male once adulthood is attained. Such males, however, can be distinguished on the basis of central nervous system (CNS) morphology.

One of the major developments within the sexual differentiation literature has been the demonstration that there is a clearly discernible sex difference in the volume of a specific nucleus within the preoptic area of several species of mammals. The size of what has been termed the sexually dimorphic nucleus of the preoptic area (SDN-POA) is larger in male than in female rats (Gorski, Gordon, Shryne, & Southam, 1978; Raisman & Field, 1971, 1973), gerbils (Commins & Yahr, 1984), guinea pigs (Hines, Davis, Coquelin, Goy, & Gorski, 1985), and humans (Swaab & Fliers, 1985). Moreover, the magnitude of this sex difference depends on the degree to which the nervous system was exposed to androgens during perinatal development (Döhler et al., 1984, 1986; Handa, Corbier, Shryne, Schoonmaker, & Gorski, 1985; Jacobson, Csernus, Shryne, & Gorski, 1981). A comparison of the area of the SDN-POA of prenatally stressed and nonstressed rats revealed a marked reduction in the stressed males (Anderson, Rhees, & Fleming, 1985; Anderson, Fleming, Rhees, & Kinghorn, 1986; Kerchner, & Ward, 1992). The SDN-POA of females was not altered by exposure to stress during fetal life.

Similar effects were obtained in the sexually dimorphic asymmetry that exists in the thickness of the rat's cerebral cortex. In some regions, the right cortex in adult male rats is significantly thicker than the corresponding area on the left. In females, it is the diameter of specific regions in the left cortex that exceeds that in the right (Diamond, 1984; Diamond, Dowling, & Johnson, 1981; Diamond, Johnson, Young, & Singh, 1983). The normal male pattern of right–left asymmetry was not found in males whose mothers had been stressed during pregnancy (Fleming, Anderson, Rhees, Kinghorn, & Bakaitis, 1986; Stewart & Kolb, 1988).

Sexually dimorphic nuclei in the spinal cord are also incompletely masculinized in prenatally stressed male rats. Specifically, there are significantly fewer motoneurons in the spinal nucleus bulbocavernosus (SNB) and the dorsolateral nucleus (DLN) of stressed compared to control males (Grisham, Kerchner, & Ward, 1991). These spinal nuclei contain neurons that innervate muscles in the perineum which produce penile reflexes active during erection and ejaculation (Breedlove & Arnold, 1980; Hart & Melese-D'Hospital, 1983; Sachs, 1982).

STRESS EFFECTS ON SEXUALLY DIMORPHIC BEHAVIORAL PATTERNS OTHER THAN COPULATION.   The effects of prenatal stress on male rodents are not limited to sexual patterns. A variety of behaviors known to differ between the sexes are feminized in males born of mothers stressed during pregnancy. For example, the tendency to kill newborn pups is much lower in prenatally stressed male rats (Miley, 1983) and mice (vom Saal, 1983) than in untreated males. Moreover, stressed males are more likely to display parental behavior when presented with young than are controls (Kinsley & Bridges, 1988; McLeod & Brown, 1988; vom Saal, 1983). Further, stressed male mice show lower levels of attack and threat behavior toward a standard adult male opponent than do control males (Harvey & Chevins, 1985; Kinsley & Svare, 1986). Group differences in the intensity of aggression persist even when the overall tendency to attack is elevated by injecting all animals with TP.

As illustrated in Figure 1, juvenile play patterns also are feminized. Stressed males, as compared to control males, spent less time playing and exhibited less of the pinning components of rough-and-tumble play (Ward & Stehm, 1991). Others have reported similar effects (Ohkawa, 1987; Takahashi, Haglin, & Talin, 1992).

It should be emphasized, however, that not all sexually dimorphic behaviors are modified by prenatal stress. For example, stressed males learn and perform an active avoidance task at the same rate as control males rather than at the accelerated

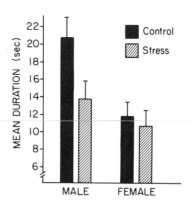

Figure 1. The average amount of time (±S.E.M.) spent in rough and tumble play by 31-day-old male and female rats derived from control mothers or from mothers stressed from days 14 to 21 of gestation. (From Ward & Stehm, 1991.)

pace shown by females (Meisel *et al.*, 1979). The effect on a second test of emotional reactivity, open-field behavior, is not as clear. Masterpasqua, Chapman, and Lore (1976) found higher levels of open-field activity in prenatally stressed male rats compared with control males, a finding interpreted to reflect feminization. Meisel *et al.* (1979) were unable to detect feminized levels of open-field behavior in stressed males, 86% of whom displayed the lordosis pattern. Thus, there is a dissociation within prenatally stressed males in the establishment of various sexually dimorphic behavior potentials that tend to be correlated in normal males and females.

PARTIAL EFFECTS OF PRENATAL STRESS ON SEXUAL DIFFERENTIATION. The data reviewed thus far suggest that males subjected to prenatal stress exhibit a partially feminized behavioral repertoire but their potential for sexually dimorphic behaviors is not completely reversed. This impression is further strengthened when the sexual patterns displayed by these animals in adulthood are examined in greater detail. Compared to untreated males, prenatally stressed males exhibit high rates of lordosis behavior when given estradiol and progesterone (Dahlöf *et al.*, 1977; Dörner, Götz, & Döcke, 1983; Götz & Dörner, 1980; Meisel *et al.*, 1979; Rhees & Fleming, 1981; Ward, 1972; Whitney & Herrenkohl, 1977) or when under the influence of high circulating levels of testosterone (Dahlöf *et al.*, 1977; Ward, 1977). The quality of the lordosis pattern, however, is not on the order of that observed in estrous female rats. Thus, as Figure 2 illustrates, prenatally stressed males display the characteristic lordosis reflex, and, like an estrous female, often hold the posture after the stud male has dismounted. However, unlike the passive lordosis response, which usually is only exhibited if the receptive animal first is mounted, estrous females also display a number of active behavioral components that are subsumed under the label of "soliciting" patterns. These include hopping and darting interspersed by episodes of freezing and ear wiggling. Such soliciting behaviors are rarely seen in prenatally stressed males, which, nevertheless, may readily show high-quality lordosis responses when mounted.

Additional evidence of a dissociation in the organization of sexually dimorphic behaviors in prenatally stressed males is seen when the behavioral potentials of individual animals are assessed. In the typical manifestation of the syndrome, most prenatally stressed males do not copulate when placed with estrous females, even when exposed to treatments that activate ejaculatory behavior in previously non-

Figure 2.   Holding component of the estrous behavior pattern shown by a prenatally stressed male rat that had been injected with estradiol benzoate and progesterone prior to testing.

copulating control males. Included in the latter category of treatments are exogenous injections of TP, mild electric skin shock (Ward, 1977), and administration of the opiate receptor blocker naloxone (Rhees, Badger, & Fleming, 1983). Nevertheless, if both male and female behaviors are assessed within the same group of animals, four distinct subgroups become evident (e.g., Ward, 1972, 1977; Ward & Reed, 1985). A few males are completely asexual, a few show only the ejaculatory pattern, and the rest either exhibit only lordosis behavior or are bisexual; that is, they mount an estrous female but show lordosis when mounted by a vigorous male. Just what causes the variability and dissociations in behavioral potentials among animals that all experienced the same prenatal treatment is puzzling, particularly since heterogeneity can even be found among males derived from the same litter.

ETIOLOGY OF THE PRENATAL STRESS SYNDROME

Any prenatal manipulation resulting in the feminization and demasculinization of adult sexual behaviors in males strongly suggests that the etiology involves an abnormality in the output of the fetal gonads. That appears to be the case with regard to the prenatal stress syndrome. Figure 3 summarizes a variety of hormonal mechanisms implicated in the process of sexual differentiation that show abnormal ontogenetic patterns in male fetuses of rats stressed during pregnancy.

A parametric study of plasma testosterone titers in fetal rats has revealed that significant sex differences normally begin to appear after day 17 post-conception, with the largest difference between males and females occurring on days 18 and 19 of gestation (Weisz & Ward, 1980; Ward & Weisz, 1984). Thereafter, males have consistently higher levels than female littermates, but the magnitude of the difference is not on the order of that seen on days 18 and 19.

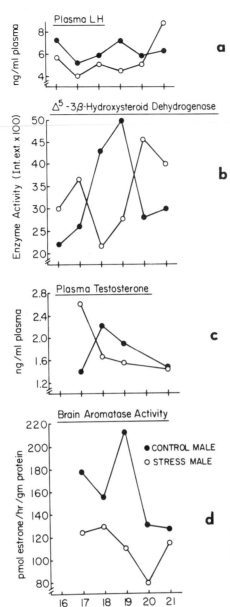

Figure 3. Ontogenetic pattern of plasma LH, of testicular Δ^5-3β-hydroxysteroid dehydrogenase activity, of plasma testosterone, and of brain aromatase activity in control and stressed male rats from days 16–21 of gestation. (Data were adapted from the following sources, beginning with the lowest panel: Weisz et al., 1982; Ward & Weisz, 1980; Orth et al., 1983; Salisbury et al., 1989.

Perusal of Figure 3 reveals that the pattern of plasma testosterone levels in stressed male fetuses is quite different from that in control male fetuses. Stressed fetuses do not exhibit the surge in plasma testosterone that occurs on days 18 and 19 of gestation in control animals. Instead, stressed fetuses exhibit a premature surge on day 17 (Ward & Weisz, 1980, 1984). A very similar profile of disrupted gonadal steroid activity is indicated by the ontogenetic pattern of Δ^5-3β-hydroxysteroid dehydrogenase (3β-HSD) in the Leydig cells of stressed fetal rats (Orth, Weisz, Ward, & Ward, 1983; Ward, Ward, Hayden, Weisz, & Orth, 1990). As shown in Figure 3, the activity of this key steroidogenic enzyme normally peaks on days 18 and 19 of gestation. Stressed fetuses lack this characteristic surge. Instead, higher than normal levels of 3β-HSD are found in stressed males both before and after the critical day-18–19 period.

The abnormally timed elevations in testicular steroidogenesis found on some days of gestation in stressed males fail to compensate for, i.e., prevent or reverse, the feminization believed to take place on days 18 and 19, the period when steroidogenic activity is conspicuously low. Disruption of normal gonadal steroidogenic activity during these particular 2 days results in low plasma levels of testosterone, a condition that has permanent consequences on sexually dimorphic behaviors exhibited at later stages of life.

Several studies support the conclusion that it is the suppression rather than the elevation of plasma testosterone titers on specific days of gestation that alters the course of sexual differentiation in stressed male rats. In the first of these (Ward *et al.*, 1990), pregnant rats were injected with saline or the opioid receptor blocker naltrexone $\frac{1}{2}$ hour before being stressed. The 3β-HSD activity was measured in the Leydig cells of their fetuses on days 17, 18, or 19 of gestation. Compared to offspring from stressed mothers given saline, stressed males from mothers pretreated with naltrexone had normal levels of 3β-HSD activity on days 18 and 19. However, naltrexone did not block the elevated 3β-HSD activity that occurs on day 17 of gestation in stressed males. Interestingly, prenatally stressed males did not exhibit high levels of lordosis behavior in adulthood if the mothers were given naltrexone before each stress session (Ward, Monaghan, & Ward, 1986). These studies suggest that a normalization of testicular steroidogenesis on days 18 and 19 of gestation is sufficient to prevent behavioral feminization in stressed rat fetuses.

The contribution of abnormally low testosterone titers to the etiology of the prenatal stress syndrome is further indicated by the finding that prenatally stressed male rats show normal patterns of sexual behavior in adulthood if they receive exogenous androgen treatment during perinatal development (Dörner, Götz, *et al.*, 1983). Specifically, when stressed mothers were given TP on days 17, 19, and 21 of gestation and their male pups were injected with TP on day 2 post-partum, normal patterns of ejaculatory behavior were observed when the offspring reached adulthood, and none displayed the lordosis pattern.

Additional links in the physiological chain of events underlying the prenatal stress syndrome have been identified. As seen in Figure 3, plasma LH levels are suppressed during days 16–20 of gestation in stressed fetuses (Salisbury, Reed, Ward, & Weisz, 1989). This finding is important because testosterone secretion by fetal testes has been shown to be stimulated by LH (Picon & Ktorza, 1976; Sanyal & Villee, 1977; Feldman & Bloch, 1978). Finally, Figure 3 demonstrates that the activity of aromatizing enzymes measured in homogenates of combined hypothalamic and amygdaloid sections was suppressed in stressed compared to control rat fetuses on days 18, 19, and 20 of gestation (Weisz *et al.*, 1982). As mentioned earlier, there is considerable evidence that in some species the process of sexual behavior differentiation involves the intracellular conversion within specific CNS targets of aromatizable androgens to estrogens. To the extent that this mechanism is functional in the rat, it may play a role in the etiology of the prenatal stress syndrome.

EFFECTS OF PRENATAL STRESS AND PREPUBERTAL SOCIALIZATION ON ADULT SEXUAL BEHAVIOR

Although prenatal stress alters the course of sexual behavior differentiation in male rodents, the data reviewed in preceeding sections of this chapter make it clear that the modifications are incomplete. Many sexually dimorphic behaviors of prenatally stressed males are clearly distinguishable from those of control males, but stressed males would not be confused with normal females. The mixture of incompletely masculinized and/or defeminized traits found in stressed males probably is the result of the partial and often transient nature of the hormonal changes that maternal stress produces in the fetuses. It appears, then, that the prenatal stress syndrome is simply another example of the principle first suggested by Phoenix *et al.* (1959), namely, that inadequate exposure to androgen during critical periods of perinatal development prevent the behaviors of genetic males from being properly masculinized. The extent of the behavioral modifications is directly related to the timing, duration, and degree of disruption of androgenic activity. What distinguishes prenatally stressed males from other commonly studied laboratory models is the procedure by which androgen levels are manipulated. Instead of surgical or chemical castration, testosterone levels of fetuses can be modulated by imposing intense adverse environmental conditions on the mother during the course of pregnancy. The extent to which this happens routinely in nature is not yet fully appreciated.

A preparation with a strong bisexual potential such as the prenatally stressed male rat is a suitable model in which to assess the potential interaction of perinatal hormonal effects with those of prepubertal socialization. The first published report utilizing this preparation was by Dunlap *et al.* (1978). Using the restraint–light procedure described earlier, these investigators produced prenatally stressed male rats. At 22 days of age, stressed and control males were placed in one of two housing conditions at the time of weaning. The first condition was the widely used standard for weaning experimental animals, which was to group subjects of the same sex and treatment condition. The second housing condition consisted of rearing males in complete social isolation. The animals were undisturbed until 55 days of age, at which time a series of tests for male sexual behavior was given. Testing was repeated at 110 days. As shown in Figure 4, the expected results were obtained. Males raised

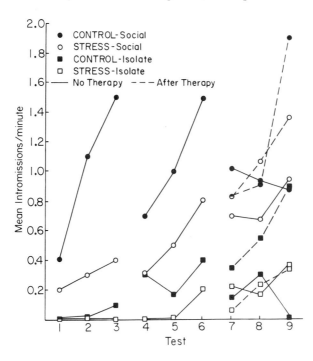

Figure 4.   Mean intromission rate shown by prenatally stressed and control male rats that from 22 days of age onward had been raised in social isolation or with other males of the same treatment group. Tests 1–3 were given beginning at 55 days of age, tests 4–6 at 100 days, and tests 7–9 at about 155 days of age. Half of the animals in each group had been rehoused (after therapy) with a prepubertal female for 2 weeks before the last three tests were given, and half (no therapy) had not. (Adapted from Dunlap *et al.*, 1978.)

in complete social isolation, regardless of prenatal treatment, failed to mate spontaneously. Also, predictably, stressed males living with other stressed males showed
impaired copulatory behavior compared to control males housed with control males.

At this point, half of the animals in each group were assigned to a cage containing prepubertal (22-day-old) females for a period of 2 weeks. Tests with estrous
females then were given to both the groups that had received "therapy" and those
for whom the original housing conditions had been continued. Exposure to a juvenile female had therapeutic consequences in two of the three groups that previously
had shown deficient copulatory behavior. As seen in Figure 4, following housing
with a female, the intromission rate increased in isolated controls and in stressed
males that had been weaned into social groups. The treatment was not effective for
prenatally stressed males that had been reared in isolation. These data show that
control males that have been normally androgenized during prenatal life show impaired copulatory behavior when raised in an impoverished social environment.
However, as had been demonstrated earlier (Gerall, Ward, *et al.*, 1967), the impairment resulting from prepubertal isolation can be partially overcome in adulthood
with an appropriate therapy such as prolonged housing with a female. Prenatally
stressed males that had developed in an abnormal androgenic milieu during fetal
life displayed impaired copulatory behavior despite being socially reared. They too,
however, were helped by therapy in adulthood. The critical group consisted of the
prenatally stressed animals in which both the prenatal hormonal milieu and the
prepubertal social environment had been impoverished. In these animals the potential for male copulatory behavior was so underdeveloped as to be impervious to
adult therapy.

We followed up this provocative study, intent both on replication and on exploring three additional issues (Ward & Reed, 1985). Specifically, we wanted to
expand the variety of social conditions under which the young males were reared to
determine the effectiveness of a different type of therapy known to promote sexual
behavior in noncopulating control males and, most importantly, to assess the effects
of early social housing on the potential for the female lordosis pattern. Accordingly,
prenatally stressed and control male rats were weaned at 16 days of age and placed
into one of the following four housing conditions: (1) paired with a same-aged
female, (2) paired with a male of the same prenatal treatment condition, (3) paired
with a male of the opposite prenatal treatment condition, or (4) placed into complete social isolation. Each isolate lived in a standard cage contained within a
wooden box with its own exhaust fan and house light. It also may be important that
no female rats were maintained in the vivaria in which the isolates or the males
housed with other males lived.

At 60 days of age, weekly tests with estrous females were given. The cumulative
percentages of males that ejaculated are shown in Figure 5. As anticipated, only a
few of the socially isolated males copulated, regardless of prenatal treatment; moreover, a large proportion of the stressed males reared with stressed males failed to
ejaculate. However, most of the prenatally stressed males housed with a control
male or with a female eventually initiated copulation.

Males that had failed to ejaculate spontaneously were given a final test with an
estrous female while exposed to mild electric skin shock. This treatment is known to
promote ejaculatory behavior in otherwise noncopulating control males (Caggiula
& Eibergen, 1969; Crowley, Popolow, & Ward, 1973). As Figure 5 illustrates, mild
skin shock was very effective in promoting copulatory behavior in stressed males

INGEBORG L.
WARD

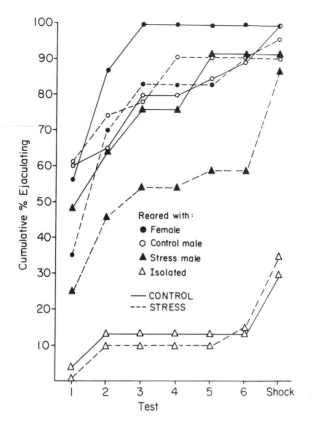

Figure 5. Cumulative percentage of control and prenatally stressed male rats ejaculating. The animals had been reared from 16 days of age onward in social isolation or with a same-aged female, control male, or prenatally stressed male. Noncopulating males were given a final test with an estrous female while receiving mild electric skin shock. (Adapted from Ward & Reed, 1985.)

reared with stressed males, a finding at odds with a previous report (Ward, 1977). A few additional isolates in both groups ejaculated when given shock.

All males then were castrated, injected weekly with estradiol benzoate and progesterone, and tested with vigorous stud males. The percentage of males that showed the female lordotic pattern on at least one of three tests is shown in Figure 6. Stressed males housed either with a stressed or a control male were more likely to exhibit lordosis behavior than were comparably housed control males. Stressed males with female partners were less likely to show lordotic behavior than were those reared with males. However, they were more responsive than control males raised with males. The high percentage of control males with female cagemates capable of showing lordosis was a particularly unexpected finding. Exposure to females seems to have precisely the reverse effect on stressed and on control males. Social isolation lowered the potential for lordosis. Fewer stressed males reared in isolation showed lordosis than did any of the socially reared stressed groups.

In summary, these data demonstrate that in male rats, the expression of both male and female sexual behaviors is affected by exposure either to prenatal stress or to various social housing conditions. Further, these two categories of variables interact to accentuate or attenuate the effect(s) expected if only one manipulation were made at a time. For example, rearing together young males that because of maternal stress had experienced an abnormal pattern of plasma testosterone during fetal development promoted abnormal male copulatory behavior coupled with a high potential for lordosis behavior. The impairment in ejaculatory behavior could be prevented by raising the stressed males with a female or with a control male. However, the abnormally high potential for the lordosis pattern could not be

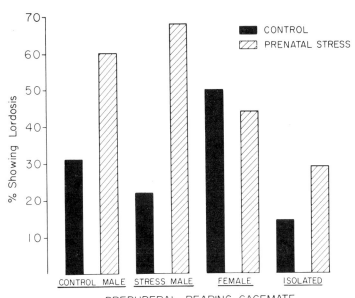

Figure 6.   Percentage of control and prenatally stressed male rats capable of showing lordosis behavior. The animals had been reared from 16 days of age onward in social isolation or with a same-aged female, control male, or prenatally stressed male. (From Ward & Reed, 1985.)

blocked by providing the stressed male with a normal male cagemate. Housing with a female lowered the expression of lordosis behavior in stressed males but did not eliminate it. The sexual behaviors shown by control males were not influenced by the type of social partner with whom the animals were reared, with one notable exception: chronic exposure to a female facilitated lordosis behavior.

Social isolation blocked the expression of whatever sexual behavior potentials a given prenatal treatment tended to promote. Few isolates exhibited the male copulatory pattern, and stressed males reared alone were less likely to show lordosis than any of the socially reared stressed groups. Regardless of prenatal treatment, the absence of adequate social stimulation resulted in a high incidence of asexual animals.

Although the modulating effects of social factors can be demonstrated readily in the prenatal stress preparation, a more recent attempt to produce an interaction between neonatal hormonal deficiencies and prepubertal socialization was less successful. As discussed earlier, in the rat the process of sexual behavior differentiation is not completed until the first few days post-partum. We postulated that favorable social conditions during prepubertal development might increase the low potential for ejaculatory behavior that neonatally castrated male rats have while lowering the capacity for lordosis behavior (Isaacson & Ward, 1986). Accordingly, male rats were castrated at 2 days of age. On day 16, control and neonatal castrates were weaned and assigned to cages containing either a same-aged female, a control male, or a neonatally castrated male. At 60 days, the control males were gonadectomized, and all animals received daily injections of TP. Tests with estrous females revealed that none of the neonatal castrates were able to ejaculate, regardless of the type of cagemate with which they had been reared. Social rearing, however, did prove important for the control group. Fewer control males reared with neonatally castrated cagemates ejaculated than did those reared with females or with control

males. Further, the few controls with neonatally castrated cagemates that did copulate required almost twice as many mounting responses to achieve their first ejaculation as did controls in the other two groups.

The animals then received weekly injections of estradiol and progesterone.
Tests with vigorous stud males revealed that the prepubertal rearing conditions did
not alter the very high levels of estrous behavior of which neonatal castrates are
capable. However, as in our previous study (Ward & Reed, 1985), female cagemates
increased the likelihood that a control male would show lordosis in adulthood.

The results of this study indicate that the extreme alterations in sexual differentiation that result from complete withdrawal of androgen during neonatal development cannot be modulated by a favorable social environment, at least not the ones
that we tested. When the organizing hormonal profile is such as to block completely
the potential for ejaculation while at the same time inducing a potential for lordosis
approximating that of normal females, insufficient plasticity remains for socialization to act on.

Paradoxically, the full androgenization that control males presumably have
undergone does provide the plasticity needed for modulation by the environment.
At this point too little is known about which elements of the social stimulus complex
critically influence behavior to predict the effects any given social situation will have
on the development of prepuberal males. For example, very different patterns of
adult sexual behavior were seen in control males that had been reared with three
different types of feminized cagemates. A female partner led to the development of
high levels of ejaculatory behavior but somehow promoted a potential for lordosis.
A neonatally castrated cagemate diminished the expression of ejaculatory behavior
but did not increase the likelihood that lordosis would occur. A prenatally stressed
male as a social partner resulted in patterns of adult sexual behavior considered to
be normal for males, i.e., ejaculatory behavior with little propensity for the lordotic
pattern. It is unclear what the unique distinguishing characteristics are that cause
these three social conditions to promote such different patterns of sexual behavior
in cagemates.

Similarly, nothing is known about the potential physiological mechanisms that
allow perinatal hormonal factors to interact with much later occurring social stimulation. It is likely that the olfactory system is involved. A number of studies have
documented that olfactory stimuli emitted by males or females alter plasma levels of
gonadal steroids in young male rats. For example, young males weaned in same-sex
groups have lower plasma testosterone levels in adulthood than males raised with
females. By the same token, exposure of young males to bedding taken from mature
males or females raises testosterone titers, with "female bedding" being more effective than "male bedding" (Dessì-Fulgheri & Lupo, 1982). Males raised in isolation
from day 21 of age have lower levels of plasma testosterone than males allowed
access to other animals (Dessì-Fulgheri, Lupo Di Prisco, & Verdarelli, 1976; Lupo
Di Prisco, Lucarini, & Dessì-Fulgheri, 1978). Thus, evidence exists that sensory
stimulation associated with specific types of social contact can alter androgen levels
in male rats.

However, knowing the effects of specific types of social stimulation on plasma
testosterone titers does not always enable one to predict their consequences on
sexual behavior potentials. For example, knowing that exposure to females raises
androgen levels in prepuberal males (Dessì-Fulgheri & Lupo, 1982) might provide
insight into why prenatally stressed males with female cagemates show more male
behavior and fewer lordosis patterns than do those with male partners. But, why

does this social condition promote a potential for lordosis in control males? Further, since social isolation reduces plasma testosterone titers, one might predict that lordosis behavior should be potentiated in isolated stressed males; instead, it was depressed.

CONCLUSIONS

The present state of information supports the conclusion that there is an interaction between the two categories of variables that traditionally have been invoked as explanatory concepts in the formation of sexual behavior potentials. Little is known, however, as to how this interaction is effected. Therefore, it is difficult to predict the consequences of specific social environments, even when the perinatal hormonal history of the individual animal is known. It is also clear that the same environment can have very different effects depending on the perinatal androgenic exposure of the animal.

In summary, it appears that the organizational theory of sexual behavior differentiation is more complicated than was thought in the recent past. Exposure to various patterns of gonadal hormones during critical periods of perinatal development creates different behavioral potentials. Which of these is realized or suppressed is influenced, in turn, by specific patterns of social stimulation experienced during prepubertal life. At puberty, those potentials that survived can be activated by gonadal hormones. The challenge facing us now is to tease out these complicated interactions and to identify the physiological mechanisms through which they are effected.

Acknowledgments

Supported by grant HD-04688 from the National Institute of Child Health and Human Development and by Research Scientist Award 2 K05 MH00049 from the National Institute of Mental Health.

REFERENCES

Anderson, D. K., Rhees, R. W., & Fleming, D. E. (1985). Effects of prenatal stress on differentiation of the sexually dimorphic nucleus of the preoptic area (SDN-POA) of the rat brain. *Brain Research, 332,* 113–118.

Anderson, D. K., Fleming, D. E., Rhees, R. W., & Kinghorn, E. (1986). Relationship between sexual activity, plasma testosterone, and the volume of the sexually dimorphic nucleus of the preoptic area in prenatally stressed and non-stressed rats. *Brain Research, 370,* 1–10.

Baum, M. J., & Tobet, S. A. (1986). Effect of prenatal exposure to aromatase inhibitor, testosterone, or antiandrogen on the development of feminine sexual behavior in ferrets of both sexes. *Physiology and Behavior, 37,* 111–118.

Beach, F. A. (1968). Coital behavior in dogs. III. Effects of early isolation on mating in males. *Behaviour, 30,* 218–238.

Beatty, W. W. (1979). Gonadal hormones and sex differences in nonreproductive behaviors in rodents: Organizational and activational influences. *Hormones and Behavior, 12,* 112–163.

Beckhardt, S., & Ward, I. L. (1983). Reproductive functioning in the prenatally stressed female rat. *Developmental Psychobiology, 16,* 111–118.

Breedlove, S. M., & Arnold, A. P. (1980). Hormone accumulation in a sexually dimorphic motor nucleus in the rat spinal cord. *Science, 210,* 564–566.

Caggiula, A. R., & Eibergen, R. (1969). Copulation of virgin male rats evoked by painful peripheral stimulation. *Journal of Comparative and Physiological Psychology, 69*, 414–419.

Commins, D., & Yahr, P. (1984). Adult testosterone levels influence the morphology of a sexually dimorphic area in the mongolian gerbil brain. *The Journal of comparative Neurology, 224*, 132–140.

Crowley, W. R., Popolow, H. B., & Ward, O. B. (1973). From dud to stud: Copulatory behavior elicited through conditioned arousal in sexually inactive male rats. *Physiology and Behavior, 10*, 391–394.

Dahlöf, L. G., Hård, E., & Larsson, K. (1977). Influence of maternal stress on offspring sexual behaviour. *Animal Behaviour, 25*, 958–963.

Dahlöf, L. G., Hård, E., & Larsson, K. (1978). Influence of maternal stress on the development of the fetal genital system. *Physiology and Behavior, 20*, 193–195.

Dessì-Fulgheri, F., & Lupo, C. (1982). Odour of male and female rats changes hypothalamic aromatase and 5α-reductase activity and plasma sex steroid levels in unisexually reared male rats. *Physiology and Behavior, 28*, 231–235.

Dessì-Fulgheri, F., Lupo Di Prisco, C., & Verdarelli, P. (1976). Effects of two kinds of social deprivation on testosterone and estradiol-17β plasma levels in the male rat. *Experientia, 32*, 114–115.

Diamond, M. C. (1984). Age, sex, and environmental influences. In N. Geschwind & A. M. Galaburda (Eds.), *Cerebral dominance: The biological foundations* (pp. 134–146). Cambridge: Harvard University Press.

Diamond, M. C., Dowling, G. A., & Johnson, R. E. (1981). Morphological cerebral cortical asymmetry in male and female rats. *Experimental Neurology, 71*, 261–268.

Diamond, M. D., Johnson, R. E., Young, D., & Singh, S. S. (1983). Age-related morphological differences in the rat cerebral cortex and hippocampus: Male–female; right–left. *Experimental Neurology, 81*, 1–13.

Döhler, K. D., Coquelin, A., Davis, F., Hines, M., Shryne, J. E., & Gorski, A. (1984). Pre- and postnatal influence of testosterone propionate and diethylstilbestrol on differentiation of the sexually dimorphic nucleus of the preoptic area in male and female rats. *Brain Research, 302*, 291–295.

Döhler, K. D., Coquelin, A., Davis, F., Hines, M., Shryne, J. E., Sickmoller, P. M., Jarzab, B., & Gorski, R. A. (1986). Pre- and postnatal influence of an estrogen antagonist and an androgen antagonist on differentiation of the sexually dimorphic nucleus of the preoptic area in male and female rats. *Neuroendocrinology, 42*, 443–448.

Dörner, G., Geier, T., Ahrens, L., Krell, Münx, G., Sieler, H., Kittner, E., & Müller, H. (1980). Prenatal stress as possible aetiogenetic factor of homosexuality in human males. *Endokrinologie, 75*, 365–368.

Dörner, G., Götz, F., & Döcke, W. D. (1983). Prevention of demasculinization and feminization of the brain in prenatally stressed male rats by perinatal androgen treatment. *Experimental and Clinical Endocrinology, 81*, 88–90.

Dörner, G., Schenk, B., Schmiedel, B., & Ahrens, L. (1983). Stressful events in prenatal life of bi- and homosexual men. *Experimental and Clinical Endocrinology, 81*, 83–87.

Duffy, J. A., & Hendricks, S. E. (1973). Influences of social isolation during development on sexual behavior of the rat. *Animal Learning and Behavior, 1*, 223–227.

Dunlap, J. L., Zadina, J. E., & Gougis, G. (1978). Prenatal stress interacts with prepuberal social isolation to reduce male copulatory behavior. *Physiology and Behavior, 21*, 873–875.

Feldman, S. C., & Bloch, E. (1978). Developmental pattern of testosterone synthesis by fetal rat testes in response to luteinizing hormone. *Endocrinology, 102*, 999–1007.

Fillion, T. J., & Blass, E. M. (1986). Infantile experience with suckling odors determines adult sexual behavior in male rats. *Science, 231*, 729–731.

Fleming, D. E., Anderson, R. H., Rhees, R. W., Kinghorn, E., & Bakaitis, J. (1986). Effects of prenatal stress on sexually dimorphic asymmetries in the cerebral cortex of the male rat. *Brain Research Bulletin, 16*, 395–398.

Folman, Y., & Drori, D. (1965). Normal and aberrant copulatory behaviour in male rats (*R. norvegicus*) reared in isolation. *Animal Behaviour, 13*, 427–429.

Friedman, M. I., & Bruno, J. P. (1976). Exchange of water during lactation. *Science, 191*, 409–410.

George, F. W., & Ojeda, S. R. (1982). Changes in aromatase activity in the rat brain during embryonic, neonatal, and infantile development. *Endocrinology, 111*, 522–529.

Gerall, A. A. (1963). An exploratory study of the effects of social isolation variables on the sexual behaviour of male guinea pigs. *Animal Behaviour, 11*, 274–282.

Gerall, A. A., & Ward, I. L. (1966). Effects of prenatal exogenous androgen on the sexual behavior of the female albino rat. *Journal of Comparative and Physiological Psychology, 62*, 370–375.

Gerall, A. A., Hendricks, S. E., Johnson, L. L., & Bounds, T. W. (1967). Effects of early castration in male rats on adult sexual behavior. *Journal of Comparative and Physiological Psychology, 64*, 206–212.

Gerall, H. D., Ward, I. L., & Gerall, A. A. (1967). Disruption of the male rat's sexual behaviour induced by social isolation. *Animal Behaviour, 15*, 54–58.

Goldfoot, D. A., & Wallen, K. (1979). Development of gender role behaviors in heterosexual and isosexual groups of infant rhesus monkeys. In D. J. Chivers & J. Herbert (Eds.), *Recent advances in primatology, Vol. 1: Behaviour* (pp. 155–159). London: Academic Press.

Goldfoot, D. A., Wallen, K., Neff, D. A., McBriar, M. C., & Goy, R. W. (1984). Social influences on the display of sexually dimorphic behavior in rhesus monkeys: Isosexual rearing. *Archives of Sexual Behavior, 13,* 395–412.

Gorski, R. A., Gordon, J. H., Shryne, J. E., & Southam, A. M. (1978). Evidence for a morphological sex difference within the medial preoptic area of the rat brain. *Brain Research, 148,* 333–346.

Götz, F., & Dörner, G. (1980). Homosexual behaviour in prenatally stressed male rats after castration and oestrogen treatment in adulthood. *Endokrinologie, 76,* 115–117.

Goy, R. W. (1981). Differentiation of male social traits in female rhesus macaques by prenatal treatment with androgens: Variation in type of androgen, duration, and timing of treatment. In M. J. Novy & J. A. Resko (Eds.), *Fetal endocrinology* (pp. 319–339). New York: Academic Press.

Goy, R. W., & Goldfoot, D. A. (1974). Experiential and hormonal factors influencing development of sexual behavior in the male rhesus monkey. In F. O. Schmitt & F. G. Worden (Eds.), *The neurosciences, third study program* (pp. 571–581). Cambridge, MA: MIT Press.

Goy, R. W., & McEwen, B. S. (1980). *Sexual differentiation of the brain.* Cambridge: MIT Press.

Goy, R. W., & Robinson, J. A. (1982). Prenatal exposure of rhesus monkeys to potent androgens: Morphological, behavioral and physiological consequences. In *Banbury report 11: Environmental effects on maturation* (pp. 1–21). New York: Cold Spring Harbor Laboratory.

Goy, R. W., & Wallen, K. (1979). Experiential variables influencing play, foot-clasp mounting and adult sexual competence in male rhesus monkeys. *Psychoneuroendocrinology, 4,* 1–12.

Goy, R. W., & Young, W. C. (1957). Somatic basis of sexual behavior patterns in guinea pigs. *Psychosomatic Medicine, 19,* 144–151.

Goy, R. W., Wallen, K., & Goldfoot, D. A. (1974). Social factors affecting the development of mounting behavior in male rhesus monkeys. In W. Montagna & W. A. Sadler (Eds.), *Reproductive behavior* (pp. 223–247). New York: Plenum Press.

Grady, K. L., Phoenix, C. H., & Young, W. C. (1965). Role of the developing rat testis in differentiation of the neural tissues mediating mating behavior. *Journal of Comparative and Physiological Psychology, 59,* 176–182.

Grisham, W., Kerchner, M., & Ward, I. L. (1991). Prenatal stress alters sexually dimorphic nuclei in the spinal cord of male rats. *Brain Research, 551,* 126–131.

Gruendel, A. D., & Arnold, W. J. (1969). Effects of early social deprivation on reproductive behavior of male rats. *Journal of Comparative and Physiological Psychology, 67,* 123–128.

Gruendel, A. D., & Arnold, W. J. (1974). Influence of preadolescent experiential factors on the development of sexual behavior in albino rats. *Journal of Comparative and Physiological Psychology, 86,* 172–178.

Gubernick, D. J., & Alberts, J. R. (1983). Maternal licking of young: Resource exchange and proximate controls. *Physiology and Behavior, 31,* 593–601.

Handa, R. J., Corbier, P., Shryne, J. E., Schoonmaker, J. N., & Gorski, R. A. (1985). Differential effects of the perinatal steroid environment on three sexually dimorphic parameters of the rat brain. *Biology of Reproduction, 32,* 855–864.

Hansen, S. (1977). Mounting behavior and receptive behavior in developing female rats and the effect of social isolation. *Physiology and Behavior, 19,* 749–752.

Hård, E., & Larsson, K. (1968). Dependence of adult mating behavior in male rats on the presence of littermates in infancy. *Brain, Behavior and Evolution, 1,* 405–419.

Harper, L. V. (1968). The effects of isolation from birth on the social behaviour of guinea pigs in adulthood. *Animal Behaviour, 16,* 58–64.

Harris, G. W., & Levine, S. (1965). Sexual differentiation of the brain and its experimental control. *Journal of Physiology, 181,* 379–400.

Hart, B. L., & Melese-D'Hospital, P. Y. (1983). Penile mechanisms and the role of striated penile muscles in penile reflexes. *Physiology and Behavior, 31,* 807–813.

Harvey, P. W., & Chevins, P. F. D. (1984). Crowding or ACTH treatment of pregnant mice affects adult copulatory behavior of male offspring. *Hormones and Behavior, 18,* 101–110.

Harvey, P. W., & Chevins, P. F. D. (1985). Crowding pregnant mice affects attack and threat behavior of male offspring. *Hormones and Behavior, 19,* 86–97.

Herrenkohl, L. R. (1979). Prenatal stress reduces fertility and fecundity in female offspring. *Science, 206,* 1097–1099.

Herrenkohl, L. R., & Politch, J. A. (1978). Effects of prenatal stress on the estrous cycle of female offspring as adults. *Experientia, 34,* 1240–1241.

Herrenkohl, L. R., & Scott, S. (1984). Prenatal stress and postnatal androgen: Effects on reproduction in female rats. *Experientia, 40,* 101–103.

Herrenkohl, L. R., & Whitney, J. B. (1976). Effects of prepartal stress on postpartal nursing behavior, litter development and adult sexual behavior. *Physiology and Behavior, 17,* 1019–1021.

Hines, M., Davis, F. C., Coquelin, A., Goy, R. W., & Gorski, R. A. (1985). Sexually dimorphic regions in the medial preoptic area and the bed nucleus of the stria terminalis of the guinea pig brain: A description and an investigation of their relationship to gonadal steroids in adulthood. *The Journal of Neuroscience, 5,* 40–47.

Hoepfner, B. A., & Ward, I. L. (1988). Prenatal and neonatal androgen exposure interact to affect sexual differentiation in female rats. *Behavioral Neuroscience, 102,* 61–65.

Huffman, L., & Hendricks, S. E. (1981). Prenatally injected testosterone propionate and sexual behavior of female rats. *Physiology and Behavior, 26,* 773–778.

Imperato-McGinley, J., Guerrero, L., Gautier, T., & Peterson, R. E. (1974). Steroid 5α-reductase deficiency in man: An inherited form of male pseudohermaphroditism. *Science, 186,* 1213–1215.

Imperato-McGinley, J., Binienda, Z., Arthur, A., Mininberg, D. T., Vaughan, E. D., Jr., & Quimby, F. W. (1985). The development of a male pseudohermaphroditic rat using an inhibitor of the enzyme 5α-reductase. *Endocrinology, 116,* 807–812.

Imperato-McGinley, J., Binienda, Z., Gedney, J., & Vaughan, E. D., Jr. (1986). Nipple differentiation in fetal male rats treated with an inhibitor of the enzyme 5α-reductase: Definition of a selective role for dihydrotestosterone. *Endocrinology, 118,* 132–137.

Isaacson, M. D., & Ward, I. L. (1986). Prepuberal social rearing conditions and sexual behavior in control and neonatally castrated male rats. *Physiology and Behavior, 37,* 469–473.

Jacobson, C. D., Csernus, V. J., Shryne, J. E., & Gorski, R. A. (1981). The influence of gonadectomy, androgen exposure, or a gonadal graft in the neonatal rat on the volume of the sexually dimorphic nucleus of the preoptic area. *The Journal of Neuroscience, 1,* 1142–1147.

Jost, A. (1985). Sexual organogenesis. In N. T. Adler, D. Pfaff, & R. Goy (Eds.), *Handbook of behavioral neurobiology: Reproduction, Vol. 7* (pp. 3–19). New York: Plenum Press.

Kerchner, M., & Ward, I. L. (1992). SDN-MPOA volume in male rats is decreased by prenatal stress but is not related to ejaculatory behavior. *Brain Research,* (in press).

Kinsley, C. H., & Bridges, R. S., (1988). Prenatal stress and maternal behavior in intact virgin rats: Response latencies are decreased in males and increased in females. *Hormones and Behavior, 22,* 76–89.

Kinsley, C., & Svare, B. (1986). Prenatal stress reduces intermale aggression in mice. *Physiology and Behavior, 36,* 783–786.

Lupo Di Prisco, C., Lucarini, N., & Dessì-Fulgheri, F. (1978). Testosterone aromatization in rat brain is modulated by social environment. *Physiology and Behavior, 20,* 345–348.

MacLusky, N. J., Philip, A., Hurlburt, C., & Naftolin, F. (1985). Estrogen formation in the developing rat brain: Sex differences in aromatase activity during early post-natal life. *Psychoneuroendocrinology, 10,* 355–361.

Mason, W. A. (1960). The effects of social restriction on the behavior of rhesus monkeys: I. Free social behavior. *Journal of Comparative and Physiological Psychology, 53,* 582–589.

Masterpasqua, F., Chapman, R. H., & Lore, R. K. (1976). The effects of prenatal psychological stress on the sexual behavior and reactivity of male rats. *Developmental Psychobiology, 9,* 403–411.

McLeod, P. J., & Brown, R. E. (1988). The effects of prenatal stress and postweaning housing conditions on parental and sexual behavior of male Long–Evans rats. *Psychobiology, 16,* 372–380.

Meisel, R. L., Dohanich, G. P., & Ward, I. L. (1979). Effects of prenatal stress on avoidance acquisition, open-field performance and lordotic behavior in male rats. *Physiology and Behavior, 22,* 527–530.

Miley, W. M. (1983). Prenatal stress suppresses hunger-induced rat-pup killing in Long–Evans rats. *Bulletin of the Psychonomic Society, 21,* 495–497.

Missakian, E. A. (1969). Reproductive behavior of socially deprived male rhesus monkeys (*Macaca mulatta*). *Journal of Comparative and Physiological Psychology, 69,* 403–407.

Moore, C. L. (1982). Maternal behavior of rats is affected by hormonal condition of pups. *Journal of Comparative and Physiological Psychology, 96,* 123–129.

Moore, C. L. (1984). Maternal contribution to the development of masculine sexual behavior in laboratory rats. *Developmental Psychobiology, 17,* 347–356.

Moore, C. L., & Morelli, G. A. (1979). Mother rats interact differently with male and female offspring. *Journal of Comparative and Physiological Psychology, 93,* 677–684.

Nadler, R. D. (1969). Differentiation of the capacity for male sexual behavior in the rat. *Hormones and Behavior, 1,* 53–63.

Ohkawa, T. (1987). Sexual differentiation of social play and copulatory behavior in prenatally stressed male and female offspring of the rat: the influence of simultaneous treatment by tyrosine during the exposure to the prenatal stress. *Folia Endocrinologica, 63,* 823–825.

Orth, J. M., Weisz, J., Ward, O. B., & Ward, I. L. (1983). Environmental stress alters the developmental pattern of Δ^5-3β-hydroxysteroid dehydrogenase activity in Leydig cells of fetal rats: A quantitative cytochemical study. *Biology of Reproduction, 28,* 625–631.

Phoenix, C. H., Goy, R. W., Gerall, A. A., & Young, W. C. (1959). Organizing action of prenatally administered testosterone propionate on the tissues mediating mating behavior in the female guinea pig. *Endocrinology, 65,* 369–382.

Picon, R., & Ktorza, A. (1976). Effects of LH on testosterone production by foetal rat testes *in vitro. FEBS Letters, 68,* 19–22.

Politch, J. A., & Herrenkohl, L. R. (1984). Effects of prenatal stress on reproduction in male and female mice. *Physiology and Behavior, 32,* 95–99.

Raisman, G., & Field, P. M. (1971). Sexual dimorphism in the preoptic area of the rat. *Science, 173,* 731–733.

Raisman, G., & Field, P. M. (1973). Sexual dimorphism in the neuropil of the preoptic area of the rat and its dependence on neonatal androgen. *Brain Research, 54,* 1–29.

Reddy, V. V. R., Naftolin, F., & Ryan, K. J. (1974). Conversion of androstenedione to estrone by neural tissues from fetal and neonatal rats. *Endocrinology, 94,* 117–121.

Rhees, R. W., & Fleming, D. E. (1981). Effects of malnutrition, maternal stress, or ACTH injection during pregnancy on sexual behavior of male offspring. *Physiology and Behavior, 27,* 879–882.

Rhees, R. W., Badger, D. S., & Fleming, D. E. (1983). Naloxone induces copulation in control but not in prenatally stressed male rats. *Bulletin of the Psychonomic Society, 21,* 498–500.

Sachs, B. D. (1982). Role of striated penile muscles in penile reflexes, copulation, and induction of pregnancy in the rat. *Journal of Reproduction and Fertility, 66,* 433–443.

Salisbury, R., Reed, J., Ward, I. L., & Weisz, J. (1989). Plasma luteinizing hormone levels in normal and prenatally stressed male and female rat fetuses and their mothers. *Biology of Reproduction, 40,* 111–117.

Sanyal, M. K., & Villee, C. A. (1977). Stimulation of androgen biosynthesis in rat fetal testes *in vitro* by gonadotropins. *Biology of Reproduction, 16,* 174–181.

Spevak, A. M., Quadagno, D. M., & Knoeppel, D. (1973). The effects of isolation on sexual and social behavior in the rat. *Behavioral Biology, 8,* 63–73.

Stewart, J., & Kolb, B. (1988). The effects of neonatal gonadectomy and prenatal stress on cortical thickness and asymmetry in rats. *Behavioral and Neural Biology, 49,* 344–360.

Swaab, D. F., & Fliers, E. (1985). A sexually dimorphic nucleus in the human brain. *Science, 228,* 1112–1115.

Takahashi, L. K., Haglin, C., & Talin, N. H. (1992). Prenatal stress potentiates stress-induced behavior and reduces the propensity to play in juvenile rats. *Physiology and Behavior,* (in press).

Thomas, D. A., McIntosh, T. K., & Barfield, R. J. (1980). Influence of androgen in the neonatal period on ejaculatory and postejaculatory behavior in the rat. *Hormones and Behavior, 14,* 153–162.

Thornton, J., & Goy, R. W. (1986). Female-typical sexual behavior of rhesus and defeminization by androgens given prenatally. *Hormones and Behavior, 20,* 129–147.

Tobet, S. A., Baum, M. J., Tang, H. B., Shim, J. H., & Canick, J. A. (1985). Aromatase activity in the perinatal rat forebrain: Effects of age, sex and intrauterine position. *Developmental Brain Research, 23,* 171–178.

Turner, C. H., Davenport, R. K., Jr., & Rogers, C. M. (1969). The effect of early deprivation on the social behavior of adolescent chimpanzees. *American Journal of Psychiatry, 125,* 1531–1536.

Valenstein, E. S., Riss, W., & Young, W. C. (1955). Experiential and genetic factors in the organization of sexual behavior in male guinea pigs. *Journal of Comparative and Physiological Psychology, 48,* 397–403.

vom Saal, F. S. (1983). Variation in infanticide and parental behavior in male mice due to prior intrauterine proximity to female fetuses: Elimination by prenatal stress. *Physiology and Behavior, 30,* 675–681.

Wallen, K., Bielert, C., & Slimp, J. (1977). Foot clasp mounting in the prepubertal rhesus monkey: Social and hormonal influences. In F. E. Poirie & S. Cevalier-Skolnikoff (Eds.), *Biosocial development among primates* (pp. 439–461). New York: Garland Publishing.

Wallen, K., Goldfoot, D. A., & Goy, R. W. (1981). Peer and maternal influences on the expression of foot-clasp mounting by juvenile male rhesus monkeys. *Developmental Psychobiology, 14,* 299–309.

Ward, I. L. (1969). Differential effect of pre- and postnatal androgen on the sexual behavior of intact and spayed female rats. *Hormones and Behavior, 1,* 25–36.

Ward, I. L. (1972). Prenatal stress feminizes and demasculinizes the behavior of males. *Science, 175,* 82–84.

Ward, I. L. (1977). Exogenous androgen activates female behavior in noncopulating, prenatally stressed male rats. *Journal of Comparative and Physiological Psychology, 91,* 465–471.

Ward, I. L. (1984). The prenatal stress syndrome: Current status. *Psychoneuroendocrinology, 9,* 3–11.

Ward, I. L., & Reed, J. (1985). Prenatal stress and prepuberal social rearing conditions interact to determine sexual behavior in male rats. *Behavioral Neuroscience, 99,* 301–309.

Ward, I. L., & Renz, F. J. (1972). Consequences of perinatal hormone manipulation on the adult sexual behavior of female rats. *Journal of Comparative and Physiological Psychology, 78,* 349–355.

Ward, I. L., & Stehm, K. E. (1991). Prenatal stress feminizes juvenile play patterns in male rats. *Physiology and Behavior, 50,* (in press).

Ward, I. L., & Ward, O. B., (1985). Sexual behavior differentiation: Effects of prenatal manipulations in rats. In N. Adler, D. Pfaff, & R. W. Goy (Eds.), *Handbook of behavioral neurobiology, Vol. 7* (pp. 77–97). New York: Plenum Press.

Ward, I. L., & Weisz, J. (1980). Maternal stress alters plasma testosterone in fetal males. *Science, 207,* 328–329.

Ward, I. L., & Weisz, J. (1984). Differential effects of maternal stress on circulating levels of corticosterone, progesterone, and testosterone in male and female rat fetuses and their mothers. *Endocrinology, 114,* 1635–1644.

Ward, I. L., Ward, O. B., Hayden, T., Weisz, J., & Orth, J. M. O. (1990). Naltrexone normalizes the suppression but not the surge of Δ^5-3β-hydroxysteroid dehydrogenase activity in Leydig cells of stressed rat fetuses. *Endocrinology, 127,* 88–92.

Ward, O. B., Monaghan, E. P., & Ward, I. L. (1986). Naltrexone blocks the effects of prenatal stress on sexual behavior differentiation in male rats. *Pharmacology, Biochemistry and Behavior, 25,* 573–576.

Weisz, J., & Ward, I. L. (1980). Plasma testosterone and progesterone titers of pregnant rats, their male and female fetuses, and neonatal offspring. *Endocrinology, 106,* 306–316.

Weisz, J., Brown, B. L., & Ward, I. L. (1982). Maternal stress decreases steroid aromatase activity in brains of male and female rat fetuses. *Neuroendocrinology, 35,* 374–379.

Whalen, R. E., & Edwards, D. A. (1967). Hormonal determinants of the development of masculine and feminine behavior in male and female rats. *Anatomical Record, 157,* 173–180.

Whitney, J. B., & Herrenkohl, L. R. (1977). Effects of anterior hypothalamic lesions on the sexual behavior of prenatally-stressed male rats. *Physiology and Behavior, 19,* 167–169.

Wilke, D. L., Tseu, S. R., Rhees, R. W., & Fleming, D. E. (1982). Effects of environmental stress or ACTH treatment during pregnancy on maternal and fetal plasma androstenedione in the rat. *Hormones and Behavior, 16,* 293–303.

Wilson, J. D. (1978). Sexual differentiation. *Annual Review of Physiology, 40,* 279–306.

Wilson, J. D., Griffin, J. E., & George, F. W. (1980). Sexual differentiation: Early hormone synthesis and action. *Biology of Reproduction, 22,* 9–17.

Wilson, J. D., Griffin, J. E., George, F. W., & Leshin, M. (1981). The role of gonadal steroids in sexual differentiation. *Recent Progress in Hormone Research, 37,* 1–34.

6

Fetal Drug Exposure and Sexual Differentiation of Males

O. BYRON WARD

INTRODUCTION

A variety of commonly abused drugs interfere with steroidogenesis in the testes of adult animals and men. These drugs suppress hormone synthesis by a direct inhibitory action on the testes, and/or by disrupting hypothalamic–pituitary stimulation of the gonads. Plasma testosterone (T) titers are attenuated after use of any of a wide variety of compounds, including licit addictive substances, such as alcohol and nicotine, as well as illicit drugs, such as marijuana and opiates. Some males are passively exposed to these substances at very early stages of development because women who have become dependent on drugs often continue to use them when pregnant. Based on a 1988 survey of 36 hospitals across the United States by the National Association for Perinatal Addiction Research and Education, it is estimated that there are over 1,000 babies born each day in the United States to women who used opiates, cocaine, phencyclidine (PCP), amphetamines, and/or marijuana during pregnancy (NIDA, 1989). That estimate does not include the large number of additional women who use only the more commonly abused substances, alcohol and/or nicotine, while pregnant. For many abused drugs there is little if any effective placental barrier, so that the developing fetus is exposed to the drug taken by the mother. Thus, many fetal males are inadvertently exposed to drugs that, among other things, suppress T levels.

There are potentially very important biological consequences of any drug-induced disruption of testicular steroidogenesis in fetal males. Normal masculinization and defeminization of the fetus are thought to be dependent on the actions of

O. BYRON WARD   Department of Psychology, Villanova University, Villanova, Pennsylvania 19085.

Sexual Differentiation, Volume 11 of *Handbook of Behavioral Neurobiology*, edited by Arnold A. Gerall, Howard Moltz, and Ingeborg L. Ward, Plenum Press, New York, 1992.

181

an androgen, T or its metabolites, during critical periods of prenatal or perinatal development. The source of this T is the testes of the fetal male. If maternal drug use impairs fetal testicular function, androgen-sensitive target structures would develop without adequate hormonal stimulation. A partial failure of masculinization as well as incomplete defeminization of the fetus could result from any drug-induced alterations in plasma T levels. For a more detailed coverage of the sexual differentiation hypothesis, see Ward and Ward (1985) and Chapters 3 and 5 by W. Beatty and I. L. Ward in this volume.

The present chapter reviews the evidence that *in utero* exposure to certain nonsteroidal drugs is associated with alterations in (1) fetal T titers, (2) morphological and/or physiological parameters that are normally masculinized by exposure to T during prenatal development, and (3) sexually dimorphic copulatory patterns as well as other behaviors that show variation related to gender. The normal differentiation of these behaviors also is dependent on the organizational action of adequate amounts of prenatal T and/or its metabolites.

As we focus on drugs of abuse, some T-attenuating drugs that are used primarily for therapeutic purposes, for example, barbiturates and cimetidine, are also covered.

There are numerous difficulties in trying to assess the effects of maternal drug use during pregnancy on human sexual differentiation. These include:

1. The drug-addicted woman is likely to be a polydrug abuser, making it difficult to attribute any abnormalities in her offspring to exposure to any one particular substance.
2. It is often difficult to determine the amount of a particular drug a woman was taking during specific stages of pregnancy. Not only is this information subject to the usual problems of retrospective reports, but there are the added complications of not knowing the purity of the substance when a street drug was taken and the fact that many women are reluctant to admit taking an illicit substance, especially while pregnant.
3. During pregnancy the drug abuser is likely to neglect proper nutrition and other aspects of good prenatal care. Poor nutritional status during pregnancy and other general health considerations may create direct or stress-mediated effects on sexual differentiation, independent of prenatal drug exposure.
4. Women who take drugs while pregnant often continue to take them after the child's birth, raising questions about the extent to which alterations in the child might be attributable to drug-related neglect during early postnatal development rather than to *in utero* drug exposure.
5. Many of these drugs act primarily as behavioral teratogens; i.e., they produce abnormalities in neurological functioning and behavior of the offspring in the absence of any gross physical defects. Although no less devastating, behavioral effects are not as easily recognized at birth and during very early development as are physical deformities. The relationship between behavioral abnormalities and the drug status of the mother during pregnancy is, therefore, less likely to be detected.

For the above reasons, nearly all studies on the effects of *in utero* exposure to nonsteroidal drugs on the sexual differentiation of males have been done with experimental animals and, more specifically, with laboratory rodents. The animal studies offer a number of advantages in terms of experimental control. Exposure to one drug, or to specific drug combinations, in known dosage, during limited stages of gestation is easily achieved. If the drug has an adverse effect on food intake, a pair-fed control group can help to rule out altered nutritional status during pregnancy as a confounding variable. Cross-fostering of the offspring at birth to non-drug mothers allows control of postnatal factors that might influence development. Another advantage is that morphological, physiological, and behavioral influences of drug exposure can be studied not only in the neonate and thereafter but also in the fetus. The obvious limitation of rat studies is that the results do not necessarily generalize directly to other species, including humans.

As each drug is discussed, the evidence that testicular steroidogenesis is altered by exposure during adulthood is covered first. Some comment is made on the prevalence of use of the substance by pregnant women. Then the experimental studies on fetal endocrine effects as well as evidence for residual effects of fetal drug exposure are examined.

ETHANOL

ACTIONS IN ADULT MALES

Either acute or chronic exposure of adult rats or men to ethanol depresses T levels. A few representative early studies from what is now a voluminous literature are presented to illustrate these findings. In rats, acute exposure achieved by intraperitoneal (IP) injection of 1.5 to 2.5 g of ethanol/kg of body weight reduced plasma T levels by over 50% for a period of 6 hours or more. Paradoxically, a smaller dose (0.75 g/kg) led to a substantial increase in plasma T levels (Cicero & Badger, 1977b). In nonalcoholic men, drinking 2.4 mL/kg of 100-proof vodka induced a 20% reduction in plasma T levels. The T fell within 30 minutes and was still depressed 3 hours after consumption. Blood alcohol levels peaked at slightly over 100 mg/dL (Mendelson, Mello, & Ellingboe, 1977). Chronic ethanol exposure in rats for 4 months, during which blood alcohol levels ranged from 60 to 170 mg/dL, depressed plasma T concentrations by about 40% compared to pair-fed controls (Van Thiel, Gavaler, Cobb, Sherins, & Lester, 1979). In nonalcoholic men, chronic consumption of ethanol [every 3 hours around the clock for 25 days (3 g/kg per day)] depressed plasma T between 29% and 55%, depending on the individual (Gordon, Altman, Southren, Rubin, & Lieber, 1976).

Ethanol appears to suppress T in adult males by acting at a variety of levels in the hypothalamic–pituitary–gonadal axis. Normally, T synthesis is governed by a sequence of steps which are initiated by a release of luteinizing hormone-releasing hormone (LHRH) from the hypothalamus. The LHRH acts on the anterior pituitary to release luteinizing hormone (LH), which in turn interacts with gonadotropin receptors in the testes to facilitate the synthesis of T by Leydig cells. Each link in this chain of events is influenced by alcohol. The primary inhibitory effect of ethanol appears to be directly on the testis. Ethanol causes a depletion of testicular gonadotropin receptors (Bhalla, Rajan, & Newman, 1983; Gantt *et al.*, 1982). Further, ethanol reduces the activity of testicular 3β-hydroxysteroid dehydrogenase/isomerase, the rate-limiting enzyme in the synthesis of T from pregnenolone (Chiao, Johnston, Gavaler, & Van Thiel, 1981). In addition to blocking synthesis at the testicular level, ethanol reduces circulating T by enhancing liver metabolism. Ethanol also attenuates synthesis of T by depressing LH levels. It seems likely that the primary central action on LH is the result of inhibition of LHRH rather than an effect exerted directly at the level of the pituitary, although this point is not yet clear (see Cicero, 1981; Van Thiel & Gavaler, 1982, for reviews).

USE DURING PREGNANCY

On a national level, no accurate estimates are available for the prevalence of alcohol use (or, for that matter, the use of any other drug of abuse) during pregnancy (Jones & Lopez, 1988). Not surprisingly, women tend to underreport their use of alcohol while pregnant (Ernhart, Morrow-Tlucak, Sokol, & Martier, 1988). However, alcohol and nicotine are the most common teratogens to which fetuses are exposed. The 1985 National Household Survey on Drug Abuse showed that 90% of women of childbearing age (15 to 44 years) had used alcohol at least once and that 61% had used alcohol within the past month (Adams, Gfroerer, & Rouse, 1989). In a study of over 7,000 women in an antenatal clinic in Cleveland, 11% tested positive on a test for alcoholism, and 3.5% were estimated to have consumed 21 or more drinks per week during the early part of their pregnancy (Ernhart, Sokol, Ager, Morrow-Tlucak, & Martier, 1989). In 1987 in a large hospital in Dallas, 1.4% of pregnant women were considered alcohol abusers, as defined as consuming four or more alcoholic drinks per day (Little, Snell, Gilstrap, Gant, & Rosenfeld, 1989). Mello, Mendelson, and Teoh (1989) found that peak blood alcohol levels in women who averaged 3.8 drinks/day ranged from 27 to 233 mg/dL.

Alcohol, like other drugs with a relatively small molecular weight, passes freely across the placental barrier into the fetus. On day 19 of gestation, ethanol levels in the blood and brain of fetuses of chronically intoxicated rats were equal to those in the mothers (Guerri, Esquifino, Sanchis, & Grisolia, 1984). In effect, there is virtually no placental barrier to ethanol (Szeto, 1989).

In about 25% of infants whose mothers are heavy drinkers, especially during early pregnancy, the effects of *in utero* exposure to ethanol are expressed as a full-blown fetal alcohol syndrome (FAS). In FAS, the most prominent features are characteristic changes in facial appearance. However, mild to moderate mental retardation, microcephaly, growth retardation, and irritability usually also occur in FAS (Clarren & Smith, 1978; Colangelo & Jones, 1982; Hanson, Jones, & Smith, 1976; Jones & Smith, 1973; Jones, Smith, Ulleland, & Streissguth, 1973). Malformations of the external genitals, including hypospadias and cryptorchidism, are occasionally noted (Lausecker, Withofs, Ritz, & Pennerath, 1976; Majewski, Bierich, Michaelis, Loser, & Leiber, 1977). The full FAS has been seen only in offspring of women who drank six or more drinks per day. However, a partial expression of the facial abnormalities of FAS is sometimes noted in offspring of women consuming three or more drinks per day (Ernhart *et al.*, 1989). Furthermore, although the facial malformations of FAS are readily recognized, cognitive impairment is a more frequent and less easily detected characteristic (Abel, 1980) and may be induced by lower alcohol consumption than are physical anomalies. Thus, although women do tend to reduce their alcohol intake during pregnancy, many continue to drink, some of them heavily, and the alcohol passes freely into the fetus, as is dramatically reflected in FAS individuals.

In rats, pregnant or otherwise, the experimental technique of choice for inducing consumption of enough alcohol so that clinically relevant blood alcohol levels are obtained is as follows (Lieber & DeCarli, 1989). A liquid diet in which ethanol (5 g/dL) provides 36% of total calories is the sole available source of nutrition. Daily intake of 10 to 14 g ethanol/kg of body weight is typically found. With the high rate of metabolic clearance of ethanol found in rats, this intake level produces blood alcohol levels in pregnant rats that are in the desired range of 125 to 200 mg/dL. The liquid ethanol diet procedure sustains clinically relevant blood alcohol levels in rats without introducing the stress of frequent intubation. Alcohol consumption reduces caloric intake; thus, appropriate controls for nutritional factors must be included in any study. Pair-fed control animals are matched with regard to daily intake by utilizing an isocaloric diet in which sucrose or dextrin-maltose is substituted for ethanol.

Effects on Fetal Testicular Function

In experimental studies, the normal ontogeny of fetal endocrine function has to be kept in mind, so that the effects of drug exposure encompassing the period of fetal T production can be evaluated. In the rat, pregnancy usually lasts about 23 days. The critical perinatal period during which testicular steroids regulate sexual differentiation begins during the second week of gestation. At approximately the 14th day of gestation, differentiation of the tissue that becomes the testes in the male takes place (Torrey, 1945). By the 16th day the testes actively begin secreting T as indicated by significant sex differences in T levels of amniotic fluid (Ward, Needham, Ward, & Hess, 1985). On days 18 and 19 of gestation, a characteristic surge in T titers is seen in the plasma of fetal males (Ward & Weisz, 1984; Weisz & Ward, 1980). A second surge of plasma T is noted shortly after birth (Corbier, Kerdelhue, Picon, & Roffi, 1978). Evidence from a recent study using a LHRH antagonist suggests that the hypothalamic–pituitary axis begins to control fetal testicular steroidogenesis on day 18 of gestation (Lalau, Aubert, Carmignac, Grégoire, & Dupouy, 1990). In the human male fetus, T levels in blood are elevated during the period from approximately the 11th week through the 17th week of gestation (Reyes, Boroditsky, Winter, & Faiman, 1974).

FETAL T LEVELS. McGivern, Raum, Salido, and Redei (1988) evaluated the effects of maternal ethanol consumption on fetal T levels. Pregnant rats were exposed to a liquid ethanol diet beginning on day 14 of gestation. The T levels were measured in the trunk tissue of their male fetuses on days 17, 18, 19, and 20 of gestation. Ethanol-exposed male fetuses failed to show the elevation in T levels on days 18 and 19 of gestation that characterized fetuses from isocaloric control animals (see Figure 1). Furthermore, testes taken from males whose mothers consumed

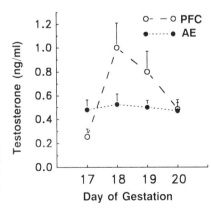

Figure 1. Testosterone levels in trunk tissue of alcohol-exposed (AE) male fetuses and male fetuses from pair-fed control (PFC) dams. Data, means ± S.E.M. (From McGivern *et al.*, 1988. Copyright 1988 by The Research Society on Alcoholism. Adapted by permission.)

14.6 g/kg of ethanol daily showed a dramatically lower production of T in response to stimulation by rat LH on day 18 and day 20 of gestation.

In contrast to the finding of McGivern *et al.* (1988), control males did not have higher levels of whole body T on days 18 and 19 of gestation than ethanol-exposed male rats in a study by Dahlgren *et al.* (1989). However, this latter study must be evaluated in terms of the abnormal T pattern shown by the males in the control group. The control male fetuses did not show the typical increase in T from day 17 to days 18 and 19 of gestation that other investigators have consistently reported (Lichtensteiger & Schlumpf, 1985; McGivern *et al.*, 1988; Ward & Weisz, 1984; Weisz & Ward, 1980). The abnormally low levels of T in control fetuses may have been caused by the stressful gastric intubation procedure to which the control as well as ethanol mothers were exposed three times daily beginning on the 15th day of gestation. Ward and Weisz (1980, 1984) have shown that stressing pregnant rats three times a day with a restraint and light procedure prevents the surge normally seen in plasma T on days 18 and 19 of gestation. Although administration of ethanol through intubation is sometimes used as a model of "binge" drinking, the gavage procedure does not produce chronically elevated blood alcohol levels and does introduce an undesirable stress component. Thus, the failure of the Dahlgren *et al.* (1989) study to replicate the finding that ethanol-exposed fetuses have lower T levels than normal seems to reflect the choice of an inappropriate technique for administering the ethanol and control solutions. Furthermore, in the Dahlgren *et al.* (1989) study, the total ethanol administered per day by gavage was 6 g/kg, less than half of that consumed by the rats in the study by McGivern *et al.* (1988).

Serum T levels have been measured on day 18 of gestation in rat fetuses whose mothers were maintained on a liquid ethanol diet beginning on day 12 of gestation. Although serum T levels tended to be somewhat lower in ethanol-exposed fetuses than in isocaloric controls, the difference was not statistically significant (Udani, Parker, Gavaler, & Van Thiel, 1985).

NEONATAL TESTICULAR FUNCTION.   Fetal exposure to ethanol seems to have a suppressive effect on the surge in plasma T that normally occurs in male rat pups a few hours after birth. Plasma T levels in male rat pups from mothers given a liquid ethanol diet throughout gestation are lower than those in male pups from isocaloric control mothers at birth and 1 and 4 hours following delivery (Redei & McGivern, 1988). Rudeen and Kappel (1985) and Rudeen, Ganjam, and Mohammed (1988) also have reported significantly lower plasma T concentrations in newborn rat pups that were exposed to ethanol *in utero*. Further, newborn rats from mothers exposed to ethanol during pregnancy have lower testicular weights (Kakihana, Butte, & Moore, 1980; Rudeen *et al.*, 1988). Ethanol exposure throughout gestation (Rudeen *et al.*, 1988) or from day 12 of gestation until parturition (Kelce, Rudeen, & Ganjam, 1989) also reduced the activity of 17α-hydroxylase, an enzyme involved in the synthesis of T in neonatal rats. Dihydrotestosterone (DHT) levels in plasma and brain measured 1 or 2 days after birth also were reduced by prenatal ethanol (Kakihana *et al.*, 1980). Although McGivern, Holcomb, and Poland (1987) found no significant effect of ethanol exposure from gestational day 14 until birth on neonatal T levels in serum, the weight of the evidence is that *in utero* ethanol exposure attenuates the early postnatal surge of T required for complete masculinization and defeminization of male rats (Redei & McGivern, 1988).

ANOGENITAL DISTANCE.   The distance from the anal opening to the base of the genitalia in the neonatal rat pup is an easily measured physical parameter that is widely believed to reflect the amount of androgen to which the newborn was exposed during the prenatal period. This so-called anogenital (A-G) distance is approximately twice as long in normal male as in female pups. The A-G distance is sensitive to decreased fetal plasma T titers as illustrated by the work of Goldman and colleagues. If the synthesis of fetal T is reduced by treating the mother during pregnancy with any one of a number of different steroidogenic enzyme inhibitors, the A-G distances in neonatal males are shorter than those in control males, and hypospadias is induced (Goldman, 1970, 1971; Goldman, Bongiovanni, & Yakovac, 1966). More extreme modification of the A-G distance is seen in males from mothers who were injected with the androgen receptor blocker flutamide. The A-G distance of neonatal males exposed *in utero* to flutamide is virtually indistinguishable from that of females (Clemens, Gladue, & Coniglio, 1978).

A number of studies have found that the A-G distance of neonatal male pups exposed *in utero* to alcohol is shorter than that in male pups from isocaloric control mothers (Rudeen, 1986; Rudeen, Kappel, & Lear, 1986; Udani et al., 1985). McGivern (1987) questioned the interpretation that the reduced A-G distance in ethanol-exposed animals was a true reflection of a reduction in prenatal T. He noted that pups from mothers given an ethanol diet during pregnancy always weigh less than those from isocaloric pair-fed control mothers, thus confounding A-G distance with body size. Pair feeding controls for calories consumed but does not take into account nutritional deficiency caused by ethanol-induced interference with placental transfer, intestinal absorption, and the metabolism of nutrients (see Colangelo & Jones, 1982, for a review). In McGivern's 1987 study, he too found shorter A-G distances in males prenatally exposed to ethanol but noted that when A-G distance was divided by body weight, the difference between ethanol-exposed and control pups was no longer significant. However, Udani et al. (1985) used an A-G length index in which A-G distance was divided by the length of the animal rather than by body weight. This A-G length index was significantly shorter in ethanol-exposed pups than in control males.

A third approach was that of Chen and Smith (1979). These investigators used a somewhat different ethanol exposure technique in which 10% ethanol in a 0.125% saccharin solution was the sole source of fluid for the mother starting on day 7 of gestation. The male offspring of these mothers exhibited a significant reduction in neonatal A-G distance but no reduction in body weight. The consensus of the data seems to support the interpretation that the reduced A-G distances seen in ethanol-exposed males reflect an attenuation of fetal T rather than simply a reduction in overall body size.

LONG-TERM EFFECTS OF *IN UTERO* EXPOSURE

REPRODUCTIVE BEHAVIOR. If ethanol suppression of fetal testosterone is sufficiently prolonged and intense, it should disrupt normal sexual differentiation of behavior in males, resulting in abnormal patterns of sexual behaviors in adulthood. Several studies support this prediction. Broida, Churchill, and McGinnis (1988) found that copulatory behaviors had not been completely defeminized in adult male rats exposed *in utero* to ethanol. Males derived from pregnant rats fed a liquid diet containing 36% of calories from ethanol (days 5 through 20 of gestation) showed the receptive posture characteristic of females (lordosis) 80% of the time when mounted by other males compared to 35% by males from mothers fed an isocaloric control diet. The male offspring from pregnant rats fed an 18% alcohol-derived caloric diet were not significantly different from either the isocaloric control males or the 36% alcohol group (see Figure 2). All males had been castrated on postpartum day 3, and this probably accounts for the relatively high level of lordosis seen in the control group. Unfortunately, the castration procedure precluded assessment of the male sexual potential of these animals. In order to avoid the effects of differences in maternal care, which might be found in mothers undergoing alcohol withdrawal, all animals in this study were fostered to a control mother at birth.

Hård et al. (1984) tried a different approach to force rats to consume alcohol. Mother rats had access to a normal dry diet, but the sole source of fluid available both before and during pregnancy contained 16% ethanol. The taste of this concentration of ethanol is quite aversive to rats, so it is not clear how much ethanol the pregnant animals actually consumed under these conditions. At birth, the ethanol-exposed male offspring showed a normal A-G distance and body weight, suggesting low ethanol exposure in utero. When tested in adulthood, there was no significant change in their potential for masculine copulatory behavior. However, 45% of the males exposed to ethanol displayed lordotic behavior, compared to only 8% of control males. Chen and Smith (1979) added 10% ethanol to a 0.125% saccharin solution as the sole fluid source available from day 7 of gestation until birth. No effects were seen in pups exposed in utero to ethanol on any parameters of masculine sexual behavior, and female behavioral potentials apparently were not tested in these males.

Dahlgren et al. (1989) found no significant differences in either masculine or feminine behavior patterns of prenatally ethanol-exposed male rats compared to controls. However, as discussed previously, their data have to be interpreted cautiously. The gavage procedure used by these investigators to administer alcohol on days 15 to 21 of gestation resulted in acute rather than sustained elevations in blood alcohol level. Further, the gavage procedure may have stressed the control mothers, thus altering the sexual behavior potentials of the comparison group, as is reflected by the relatively high level of lordosis in their control males. Ward (1972) and others have shown that prenatal stress increases the potential of the male offspring to display lordosis.

Evidence for a partial failure of masculinization in rats perinatally exposed to ethanol was reported by Parker, Udani, Gavaler, and Van Thiel (1984). Male offspring of mothers given a liquid ethanol diet starting on the 12th day of pregnancy and continuing until the 10th day after parturition showed significantly fewer intromission responses and a longer latency to mount than control males. This is the only study in which deficits in masculine behavior patterns were noted in males exposed to ethanol during the perinatal period. The relationship between fetal ethanol ex-

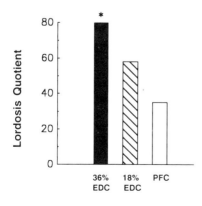

Figure 2. Lordosis quotient in male rats from dams given a diet containing 36% or 18% ethanol-derived calories (EDC) during days 5–20 of pregnancy or from pair-fed control (PFC) dams. *$p < 0.005$ compared to PFC males. (From Broida et al., 1988. Adapted by permission of the authors.)

posure and male sexual behavior is clouded by the fact that the mothers of the males
in this study continued to consume the ethanol diet during the first 10 days following parturition.

The weight of available evidence seems to suggest that prenatal ethanol exposure increases the potential for female behavior in male rats. The effect on male
behavior is more equivocal. Unfortunately, the Broida *et al.* (1988) study is the only
one to date in which alterations in reproductive behavior potentials were investigated under the following important conditions: (1) the liquid ethanol diet used was
likely to sustain clinically relevant blood alcohol levels, and (2) exposure to ethanol
was restricted to the prenatal period.

NONREPRODUCTIVE BEHAVIORS.   Sexually dimorphic nonreproductive behaviors also show a tendency toward feminization in males exposed *in utero* to ethanol.
Female rats normally show a greater preference for saccharin solutions than males
(Valenstein, Cox, & Kakolewski, 1967; Valenstein, Kakolewski, & Cox, 1967). Mc-
Givern and co-workers have shown that saccharin intake is greater in males exposed
to ethanol during the last week of fetal development than in isocaloric control
males. This finding held regardless of whether saccharin intake was expressed as a
percentage of total fluid intake (McGivern *et al.*, 1987) or was adjusted for group
differences in body weight (McGivern, Clancy, Hill, & Noble, 1984). Abel and Dintcheff (1986) were unable to replicate the finding of increased saccharin intake in
males exposed to ethanol *in utero*. However, Abel and Dintcheff used an intubation
procedure to administer 7 g of ethanol/kg per day to the pregnant dams, a dosage
approximately half of that consumed by the animals in McGivern's study.

Another nonreproductive sexually dimorphic behavior is the amount of play
fighting shown by juvenile rats. Males normally exhibit more play fighting than
females (Olioff & Stewart, 1978; Poole & Fish, 1976). Meyer and Riley (1986) found
that play levels in male rats exposed to ethanol during prenatal development were
not higher than in control females and were significantly lower than in males from
isocaloric control mothers. Dams in this study consumed an average of 13.2 g
ethanol/kg per day from gestational days 6 to 20. Collectively, these data support
the idea that alcohol-induced alterations in the fetal hormonal environment may
have important consequences on a variety of sexually dimorphic behaviors exhibited
by juvenile and adult males.

BRAIN DIFFERENCES.   There are structural differences in the brains of males
and females. The most prominent of these, the sexually dimorphic nucleus of the
preoptic area (SDN-POA), is several times larger in the brain of male rats (Gorski,
Gordon, Shryne, & Southam, 1978) and humans (Swaab & Fliers, 1985) than in
females. In rats, development of this neural sexual dimorphism is dependent on
perinatal exposure to testosterone or its metabolites (Döhler *et al.*, 1984, 1986).
Barron, Tieman, and Riley (1988) have shown (see Figure 3) that the adult volume of
the SDN-POA is reduced by approximately 50% in male rats whose mothers consumed an average of 13.2 g ethanol/kg per day from days 6 to 20 of gestation.
Further, the mean size of individual neurons within the SDN-POA was also dramatically reduced in males exposed *in utero* to ethanol. The effect on the central nervous
system (CNS) was restricted to sexually dimorphic structures. The volume of another nucleus, the nucleus of the anterior commissure, which is not sexually dimorphic, was not altered by prenatal ethanol exposure. Two other studies (Rudeen,
1986; Rudeen *et al.*, 1986) also found the volume of the SDN-POA to be signifi-

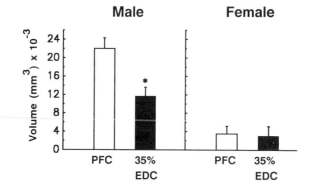

Figure 3. Volume of the SDN-POA in adult rats derived from dams given a diet containing 35% ethanol-derived calories (35% EDC) during days 6–20 of pregnancy or from pair-fed control (PFC) dams. Error bars represent S.E.M.s. *$p < 0.01$ compared to PFC of the same sex. (From Barron et al., 1988. Copyright 1988 by The Research Society on Alcoholism. Adapted by permission.)

cantly smaller in ethanol-exposed male rats than in control males. However, in the studies by Rudeen and co-workers, ethanol exposure started during gestation and then continued for the first 2 weeks of postnatal development.

Another sex difference in the brains of rats and humans is the size of the corpus callosum. Adult males have larger callosums than adult females. Zimmerberg and Scalzi (1989) have demonstrated that this sex difference is already present on the third day of postnatal life in rats. However, 3-day-old males from mothers given a liquid ethanol diet from days 5 to 20 of gestation are not different from females; i.e., no sex difference occurs in the size of the callosum. Since the ethanol treatment did not alter the size of the corpus callosum in females, the authors suggest that this neurological alteration is hormonally mediated.

SUMMARY

In summary, sustained ethanol exposure of the pregnant rat suppresses the surge in fetal testosterone necessary for total masculinization and defeminization of sexually dimorphic behavior, physiology, and morphology in their male offspring. Reflections of this suppression are seen in (1) reduced neonatal T levels, (2) shortened A-G distances in neonates, (3) an increased potential for the female lordosis pattern in adulthood, (4) a feminized intake pattern for saccharin solution, (5) female levels of rough and tumble play exhibited by juveniles, and (6) incomplete masculinization of the SDN-POA, a sexually dimorphic diencephalic structure. No gross changes in external genitalia were noted in experimental animals.

For boys exposed to ethanol *in utero,* no attempts to assess any possible alterations in sexually dimorphic behaviors have been reported.

NICOTINE

ACTION IN ADULT MALES

Nicotine has a direct inhibitory effect on the Leydig cells of the testes, suppressing their production of T. Nicotine reduced production of T in a culture of rat testicular cells in which approximately 5–10% of the cells were Leydig cells (Kasson & Hsueh, 1985) and in cultures where Leydig cells were isolated and incubated (Yeh, Barbieri, & Friedman, 1989). One or more steps in the biosynthetic pathway for T were inhibited. Furthermore, both nicotine and cotinine, the main metabolite of nicotine, suppress LH-stimulated T production by isolated Leydig cells of mice (Patterson, Stringham, & Meikle, 1990).

The strongest evidence *in vivo* that nicotine inhibits T production comes from a study with tracheostomized dogs. In adult male beagles that "smoked" approximately 12 cigarettes daily (1.5 mg nicotine/cigarette) for 2 years or more, testosterone levels were reduced to 54% of the level of control dogs (Mittler, Pogach, & Ertel, 1983).

In humans, similar findings have been reported for heavy smokers. Twenty-five young men who smoked more than 20 cigarettes per day had serum T levels that were 57% of the T levels in a group of 20 nonsmokers (Shaarawy & Mahmoud, 1982). In a smaller study, Briggs (1973) also found reduced T levels in the serum of heavy smokers compared to nonsmokers.

In contrast, smokers had higher T levels than nonsmokers when inclusion in the category of smoker was dependent on smoking five or more cigarettes a day (Deslypere & Vermeulen, 1984). Perhaps the difference in findings is related to dosage, i.e., the amount of nicotine consumed.

It is difficult to study the acute effects of smoking on hormone levels in nonsmokers. Nonsmokers frequently show a nonspecific stress response, becoming temporarily ill if they smoke a number of cigarettes within several hours. Noting this problem, Winternitz and Quillen (1977) used habitual smokers in their study of the acute effects of cigarette smoking on T levels. Their subjects smoked eight cigarettes (2.5 mg nicotine/cigarette) over a 2-hour period. No significant fluctuations from baseline T levels were noted. In contrast, in the study by Briggs (1973), 7 days of abstinence from smoking produced a significant rise in plasma T levels of heavy smokers.

USE DURING PREGNANCY

Several studies have indicated that about one out of every three women (range 28% to 41%) smokes cigarettes before becoming pregnant and an even higher percentage drank alcohol (range 62% to 96%) prior to pregnancy (Condon & Hilton, 1988; Fried, Barnes, & Drake, 1985; Johnson, McCarter, & Ferencz, 1987; Waterson & Murray-Lyon, 1989). However, women have more difficulty reducing their nicotine than their ethanol intake while pregnant. Thus, approximately 23% continued to smoke while pregnant, with 6% to 12% being considered heavy smokers. Heavy smoking was defined either as smoking 20 or more cigarettes per day or as smoking cigarettes containing greater than 16 mg of nicotine per day. Heavy smoking during pregnancy reduces the birth weight of the child and increases the chances of perinatal mortality.

Nicotine crosses the placenta readily and enters the developing fetus. Nicotine appeared in the fetal circulation within 5 minutes of maternal administration when pregnant rats were injected with 0.02 mg/kg of [^{3}H]nicotine on day 19 of gestation. The level of radioactivity was higher in fetal plasma than in maternal plasma from 30 minutes until 20 hours following injection (Mosier & Jansons, 1972). In pregnant rhesus monkeys, fetal plasma nicotine levels reached maternal levels within 20 minutes following a 1 mg/kg IV injection of labeled nicotine to the mother. Nicotine levels in the fetal circulation remained higher than in the maternal circulation for at least 2 hours (Suzuki *et al.*, 1974). In humans, nicotine levels in amniotic fluid and in umbilical vein serum at birth exceed those in maternal vein serum (Luck, Nau, Hansen, & Steldinger, 1985). Also in human neonates, cord blood levels of continine are approximately 40 times higher than normal when the mother smoked during pregnancy (Sorsa & Husgafvel-Pursiainen, 1988). Nicotine and continine were found in amniotic fluid samples taken between 14 and 20 weeks of gestation in women who smoked (Van Vunakis, Langone, & Milunsky, 1974).

EFFECTS ON FETAL TESTICULAR FUNCTION

Nicotine exposure blocks the normal elevation in plasma T found in male rat fetuses on the 18th day of gestation. In order to provide sustained exposure to nicotine, Lichtensteiger and Schlumpf (1985) implanted osmotic minipumps into rats on the 12th day of pregnancy. These pumps delivered either nicotine (25 μg/100 g per hour) or saline. The total daily nicotine dose on the 18th day of gestation was approximately 4.5 to 5 mg/kg of maternal body weight. As expected, plasma T levels were elevated on day 18 in saline-treated male fetuses compared to day 17 of gestation. Nicotine-treated males did not show this rise in T levels on day 18 of gestation.

LONG-TERM EFFECTS OF *IN UTERO* EXPOSURE

REPRODUCTIVE BEHAVIOR. Prenatal nicotine exposure leads to a marked reduction in the number of male rats that exhibit masculine copulatory patterns in adulthood. Segarra and Strand (1989) injected pregnant rats twice daily with 0.25 mg/kg nicotine or saline from gestational days 3 through 21. The males exposed as fetuses to nicotine were less likely than control males to copulate and ejaculate (see Figure 4). It should be noted that testosterone levels in adulthood were almost 50% lower in the males exposed to nicotine during gestation than in controls.

Prenatal exposure to nicotine also has been reported to increase the potential of male rats for feminine copulatory behavior. Males derived from dams implanted with osmotic minipumps that delivered a high dose of nicotine from days 12 to 19 of pregnancy showed significantly more lordotic responses than control males (Rodriguez-Sierra & Hendricks, 1984).

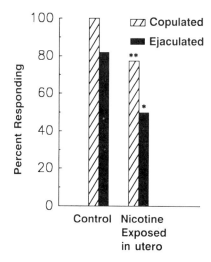

Figure 4. Percentage of adult males exposed to nicotine *in utero* that copulated and ejaculated. Nicotine (0.25 mg/kg) was administered twice daily to pregnant dams on days 3–21 of gestation. Control dams were injected with saline. *$p < 0.025$, **$p < 0.01$, compared with control males (chi-square tests). (From Segarra & Strand, 1989. Copyright 1989 by Elsevier Science Publishers. Adapted by permission.)

NONREPRODUCTIVE BEHAVIOR.   Adult male rats exposed to nicotine *in utero* exhibit a feminized level of preference for saccharin solutions. Lichtensteiger and Schlumpf (1985) found that normal females showed a significantly higher preference for 0.06, 0.125, and 0.25% solutions of saccharin than normal males. Exposure to nicotine on days 12 through approximately 20 of gestation abolished this sexual dimorphism in behavior. Specifically, prenatal nicotine exposure had no effect on the saccharin preference of female offspring but increased the preference of male offspring to the level of females.

ANATOMIC MODIFICATIONS.   There have been no reports of a high incidence of genital anomalies, such as hypospadias, that might result from fetal androgen deficiency in children exposed *in utero* to nicotine. However, data from the Collaborative Perinatal Project of the National Institute of Neurological and Communicative Disorders and Stroke suggest a small increase in hypospadias. Of the 929 male children whose mothers smoked 30 or more cigarettes per day while pregnant, 1.5% had hypospadias compared to 0.70% of the 24,613 male children from mothers who never smoked or smoked less than 30 cigarettes per day (Heinonen, Slone, & Shapiro, 1977).

SUMMARY

Prenatal nicotine exposure suppresses the surge in testosterone normally found on day 18 of gestation in male rats. The functional significance of this hormonal alteration during fetal development is reflected in an incomplete masculinization of adult copulatory patterns and a feminization of saccharin solution preference levels. The possible effects of this drug on the host of other sexually dimorphic behaviors have not been investigated.

MARIJUANA

Marijuana is a crushed mixture of the flowering tops and leaves of the plants of *Cannabis sativa*. The principal psychoactive ingredient of marijuana is Δ^9-tetrahydrocannabinol (THC). In experimental studies, THC or cannabinol (CBN), rather than the crude extracts of the entire plant, are typically used. Cannabinol is another purified constituent of marijuana that has biological but not psychoactive effects.

ACTIONS IN ADULT MALES

The effects of chronic marijuana smoking on plasma T levels in humans are equivocal. In one study, men who smoked, on the average, 5 to 9 marijuana joints per week had T levels that were one-third lower than those of men in the control group who had never used marijuana (Kolodny, Masters, Kolodner, & Toro, 1974). The control men were matched to the marijuana users in a number of dimensions including age and cigarette-smoking habits. Men who smoked over 10 joints per week had a 58% reduction in T levels. The T levels rose dramatically in a few men who discontinued marijuana use for a 2-week period. However, T levels did not show fluctuations related to marijuana use in two studies in which chronic marijuana users abstained from use for 5 days, then smoked marijuana for 3 weeks, and then abstained from use for an additional 5-day period (Mendelson, Kuehnle, Ellingboe, & Babor, 1974; Mendelson, Ellingboe, Kuehnle, & Mello, 1978).

Acute reductions in plasma T have been found to occur after smoking even a single marijuana cigarette. Men who had been regular marijuana users in the past but who had abstained from use for 11 days showed an average decrease of 35% in plasma T concentration 3 hours after smoking a cigarette containing 900 mg of marijuana (Kolodny, Lessin, Toro, Masters, & Cohen, 1976). Similarly, after a 50-minute infusion of 10 mg of THC, plasma T levels fell 5% per hour for the 6 hours observed (Wall, Brine, & Perez-Reyes, 1978, as cited in Barnett & Chiang, 1983).

A pronounced suppression of plasma T is observed in adult male rodents following exposure to a substantial dose of THC (Dalterio, Bartke, & Mayfield, 1981; Kumar & Chen, 1983; Symons, Teale, & Marks, 1976). For example, 4 hours after a single oral dose of 50 mg/kg of THC, plasma T concentrations of adult mice were 22% of control values (Dalterio Bartke, Roberson, Watson, & Burstein, 1978). Moderate to large doses of THC produced a transient increase in plasma T within 10 minutes followed by a prolonged suppression within 20 minutes (Dalterio, Bartke, & Mayfield, 1981). Similar biphasic effects of THC on testicular T levels have been noted (Dalterio, Mayfield, Bartke, and Morgan, 1985). Small doses of THC may produce increases in plasma T concentrations that last at least 60 minutes (Dalterio, Bartke, & Mayfield, 1981).

The mechanism(s) through which THC suppresses plasma T levels may involve action at a central level, on hypothalamic releasing factors, which, in turn, alter pituitary gonadotropins, and/or direct effects on pituitary production of gonadotropins. A central action at the level of the hypothalamus and/or pituitary is supported by numerous *in vivo* demonstrations that THC suppresses plasma titers of LH (Chakravarty, Sheth, Sheth, & Ghosh, 1982; Collu, Letarte, Leboeuf, & Ducharme, 1975; Dalterio *et al.*, 1978; Symons *et al.*, 1976). There is also indirect evidence that THC reduces synthesis of LHRH in the hypothalamus of intact male rats (Kumar & Chen, 1983). However, *in vitro* studies utilizing isolated hypothalamic tissue from adult male mice have found that LHRH secretion from the median eminence in response to noradrenergic stimulation was increased in the presence of THC (Dalterio, Steger, Peluso, & DePaolo, 1987). Further, THC treatment enhanced the release of LH from isolated pituitary tissue in response to LHRH. It is difficult to reconcile the *in vitro* findings of Dalterio *et al.* (1987) of enhanced responsivity of the hypothalamic–pituitary axis with frequent *in vivo* reports that plasma LH levels are reduced by THC. Thus, the nature of the central role exerted by THC that leads to alterations in gonadal steroidogenesis in male rodents is still unclear.

On the other hand, there is general agreement on the peripheral effects exerted by cannabinoids directly on the testes. In a number of *in vitro* studies, THC and CBN inhibited steroidogenesis in isolated testes and in cultures of Leydig cells. In mice, THC completely blocked the increase in T production normally induced by adding gonadotropins to such preparations, and CBN blocked both basal and gonadotropin-stimulated T production (Dalterio, Bartke, & Burstein, 1977; Dalterio *et al.*, 1978; Dalterio, Bartke, & Mayfield, 1985). Very low concentrations of a number of cannabinoids inhibit gonadotrophin-stimulated T production of Leydig cell preparations from rats. In fact, cannabinoid concentrations that approximated those found in testes after 15 minutes of exposure of rats to the smoke of a single marijuana cigarette were effective in inhibiting steroidogenesis (Jakubovic, McGeer, & McGeer, 1979).

Burstein, Hunter, and Shoupe (1979) and Burstein, Hunter, Shoupe, and Taylor (1978) have presented evidence that the primary peripheral inhibitory effect of THC on testicular production of T is by suppressing the release of cholesterol from its ester storage form. There seems to be no interference by THC with later steps in the synthesis pathway, i.e., the conversion of either pregnenolone or progesterone to T. Further, a dose of THC that reduces T by 75% does not interfere with gonadotropin binding to Leydig cells.

There also is evidence that THC acts as an androgen antagonist directly at peripheral target tissues, e.g., the prostate (Ghosh, Chatterjee, & Ghosh, 1981; Purohit, Singh, & Ahluwalia, 1979; Purohit, Ahluwahlia, & Vigersky, 1980).

USE DURING PREGNANCY

Marijuana is the most frequently used illicit drug in the United States (Nahas & Frick, 1986). Approximately 44% of all women of childbearing age have taken marijuana, and an estimated 6 million are current users (Adams *et al.*, 1989). Marijuana use declines with pregnancy. It is estimated that approximately 13% of pregnant women smoke marijuana, with greater frequency during the first trimester than at later stages (Fried *et al.*, 1985; Fried, Buckingham, & Von Kulmiz, 1983). However, heavy users (over five joints per week) of marijuana are less likely to reduce consumption during pregnancy than are heavy users of alcohol or nicotine. Specifically, heavy users of marijuana declined from 5% prior to pregnancy to a steady 3% throughout pregnancy (Fried *et al.*, 1985).

Marijuana smoked by the mother crosses the placenta and accumulates in the developing fetus, and THC is present in cord-blood samples taken at delivery from fetuses whose mothers smoked marijuana daily during the last trimester of pregnancy (Blackard & Tennes, 1984). In experimental studies with animals, THC has been detected in fetal hamsters (Idänpään-Heikkila, Fritchie, Englert, Ho, & McIsaac, 1969), mice (Harbison & Mantilla-Plata, 1972), rats (Hutchings, Martin, Gamagaris, Miller, & Fico, 1989), and dogs (Martin, Dewey, Harris, & Beckner, 1977) after maternal administration. The most complete study demonstrating the speed with which THC crosses the placenta and enters the fetus was conducted using rhesus monkeys (Bailey, Cunny, Paule, & Slikker, 1987). Simultaneous maternal and fetal blood samples were taken at 1, 3, 6, 15, 30, 60, 120, and 180 minutes following a small (0.03 mg/kg) maternal IV dose of THC. The plasma concentration of THC following this IV dose was approximately the same as that produced by exposure of the monkey to the smoke from 15 puffs on a marijuana cigarette. Within 15 minutes of maternal administration, fetal plasma THC levels peaked, being 27.6% of the levels found in maternal plasma at that point in time. Fetal plasma THC levels fell more slowly than maternal levels, so that 60 minutes after administration the fetus had retained 40% of the concentration found in the mother. At two hours, the fetal level was 71% of the mother's level, and concentrations were equal after 3 hours.

EFFECTS ON FETAL TESTICULAR FUNCTION

The experimental animal of choice for investigators studying the effects of THC on sexual differentiation has been the mouse. Pregnancy last 19 or 20 days in this species. In mice, fetal testes begin secreting T on the 12th day of gestation. Maximal sex differences in whole-body T levels are seen on the 16th day of gestation (Dalterio & Bartke, 1981).

FETAL ANDROGEN LEVELS.   Maternal consumption of cannabinoids during pregnancy interferes with testicular steroidogenesis in the mouse fetus. Both T and DHT levels were reduced in male mice on the 16th day of gestation following maternal treatment with either THC or CBN (Dalterio & Bartke, 1981). Specifically, pregnant mice were given 50 mg/kg of THC or CBN orally, once per day, starting on the 12th day of gestation. Levels of THC in the blood of pregnant mice following this dose are only slightly higher than those found in humans after they have smoked two marijuana cigarettes containing 13 mg THC each (Abel, Tan, & Subramanian, 1987). Four hours after the drug administration on day 16 of gestation, the mothers were killed, and T levels were determined in the whole bodies of their fetuses. As shown in Figure 5, T levels were reduced by 23% in THC-exposed fetuses compared to male fetuses from control mothers. The T levels in female fetuses were not altered by maternal cannabinoid use.

NEONATAL AND JUVENILE TESTICULAR FUNCTION.   There are residual effects of prenatal THC exposure on testicular functioning during early postnatal development in rats. Serum T levels on the day of birth were nearly 60% lower in males derived from mothers given 6 mg/kg of THC from day 14 through 19 of gestation, compared to neonates from oil-injected control mothers. Luteinizing hormone levels were also suppressed in neonates exposed *in utero* to THC. The suppression of both T and LH was still present when the pups were 5 and 25 days old but had disappeared by 60 days of age (Ahluwahlia, Rajguru, & Nolan, 1985).

LONG-TERM EFFECTS OF *IN UTERO* EXPOSURE

There are virtually no well-controlled studies of sexually dimorphic behaviors in males in whom exposure to cannabinoids was restricted entirely to the prenatal period of development. In one study only 40% of male rats exposed *in utero* to THC were capable, as adults, of impregnating females as compared to approximately 73% of control males (Ahluwahlia *et al.*, 1985). Unfortunately, the nature of the deficit underlying such infertility is not clear, since only the end product of reproductive behavior was observed. For example, the ventral prostate glands of these THC-exposed males showed morphological abnormalities in adulthood. Thus, the infertility of THC males could have been caused either by a behavioral deficit or by the

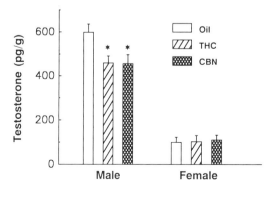

Figure 5.   Testosterone concentrations in fetal mice (16th day of gestation) taken from pregnant dams that received 50 mg/kg of Δ^9-tetrahydrocannabinol (THC) or cannabinol (CBN) or the carrier (OIL) daily, starting on the 12th day of gestation. Values are means ± S.E.M. *$p < 0.05$, compared to oil-treated fetuses of the same sex. (From Dalterio & Bartke, 1981. Copyright 1981 by *Journal of Endocrinology*, Ltd. Adapted by permission.)

malfunctioning of some vital physical component of the reproductive apparatus. Similar reductions in fertility as well as reduced seminal vesicle weights were noted in male rats derived from dams exposed to smoke from marijuana while pregnant (Fried & Charlebois, 1979).

In a study conducted on mice, mothers were given 50 mg/kg of THC or CBN beginning on day 20 of pregnancy and continuing for the first 6 days following parturition of the litter. Thus, fetal exposure to THC was combined with the post-natal THC exposure resulting from nursing by the lactating mother. As adults, fewer than half of the males exposed to THC during the perinatal period attempted to mount when given access to a receptive female (Dalterio, 1980; Dalterio & Bartke, 1979). Reductions were also seen in the male behavior of CBN-exposed males compared to the control males, all of whom were sexually active. Plasma T levels in adulthood were not significantly reduced in males exposed to THC or CBN perinatally. In a later study, Dalterio, Bartke, Blum, and Sweeney (1981) did report a significant reduction in plasma T concentrations of adult male mice exposed to the same perinatal THC treatment as used in the earlier study.

SUMMARY

There is a large volume of research demonstrating that THC and other constituents of marijuana suppress testicular steroidogenesis in adult males of several species, including man. Further, Dalterio and Bartke (1981) reported that THC suppressed T in fetal male mice. However, there is a surprising sparcity of studies investigating the long-term effects of fetal THC exposure on sexually dimorphic behaviors. In the only study of reproductive potentials in males in whom THC exposure was restricted to the prenatal period, copulatory behaviors were not directly observed, and thus the nature of the deficit resulting in the observed infertility remains unclear. There are apparently no studies assessing a potential enhancement of feminine copulatory patterns in males exposed *in utero* to THC, and this reviewer also found no studies of the effects of prenatal THC exposure on sexually dimorphic nonreproductive behaviors.

OPIATES

The opiate narcotics of primary interest for this review include (1) heroin, a semisynthetic derivative of opium that is the narcotic most frequently used by opiate addicts (heroin is highly lipid soluble and thus passes through both the placental and blood–brain barriers with ease), (2) morphine, a natural derivative of opium (heroin is rapidly metabolized to morphine in the body), and (3) methadone, a synthetic narcotic that acts on opiate receptors. Methadone is a long-acting opioid in humans, having a plasma half-life of 24 to 36 hours. In the rat the plasma half-life is only 70 to 90 minutes (Kreek, 1982). Methadone is used in the United States as a legally controlled substitute in the treatment of heroin addicts.

ACTIONS IN ADULT MALES

HUMANS.   Male opiate addicts frequently report impotence and a loss of libido. These disturbances in sexual function may be related to a depressed concentration in serum T, which is usually (Azizi, Vagenakis, Longcope, Ingbar, & Braverman, 1973; Mendelson, Mendelson, & Patch, 1975; Mendelson, Meyer, Ellingboe, Mirin, & McDougle, 1975) but not always (Ragni, De Lauretis, Bestetti, Sghedoni, & Gambaro, 1988) found in both heroin- and methadone-maintained addicts. Cicero, Bell, *et al.* (1975) noted a 43% reduction in plasma T in men on methadone but found no decrease in T levels in heroin addicts. However, the "street" heroin used by the addicts in the studies of both Ragni *et al.* (1988) and Cicero, Bell, *et al.* (1975) was of unknown purity. As a general rule, the concentration of heroin in drugs obtained on the street in the United States is relatively low and may vary considerably. In order to investigate the effects of heroin of known dosage on plasma T in males, Mendelson, Mendelson, *et al.* (1975) studied men in a heroin maintenance clinic in London, England. Testosterone levels were depressed in most of these addicts. Three additional studies on the effects of opiates on neuroendocrine function in men have examined heroin addicts in Hong Kong (Mendelson & Mello, 1975, 1982; Wang, Chan, & Yeung, 1978). In Hong Kong a high-quality relatively pure heroin is readily available on the illicit drug market. Further, polydrug use is not a problem as it is in many opiate addicts in the United States. Thus, the effects of known exposure to heroin alone on plasma T levels can be evaluated. In all three studies of Hong Kong addicts, plasma T levels were significantly depressed compared to men in the control groups. On abstinence from heroin (Mendelson & Mello, 1975) or methadone (Mendelson, Mendelson, *et al.*, 1975; Woody *et al.*, 1988), T concentrations return to normal.

Luteinizing hormone levels are also depressed by methadone (Martin *et al.*, 1973; Woody *et al.*, 1988) and heroin (Mendelson & Mello, 1982; Mendelson, Ellingboe, Kuehnle, & Mello, 1980). When blood samples were taken every 30 minutes preceding and following IV heroin administration, LH levels were suppressed prior to the reduction in T, suggesting a central site of action for this opiate (Mirin, Mendelson, Ellingboe, & Meyer, 1976). However, opiates also may have an inhibitory effect directly on the gonads, as is suggested by the finding that, *in vitro*, human testicular tissue pretreated with methadone does not respond to gonadotropins (Purohit, Singh, & Ahluwalia, 1978).

EXPERIMENTAL ANIMALS.　In a series of studies, Cicero and his co-workers have demonstrated that morphine has very potent effects on the hypothalamic–pituitary–gonadal axis of the adult male rat. A very small (0.84 mg/kg) dose of morphine reduces serum T levels by 50% within 4 hours of administration (Cicero, Wilcox, Bell, & Meyer, 1976). It required 1.6 mg/kg of methadone to produce the same 50% reduction on serum T levels. A larger dose of morphine (20 mg/kg) reduced T levels by approximately 85%, with the suppression lasting for about 6 hours. Chronic treatment with morphine by pellet implantation for 3 days produced an 87% reduction in serum T levels and reduced the weight of such T-dependent secondary sex structures as the seminal vesicles and prostate (Cicero, Meyer, Wiest, Olney, & Bell, 1975).

In response to acute morphine administration, serum LH levels fall within the first 30 to 60 minutes, followed by a reduction in T levels 1 to 2 hours later (Cicero, Wilcox, *et al.*, 1976). This time course suggests that the suppression of testicular steroidogenesis may be secondary to an inhibitory effect of morphine on hypothalamic–pituitary production of LH. Morphine does not inhibit the release of LH in response to exogenous LHRH either *in vivo* or *in vitro* (Cicero, Badger, Wilcox, Bell, & Meyer, 1977). However, morphine does block potassium-stimulated release of LHRH from slices of mediobasal hypothalamus (Drouva, Epelbaum, Tapia-

Arancibia, Laplante, & Kordon, 1981). This combination of findings suggests that morphine exerts its inhibitory effect at the hypothalamic rather than the pituitary level.

Morphine does not seem to have a direct inhibitory effect at the peripheral (gonadal) level in experimental animals. In isolated testes, morphine did not alter the basal rate of T production, nor did it inhibit gonadotropin-stimulated release of T (Cicero, Bell, Meyer, & Schweitzer, 1977). Further, morphine does not produce any atrophy of the secondary sex structures if rats are given exogenous gonadotropins (Cicero, Meyer, Bell, & Koch, 1976).

USE DURING PREGNANCY

Zagon (1985) projected that there may be 6,000 to 9,000 births each year to opiate-consuming women in the United States. Zagon's estimate was based on a conjecture by Carr (1975) that there are approximately 3,000 pregnant users of opiates in New York City and on the belief that one-third to one-half of the chronic heroin and methadone users in the United States live in New York City. More recently, Hans (1989) has estimated that the number of pregnancies in opiate-dependent women in the United States may be as high as 10,000 per year. Since the early 1970s, opiate-dependent women who seek prenatal care frequently have been placed in a methadone maintenance program. Presumably, one of the benefits of having the addicted pregnant woman consume a consistent daily dose of the long-acting opiate methadone is that the fetus is protected from the repeated *in utero* drug withdrawals that might accompany erratic maternal use of street drugs of varying purity (Kaltenbach & Finnegan, 1989).

Methadone and morphine are transferred across the placenta in mice (Choi, Sperling, & On'gele, 1976; Shah, Donald, Bertolatus, & Hixson, 1976), rats (Gabrielsson & Paalzow, 1983; Jóhannesson, Steele, & Becker, 1972; Peters, Turnbow, & Buchenauer, 1972), and sheep (Golub, Eisele, & Anderson, 1986; Szeto, Umans, & McFarland, 1982). For methadone, in particular, the transfer is rapid, and detectable levels are present for a considerable period of time in the fetus. Peak levels of methadone in fetal mice are reached within 15 minutes following maternal administration (Shah *et al.*, 1976). Small amounts of methadone continue to be detectable for the first few weeks of postnatal development in rats exposed *in utero* to methadone and cross-fostered at birth to nondrug mothers (Levitt, Hutchings, Bodnarenko, & Leicach, 1982). Opiate concentrations in fetal brains are two to four times the level found in the maternal brain in rats, mice, and guinea pigs (Gabrielsson & Paalzow, 1983; Jóhannesson *et al.*, 1972; Pak & Ecobichon, 1982; Shah *et al.*, 1976). Although an effective blood–brain barrier partially blocks opiate entry into the adult brain, the fetus has not developed this screening mechanism. Fetal opiate concentrations in plasma and in tissues other than brain are lower than in the mother, suggesting that there is some partial placental barrier.

The nature of the partial barrier to the transfer of opiates from the mother to the fetus differs for methadone and morphine (Szeto *et al.*, 1982). At least 75% of methadone in maternal plasma is protein-bound, leaving only 25% free for transplacental transfer. In contrast, a very low percentage of morphine is protein bound. However, lipid-soluble drugs (e.g., heroin and methadone) move across the placental barrier much faster than non-lipid-soluble drugs (e.g., morphine). When placental transfer of morphine and methadone were directly compared in pregnant ewes, morphine moved across the placental barrier from mother to fetus 15 times more slowly than methadone (Szeto *et al.*, 1982).

There are also species differences in placental transfer and disposition of opiates. Morphine transferred readily from mother to fetus in near-term rhesus monkeys, with fetal and maternal plasma levels being equal within 60 minutes (Golub *et al.*, 1986). In the same study, transplacental transfer of morphine in the pregnant ewe was much more limited. Further species differences are apparent when one examines the effects of pregnancy on methadone metabolism. In guinea pigs, the plasma half-life of methadone was approximately doubled during pregnancy (Pak & Ecobichon, 1982). In contrast, in pregnant women as they approach term, the plasma half-life of methadone is approximately one-third that of the nonpregnant state (Kreek, 1982).

Methadone crosses the placenta into the human fetus, as indicated by its presence in umbilical cord plasma and amniotic fluid at birth in fetuses from women on a methadone maintenance program (Blinick, Inturrisi, Jerez, & Wallach, 1975; Harper, Solish, Feingold, Gersten-Woolf, & Sokal, 1977; Inturrisi & Blinick, 1973). Methadone concentrations are typically somewhat lower in umbilical cord plasma and amniotic fluid than in maternal plasma, again indicating some limitation on transplacental transfer. For example, in the Blinick *et al.* (1975) study, umbilical cord plasma and amniotic fluid concentrations were 57% and 73%, respectively, of the maternal plasma concentration. Methadone has been detected in amniotic fluid as early as the 14th week of gestation (Blinick *et al.*, 1975).

Morphine is also transferred across the placenta in the human, as indicated by the detection of small amounts in the umbilical vein and artery following intrathecal administration 1 to 5 hours prior to delivery (Bonnardot, Maillet, Colau, Millot, & Deligne, 1982).

The fetus is passively exposed to opiates through placental transfer, as is dramatically demonstrated by the opiate withdrawal symptoms that the neonates from addicted women undergo following delivery. With the abrupt withdrawal of the drug source and the continuing metabolism and excretion of the drug by the infant, drug levels fall, and the infant displays a "neonatal abstinence syndrome" (Desmond & Wilson, 1975; Finnegan, Kron, Connaughton, & Emich, 1975b). In the case of heroin or methadone addiction, the symptoms usually appear within 72 hours of birth and may include restlessness, tremors, hyperirritability, vague autonomic symptoms, gastrointestinal dysfunction, respiratory distress, increased muscle tone, and a high-pitched cry (Finnegan, 1985; Finnegan, Kron, Connaughton, & Emich, 1975a; Perlmutter, 1974).

EFFECTS ON FETAL TESTICULAR FUNCTION

In utero exposure to methadone suppresses T titers in the blood of fetal male rats (Singh, Purohit, & Ahluwalia, 1980). As little as 5 mg/kg per day of methadone, administered to the dam on days 14 through 19, reduced T concentrations in the serum of male fetuses on day 20 of gestation by 42%. The reduction in fetal T induced by methadone could be prevented by simultaneous treatment with the opiate antagonist naloxone. In the fetus, as in the adult male rat, the inhibition did not seem to involve a direct effect at the level of the gonads. Methadone exposure did not alter the synthesis of T by isolated fetal testicular tissue (Singh *et al.*, 1980).

Morphine also disrupts testicular steroidogenesis in fetal male rats. The activity of 3β-hydroxysteroid dehydrogenase (3β-HSD), a key steroidogenic enzyme in the synthesis pathway for T, was quantified in individual Leydig cells of testes taken on days 18–21 of gestation (Badway, Orth, & Weisz, 1981). Opiate-treated dams received an initial dose of 4 mg/kg per day of morphine beginning on day 16 of gestation. Injections were given every 8 hours. The dose was increased by 4 mg/kg per day on each subsequent day. The 3β-HSD activity was significantly lower on days 19 and 20 of gestation in testes from morphine-exposed fetuses than in fetuses from saline-injected controls.

LONG-TERM EFFECTS OF *IN UTERO* EXPOSURE

EXPERIMENTAL ANIMALS.   *In utero* exposure of male rats to morphine is associated with feminized and incompletely masculinized sexual behavior patterns in adulthood (Ward, Orth, & Weisz, 1983). Chronic fetal opiate exposure was achieved by injecting the pregnant dam with morphine every 8 hours during days 15 through 22 of gestation. Low-dose morphine mothers received 1 mg/kg per day, with the dosage level doubled every other day. A high-dose group started at 4 mg/kg per day, with the dose increased by 4 mg/kg per day each day. The morphine-exposed pups from both the low- and high-dose mothers were cross-fostered at birth to control mothers. A series of tests given to the male offspring at 70–90 days of age revealed that exposure to either the low or high dose of morphine during fetal development significantly reduced the percentage of males that showed the ejaculatory pattern in adulthood (see Figure 6). Furthermore, males exposed *in utero* to the high, but not the low, dose of morphine had an enhanced potential for female copulatory patterns, i.e., displayed lordosis when mounted by a stud male (see Figure 6). There were no significant differences caused by prenatal morphine exposure in adult levels of plasma LH or T or in body weight, anogenital distance, penile length, or testes weight. Thus, *in utero* exposure to morphine induced behavioral feminization

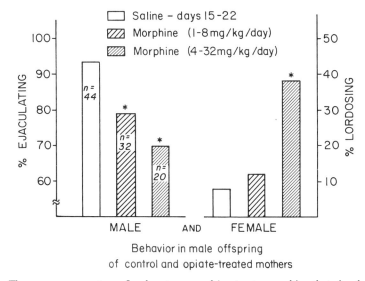

Behavior in male offspring
of control and opiate-treated mothers

Figure 6.   The average percentage of male rats exposed *in utero* to morphine that ejaculated in adulthood in four tests is shown in the left panel. The percentage of these males that showed lordosis on at least one of five tests for female behavior is shown in the right panel. *$p < 0.005$, compared with control males (binomial tests). (From Ward *et al.*, 1983. Copyright 1983 by S. Karger. Adapted by permission.)

and a partial failure of masculinization without readily apparent concomitant alterations in adult endocrine state or anatomy (Ward *et al.*, 1983).

Male rats exposed to morphine during the period of gestation immediately preceding the surge of plasma T show normal male copulatory patterns (Vathy, Etgen, & Barfield, 1985). In the Vathy *et al.* (1985) study, twice-daily injections of 10 mg/kg were initiated on day 12 of gestation and were terminated on the day when plasma T levels first increase, i.e., day 18 of gestation. Thus, morphine exposure overlapped to a minimal extent with the period of maximal fetal testicular steroidogenesis, and behavior was not altered.

HUMANS. The effects of *in utero* exposure to drugs on the sexual differentiation of human males has been investigated in two studies of the children from women addicted to opiates during pregnancy. In one study (Ward, Kopertowski, Finnegan, & Sandberg, 1989), gender-related behaviors in 5- to 7-year-old boys who were the progeny of women from a methadone maintenance clinic in Philadelphia were directly assessed. At this age, drug-exposed boys showed some feminization on two tests that are widely used in the assessment of gender identity. In one, the It Scale for Children, the child plays with a stick figure ("It") that has no distinct gender clues. The assumption is that children project their own sex role preference onto the stick figure.

In one part of this test, the child is shown four pictures representing a girl, a girl dressed as a boy, a boy dressed as a girl, and a boy. When asked "Which child would 'It' like to be?" 35% of the drug-exposed boys chose the girl or an ambiguous figure, compared to 8% of the control boys. Regression analyses revealed that this feminization of the boy's choices was significantly associated with prenatal exposure to drugs but was not associated to variations in certain postnatal rearing conditions (i.e., being raised without a father figure or by someone other than the biological mother).

In a second test of gender identity, 39% of drug-exposed boys, compared to 15% of control boys, drew a female when asked to "Draw a Person." It is assumed that the child projects his own self-image onto the drawing. Although these results are congruent with the hypothesis that *in utero* exposure to opiates may lead to some feminization of gender identity in boys, it should be noted that no profound disturbances in gender identity were found in this study, i.e., no reports of dressing in feminine clothing, expressing a wish to be a girl, or showing a preference for feminine over masculine toys. During pregnancy, the mothers received an average of 35 mg of methadone daily, and many were polydrug users, with the majority continuing to use some heroin.

Sandberg, Meyer-Bahlburg, Rosen, and Johnson (1990) have reported that 6- to 8-year-old boys exposed to methadone *in utero* were rated by their primary caretakers as engaging in more stereotypically feminine play activities, such as doll and house play, than were control boys. This increase in cross-gender behavior was significant both for boys whose mothers were polydrug users during pregnancy and for those who used only methadone. The mothers of these children were from the New York City area and received an average of 39 mg of methadone per day during pregnancy.

SUMMARY

In rodents, prenatal exposure to opiates disrupts testosterone production by the male fetus. This attenuation in fetal testosterone levels is associated with feminized and incompletely masculinized copulatory behavior potentials in adulthood. These residual behavioral effects are dissociated from gross genital morphology and endocrine status, which appear to be normal in adult rats that have been exposed to opiates *in utero*.

Although there is no direct evidence as to whether or not opiate exposure of the human attenuates fetal testosterone titers, it has been shown that boys who were exposed *in utero* to methadone show some subtle evidence of feminization of gender identity and an increased preference for stereotypically feminine play activities. As in the animal work, no abnormalities of reproductive organs have been reported in opiate-exposed human males.

Barbiturates and Phenytoin

The coverage of this broad category of drugs focuses only on those substances that often are used chronically throughout gestation. Phenobarbital (phenobarbitone), a long-acting drug, is the barbiturate of choice in the treatment of epilepsy. In epileptics anticonvulsant drugs typically are continued throughout pregnancy, as the prevention of seizures is considered to be more important than the small risk of drug-induced teratological effects on the fetus. Pentobarbital (Nembutal®) and secobarbital (Seconal®) are short-acting barbiturates with considerable abuse potential (Wesson & Smith, 1977). Addiction to these barbiturates also might lead to sustained exposure throughout gestation for some fetuses. Finally, in addition to the barbiturates, phenytoin (diphenylhydantoin, Dilantin®), a hydantoin, also is included because of its very common use as an anticonvulsant drug for pregnant epileptic women.

Actions in Adult Males

A 40% reduction in serum T levels of adult male rats is seen 2 hours following an anesthetic-level dose (40 mg/kg) of pentobarbital (Cicero & Badger, 1977a). The suppression of T lasted for 3 to 4 hours and was preceded by a fall in LH, suggesting it to be secondary to an inhibitory effect exerted at the hypothalamic–pituitary level. Pentobarbital also suppressed LH within 30 minutes in castrated male rats (Beattie, Campbell, Nequin, Soyka, & Schwartz, 1973). A single dose of phenobarbital (60 mg/kg) or phenytoin (50 mg/kg) in rats suppressed T levels in plasma samples taken 4 hours after drug administration (Kaneyuki, Kohsaka, & Shohmori, 1979). The dosage of phenobarbital used by Kaneyuki *et al.* (1979) is probably within a clinically relevant range, since an 80 mg/kg dose of phenobarbital in mice produced plasma concentrations of the drug that were well within the human anticonvulsant therapeutic range (Middaugh, 1986). Phenobarbital also reduced the weight of the seminal vesicles in immature and mature rats (Conney, Levin, Jacobson, & Kuntzman, 1973).

In addition to reducing T by suppressing LH, barbiturates may exert an inhibitory action directly on the gonads. Phenobarbital inhibits 3β-HSD, at least in ovarian tissue of immature rats (Gupta & Karavolas, 1973). Further reduction in circulating T may result from barbiturate-induced increases in the rate of liver metabolism of steroids. Phenobarbital increases the activity of liver enzymes that hydroxylate T (Conney *et al.*, 1973).

In men, exposure to anticonvulsant drugs seems to reduce androgen levels. A substantial percentage of male epileptics treated with phenytoin plus other anticonvulsant drugs report impaired potency, which was correlated with abnormally low levels of urinary androsterone, an androgen metabolite (Christiansen & Lund, 1975).

USE DURING PREGNANCY

It is estimated that there are over 18,000 births per year in the United States alone to women taking anticonvulsant drugs to control epilepsy (Vorhees, 1986). However, the use of barbiturates to control seizures accounts for only a small fraction of all women who have been exposed to these drugs during pregnancy. Licit uses have included treatment of morning sickness, as tranquilizers, for anesthesia, and to control insomnia and hypertension. In recent years, benzodiazepenes have largely replaced barbiturates as hypnotic and antianxiety agents. For a more complete coverage of the numerous reasons that women take barbiturates during pregnancy, see the review by Reinisch and Sanders (1982).

Barbiturates are also the most commonly abused prescription drug (Wesson & Smith, 1977). Secobarbital and pentobarbital in particular produce an intoxication similar to alcohol. The tranquilizing and euphoric effects that result from these short-acting barbiturates have made them very common drugs of abuse. Both physical and psychological addiction to barbiturates occurs, thus resulting in sustained exposure to the drug. As a result of the many licit as well as illicit uses of barbiturates, surveys done in the 1970s found that approximately one out of every four women was exposed to barbiturates during her pregnancy (Forfar & Nelson, 1973; Heinonen *et al.*, 1977; Hill, 1973).

Barbiturates and phenytoin pass across the placenta rapidly. In 19 women given pentobarbital IV within 5 minutes of delivery as an obstetric analgesic, levels of the drug in umbilical cord blood were 77% of the levels in blood taken from a maternal vein (Fealy, 1958). Similar rapid placental transfer of secobarbital just before delivery has been reported (Root, Eichner, & Sunshine, 1961). Barbiturates administered to the mother also enter the fetus readily in earlier stages of development. In 35 women who had legal abortions performed in Sweden some time between the fourth and seventh month, amobarbital or phenobarbital was administered IM at time intervals that varied from 30 minutes to 50 hours before the procedure. Barbiturate levels in umbilical cord blood were at least equal to maternal vein levels at all time intervals up to 24 hours (Ploman & Persson, 1957).

In neonates from women who were treated with phenobarbital to control epilepsy during pregnancy, umbilical cord levels of drug were 95% of maternal vein levels (Melchior, Svensmark, & Trolle, 1967). In neonates whose mothers received phenobarbital for a short period before delivery to control hypertension, drug levels in cord blood were equal to maternal levels. The plasma half-life of phenobarbital in the neonates varied from 77 to 404 hours (Jalling, Boréus, Kållberg, & Agurell, 1973).

Barbiturate withdrawal symptoms have been noted in infants whose mothers were barbiturate addicted during pregnancy, taking amobarbital or secobarbital, as well as in infants whose mothers were taking phenobarbital to control seizures or for other medical reasons (Desmond, Schwanecke, Wilson, Yasunaga, & Burgdorff, 1972). The symptoms resemble those seen in infants whose mothers are heroin addicts, except that the onset of the symptoms for those exposed to phenobarbital is later, usually about 7 days after birth. The delayed onset is probably related to the long half-life of phenobarbital. Symptoms similar to those seen following abrupt withdrawal of barbiturates in adults, including seizures, have been reported in an infant passively addicted through his mother's secobarbital abuse during pregnancy (Bleyer & Marshall, 1972).

Maternally administered phenytoin also passes into the fetus. Phenytoin levels in umbilical artery and vein were virtually identical to maternal vein levels at delivery in women given the drug throughout pregnancy to control their seizure disorder (Mirkin, 1971a, 1971b). A widespread distribution in fetal tissue was found when phenytoin was injected immediately prior to therapeutic abortion (Mirkin, 1971a). In rats, on days 18 or 19 of gestation, phenytoin levels in fetal liver and brain exceed those in the mother 1 hour after IV administration to the mother (Mirkin, 1971a). After oral phenytoin administration to rat mothers on the 19th day of gestation, fetal brain levels equaled maternal brain levels at 4, 8, 16, 24, and 48 hours after administration (Gabler & Falace, 1970). Similarly, 6 hours following an IV infusion of phenytoin to rhesus monkeys on the 150th day of gestation, fetal (umbilical) artery and vein drug levels were equal to maternal artery levels. The tissue distribution of the drug was similar for fetus and mother in the case of most organs (Gabler & Hubbard, 1972).

EFFECTS ON FETAL TESTICULAR FUNCTION

FETAL T LEVELS.   Maternal phenobarbital administration during the period of maximal prenatal steroidogenesis suppresses fetal T levels in rats. Testosterone concentrations in the brains and plasma of fetal rats on day 20 of gestation were significantly depressed when the mother was given 40 mg/kg per day of phenobarbital on days 17, 18, and 19 of gestation (Gupta, Yaffe, & Shapiro, 1982). Control dams received saline injections.

NEONATAL TESTICULAR FUNCTION.   Testosterone concentrations in the plasma of rats whose mothers received 40 mg/kg per day of phenobarbital on days 17, 18, 19, and 20 of gestation were approximately 50% of control values on the day of birth. The T concentrations in the brains of phenobarbital-exposed male rats were less than one-third of those in control neonates. Further, the capacity to synthesize T was significantly impaired in testes taken on the day of birth from phenobarbital-exposed male rats compared to that of control males (Gupta *et al.*, 1982; Sonawane & Yaffe, 1983).

ANOGENITAL DISTANCE.   At birth, the A-G distance in male rats derived from dams that received 40 mg/kg per day of phenobarbital on days 12–19 of gestation is significantly shorter than that of males from saline-injected dams (Gupta, Shapiro, & Yaffe, 1980). The body weight of the neonates was not reduced by the prenatal phenobarbital. On day 18 of gestation the A-G distance of fetal male mice from mothers given 50 mg/kg of phenytoin on days 11–14 of gestation was reduced by 44% and was not different from that in females (Gupta & Goldman, 1986).

HUMAN NEONATAL GENITAL ANOMALIES.   Zellweger (1974), from a review of the literature, concluded that the use of anticonvulsant drugs during pregnancy was associated with a small increase in the incidence of hypospadias. This point is of particular interest for this review, since hypospadias is the result of fetal androgen deficiency (Angerpointner, 1984). Not all reviewers have noted an increased incidence of hypospadias. Janz (1975) and Hill, Verniaud, Horning, McCulley, and

Morgan (1974) found that the incidence of hypospadias associated with the use of phenobarbital, phenytoin, etc. in pregnancy was 0.3% to 0.4%, which is essentially the same as the 0.37% incidence seen in the infant population at large (Heinonen *et al.*, 1977). However, in a study not included in the 27 studies reviewed by Hill *et al.* (1974) and Janz (1975), Fedrick (1973) found an incidence of 1.4% hypospadias in 217 infants delivered to epileptic mothers in Britain compared to 0.3% in his control mothers. A similar incidence of hypospadias (1.5%) was found in the offspring of women on anticonvulsant drugs in Finland (Granström & Hiilesmaa, 1982).

Recently, material was compiled from Italy, Sweden, and Lyon, France, in which detailed diagnoses were available about the malformations associated with the use of a single anticonvulsant drug during pregnancy. Of the 352 infants studied, 1.4% had hypospadias or micropenis (Bertollini, Källen, Mastroiacovo, & Robert, 1987). Further, in a large study of infants selected because they had congenital abnormalities, Nelson and Forfar (1971) noted that prenatal barbiturate exposure was associated with "intersex" and hypospadias in children. No detailed data on this point were provided. Since the 1974 and 1975 reviews, hypospadias, undescended testes, and other genital anomalies have been noted in a number of small sample case reports on the children of women who took anticonvulsants while pregnant (Barr, Poznanski, & Schmickel, 1974; DiLiberti, Farndon, Dennis, & Curry, 1984; Finnell & DiLiberti, 1983; Hanson & Smith, 1975; Hanson, Myrianthopoulos, Harvey, & Smith, 1976; Owens-Stively, Orson, & Martin, 1985; Pinto, Gardner, & Rosenbaum, 1977; Zackai, Mellman, Neiderer, & Hanson, 1975). Thus, there is considerable evidence suggesting that exposure to anticonvulsant medication during pregnancy is associated with a small increase in the incidence of genital malformations, especially hypospadias, in male offspring.

LONG-TERM EFFECTS OF *IN UTERO* EXPOSURE

Residual effects of prenatal barbiturate exposure on adult endocrine status and reproductive anatomy have been reported. As adults, male rats exposed to phenobarbital during gestation have significantly lower plasma LH and T levels, reduced seminal vesicle weights, and an impaired capacity of the testes to synthesize T (Gupta *et al.*, 1980, 1982).

Reduced fertility also has been found in male rats exposed to phenobarbital during prenatal development. Only 64% of phenobarbital-exposed males successfully impregnated females as compared to 100% of the control group (Gupta *et al.*, 1982). Interestingly, the females paired with the phenobarbital-exposed males were found to have sperm plugs, although they did not become pregnant. However, when 80 mg/kg of phenobarbital was combined with 80 mg/kg of phenytoin on days 5–20 of pregnancy, only 52% of the male offspring mated, as assessed by the presence of sperm plugs in the paired females, compared to 86% of control males (Takagi, Alleva, Seth, & Balazs, 1986).

In open-field tests, the activity level of adult male mice exposed from the 13th day of gestation until parturition to 40 mg/kg per day of phenobarbital was increased to levels characteristic of females (Zemp & Middaugh, 1975). Further, phenobarbital-exposed males, like normal females, exhibited short latencies in a passive avoidance task. As Reinisch and Sanders (1982) have pointed out, these data can be interpreted to reflect a demasculinization of nonreproductive behavior resulting from prenatal barbiturate exposure.

Exposure of pregnant experimental animals to levels of phenobarbital that fall within a clinically relevant anticonvulsant range suppresses circulating T titers in developing male fetuses. Prenatal exposure of male rats to barbiturates has long-term consequences not only on sexually dimorphic behaviors but also on adult endocrine status. The weight of the evidence also suggests a small increase in the incidence of hypospadias in human males exposed prenatally to anticonvulsant drugs.

Miscellaneous Drugs

Cimetidine

Cimetidine (Tagamet®) is a drug that blocks type 2 histamine receptors. It is widely used in the control of duodenal and peptic ulcers and has been used during pregnancy for the relief of a variety of gastrointestinal symptoms, including those commonly referred to as "morning sickness." In high concentrations cimetidine acts as a weak antiandrogen by competitively binding to cytosol androgen receptors, as has been demonstrated in rat ventral prostate (Foldesy, Vanderhoof, & Hahn, 1985; Sivelle, Underwood, & Jelly, 1982) and mouse kidney tissue (Funder & Mercer, 1979). *In vivo*, cimetidine, in high dose levels, causes reductions in prostate and seminal vesicle weights in male rats (Foldesy *et al.*, 1985; Leslie & Walker, 1977; Sivelle *et al.*, 1982). After 6 weeks of daily cimetidine administration to male rats, reduced weights of accessory sexual organs were accompanied by elevated gonadotropin levels (Baba, Paul, Pollow, Janetschek, & Jacobi, 1981). At therapeutic levels in men, cimetidine either has no effect on plasma T levels (Spona *et al.*, 1987; Stubbs *et al.*, 1983) or causes small increases in T (Peden, Boyd, Browning, Saunders, & Wormsley, 1981; Van Thiel, Gavaler, Smith, & Paul, 1979; Wang, Lai, Lam, & Yeung, 1982). The increases in T have been attributed to cimetidine's antagonism of the normal negative feedback that androgens exert on gonadotropin secretion (Peden, Cargill, Browning, Saunders, & Wormsley, 1979). Gynecomastia and even loss of libido that progressed to impotence have occasionally been reported in men taking cimetidine (Peden *et al.*, 1979; Spence & Celestin, 1979), but the occurrence of these disorders is very rare (Gifford, Aeugle, Myerson, & Tannenbaum, 1980). In one survey, gynecomastia, the most frequent endocrine-related complaint, was reported in only 0.2% of over 9,000 patients taking cimetidine (Gifford *et al.*, 1980).

In view of the weak antiandrogenic properties of cimetidine and the placental transfer of the drug in humans (Howe, McGowan, Moore, McCaughey, & Dundee, 1981; Schenker, Dicke, Johnson, Mor, & Henderson, 1987), the possibility that cimetidine use during pregnancy might disrupt sexual differentiation of males has been investigated. The results of these studies are contradictory. Cimetidine mixed in the drinking water of pregnant rats starting on day 12 of gestation at a dosage level that averaged 137 mg/kg per day reduced the A-G distance of the male offspring on the day of birth (Anand & Van Thiel, 1982). Even a much smaller dose (17.1 mg/kg per day) of cimetidine during the same prenatal period significantly reduced the A-G distance at 1 day of age (Parker, Schade, Pohl, Gavaler, & Van Thiel, 1984). McGivern (1987) also found that A-G distance on the day of birth was reduced in male rats exposed *in utero* to cimetidine. In contrast, neither Walker, Bott, and Bond (1987) nor Shapiro, Hirst, Babalola, and Bitar (1988) were able to demonstrate any effect of prenatal cimetidine on neonatal A-G distance despite using large doses of cimetidine.

In all of these studies, exposure to cimetidine was continued for at least the first week of postnatal development, and various anatomic, hormonal, and/or behavioral measures were made in adulthood. Studies finding an effect on day 1 of postnatal life also found alterations in adult anatomic, endocrine, and/or behavioral measures, whereas those not finding a neonatal effect also found no effect in adulthood. Since cimetidine exposure was continued well beyond the period of fetal development, these findings are not covered in detail.

COCAINE

In recent years, cocaine use has increased dramatically. In a recent survey (Frank *et al.*, 1988) of 679 women who enrolled for routine prenatal care in a Boston clinic, 17% used cocaine at least once while pregnant. Placental transfer of cocaine has been demonstrated in rats (Wiggins, Rolsten, Ruiz, & Davis, 1989). A single injection of 30 mg/kg or 40 mg/kg of cocaine to adult male rats induces a biphasic effect on plasma T concentrations. A transient increase is followed by a depression in T levels (Berul & Harclerode, 1989; Gordon, Mostofsky, & Gordon, 1980). Chronic cocaine administration for 15 days at 40 mg/kg per day to adult rats depressed T levels by 77% and significantly reduced epididymis and seminal vesicle weights (Berul & Harclerode, 1989).

There is very little information available on the effects of fetal exposure to cocaine on sexual differentiation. Raum, McGivern, Peterson, Shryne, and Gorski (1990) examined the effects of cocaine injections to pregnant rats on days 15–20 of gestation on a number of reproductive indices in the male offspring. The A-G distance on the day of birth was not influenced by prenatal cocaine exposure, and the adult levels of plasma T and weights of the testes, seminal vesicles, prostate, and epididymis were not altered. However, adult scent marking, a behavior whose normal organization in males is dependent on early androgen exposure, was depressed in male rats exposed to as little as 3 mg/kg per day of cocaine during gestation. There also was some suggestion of an impairment in copulatory behavior in males exposed *in utero* to larger doses of cocaine (Raum *et al.*, 1990). Male mice exposed to cocaine through maternal injection on day 8, 10, or 11 of gestation showed an increased incidence of cryptorchidism (Mahalik, Gautieri, & Mann, 1980). Chasnoff, Chisum, and Kaplan (1988) noted an unusually high incidence (four out of 50) of hypospadias in male infants from cocaine-using women.

CONCLUSIONS

Within the past 10 years studies have appeared that found that T concentrations in male fetuses are suppressed when pregnant animals are exposed to such commonly used drugs of abuse as ethanol, nicotine, marijuana, opiates, and barbiturates. At birth, male rodents exposed *in utero* to these drugs had no gross deviation of their external genitalia, but on close examination of ethanol and barbiturate males, a small but significant reduction (feminization) of the A-G distance was noted. Functional alterations, rather than anatomic anomalies, were the primary consequence of a drug-induced attenuation of normal gonadal hormone stimulation during critical prenatal stages of development in animals. These functional changes typically were not detected until adolescence or adulthood. A variety of sexually dimorphic behaviors have been demonstrated to be feminized and/or incompletely masculinized in drug-exposed males. In the case of exposure to certain drugs (e.g., barbiturates), deviations in T concentrations persisted into adulthood, whereas prenatal exposure to other drugs (e.g., opiates) led to behavioral variations despite apparently normal adult gonadal hormone levels.

If fetal exposure of human males to drugs of abuse also causes modifications in gender-related behaviors without concomitant changes in the gross appearance of the genitalia, it should not come as a surprise that this type of behavioral teratogenic effect has not been easily recognized. As Vorhees (1989) has pointed out, linkage of problems that first appear in adulthood to the mother's drug history during pregnancy is problematic. The detection of a relationship between a physical defect that is clearly visible at birth and maternal drug use is much more likely than a behavioral effect that appears years after the pregnancy. Indeed, it will take very careful and well-controlled studies to tease out any prenatal drug effects on sexually dimorphic behaviors in humans. There are many confounding variables that make it very difficult to select an appropriate contrast group for drug-addicted mothers. However, in view of the preliminary findings of Ward *et al.* (1989) and Sandberg *et al.* (1990) that gender orientation in boys exposed *in utero* to opiates is somewhat feminized, further investigation is warranted into possible endocrine-mediated effects of maternal drug abuse on sexual behavior differentiation of human males.

REFERENCES

Abel, E. L. (1980). Fetal alcohol syndrome: Behavioral teratology. *Psychological Bulletin, 87*, 29–50.

Abel, E. L., & Dintcheff, B. A. (1986). Saccharin preference in animals prenatally exposed to alcohol: No evidence of altered sexual dimorphism. *Neurobehavioral Toxicology and Teratology, 8*, 521–523.

Abel, E. L., Tan, S. E., & Subramanian, M. (1987). Effects of Δ^9-tetrahydrocannabinol, phenobarbital, and their combination on pregnancy and offspring in rats. *Teratology, 36*, 193–198.

Adams, E. H., Gfroerer, J. C., & Rouse, B. A. (1989). Epidemiology of substance abuse including alcohol and cigarette smoking. In D. E. Hutchings (Ed.), *Prenatal abuse of licit and illicit drugs: Annals of the New York Academy of Sciences, Vol. 562* (pp. 14–20). New York: New York Academy of Sciences.

Ahluwalia, B. S., Rajguru, S. U., & Nolan, G. H. (1985). The effect of Δ^9-tetrahydrocannabinol *in utero* exposure on rat offspring fertility and ventral prostate gland morphology. *Journal of Andrology, 6*, 386–391.

Anand, S., & Van Thiel, D. H. (1982). Prenatal and neonatal exposure to cimetidine results in gonadal and sexual dysfunction in adult males. *Science, 218*, 493–494.

Angerpointner, T. A. (1984). Hypospadias—Genetics, epidemiology and other possible aetiological influences. *Zeitschrift für Kinderchirurgie, 39*, 112–118.

Azizi, F., Vagenakis, A. G., Longcope, C., Ingbar, S. H., & Braverman, L. E. (1973). Decreased serum testosterone concentration in male heroin and methadone addicts. *Steroids, 22*, 467–472.

Baba, S., Paul, H. J., Pollow, K., Janetschek, G., & Jacobi, G. H. (1981). *In vivo* studies on the antiandrogenic effects of cimetidine versus cyproterone acetate in rats. *The Prostate, 2*, 163–174.

Badway, D., Orth, J., & Weisz, J. (1981). Effect of morphine on Δ^5-3β-OL steroid dehydrogenase (3β-OHD) in Leydig cells of fetal rats: A quantitative cytochemical study. *Anatomical Record, 199*, 15a.

Bailey, J. R., Cunny, H. C., Paule, M. G., & Slikker, W. (1987). Fetal disposition of Δ^9-tetrahydrocannabinol (THC) during late pregnancy in the rhesus monkey. *Toxicology and Applied Pharmacology, 90*, 315–321.

Barnett, G., & Chiang, C.-W. N. (1983). Effects of marijuana on testosterone in male subjects. *Journal of Theoretical Biology, 104*, 685–692.

Barr, M., Poznanski, A. K., & Schmickel, R. D. (1974). Digital hypoplasia and anticonvulsants during gestation: A teratogenic syndrome. *The Journal of Pediatrics, 84*, 254–256.

Barron, S., Tieman, S. B., & Riley, E. P. (1988). Effects of prenatal alcohol exposure on the sexually dimorphic nucleus of the preoptic area of the hypothalamus in male and female rats. *Alcoholism: Clinical and Experimental Research, 12,* 59–64.

Beattie, C. W., Campbell, C. S., Nequin, L. G., Soyka, L. F., & Schwartz, N. B. (1973). Barbiturate blockade of tonic LH secretion in the male and female rat. *Endocrinology, 92,* 1634–1638.

Bertollini, R., Källen B., Mastroiacovo, P., & Robert, E. (1987). Anticonvulsant drugs in monotherapy. Effect on the fetus. *European Journal of Epidemiology, 3,* 164–171.

Berul, C. I., & Harclerode J. E. (1989). Effects of cocaine hydrochloride on the male reproductive system. *Life Sciences, 45,* 91–95.

Bhalla, V. K., Rajan, V. P., & Newman, M. E. (1983). Alcohol-induced luteinizing hormone receptor deficiency at the testicular level. *Alcoholism: Clinical and Experimental Research, 7,* 153–162.

Blackard, C., & Tennes, K. (1984). Human placental transfer of cannabinoids. *The New England Journal of Medicine, 311,* 797.

Bleyer, W. A., & Marshall, R. E. (1972). Barbiturate withdrawal syndrome in a passively addicted infant. *Journal of the American Medical Association, 221,* 185–186.

Blinick, G., Inturrisi, C. E., Jerez, E., & Wallach, R. C. (1975). Methadone assays in pregnant women and progeny. *American Journal of Obstetrics and Gynecology, 121,* 617–621.

Bonnardot, J. P., Maillet, M., Colau, C., Millot, F., & Deligne, P. (1982). Maternal and fetal concentration of morphine after intrathecal administration during labour. *British Journal of Anaesthesiology, 54,* 487–489.

Briggs, M. H. (1973). Cigarette smoking and infertility in men. *The Medical Journal of Australia, 63,* 616–617.

Broida, J. P., Churchill, J., & McGinnis, M. (June, 1988). Prenatal exposure to alcohol facilitates lordosis in male rats. Poster presented at the *Annual Conference on Reproductive Behavior,* Omaha, Nebraska.

Burstein, S., Hunter, S. A., Shoupe, T. S., & Taylor, P. (1978). Cannabinoid inhibition of testosterone synthesis by mouse Leydig cells. *Research Communications in Chemical Pathology and Pharmacology, 19,* 557–560.

Burstein, S., Hunter, S. A., & Shoupe, T. S. (1979). Site of inhibition of Leydig cell testosterone synthesis by Δ^1-tetrahydrocannabinol. *Molecular Pharmacology, 15,* 633–640.

Carr, J. N. (1975). Drug patterns among drug-addicted mothers: Incidence, variance in use, and effects on children. *Pediatric Annals, 4,* 408–417.

Chakravarty, I., Sheth, P. R., Sheth, A. R., & Ghosh, J. J. (1982). Delta-9-tetrahydrocannabinol: Its effect on hypothalamo-pituitary system in male rats. *Archives of Andrology, 8,* 25–27.

Chasnoff, I. J., Chisum, G. M., & Kaplan, W. E. (1988). Maternal cocaine use and genitourinary tract malformations. *Teratology, 37,* 201–204.

Chen, J. J., & Smith, E. R. (1979). Effects of perinatal alcohol on sexual differentiation and open-field behavior in rats. *Hormones and Behavior, 13,* 219–231.

Chiao, Y.-B., Johnston, D. E., Gavaler, J. S., & Van Thiel, D. H. (1981). Effects of chronic ethanol feeding on testicular content of enzymes required for testosteronogenesis. *Alcoholism: Clinical and Experimental Research, 5,* 230–236.

Choi, Y. S., Sperling, F., & On'gele, N. A. (1976). Morphine distribution in pregnant mice and their conceptuses following daily dosing. *Toxicology and Applied Pharmacology, 37,* 127.

Christiansen, P., & Lund, M. (1975). Sexual potency, testicular function and excretion of sexual hormones in male epileptics. In D. Janz (Ed.), *Epileptology: Proceedings of the seventh international symposium on epilepsy* (pp. 190–191). Berlin: Thieme Edition Publishing Sciences Group.

Cicero, T. J. (1981). Neuroendocrinological effects of alcohol. *Annual Reviews of Medicine, 32,* 123–142.

Cicero, T. J., & Badger, T. M. (1977a). A comparative analysis of the effects of narcotics, alcohol and the barbiturates on the hypothalamic–pituitary–gonadal axis. *Advances in Experimental Medicine and Biology, 85B,* 95–115.

Cicero, T. J., & Badger, T. M. (1977b). Effects of alcohol on the hypothalamic–pituitary–gonadal axis in the male rat. *The Journal of Pharmacology and Experimental Therapeutics, 201,* 427–433.

Cicero, T. J., Bell, R. D., Wiest, W. G., Allison, J. H., Polakoski, K., & Robins, E. (1975). Function of the male sex organs in heroin and methadone users. *The New England Journal of Medicine, 292,* 882–887.

Cicero, T. J., Meyer, E. R., Wiest, W. G., Olney, J. W., & Bell, R. D. (1975). Effects of chronic morphine administration on the reproductive system of the male rat. *The Journal of Pharmacology and Experimental Therapeutics, 192,* 542–548.

Cicero, T. J., Meyer, E. R., Bell, R. D., & Koch, G. A. (1976). Effects of morphine and methadone on serum testosterone and luteinizing hormone levels and on the secondary sex organs of the male rat. *Endocrinology, 98,* 367–372.

Cicero, T. J., Wilcox, C. E., Bell, R. D., & Meyer, E. R. (1976). Acute reductions in serum testosterone levels by narcotics in the male rat: Stereospecificity, blockade by naloxone and tolerance. *The Journal of Pharmacology and Experimental Therapeutics, 198,* 340–346.

Cicero, T. J., Badger, T. M., Wilcox, C. E., Bell, R. D., & Meyer, E. R. (1977). Morphine decreases luteinizing hormone by an action on the hypothalamic–pituitary axis. *The Journal of Pharmacology and Experimental Therapeutics, 203,* 548–555.

Cicero, T. J., Bell, R. D., Meyer, E. R., & Schweitzer, J. (1977). Narcotics and the hypothalamic–pituitary–gonadal axis: Acute effects on luteinizing hormone, testosterone and androgen-dependent systems. *The Journal of Pharmacology and Experimental Therapeutics, 201,* 76–83.

Clarren, S. K., & Smith, D. W. (1978). The fetal alcohol syndrome. *The New England Journal of Medicine, 298,* 1063–1067.

Clemens, L. G., Gladue, B. A., & Coniglio, L. P. (1978). Prenatal endogenous androgenic influences on masculine sexual behavior and genital morphology in male and female rats. *Hormones and Behavior, 10,* 40–53.

Colangelo, W. & Jones, D. G. (1982). The fetal alcohol syndrome: A review and assessment of the syndrome and its neurological sequelae. *Progress in Neurobiology, 19,* 271–314.

Collu, R., Letarte, J., Leboeuf, G., & Ducharme, J. R. (1975). Endocrine effects of chronic administration of psychoactive drugs to prepuberal male rats. I: Δ^9-Tetrahydrocannabinol. *Life Sciences, 16,* 533–542.

Condon, J. T., & Hilton, C. A. (1988). A comparison of smoking and drinking behaviours in pregnant women: Who abstains and why. *The Medical Journal of Australia, 148,* 381–385.

Conney, A. H., Levin, W., Jacobson, M., & Kuntzman, R. (1973). Effects of drugs and environmental chemicals on steroid metabolism. *Clinical Pharmacology and Therapeutics, 14,* 727–741.

Corbier, P., Kerdelhue, R., Picon, R., & Roffi, J. (1978). Changes in testicular weight and serum gonadotropin and testosterone levels before, during, and after birth in the perinatal rat. *Endocrinology, 103,* 1985–1990.

Dahlgren, I. L., Eriksson, C. J. P., Gustafsson, B., Harthon, C., Hård, E., & Larsson, K. (1989). Effects of chronic and acute ethanol treatment during prenatal and early postnatal ages on testosterone levels and sexual behaviors in rats. *Pharmacology Biochemistry and Behavior, 33,* 867–873.

Dalterio, S. L. (1980). Perinatal or adult exposure to cannabinoids alters male reproductive functions in mice. *Pharmacology, Biochemistry and Behavior, 12,* 143–153.

Dalterio, S., & Bartke, A. (1979). Perinatal exposure to cannabinoids alters male reproductive function in mice. *Science, 205,* 1420–1422.

Dalterio, S., & Bartke, A. (1981). Fetal testosterone in mice: Effect of gestational age and cannabinoid exposure. *Journal of Endocrinology, 91,* 509–514.

Dalterio, S., Bartke, A., & Burstein, S. (1977). Cannabinoids inhibit testosterone secretion by mouse testes *in vitro. Science, 196,* 1472–1473.

Dalterio, S., Bartke, A., Roberson, C., Watson, D., & Burstein, S. (1978). Direct and pituitary-mediated effects of Δ^9-THC and cannabinol on the testis. *Pharmacology, Biochemistry and Behavior, 8,* 673–678.

Dalterio, S., Bartke, A., Blum, K., & Sweeney, C. (1981). Marihuana and alcohol: Perinatal effects on development of male reproductive functions in mice. *Progress in Biochemical Pharmacology, 18,* 143–154.

Dalterio, S., Bartke, A., & Mayfield, D. (1981). Δ^9-Tetrahydrocannabinol increases plasma testosterone concentrations in mice. *Science, 213,* 581–583.

Dalterio, S., Bartke, A., & Mayfield, D. (1985). Effects of Δ^9-tetrahydrocannabinol on testosterone production *in vitro:* Influence of Ca^{++}, Mg^{++} or glucose. *Life Sciences, 37,* 1425–1433.

Dalterio, S., Mayfield, D., Bartke, A., & Morgan, W. (1985). Effects of psychoactive and non-psychoactive cannabinoids on neuroendocrine and testicular responsiveness in mice. *Life Sciences, 36,* 1299–1306.

Dalterio, S., Steger, R., Peluso, J., & De Paolo, L. (1987). Acute Δ^9-tetrahydrocannabinol exposure: Effects on hypothalamic–pituitary–testicular activity in mice. *Pharmacology, Biochemistry and Behavior, 26,* 533–537.

Deslypere, J. P., & Vermeulen, A. (1984). Leydig cell function in normal men: Effects of age, life-style, residence, diet and activity. *Journal of Clinical Endocrinology and Metabolism, 59,* 955–962.

Desmond, M. M., & Wilson, G. S. (1975). Neonatal abstinence syndrome: Recognition and diagnosis. *Addictive Diseases: An International Journal, 2,* 113–121.

Desmond, M. M., Schwanecke, R. P., Wilson, G. S., Yasunaga, S., & Burgdorff, I. (1972). Maternal barbiturate utilization and neonatal withdrawal symptomatology. *The Journal of Pediatrics, 80,* 190–197.

DiLiberti, J. H., Farndon, P. A., Dennis, N. R., & Curry, C. J. R. (1984). The fetal valproate syndrome. *American Journal of Medical Genetics, 19,* 473–481.

Döhler, K.-D., Coquelin, A., Davis, F., Hines, M., Shryne, J. E., & Gorski, R. A. (1984). Pre- and postnatal influence of testosterone propionate and diethylstilbestrol on differentiation of the sexually dimorphic nucleus of the preoptic area in male and female rats. *Brain Research, 302,* 291–295.

Döhler, K. D., Coquelin, A., Davis, F., Hines, M., Shryne, J. E., Sickmöller, P. M., Jarzab, B., & Gorski, R. A. (1986). Pre- and postnatal influence of an estrogen antagonist and an androgen antagonist on differentiation of the sexually dimorphic nucleus of the preoptic area in male and female rats. *Neuroendocrinology, 42,* 443–448.

Drouva, S. V., Epelbaum, J., Tapia-Arancibia, L., Laplante, E., & Kordon, C. (1981). Opiate receptor modulate LHRH and SRIF release from mediobasal hypothalamic neurons. *Neuroendocrinology, 32,* 163–167.

Ernhart, C. B., Morrow-Tlucak, M., Sokol, R. J., & Martier, S. (1988). Underreporting of alcohol use in pregnancy. *Alcoholism: Clinical and Experimental Research 12,* 506–511.

Ernhart, C. B., Sokol, R. J., Ager, J. W., Morrow-Tlucak, M., & Martier, S. (1989). Alcohol-related birth defects: Assessing the risk. In D. E. Hutchings (Ed.), *Prenatal abuse of licit and illicit drugs: Annals of the New York Academy of Sciences, Vol. 562* (pp. 159–172). New York: New York Academy of Sciences.

Fealy, J. (1958). Placental transmission of pentobarbital sodium. *Obstetrics and Gynecology, 11,* 342–349.

Fedrick, J. (1973). Epilepsy and pregnancy: A report from the Oxford record linkage study. *British Medical Journal, 2,* 442–448.

Finnegan, L. P. (1985). Effects of maternal opiate abuse on the newborn. *Federation Proceedings, 44,* 2314–2317.

Finnegan, L. P., Kron, R. E., Connaughton, J. F., & Emich, J. P. (1975a). Assessment and treatment of abstinence in the infant of the drug-dependent mother. *International Journal of Clinical Pharmacology, 12,* 19–32.

Finnegan, L. P., Kron, R. E., Connaughton, J. F., & Emich, J. P. (1975b). Neonatal abstinence syndrome: Assessment and management. *Addictive Diseases: An International Journal, 2,* 141–158.

Finnell, R. H., & DiLiberti, J. H. (1983). Hydantoin-induced teratogenesis: Are arene oxide intermediates really responsible? *Helvetica Paediatrica Acta, 38,* 171–177.

Foldesy, R. G., Vanderhoof, M. M., & Hahn, D. W., (1985). *In vitro* and *in vivo* comparisons of antiandrogenic potencies of two histamine H_2-receptor antagonists, cimetidine and entintidine-HCl (42087). *Proceedings of the Society for Experimental Biology and Medicine, 179,* 206–210.

Forfar, J. O., & Nelson, M. M. (1973). Epidemiology of drugs taken by pregnant women: Drugs that may affect the fetus adversely. *Clinical Pharmacology and Therapeutics, 14,* 632–642.

Frank, D. A., Zuckerman, B. S., Amaro, H., Aboagye, K., Bauchner, H., Cabral, H., Fried, L., Hingson, R., Kayne, H., Levenson, S. M., Parker, S., Reece, H., & Vinci, R. (1988). Cocaine use during pregnancy: Prevalence and correlates. *Pediatrics, 82,* 888–895.

Fried, P. A., & Charlebois, A. T. (1979). Cannabis administered during pregnancy: First- and second-generation effects in rats. *Physiological Psychology, 7,* 307–310.

Fried, P. A., Buckingham, M., & Von Kulmiz, P. (1983). Marijuana use during pregnancy and perinatal risk factors. *American Journal of Obstetrics and Gynecology, 146,* 992–994.

Fried, P. A., Barnes, M. V., & Drake, E. R. (1985). Soft drug use after pregnancy compared to use before and during pregnancy. *American Journal of Obstetrics and Gynecology, 151,* 787–792.

Funder, J. W., & Mercer, J. E. (1979). Cimetidine, a histamine H_2 receptor antagonist, occupies androgen receptors. *Journal of Clinical Endocrinology and Metabolism, 48,* 189–191.

Gabler, W. L., & Falace, D. (1970). The distribution and metabolism of dilantin in non-pregnant, pregnant and fetal rats. *Archives Internationales de Pharmacodynamie et de Therapie, 184,* 45–58.

Gabler, W. L., & Hubbard, G. L. (1972). The distribution of 5,5-diphenylhydantoin (DPH) and its metabolites in maternal and fetal rhesus monkey tissues. *Archives Internationales de Pharmacodynamie et de Therapie, 200,* 222–230.

Gabrielsson, J. L., & Paalzow, L. K. (1983). A physiological pharmacokinetic model for morphine disposition in the pregnant rat. *Journal of Pharmacokinetics and Biopharmaceutics, 11,* 147–163.

Gantt, P. A., Tho, P. T., Bhalla, V. K., McDonough, P. G., Costoff, A., & Mahesh, V. B. (1982). Effect of ethanol-containing liquid diet upon gonadotropin receptor depletion in rat testis. *The Journal of Pharmacology and Experimental Therapeutics, 223,* 848–853.

Ghosh, S. P., Chatterjee, T. K., & Ghosh, J. J. (1981). Antiandrogenic effect of delta-9-tetrahydrocannabinol in adult castrated rats. *Journal of Reproduction and Fertility, 62,* 513–517.

Gifford, L. M., Aeugle, M. E., Myerson, R. M., & Tannenbaum, P. J. (1980). Cimetidine postmarket outpatient surveillance program. *Journal of the American Medical Association, 243,* 1532–1535.

Goldman, A. S. (1970). Production of congenital lipoid adrenal hyperplasia in rats and inhibition of cholesterol side-chain cleavage. *Endocrinology, 86,* 1245–1251.

Goldman, A. S. (1971). Production of hypospadias in the rat by selective inhibition of fetal testicular 17α-hydroxylase and C_{17-20}-lyase. *Endocrinology, 88,* 527–531.

Goldman, A. S., Bongiovanni, A. M., & Yakovac, W. C. (1966). Production of congenital adrenal cortical hyperplasia, hypospadias, and clitoral hypertrophy (adrenogenital syndrome) in rats by inactivation of 3β-hydroxysteroid dehydrogenase. *Proceedings of the Society for Experimental Biology and Medicine, 121,* 757–766.

Golub, M. S., Eisele, J. H., & Anderson, J. H. (1986). Maternal–fetal distribution of morphine and alfentanil in near-term sheep and rhesus monkeys. *Developmental Pharmacology, 9,* 12–22.

Gordon, G. G., Altman, K., Southren, A. L., Rubin, E., & Lieber, C. S. (1976). Effect of alcohol (ethanol) administration on sex-hormone metabolism in normal men. *The New England Journal of Medicine, 295,* 793–797.

Gordon, L. A., Mostofsky, D. I., & Gordon, G. G. (1980). Changes in testosterone levels in the rat following intraperitoneal cocaine HCl. *International Journal of Neuroscience, 11,* 139–141.

Gorski, R. A., Gordon, J. H., Shryne, J. E., & Southam, A. M. (1978). Evidence for a morphological sex difference within the medial preoptic area of the rat brain. *Brain Research, 148,* 333–346.

Granström, M.-L., & Hiilesmaa, V. K. (1982). Malformations and minor anomalies in the children of epileptic mothers: Preliminary results of the prospective Helsinki study. In D. Janz, M. Dam, A. Richens, L. Bossi, H. Helge, & D. Schmidt (Eds.), *Epilepsy, pregnancy, and the child* (pp. 303–307). New York: Raven Press.

Guerri, C., Esquifino, A., Sanchis, R., & Grisolia, S. (1984). Growth, enzymes and hormonal changes in offspring of alcohol-fed rats. *Ciba Foundation Symposium, 105,* 85–102.

Gupta, C., & Goldman, A. S. (1986). The arachidonic acid cascade is involved in the masculinizing action of testosterone on embryonic external genitalia in mice. *Proceedings of the National Academy of Sciences, U.S.A., 83,* 4346–4349.

Gupta, C., & Karavolas, H. J. (1973). Lowered ovarian conversion of ^{14}C-pregnenolone to progesterone and other metabolites during phenobarbital (PB) block of PMS-induced ovulation in immature rats: Inhibition of 3β-hydroxysteroid dehydrogenation. *Endocrinology, 92,* 117–124.

Gupta, C., Shapiro, B. H., & Yaffe, S. J. (1980). Reproductive dysfunction in male rats following prenatal exposure to phenobarbital. *Pediatric Pharmacology, 1,* 55–62.

Gupta, C., Yaffe, S. J., & Shapiro, B. H. (1982). Prenatal exposure to phenobarbital permanently decreases testosterone and causes reproductive dysfunction. *Science, 216,* 640–642.

Hans, S. L. (1989). Developmental consequences of prenatal exposure to methadone. In D. E. Hutchings (Ed.), *Prenatal abuse of licit and illicit drugs.* Annals of The New York Academy of Sciences, Vol. 562 (pp. 195–207). New York: New York Academy of Sciences.

Hanson, J. W., & Smith, D. W. (1975). The fetal hydantoin syndrome. *The Journal of Pediatrics, 87,* 285–290.

Hanson, J. W., Jones, K. L., & Smith, D. W. (1976). Fetal alcohol syndrome: Experience with 41 patients. *Journal of the American Medical Association, 235,* 1458–1460.

Hanson, J. W., Myrianthopoulos, N. C., Harvey, M. A. S., & Smith, D. W. (1976). Risks to the offspring of women treated with hydantoin anticonvulsants, with emphasis on the fetal hydantoin syndrome. *The Journal of Pediatrics, 89,* 662–668.

Harbison, R. D., & Mantilla-Plata, B. (1972). Prenatal toxicity, maternal distribution and placental transfer of tetrahydrocannabinol. *The Journal of Pharmacology and Experimental Therapeutics, 180,* 446–453.

Hård, E., Dahlgren, I. L., Engel, J., Larsson, K., Liljequist, S., Lindh, A. S., & Musi, B. (1984). Development of sexual behavior in prenatally ethanol-exposed rats. *Drug and Alcohol Dependence, 14,* 51–61.

Harper, R. G., Solish, G., Feingold, E., Gersten-Woolf, N. B., & Sokal, M. M. (1977). Maternal ingested methadone, body fluid methadone, and the neonatal withdrawal syndrome. *American Journal of Obstetrics and Gynecology, 129,* 417–424.

Heinonen, O. P., Slone, D., & Shapiro, S. (1977). *Birth defects and drugs in pregnancy.* Littleton, MA: Publishing Sciences Group.

Hill, R. M. (1973). Drugs ingested by pregnant women. *Clinical Pharmacology and Therapeutics, 14,* 654–659.

Hill, R. M., Verniaud, W. M., Horning, M. G., McCulley, L. B., & Morgan, N. F. (1974). Infants exposed *in utero* to antiepileptic drugs. *American Journal of Diseases of Children, 127,* 645–653.

Howe, J. P., McGowan, W. A. W., Moore, J., McCaughey, W., & Dundee, J. W. (1981). The placental transfer of cimetidine. *Anaesthesia, 36,* 371–375.

Hutchings, D. E., Martin, B. R., Gamagaris, Z., Miller, N., & Fico, T. (1989). Plasma concentrations of delta-9-tetrahydrocannabinol in dams and fetuses following acute or multiple prenatal dosing in rats. *Life Sciences, 44,* 697–701.

Idänpään-Heikkila, J., Fritchie, G. E., Englert, L. F., Ho, B. T., & McIsaac, W. M. (1969). Placental transfer of tritiated-1-Δ⁹-tetrahydrocannabinol. *The New England Journal of Medicine, 281,* 330.

Inturrisi, C. E., & Blinick, G. (1973). The quantitation of methadone in human amniotic fluid. *Research Communications in Chemical Pathology and Pharmacology, 6*, 353–356.

Jakubovic, A., McGeer, E. G., & McGeer, P. L. (1979). Effects of cannabinoids on testosterone and protein synthesis in rat testis Leydig cells *in vitro. Molecular and Cellular Endocrinology, 15*, 41–50.

Jalling, B., Boréus, L. O., Kållberg, N., & Agurell, S. (1973). Disappearance from the newborn of circulating prenatally administered phenobarbital. *European Journal of Clinical Pharmacology, 6*, 234–238.

Janz, D. (1975). The teratogenic risk of antiepileptic drugs. *Epilepsia, 16*, 159–169.

Jóhannesson, T., Steele, W. J., & Becker, B. A. (1972). Infusion of morphine in maternal rats at near-term: Maternal and foetal distribution and effects on analgesia, brain DNA, RNA and protein. *Acta Pharmacologica et Toxicologica, 31*, 353–368.

Johnson, S. F., McCarter, R. J., & Ferencz, C. (1987). Changes in alcohol, cigarette and recreational drug use during pregnancy: Implications for intervention. *American Journal of Epidemiology, 126*, 695–702.

Jones, C. L., & Lopez, R. E. (1988). Direct and indirect effects on the infant of maternal drug abuse. In *Component Report on Drug Abuse, Expert Panel on Prenatal Care*. Washington, DC: Department of Health and Human Services/National Institutes of Health.

Jones, K. L., & Smith, D. W. (1973). Recognition of the fetal alcohol syndrome in early infancy. *Lancet, 2*, 999–1001.

Jones, K. L., Smith, D. W., Ulleland, C. N., & Streissguth, A. P. (1973). Pattern of malformation in offspring of chronic alcoholic mothers. *Lancet, 1*, 1267–1271.

Kakihana, R., Butte, J. C., & Moore, J. A. (1980). Endocrine effects of maternal alcoholization: Plasma and brain testosterone, dihydrotestosterone, estradiol, and corticosterone. *Alcoholism: Clinical and Experimental Research, 4*, 57–61.

Kaltenbach, K. A., & Finnegan, L. P. (1989). Prenatal narcotic exposure: Perinatal and developmental effects. *NeuroToxicology, 10*, 597–604.

Kaneyuki, T., Kohsaka, M., & Shohmori, T. (1979). Sex hormone metabolism in the brain: Influence of central acting drugs on 5α-reduction in rat diencephalon. *Endocrinologica Japonica, 26*, 345–351.

Kasson, B. G., & Hsueh, A. J. W. (1985). Nicotinic cholinergic agonists inhibit androgen biosynthesis by cultured rat testicular cells. *Endocrinology, 117*, 1874–1880.

Kelce, W. R., Rudeen, P. K., & Ganjam, V. K. (1989). Prenatal ethanol exposure alters steroidogenic enzyme activity in newborn rat testes. *Alcoholism: Clinical and Experimental Research, 13*, 617–621.

Kolodny, R. C., Masters, W. H., Kolodner, R. M., & Toro, G. (1974). Depression of plasma testosterone levels after chronic intensive marihuana use. *The New England Journal of Medicine, 290*, 872–874.

Kolodny, R. C., Lessin, P., Toro, G., Masters, W. H., & Cohen, S. (1976). Depression of plasma testosterone with acute marihuana administration. In M. C. Braude & S. Szara (Eds.), *The Pharmacology of Marihuana* (pp. 217–225). New York: Raven Press.

Kreek, M. J. (1982). Opioid disposition and effects during chronic exposure in the perinatal period in man. *Advances in Alcohol and Substance Abuse, 1*, 21–53.

Kumar, M. S. A., & Chen, C. L. (1983). Effect of an acute dose of Δ^9-THC on hypothalamic luteinizing hormone releasing hormone and Met-enkephalin content and serum levels of testosterone and corticosterone in rats. *Substance and Alcohol Actions/Misuse, 4*, 37–43.

Lalau, J. D., Aubert, M. L., Carmignac, D. F., Grégoire, I., & Dupouy, J. P. (1990). Reduction in testicular function in rats. I. Reduction by a specific gonadotropin-releasing hormone antagonist in fetal rats. *Neuroendocrinology, 51*, 284–288.

Lausecker, C., Withofs, L., Ritz, N., & Pennerath, A. (1976). A propos du syndrome dit "d'alcoolisme foetal." *Pediatrie, 31*, 741–747.

Leslie, G. B., & Walker, T. F. (1977). A toxicological profile of cimetidine. In W. L. Burland & M. A. Simkins (Eds.), *Cimetidine: Proceedings of the Second International Symposium on Histamine H_2-Receptor Antagonists* (pp. 24–33). Amsterdam: Excerpta Medica.

Levitt, M., Hutchings, D. E., Bodnarenko, S. R., & Leicach, L. L. (1982). Postnatal persistence of methadone following prenatal exposure in the rat. *Neurobehavioral Toxicology and Teratology, 4*, 383–385.

Lichtensteiger, W., & Schlumpf, M. (1985). Prenatal nicotine affects fetal testosterone and sexual dimorphism of saccharin preference. *Pharmacology Biochemistry and Behavior, 23*, 439–444.

Lieber, C. S., & DeCarli, L. M. (1989). Liquid diet technique of ethanol administration: 1989 update. *Alcohol and Alcoholism, 24*, 197–211.

Little, B. B., Snell, L. M., Gilstrap, L. C., Gant, N. F., & Rosenfeld, C. R. (1989). Alcohol abuse during pregnancy: Changes in frequency in a large urban hospital. *Obstetrics and Gynecology, 74*, 547–550.

Luck, W., Nau, H., Hansen, R., & Steldinger, R. (1985). Extent of nicotine and cotinine transfer to the human fetus, placenta and amniotic fluid of smoking mothers. *Developmental Pharmacology and Therapeutics, 8*, 384–395.

Mahalik, M. P., Gautieri, R. F., & Mann, D. E., Jr. (1980). Teratogenic potential of cocaine hydrochloride in CF-1 mice. *Journal of Pharmaceutical Sciences, 69*, 703–706.

Majewski, F., Bierich, J. R., Michaelis, R., Löser, H., & Leiber, B. (1977). Über die Alkohol-Embryopathie, eine häufige intrauterine Schädigung. *Monatsschrift für Kinderheilkunde, 125*, 445–446.

Martin, B. R., Dewey, W. L., Harris, L. S., & Beckner, J. S. (1977). ^{3}H-Δ^9-Tetrahydrocannabinol distribution in pregnant dogs and their fetuses. *Research Communications in Chemical Pathology and Pharmacology, 17*, 457–470.

Martin, W. R., Jasinski, D. R., Haertzen, C. A., Kay, D. C., Jones, B. E., Mansky, P. A., & Carpenter, R. W. (1973). Methadone—a reevaluation. *Archives of General Psychiatry, 28*, 286–295.

McGivern, R. F. (1987). Influence of prenatal exposure to cimetidine and alcohol on selected morphological parameters of sexual differentiation: A preliminary report. *Neurotoxicology and Teratology, 9*, 23–26.

McGivern, R. F., Clancy, A. N., Hill, M. A., & Noble, E. P. (1984). Prenatal alcohol exposure alters adult expression of sexually dimorphic behavior in the rat. *Science, 224*, 896–898.

McGivern, R. F., Holcomb, C., & Poland, R. E. (1987). Effects of prenatal testosterone propionate treatment on saccharin preference of adult rats exposed to ethanol *in utero. Physiology and Behavior, 39*, 241–246.

McGivern, R. F., Raum, W. J., Salido, E., & Redei, E. (1988). Lack of prenatal testosterone surge in fetal rats exposed to alcohol: Alterations in testicular morphology and physiology. *Alcoholism: Clinical and Experimental Research, 12*, 243–247.

Melchior, J. C., Svensmark, O., & Trolle, D. (1967). Placental transfer of phenobarbitone in epileptic women, and elimination in newborns. *The Lancet, 2*, 860–861.

Mello, N. K., Mendelson, J. H., & Teoh, S. K. (1989). Neuroendocrine consequences of alcohol abuse in women. In D. E. Hutchings (Ed.), *Prenatal abuse of licit and illicit drugs: Annals of the New York Academy of Sciences, Vol. 562* (pp. 211–240). New York: The New York Academy of Sciences.

Mendelson, J. H., & Mello, N. K. (1975). Plasma testosterone levels during chronic heroin use and protracted abstinence: A study of Hong Kong addicts. *Clinical Pharmacology and Therapeutics, 17*, 529–533.

Mendelson, J. H., & Mello, N. K. (1982). Hormones and psychosexual development in young men following chronic heroin use. *Neurobehavioral Toxicology and Teratology, 4*, 441–445.

Mendelson, J. H., Kuehnle, J., Ellingboe, J., & Babor, T. F. (1974). Plasma testosterone levels before, during and after chronic marihuana smoking. *The New England Journal of Medicine, 291*, 1051–1055.

Mendelson, J. H., Mendelson, J. E., & Patch, V. D. (1975). Plasma testosterone levels in heroin addiction and during methadone maintenance. *The Journal of Pharmacology and Experimental Therapeutics, 192*, 211–217.

Mendelson, J. H., Meyer, R. E., Ellingboe, J., Mirin, S. M., & McDougle, M. (1975). Effects of heroin and methadone on plasma cortisol and testosterone. *The Journal of Pharmacology and Experimental Therapeutics, 195*, 296–302.

Mendelson, J. H., Mello, N. K., & Ellingboe, J. (1977). Effects of acute alcohol intake on pituitary–gonadal hormones in normal human males. *The Journal of Pharmacology and Experimental Therapeutics, 202*, 676–682.

Mendelson, J. H., Ellingboe, J., Kuehnle, J. C., & Mello, N. K. (1978). Effects of chronic marihuana use on integrated plasma testosterone and luteinizing hormone levels. *The Journal of Pharmacology and Experimental Therapeutics, 207*, 611–617.

Mendelson, J. H., Ellingboe, J., Kuehnle, J. C., & Mello, N. K. (1980). Heroin and naltrexone effects on pituitary–gonadal hormones in man: Interaction of steroid feedback effects, tolerance and supersensitivity. *The Journal of Pharmacology and Experimental Therapeutics, 214*, 503–506.

Meyer, L. S., & Riley, E. P. (1986). Social play in juvenile rats prenatally exposed to alcohol. *Teratology, 34*, 1–7.

Middaugh, L. D. (1986). Phenobarbital during pregnancy in mouse and man. *NeuroToxicology, 7*, 287–302.

Mirin, S. M., Mendelson, J. H., Ellingboe, J., & Meyer, R. E. (1976). Acute effects of heroin and naltrexone on testosterone and gonadotropin secretion: A pilot study. *Psychoneuroendocrinology, 1*, 359–369.

Mirkin, B. L. (1971a). Diphenylhydantoin: Placental transport, fetal localization, neonatal metabolism and possible teratogenic effects. *Journal of Pediatrics, 78,* 329–337.

Mirkin, B. L. (1971b). Placental transfer and neonatal elimination of diphenylhydantoin. *American Journal of Obstetrics and Gynecology, 109,* 930–933.

Mittler, J. C., Pogach, L., & Ertel, N. H. (1983). Effects of chronic smoking on testosterone metabolism in dogs. *Journal of Steroid Biochemistry, 18,* 759–763.

Mosier, H. D., & Jansons, R. A. (1972). Distribution and fate of nicotine in the rat fetus. *Teratology, 6,* 303–312.

Nahas, G., & Frick, H. C. (1986). Developmental effects of cannabis. *Neurotoxicology, 7,* 381–396.

Nelson, M. M., & Forfar, J. O. (1971). Associations between drugs administered during pregnancy and congenital abnormalities of the fetus. *British Medical Journal, 1,* 523–527.

NIDA. (1989). *Drug abuse and pregnancy. NIDA capsules.* Rockville, MD: Press Office of the National Institute on Drug Abuse/U.S. Department of Health and Human Services.

Olioff, M., & Stewart, J. (1978). Sex differences in the play behavior of prepubescent rats. *Physiology and Behavior, 20,* 113–115.

Owens-Stively, J. A., Orson, J., & Martin, H. F. (1985). The effect of phenytoin on testosterone metabolism *in vitro. Pediatric Pharmacology, 5,* 123–129.

Pak, R. C. K., & Ecobichon, D. J. (1982). Disposition of maternally-administered methadone and its effects on hepatic drug-metabolizing functions in perinatal guinea pigs. *Biochemical Pharmacology, 31,* 2941–2947.

Parker, S., Schade, R. R., Pohl, C. R., Gavaler, J. S., & Van Thiel, D. H. (1984). Prenatal and neonatal exposure of male rat pups to cimetidine but not ranitidine adversely affects subsequent adult sexual functioning. *Gastroenterology, 86,* 675–680.

Parker, S., Udani, M., Gavaler, J. S., & Van Thiel, D. H. (1984). Adverse effects of ethanol upon the adult sexual behavior of male rats exposed *in utero. Neurobehavioral Toxicology and Teratology, 6,* 289–293.

Patterson, T. R., Stringham, J. D., & Meikle, A. W. (1990). Nicotine and cotinine inhibit steroidogenesis in mouse Leydig cells. *Life Sciences, 46,* 265–272.

Peden, N. R., Cargill, J. M., Browning, M. C. K., Saunders, J. H. B., & Wormsley, K. G. (1979). Male sexual dysfunction during treatment with cimetidine. *British Medical Journal, 1,* 659.

Peden, N. R., Boyd, E. J. S., Browning, M. C. K., Saunders, J. H. B., & Wormsley, K. G. (1981). Effects of two histamine H_2-receptor blocking drugs on basal levels of gonadotrophins, prolactin, testosterone and oestradiol-17β during treatment of duodenal ulcer in male patients. *Acta Endocrinologica, 96,* 564–568.

Perlmutter, J. F. (1974). Heroin addiction and pregnancy. *Obstetrics and Gynecological Survey, 29,* 439–446.

Peters, M. A., Turnbow, M., & Buchenauer, D. (1972). The distribution of methadone in the nonpregnant, pregnant and fetal rat after acute methadone treatment. *The Journal of Pharmacology and Experimental Therapeutics, 181,* 273–278.

Pinto, W., Gardner, L., & Rosenbaum, P. (1977). Abnormal genitalia as a presenting sign in two male infants with hydantoin embryopathy syndrome. *American Journal of Diseases in Children, 131,* 452–455.

Ploman, L., & Persson, B. H. (1957). On the transfer of barbiturates to the human foetus and their accumulation in some of its vital organs. *Journal of Obstetrics and Gynecology, 64,* 706–711.

Poole, T. B., & Fish, J. (1976). An investigation of individual, age and sexual differences in the play of *Rattus norvegicus* (Mammalia: Rodentia). *Journal of Zoology, 179,* 249–260.

Purohit, V., Singh, H. H., & Ahluwalia, B. S. (1978). Failure of methadone-treated human testes to respond to the stimulatory effect of human chorionic gonadotrophin on testosterone biosynthesis *in vitro. Journal of Endocrinology, 78,* 299–300.

Purohit, V., Singh, H. H., & Ahluwalia, B. S. (1979). Evidence that the effects of methadone and marihuana on male reproductive organs are mediated at different sites in rats. *Biology of Reproduction, 20,* 1039–1044.

Purohit, V., Ahluwahlia, B. S., & Vigersky, R. A. (1980). Marihuana inhibits dihydrotestosterone binding to the androgen receptor. *Endocrinology, 107,* 848–850.

Ragni, G., De Lauretis, L., Bestetti, O., Sghedoni, D., & Gambaro, V. (1988). Gonadal function in male heroin and methadone addicts. *International Journal of Andrology, 11,* 93–100.

Raum, W. J., McGivern, R. F., Peterson, M. A., Shryne, J. H., & Gorski, R. A. (1990). Prenatal inhibition of hypothalamic sex steroid uptake by cocaine: Effects on neurobehavioral sexual differentiation in male rats. *Developmental Brain Research, 53,* 230–236.

Redei, E., & McGivern, R. (1988). Attenuation of postnatal testosterone surge and decreased response to LH in fetal alcohol exposed males. *Alcoholism: Clinical and Experimental Research, 12,* 341.

Reinisch, J. M., & Sanders, S. A. (1982). Early barbiturate exposure: The brain, sexually dimorphic behavior and learning. *Neuroscience and Biobehavioral Reviews, 6,* 311–319.

Reyes, F. I., Boroditsky, R. S., Winter, J. S. D., & Faiman, C. (1974). Studies on human sexual development. II. Fetal and maternal serum gonadotropin and sex steroid concentrations. *Journal of Clinical Endocrinology and Metabolism, 38,* 612–617.

Rodriguez-Sierra, J. F., & Hendricks, S. E. (1984). Prenatal exposure to nicotine alters sexual behavior of male rats at adulthood. *Society for Neuroscience Abstracts, 10,* 926.

Root, B., Eichner, E., & Sunshine, I. (1961). Blood secobarbital levels and their clinical correlation in mothers and newborn infants. *American Journal of Obstetrics and Gynecology, 81,* 948–956.

Rudeen, P. K. (1986). Reduction of the volume of the sexually dimorphic nucleus of the preoptic area by *in utero* ethanol exposure in male rats. *Neuroscience Letters, 72,* 363–368.

Rudeen, P. K., & Kappel, C. A. (1985). Blood and brain testosterone and estradiol levels in neonatal male rats exposed to alcohol *in utero. Biology of Reproduction 32,* 75.

Rudeen, P. K., Kappel, C. A., & Lear, K. (1986). Postnatal or *in utero* ethanol exposure reduction of the volume of the sexually dimorphic nucleus of the preoptic area in male rats. *Drug and Alcohol Dependence, 18,* 247–252.

Rudeen, P. K., Ganjam, V. K., & Mohammed, I. H. (1988). Neonatal testicular steroidogenesis inhibition by fetal ethanol exposure in the rat: A mechanism for brain defects. *Alcoholism: Clinical and Experimental Research, 12,* 342.

Sandberg, D. E., Meyer-Bahlburg, H. F. L., Rosen, T. S., & Johnson, H. L. (1990). Effects of prenatal methadone exposure on sex-dimorphic behavior in early school-age children. *Psychoneuroendocrinology, 15,* 77–82.

Schenker, S., Dicke, J., Johnson, R. F., Mor, L. L., & Henderson, G. I. (1987). Human placental transport of cimetidine. *The Journal of Clinical Investigation, 80,* 1428–1434.

Segarra, A. C., & Strand, F. L. (1989). Perinatal administration of nicotine alters subsequent sexual behavior and testosterone levels of male rats. *Brain Research, 480,* 151–159.

Shaarawy, M., & Mahmoud, K. Z. (1982). Endocrine profile and semen characteristics in male smokers. *Fertility and Sterility, 38,* 255–257.

Shah, N. S., Donald, A. G., Bertolatus, J. A., & Hixson, B. (1976). Tissue distribution of *levo*-methadone in nonpregnant and pregnant female and male mice: Effect of SKF 525-A. *The Journal of Pharmacology and Experimental Therapeutics, 199,* 103–116.

Shapiro, B. H., Hirst, S. A., Babalola, G. O., & Bitar, M. S. (1988). Prospective study on the sexual development of male and female rats perinatally exposed to maternally administered cimetidine. *Toxicology Letters, 44,* 315–329.

Singh, H. H., Purohit, V., & Ahluwalia, B. S. (1980). Effect of methadone treatment during pregnancy on the fetal testes and hypothalamus in rats. *Biology of Reproduction, 22,* 480–485.

Sivelle, P. C., Underwood, A. H., & Jelly, J. A. (1982). The effects of histamine H_2 receptor antagonists on androgen action *in vivo* and dihydrotestosterone binding to the rat prostate and androgen receptor *in vitro. Biochemical Pharmacology, 31,* 677–684.

Sonawane, B. R., & Yaffe, S. J. (1983). Delayed effects of drug exposure during pregnancy: reproductive function. *Biological Research in Pregnancy, 4,* 48–55.

Sorsa, M., & Husgafvel-Pursiainen, K. (1988). Assessment of passive and transplacental exposure to tobacco smoke. *IARC Scientific Publications, 89,* 129–132.

Spence, R. W., & Celestin, L. R. (1979). Gynaecomastia associated with cimetidine. *Gut, 20,* 154–157.

Spona, J., Weiss, W., Rudiger, E., Hentschel, E., Shütze, K., Reichel, W., Kerstan, E., Potzi, R. R., & Lochs, H. (1987). Effects of low and high dose oral cimetidine on hormone serum levels in patients with peptic ulcers. *Endocrinologia Experimentalis, 21,* 149–157.

Stubbs, W. A., Delitala, G., Besser, G. M., Edwards, C. R. W., Labrooy, S., Taylor, R., Misiewicz, J. J., & Alberti, K. G. M. M. (1983). The endocrine and metabolic effects of cimetidine. *Clinical Endocrinology, 18,* 167–178.

Suzuki, K., Horiguchi, T., Comas-Urrutia, A. C., Mueller-Heubach, E., Morishima, H. O., & Adamsons, K. (1974). Placental transfer and distribution of nicotine in the pregnant rhesus monkey. *American Journal of Obstetrics and Gynecology, 119,* 253–262.

Swaab, D. F., & Fliers, E. (1985). A sexually dimorphic nucleus in the human brain. *Science, 228,* 1112–1115.

Symons, A. M., Teale, J. D., & Marks, V. (1976). Effects of Δ^9-tetrahydrocannabinol on the hypothalamic–pituitary–gonadal system in the maturing male rat. *The Journal of Endocrinology, 68,* 43–44P.

Szeto, H. H. (1989). Maternal–fetal pharmacokinetics and fetal dose–response relationships. In D. E. Hutchings (Ed.), *Prenatal abuse of licit and illicit drugs: Annals of the New York Academy of Sciences, Vol. 562* (pp. 42–55). New York: New York Academy of Sciences.

Szeto, H. H., Umans, J. G., & McFarland, J. (1982). A comparison of morphine and methadone disposition in the maternal–fetal unit. *American Journal of Obstetrics and Gynecology, 143,* 700–706.

Takagi, S., Alleva, F. R., Seth, P. K., & Balazs, T. (1986). Delayed development of reproductive functions and alteration of dopamine receptor binding in hypothalamus of rats exposed prenatally to phenytoin and phenobarbital. *Toxicology Letters, 34,* 107–113.

Torrey, T. W. (1945). The development of the urinogenital system of the albino rat II. The gonads. *American Journal of Anatomy, 26,* 375–397.

Udani, M., Parker, S., Gavaler, J., & Van Thiel, D. H. (1985). Effects of *in utero* exposure to alcohol upon male rats. *Alcoholism: Clinical and Experimental Research, 9,* 355–359.

Valenstein, E. S., Cox, V. C., & Kakolewski, J. W. (1967). Further studies of sex differences in taste preferences with sweet solutions. *Psychological Reports, 20,* 1231–1234.

Valenstein, E. S., Kakolewski, J. W., & Cox, V. C. (1967). Sex differences in taste preference for glucose and saccharin solutions. *Science, 156,* 942–943.

Van Thiel, D. H., & Gavaler, J. S. (1982). The adverse effects of ethanol upon hypothalamic–pituitary–gonadal function in males and females compared and contrasted. *Alcoholism: Clinical and Experimental Research, 6,* 179–185.

Van Thiel, D. H., Gavaler, J. S., Cobb, C. F., Sherins, R. J., & Lester, R. (1979). Alcohol-induced testicular atrophy in the adult male rat. *Endocrinology, 105,* 888–895.

Van Thiel, D. H., Gavaler, J. S., Smith, W. I., & Paul, G. (1979). Hypothalamic–pituitary–gonadal dysfunction in men using cimetidine. *The New England Journal of Medicine, 300,* 1012–1015.

Van Vunakis, H., Langone, J. J., & Milunsky, A. (1974). Nicotine and continine in the amniotic fluid of smokers in the second trimester of pregnancy. *American Journal of Obstetrics and Gynecology, 120,* 64–66.

Vathy, I. U., Etgen, A. M., & Barfield, R. J. (1985). Effects of prenatal exposure to morphine on the development of sexual behavior in rats. *Pharmacology, Biochemistry and Behavior, 22,* 227–232.

Vorhees, C. V. (1986). Developmental effects of anticonvulsants. *NeuroToxicology, 7,* 235–244.

Vorhees, C. V. (1989). Concepts in teratology and developmental toxicology derived from animal research. In D. E. Hutchings (Ed.), *Prenatal abuse of licit and illicit drugs: Annals of the New York Academy of Sciences, Vol. 562* (pp. 31–41). New York: New York Academy of Sciences.

Walker, T. F., Bott, J. H., & Bond, B. C. (1987). Cimetidine does not demasculinize rat offspring exposed *in utero. Fundamental and Applied Toxicology, 8,* 188–197.

Wang, C., Chan, V., & Yeung, R. T. T. (1978). The effect of heroin addiction on pituitary–testicular function. *Clinical Endocrinology, 9,* 455–461.

Wang, C., Lai, C. L., Lam, K. C., & Yeung, K. K. (1982). Effect of cimetidine on gonadal function in man. *British Journal of Clinical Pharmacology, 13,* 791–794.

Ward, I. L. (1972). Prenatal stress feminizes and demasculinizes the behavior of males. *Science, 175,* 82–84.

Ward, I. L., & Ward, O. B. (1985). Sexual behavior differentiation: Effects of prenatal manipulations in rats. In N. Adler, D. Pfaff, & R. W. Goy (Eds.), *Handbook of behavioral neurobiology, Vol 7* (pp. 77–98). New York: Plenum Press.

Ward, I. L., & Weisz, J. (1980). Maternal stress alters plasma testosterone in fetal males. *Science, 207,* 328–329.

Ward, I. L., & Weisz, J. (1984). Differential effects of maternal stress on circulating levels of corticosterone, progesterone, and testosterone in male and female rat fetuses and their mothers. *Endocrinology, 114,* 1635–1644.

Ward, O. B., Orth, J. M., & Weisz, J. (1983). A possible role of opiates in modifying sexual differentiation. In M. Schlumpf & W. Lichtensteiger (Eds.), *Monographs in neural sciences, Vol. 9: Drugs and hormones in brain development* (pp. 194–200). Basel: S. Karger.

Ward, O. B., Needham, S. L., Ward, I. L., & Hess, D. L. (1985). Ontogeny of sex differences in amniotic fluid testosterone and estradiol levels in rats. presented at the *Annual Conference on Reproductive Behavior,* Omaha, Nebraska.

Ward, O. B., Kopertowski, D. M., Finnegan, L. P., & Sandberg, D. E. (1989). Gender-identity variation in boys prenatally exposed to opiates. In D. E. Hutchings (Ed.), *Prenatal abuse of licit and illicit drugs:*

Annals of the New York Academy of Sciences, Vol. 562 (pp. 365–366). New York: New York Academy of Sciences.

Waterson, E. J., & Murray-Lyon, I. M. (1989). Drinking and smoking patterns amongst women attending an antenatal clinic—II. During pregnancy. *Alcohol and Alcoholism, 24,* 163–173.

Weisz, J., & Ward, I. L. (1980). Plasma testosterone and progesterone titers of pregnant rats, their male and female fetuses, and neonatal offspring. *Endocrinology, 106,* 306–316.

Wesson, D. R., & Smith, D. E. (1977). *Barbiturates: Their use, misuse, and abuse.* New York: Human Sciences Press.

Wiggins, R. C., Rolsten, C., Ruiz, B., & Davis, C. M. (1989). Pharmacokinetics of cocaine: Basic studies of route, dosage, pregnancy and lactation. *NeuroToxicology, 10,* 367–382.

Winternitz, W. W., & Quillen, D. (1977). Acute hormonal response to cigarette smoking. *The Journal of Clinical Pharmacology, 17,* 389–397.

Woody, G., McLellan, A. T., O'Brien, C., Persky, H., Stevens, G., Arndt, I., & Carroff, S. (1988). Hormone secretion in methadone-dependent and abstinent patients. *NIDA Research Monograph, 81,* 216–223.

Yeh, J., Barbieri, R. L., & Friedman, A. J. (1989). Nicotine and cotinine inhibit rat testis androgen biosynthesis *in vitro. Journal of Steroid Biochemistry, 33,* 627–630.

Zackai, E. H., Mellman, W. J., Neiderer, B., & Hanson, J. W. (1975). The fetal trimethadione syndrome. *The Journal of Pediatrics, 87,* 280–284.

Zagon, I. S. (1985). Opioids and development: New lessons from old problems. In *National Institute on Drug Abuse, Research Monograph 60* (pp. 58–77). Rockville, MD: U.S. Department of Health and Human Services.

Zellweger, H. (1974). Anticonvulsants during pregnancy: A danger to the developing fetus? *Clinical Pediatrics, 13,* 338–346.

Zemp, J. W., & Middaugh, L. D. (1975). Some effects of prenatal exposure to *d*-amphetamine sulfate and phenobarbital on developmental neurochemistry and on behavior. *Addictive Diseases: An International Journal, 2,* 307–331.

Zimmerberg, B., & Scalzi, L. V. (1989). Commissural size in neonatal rats: Effects of sex and prenatal alcohol exposure. *International Journal of Developmental Neuroscience, 7,* 81–86.

7

Prenatal Hormonal Contributions to Sex Differences in Human Cognitive and Personality Development

JUNE MACHOVER REINISCH AND STEPHANIE A. SANDERS

INTRODUCTION

This chapter presents an overview of behavioral sexual differentiation in humans and presents data from our laboratory and other laboratories that highlight the importance of early (prenatal and/or postnatal) hormonal stimulation by androgenic substances in this process. Data derived from the study of individuals who were exposed to atypical levels of endogenous hormones or to exogenous compounds with an androgenic potential are presented to provide insight into the normal process underlying the emergence of sex differences (see also Reinisch, Ziemba-Davis, & Sanders, 1991).

The interplay among somatic, hormonal, behavioral, and environmental variables that shape psychosexual development is portrayed in Figure 1. At conception and during embryonic development, both males and females are essentially identical except that the cells of a normal male contain an X and a Y sex chromosome, whereas the female has two X chromosomes. The Y chromosome directs the produc-

JUNE MACHOVER REINISCH   The Kinsey Institute for Research in Sex, Gender, and Reproduction, and Department of Psychology, Indiana University, Bloomington, Indiana 47405; and Department of Psychiatry, Indiana University School of Medicine, Indianapolis, Indiana 46202.   STEPHANIE A. SANDERS   The Kinsey Institute for Research in Sex, Gender, and Reproduction, Indiana University, Bloomington, Indiana 47405.

Sexual Differentiation, Volume 11 of *Handbook of Behavioral Neurobiology*, edited by Arnold A. Gerall, Howard Moltz, and Ingeborg L. Ward, Plenum Press, New York, 1992.

JUNE MACHOVER
REINISCH AND
STEPHANIE A.
SANDERS

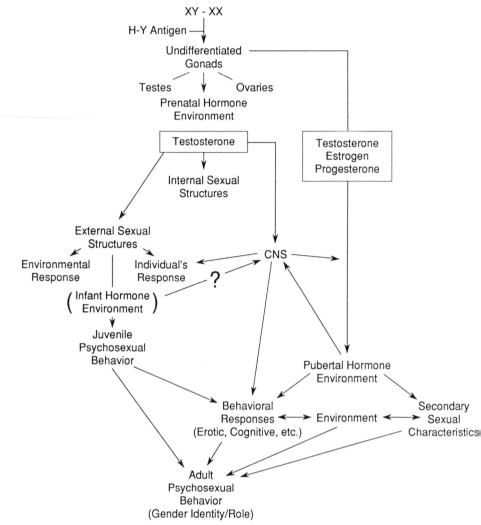

Figure 1. The interplay among somatic, hormonal, behavioral, and environmental variables that shape psychosexual development. (Derived from Money & Ehrhardt, 1972.)

tion of H-Y antigen, which in turn induces the undifferentiated gonads to become testes. In the absence of H-Y antigen or its specific cellular gonadal receptor, the embryonic gonad will eventually become an ovary.

If testes have developed, testosterone production will begin at approximately 7 to 8 weeks post-conception. It is the presence or absence of fetal androgens that determines the differentiation of the internal and external sexual structures into male or female types, respectively. The mammalian central nervous system also appears to become sexually dimorphic as a result of the relative amounts of gonadal hormones in the prenatal and/or early postnatal environment (De Vries, De Bruin, Uylings, & Corner, 1984; Gorski, 1987).

It may be said that nature's first intention is to produce a female and that something must be added to produce a phenotypic male. Concomitant to the more complex process of male differentiation is the increased possibility of developmental anomalies (Gualtieri & Hicks, 1985). As Alfred Jost so eloquently stated, "Becoming a male is a prolonged, uneasy, and risky venture. It's a kind of struggle against inherent trends toward femaleness" (1972, p. 3).

As a consequence of the sexual dimorphism of the genitalia and the brain at birth, it is likely that male and female children both receive different responses from the social environment and differentially perceive and respond to environmental stimuli (Reinisch, Gandelman, & Spiegel, 1979; Reinisch & Sanders, 1987). Puberty marks the second period of gonadal hormonal impact on behavioral development with the differentiation of secondary sexual characteristics and the release of adult levels of circulating hormones. During this period sex differences are also magnified as the adolescent responds to societal pressures to conform to sex-role stereotypes and to his or her internal predilection to acquire certain sex-identified behaviors and to excel at specific skills characterized as either masculine or feminine.

The spectrum of perceptions, the ease with which certain cognitive skills are acquired, temperament, and the resultant personality are influenced but not determined by the prenatal hormonal milieu. The social environment enhances the divergence of the sexes, with the magnitude of societally encouraged adult sex differences depending on the culture's stereotypes regarding the role behaviors of males and females. In the absence of sex-role stereotypes, however, small but consistent sex differences in cognitive ability, personality, and behavior would still emerge as a result of the divergent biological influences on brain development during gestation.

The purpose of this chapter is to illustrate the role of prenatal androgenic stimulation in the development of postnatally expressed sexual dimorphisms. Studies are cited that relate, respectively, to human endocrine syndromes, subjects whose mothers were prescribed hormonal medication during pregnancies, and sex differences in behavioral milestones during the first year of life.

Before turning to these data it is important for the reader to note certain caveats relevant to their interpretation (see also Reinisch & Gandelman, 1978). First, the empirical study of human behavioral endocrinology is subject to ethical, moral, and legal constraints. True experimental design is precluded when dealing with prenatal treatment of humans. Our confidence in the conclusions derived from human studies stems from the concordance between such findings and those of laboratory experiments in which animals were assigned randomly to treatment and control conditions. (Extensive reviews of this literature include Ellis, 1982; Hines, 1982; Reinisch, 1974, 1983).

A second caveat in the interpretation of human studies relates to the fact that almost all behavioral sex differences are *quantitative* in their expression and *not qualitative*. That is, the differences between males and females reside in the frequency with which selected behaviors are exhibited, not in whether these behaviors are expressed. Ejaculation, menstruation, gestation, and lactation are the only responses that are ordinarily expressed as qualitative sex differences.

Additionally, when interpreting and drawing conclusions regarding sexual dimorphism, it is important to note that masculinity and femininity are no longer conceptualized as opposite ends of a unidimensional bipolar continuum (Figure 2A) but are now understood to be independent or semi-independent factors. The orthogonal or *independent* model (Figure 2B: Bem, 1974; Berzins, Welling, & Wetter, 1978; Constantinople, 1973; Heilbrun, 1976; Spence, Helmreich, & Stapp, 1974; Whalen, 1974) and the *oblique* or correlated model (Figure 2C: Reinisch, 1976; Reinisch & Sanders, 1987) suggest that masculinity and femininity are, with regard to many dimensions, relatively independent. These latter two models imply that an increase in the frequency or strength of masculine behavior (masculinization) is not

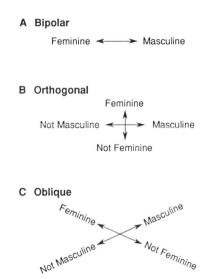

Figure 2.   Models of masculinity and femininity. (Reprinted from Reinisch and Sanders, 1987.)

necessarily associated with a concomitant decrease in feminine behavior (defeminization). Conversely, feminization is not synonymous with demasculinization. Accordingly, an individual could exhibit high frequencies of both masculine and feminine behavior (androgynous), could exhibit low levels of both kinds of behavior (undifferentiated), or could exhibit high levels of one type of behavior and low levels of the other (masculine or feminine). The difference between the orthogonal and oblique models resides in the degree of independence understood to exist between masculinity and femininity. The orthogonal model proposes complete independence, and the oblique model suggests that there is some degree of correlation between feminization and demasculinization and between masculinization and defeminization.

With these guidelines in mind let us turn to some data that suggest a biological contribution to the development of behavioral sex differences in humans.

SEX DIFFERENCES IN NONCLINICAL SAMPLES

Human behavioral sex differences are observed along many dimensions, including cognitive abilities, brain laterality, aggression, maternalism, and toy preference. A number of these behavioral differences have been related to genetic or neurological differences between the sexes. As stated earlier, there is strong evidence from the animal laboratory with confirming evidence from human studies that sex differences in neurological development and the expression of the genome are mediated by the type and quantity of androgens present in the prenatal environment, particularly the quantity of testosterone.

SEX DIFFERENCES IN THE CENTRAL NERVOUS SYSTEM

Because the brain provides the biological foundation for behavior, it seems logical to assume that sex differences in the structure and function of neural tissue may be related to at least some of the identified sex differences in behavior. At the present time, however, the relationship between structural or physiological differences on the one hand and behavioral differences on the other is speculative.

The central nervous system appears to be sexually dimorphic in a number of species (De Vries *et al.*, 1984; Gorski, 1987; Goy & McEwen, 1980; MacLusky & Naftolin, 1981). The list of known morphological and physiological brain sex differences predicated on early hormonal differences has been steadily growing during the past decade and a half (Gorski, 1987; Reinisch, 1983). Sex differences have been identified in various nuclei of the hypothalamus, cerebellum, hippocampus, amygdala, and cerebral cortex. Similarly, differences in biochemical and physiological parameters such as oxidative metabolism, protein content, serotonin levels, RNA metabolism, and cholinesterase activity have been revealed. Male and female brains also differ with regard to morphology, including such dimensions as neural connections, volume of cell nuclei, dendritic field patterns, and size of brain nuclei. Most of this research has been conducted with rodents; however, there is reason to believe that similar sex differences, determined by a comparable prenatal hormonal milieu, exist in the human brain.

In the past 5 years, several laboratories have identified morphological sex differences in the human brain. A sexually dimorphic cell group in the preoptic area of the hypothalamus, analogous to the sexually dimorphic nucleus (SDN) in rats (Gorski, Gordon, Shryne, & Southam, 1978), has been reported in humans (Swaab & Fliers, 1985). The volume of the SDN in men was found to be 2.5 times as large as that in women, and it contained 2.2 times as many cells. A sex difference in the suprachiasmatic nucleus of the human hypothalamus was also observed. This difference does not appear to be associated with differences in overall volume, cell density, or total cell number (D. F. Swaab, E. Fliers, & T. Partiman, unpublished observation, reported in Swaab & Hofman, 1984). A few years ago deLacoste-Utamsing and Holloway (1982) reported that, although human females have lower brain weights than do males, the posterior fifth of the female corpus callosum was larger, despite the fact that there was no difference in total callosal area. However, Witelson (1985) failed to confirm these findings and suggested "the possibility of a complex sex factor which, in interaction with hand preference, may be related to the size of some segment of the posterior half of the callosum" (p. 666). Analysis of larger numbers of human brains will be necessary before the relationship between sex and the corpus callosum can be clarified. However, if an anatomic difference in the corpus callosum were confirmed, it would be consistent with the other reports of sex differences in brain organization and hemispheric laterality between males and females (see below; DeVries *et al.*, 1984). Current theory holds that there is more communication between the two cerebral hemispheres in females than in males. Differences in size or function of the corpus callosum suggest an anatomic basis for functional sex differences in hemispheric laterality.

In the upcoming years it is likely that there will be an increasing number of reports of both anatomic and functional sex differences in the human brain.

COGNITIVE DIFFERENCES

Numerous studies have examined cognitive functioning in men and women. Because the brain is the organ responsible for intellectual activity, cognitive performance (or almost any behavior for that matter) may be viewed as a bioassay of

JUNE MACHOVER
REINISCH AND
STEPHANIE A.
SANDERS

underlying differences in brain functioning. Thus, sex differences in human behavior may reflect morphological and/or functional sex differences in the human brain.

In general, human males achieve higher scores than females in visuospatial tasks and mathematics, whereas females demonstrate stronger verbal skills (Benbow & Stanley, 1980, 1981, 1983; Burstein, Bank, & Jarvik, 1980; Halpern, 1986; Maccoby & Jacklin, 1974; Wilson & Vandenberg, 1979). With respect to specific types of cognitive functioning, hormones appear to be influential in two ways. (1) Differences between males and females appear to be related to prenatal hormone exposure. (2) *Within* each sex, there are relationships between cognitive functioning and adult hormone levels.

Two categories of cognitive style that have been studied in terms of sex differences and relationship to adult hormone levels are automatization and perceptual restructuring (Broverman *et al.*, 1981). Automatized behaviors are those that require little conscious attention because they are so highly practiced. Measures that reflect automatization ability include speed of reading, color naming, and object naming (i.e., tests involving associative memory, perceptual speed, and verbal abilities). Females, in general, are superior in the performance of automatization tasks that require quick, accurate responses to obvious stimulus properties and well-learned tasks. In contrast, males perform better on perceptual-restructuring tasks. These are tasks that require suppression of immediate "automatized" responses to the obvious properties of the stimulus in order to respond appropriately to hidden or embedded characteristics. Examples include the Embedded Figures Test and backward counting.

Males also appear to be more likely than females to be extremely mathematically gifted (Benbow & Benbow, 1984; Benbow & Stanley, 1980, 1981, 1983). In a study of 100,000 scholastically gifted 12-year-olds given the college entrance examination (usually taken at 17 years of age), 292 children scored 700 or more in mathematics. The boy-to-girl ratio was 13 to 1. These mathematically precocious students were also more likely than the general population to (1) be left-handed or have familial histories of left-handedness, (2) exhibit symptomatic atopic diseases such as asthma and other allergies, and (3) be myopic. Although social factors certainly can and probably do contribute to mathematical ability, there appears to be at least some biological basis for this finding of a sex difference. Genetics, brain laterality, and the effects of the prenatal hormone environment have been discussed in relationship to such mathematical precocity (Benbow & Benbow, 1984). Specifically, it has been suggested that males' superior performance on these tasks is related to testosterone exposure during prenatal development (Geschwind & Galaburda, 1985).

Interestingly, there are also more male than female retardates. A genetic/chromosomal explanation has recently been proposed linking the preponderance of mental retardation in males to the so-called "fragile X syndrome," the presence of a "fragile" site on the long arm of the X chromosome (Vandenberg, 1987; Wilson, 1987). (The term "fragile" describes its appearance and does not imply that the chromosome is more likely to break at this site.) X-linked genetic traits such as baldness and color-blindness are expressed more frequently in males than females because in females the presence of a genetic or chromosomal anomaly on one X chromosome will be attenuated by the presence of a second normal X chromosome. Additionally, it is noteworthy that left-handedness, atopic disorders, and learning disabilities frequently appear to coexist and are more prevalent among males (Geschwind & Behan, 1982; Geschwind & Galaburda, 1985; Gualtieri & Hicks, 1985).

Thus, there is now strong evidence that genetic and hormonal factors that are sex-related may underlie some of the sex differences in cognitive ability.

LOCALIZATION OF BRAIN FUNCTION AND BRAIN LATERALITY

It has been generally accepted that some degree of localization of function exists within the cerebral hemispheres (e.g., Galaburda, LeMay, Kemper, & Geschwind, 1978; Gazzaniga, 1972; Geschwind, 1970; Kimura, 1961). The left hemisphere is thought to be responsible for verbal fluency, verbal reasoning, language production, speech production and understanding, sequential processing, computational mathematics, and analytical–logical thinking. The right hemisphere is believed to be the site of intuitive approaches to problem solving, spatial abilities, musical abilities, artistic abilities, facial recognition, mathematical abilities involving abstract symbolic global concepts related to mathematical theories, and emotion. For example, it has been demonstrated that the left side of the face, which is controlled by the right hemisphere of the brain, expresses more emotion than the right (see Bruyer & Craps, 1985).

Males and females differ with respect to the degree to which the cerebral hemispheres are specialized for different types of cognitive processing. It has been suggested that sex differences in brain hemispheric lateralization underlie sex differences in cognitive functioning (Benbow & Benbow, 1984; Geschwind & Galaburda, 1985; Levy & Levy, 1978; Witelson, 1976). For example, Shucard, Shucard, Cummins, and Campos (1981) have reported sex differences in auditory evoked potentials as early as 3 months of age. Females produced higher-amplitude responses from the left hemisphere and males from the right hemisphere during both verbal and nonverbal presentations. These findings are consistent with sex differences in the later development of language and spatial functions.

With respect to sex differences in brain laterality, males appear to be more lateralized (see J. Levy & W. Heller, Chapter 8, this volume). Their superior spatial abilities, the fact that there are more male mathematical geniuses, and their enhanced degree of facial expression on the left side of the face is taken as evidence of greater right hemispheric dominance. With regard to spatial abilities, for example, males tend to show greater right hemisphere specialization for the processing of spatial stimuli than do females (Levy, 1972; McGee, 1979; McGlone, 1980). On a dichhaptic shapes task (in which two differently shaped stimuli are presented to each hand and subjects are asked to identify them by touch) given to right-handers, there are no sex differences in overall number of correct responses, but males give more correct answers for stimuli presented to their left hand (right hemisphere), whereas correct answers for females are equally distributed across both hands (hemispheres) (Witelson, 1976).

In comparison to males, females appear to be less lateralized. They show greater recovery from aphasia after stroke, suggesting that function lost in one hemisphere can be acquired in the other. Further support for this notion can be found in the differential response to unilateral brain lesions. Following brain lesions females show no difference in Verbal or Performance scores on the Wechsler Adult Intelligence Scales (WAIS), whereas males have depressed Verbal IQ scores with left-brain lesions and performance deficits with right-brain lesions (Inglis, Ruckman, Lawson, MacLean, & Monga, 1983). It is possible that these sex differences in the loss and recovery of function have more to do with sex differences in the location of functions within a hemisphere than with differences in laterality. It is possible that in males the areas specialized for particular verbal and spatial functions are located in sites that are more vulnerable to damage by stroke or trauma than they are in females (Kimura, 1983). Additional evidence of sexual dimorphism in intrahemispheric brain organization in humans is provided by studies investigating (1) the localization of various language functions by using stimulation techniques (Ojemann, 1983) and (2) the more frequent reporting of sexual arousal or orgasm related to temporal lobe epileptic seizures by women than by men (Remillard *et al.*, 1983).

In addition to differences in laterality and localization of function, sex differences in patterns of cognitive development may be related to differences in the maturation of the cerebral cortex (Waber, 1979). Data exist supporting this hypothesis. First, spatial ability shows a consistent pattern of male superiority, and this ability is related to the rate of physical maturation and the degree of androgenization in both sexes. Secondly, it appears that no relationship exists between maturational rate and performance on the verbal tasks at which females excel. Thus, the ontogeny of linguistic and spatial abilities appears to follow different developmental pathways, and sex differences in the development of cognitive abilities may be confounded with maturational differences between the sexes.

In summary, on a number of dimensions, human males and females show different behavioral proclivities, which may be related to underlying sex differences in the morphology and/or function of the brain. Currently, studies of sex differences in cognition and localization of function within the brain provide the strongest evidence for brain–behavior associations related to sex. As stated earlier, a number of researchers attribute these sex differences to early hormonal influences on development rather than to genetic factors *per se*. Let us now turn to a review of findings that provide insight into the role that prenatal hormones play in the development of sex differences.

Human Quasiexperiments

The first studies examining the effects of prenatal hormones on human behavioral development involved observations of patients with clinical endocrine syndromes that may be viewed as being somewhat comparable to existing animal experiments.

Clinical Endocrine Syndromes

Adrenogenital syndrome (AGS) or congenital adrenal hyperplasia is a genetic anomaly of one of the autosomes that results in the absence of or a deficiency in an essential enzyme needed for the synthesis of cortisol. The inability to synthesize cortisol leads to the buildup of precursors that have androgenic activity. This syndrome is similar to laboratory experiments in which female laboratory animals are exposed to androgen during the critical period of sexual development. Genetic females with this anomaly are born with masculinized genitalia, including labioscrotal fusion and clitoral hypertrophy. The degree of masculinization is dependent on the severity of the enzymatic deficiency. When the condition is identified early in life, the infant is treated with cortisone, which interrupts the production of androgenic precursors, and is given feminizing plastic surgery of the genitalia. If left untreated, the continued production of androgenic cortisol precursors will continue to virilize these girls and lead to precocious puberty and short stature.

Several studies of AGS patients revealed clear evidence of behavioral masculinization (Baker & Ehrhardt, 1974; Ehrhardt & Baker, 1974; Ehrhardt, Epstein, & Money, 1968; Ehrhardt, Evers, & Money, 1968; Money & Schwartz, 1977; Money, Schwartz, & Lewis, 1984). Even with early treatment, these girls described themselves and were described by others as more tomboyish and higher in energy expenditure than their unaffected sisters. They preferred outdoor activity, boys' toys, male playmates, and utilitarian clothing. Adrenogenital girls also demonstrated a low interest in marriage and maternal behavior coupled with early interest in careers and sex-role ambivalence. In earlier studies, when compared to "matched controls," these subjects appeared to exhibit increased IQ scores. When compared to unaffected siblings, however, no such effect was evident.

The human clinical endocrine disorder that is the closest counterpart to animal studies in which males are deprived of testosterone stimulation during early development is the androgen insensitivity or testicular feminization syndrome. This, too, is a genetically induced endocrine disorder. Although patients with this syndrome usually produce normal amounts of androgen, their somatic target cells lack the normal complement of androgen receptors and, thus, are "insensitive" and are not masculinized. Males with this syndrome are born with normal-appearing female genitalia and experience a feminizing puberty stimulated by testicular estrogen. Parallel to the behavioral findings in androgen-deprived male laboratory animals, these patients are feminized on many of the same behavioral dimensions listed above for females with the adrenogenital syndrome (Money, Ehrhardt, & Masica, 1968; Money & Ogunro, 1974).

In summary, clinical data indicate that alterations in normal levels of endogenous hormone production or hormonal responsiveness resulting from genetic anomalies affect, either directly or indirectly, the development and expression of sexually dimorphic behavior. Prenatal androgens lead to behavioral masculinization, whereas feminization occurs in the absence of androgenic stimulation.

Exogenously Hormone-Exposed Subjects

Studies of the next group of human subjects are more similar to laboratory animal experiments in that no genetic anomalies are involved in the abnormal hormonal stimulation that the fetuses experience. These individuals were exposed to steroids from an exogenous rather than endogenous source. These children were exposed to hormones during gestation as a result of medical treatment of mothers judged to be at risk for miscarriage. Detailed reviews of this literature can be found by Hines (1982), Reinisch (1983), and Sanders and Reinisch (1985).

Androgen-Based and Progesterone-Based Synthetic Progestins. A number of the synthetic progestins used by physicians in antenatal care are compounds based not on progesterone but on androgen. Many of these compounds have been shown to have a virilizing effect that in extreme cases could masculinize the genitalia almost completely. Approximately 18% of females exposed to androgen-based synthetic progestins were born with virilized genitalia (Jacobsen, 1962). An evaluation of 10 such girls (that did not include comparison to a control group) found them to be very similar to female subjects with the adrenogenital syndrome (Ehrhardt & Money, 1967).

Females (who were not genitally virilized) and males who were exposed to synthetic progestins prenatally have been compared to unexposed sibling controls on personality measures, IQ tests, and interviews (Reinisch, 1977). They were found to be more independent, individualistic, self-assured, and self-sufficient in comparison with their unexposed siblings. There were no significant differences in IQ between the exposed and unexposed siblings, contrary to an earlier finding using "matched controls" that suggested that IQ might be increased.

Another study assessed the aggression potential of 17 females and 8 males exposed to synthetic progestins with unexposed same-sex sibling controls (Reinisch, 1981). All index cases (exposed subjects) had been exposed to 19-NET (19-nor-17α-ethynyltestosterone) for at least 4 weeks with treatment starting prior to the 10th week of gestation. The aggressive potential was measured using the Leifer–Roberts Response Hierarchy—Reinisch Revision (Reinisch & Sanders, 1986). This projective test presents a series of six hypothetical moderate-level interpersonal conflict situations derived from stories of children and adolescents. Physical aggression, verbal aggression, nonaggressive coping, and withdrawal response options are paired in all possible combinations and presented as a series of forced-choice alternatives following each story. In this way a hierarchy of responses to interpersonal conflict is generated. We have demonstrated consistent and robust sex differences in physical aggression scores across samples drawn from young children to college students (Reinisch & Sanders, 1986). With same-sex siblings used as the comparison group, a strong effect of prenatal exposure to synthetic progestins indicated by higher physical aggression scores was demonstrated not only in females but also in males. Specifically, seven exposed males had greater physical aggression scores than their brothers; only one had a lower score. Twelve exposed females had higher physical aggression scores than their sisters, three had equal scores, and two had lower scores. Confirming the general sex difference on this measure, unexposed males had significantly higher physical aggression scores than unexposed females. Birth order and age had no effect. These data provide further evidence that prenatal androgen levels are behaviorally masculinizing.

Of particular interest is the unexpected demonstration of increased levels of masculine behavior among exposed males. Goy, Wolf, and Eisele (1977) found no evidence of hypermasculinization in studies of rhesus males treated prenatally with testosterone. There is, however, some supportive evidence for this effect from human clinical syndromes. For example, when nine adrenogenital syndrome boys who experienced premature puberty as a result of excess adrenal androgens were compared to their unaffected brothers, they evidenced a higher level of energy expenditure in sports and rough outdoor activities (Ehrhardt & Baker, 1974).

Other data from our laboratory provide additional evidence of increased masculinization of human males in response to higher prenatal levels of androgen. Preliminary analyses of data from 10 19-NET-exposed boys compared to their unexposed brothers revealed hypermasculinization across a variety of dimensions. Exposed males achieved higher scores on Number Facility from the Primary Mental Abilities Test. On the Cattell 16 Personality Factors, index cases scored higher on the bipolar dimensions (1) assertive/aggressive/stubborn/competitive (dominance) versus humble/mild/easily led/docile/accommodating (submissive) and (2) self-sufficient/resourceful/prefers own decisions (self-sufficiency) versus group-dependent/a "joiner"/sound follower (group adherence). Males typically attain higher scores than females on these measures.

With respect to personality factors related to sex-role, exposed males had higher masculinity scores on both the M and M–F scales of the Personal Attributes Questionnaire (Spence *et al.*, 1974). The M scale contains items that are considered to be socially desirable in both sexes but that are believed to be more typical of males than females (e.g., independence). The M scale measures masculinity independent of femininity. The M–F scale, on the other hand, consists of characteristics whose social desirability appears to vary by sex and for which masculinity and femininity are bipolar opposites (e.g., aggressiveness is judged desirable in males, and nonaggressiveness is desirable for females).

Exposed males also scored higher on the Texas Social Behavior Inventory (Spence & Helmreich, 1978), a measure of self-esteem. Males were found to generate higher scores on most measures of self-esteem. With respect to the Work and Family Orientation Questionnaire (Spence & Helmreich, 1978), index cases (1) scored higher on the Mastery dimension—affirming to a greater degree than their brothers such statements as, "If I'm not good at something I would rather keep struggling to master it than to move on to something I may be good at"; (2) had higher scores on the Work Orientation dimension—endorsing statements such as "It is very important for me to do my work as well as I can even if it isn't popular with my co-workers"; and (3) showed less fear of success. All of these variables are sex differentiated, with males characteristically receiving higher scores than females. Thus, based on measures evaluating mental abilities, masculinity/femininity, self-esteem, and attitudes toward work and career, males exposed to additional androgenic hormones during gestation appeared to exhibit enhanced levels of masculine responding.

DIETHYLSTILBESTROL, A NONSTEROIDAL SYNTHETIC ESTROGEN.   Another hormonal preparation widely used to prevent miscarriage that appears to interfere with normal hormonal action *in utero* is diethylstilbestrol (DES), a nonsteroidal synthetic estrogen. Findings on DES-treated individuals are particularly relevant to understanding the influence of the prenatal hormonal environment on the development of brain laterality and spatial ability.

A study of women exposed *in utero* to DES in combination with synthetic progestins used a dichotic listening task to reveal that this treatment appears to have an effect on the development of brain organization (Hines, 1981; Hines & Shipley, 1984). Such DES-exposed women were found to be more lateralized than their unexposed sisters, a trait characteristic of males. This finding provided evidence that brain hemispheric laterality could be influenced by altering prenatal hormones in humans.

Recent findings from our laboratory (Reinisch & Sanders, in press) provide what we believe is the first evidence of an association among prenatal hormone exposure, brain laterality, and spatial ability. Ten males exposed to DES during gestation were compared to their unexposed brothers on a measure of brain hemispheric specialization for processing nonlinguistic spatial information and on a test of cognitive abilities. The DES exposure was associated with reduced hemispheric laterality and lowered spatial ability. The reduction of laterality may be thought of as representing "feminization," and the lowered spatial ability as "demasculinization." The reason for our differential use of the terms "feminize" and "demasculinize" is that brain laterality can best be represented by a bipolar model with one end of the continuum labeled "not lateralized (feminized)" and the other "lateralized (masculinized)." It is unlikely that an individual can be simultaneously lateralized and not lateralized for a particular type of cognitive processing.

Cognitive abilities, on the other hand, are often divided into spatial and verbal abilities and may best be represented by an orthogonal or oblique model. Spatial abilities, in which males tend to excel, might be seen as a "masculine" dimension (high spatial ability = masculinized, low spatial ability = demasculinized), and verbal abilities, in which females tend to excel, might be viewed as a "feminine" dimension (high verbal ability = feminized, low verbal ability = defeminized).

Although the effects of DES on the development of laterality in males and females appear to be contradictory (females were masculinized, but males were feminized), research with laboratory animals has confirmed that estrogen exposure affects the sexes paradoxically, resulting in the behavioral feminization and/or demasculinization of males (e.g., Diamond, Llacuna, & Wong, 1973; Gray, Levine, & Broadhurst, 1965) and in behavioral masculinization and/or defeminization of females (Dörner, Döcke, & Hinz, 1971; Harris, 1964; Hines, Alsum, Gorski, & Goy, 1982).

Data from the few studies that evaluated behavioral effects in humans, other than those cited above, for the most part appear to agree with these paradoxical findings. For example, DES-exposed boys were found to be less assertive/aggressive and to have lower athletic ability (Yalom, Green, & Fisk, 1973). Reinisch (1977; Reinisch & Karow, 1977) found estrogen exposure to be associated with higher scores on group dependence and group orientation, traits more characteristic of females. There is some evidence of increased bisexuality or homosexuality among DES-exposed women (Ehrhardt et al., 1985). As Kester, Green, Finch, and Williams (1980) point out, however, the nature of the effects of prenatal DES is influenced by the timing of exposure and the presence of other hormones in the treatment regimen. More data are needed to clarify these interactions and enhance our understanding of the effects of prenatal exposure to DES.

The data reported above support the hypothesis that the development of sex differences in hemispheric lateralization and cognitive abilities is influenced by the normal dimorphism in the prenatal hormonal environment of human males and females rather than by sex differences in the genotype. A number of researchers have suggested that differences in hemispheric lateralization may mediate sex differences in cognitive ability. Further research should clarify whether prenatal exposure to sex hormones influences brain organization, which in turn produces sexually dimorphic cognitive abilities, or whether hormone exposure influences the development of each of these two processes independently. Further, it remains possible that other causative relationships exist between cognitive factors and personality/gender role alterations associated with intrauterine exposure to hormones (Reinisch, 1977; Reinisch & Karow, 1977).

SUBASYMPTOTIC PRENATAL ANDROGEN STIMULATION

We have hypothesized that most human males normally do not fully express the potential of their masculine genotype along a number of behavioral dimensions (Reinisch & Sanders, 1984). There is little doubt that testosterone produced by the fetal or neonatal testes is primarily responsible for masculine differentiation of the genitalia, hypothalamus, and behavior in mammals either by a direct effect or, in some nonprimate species, as a precursor of estradiol (Gorski, 1987; McEwen, 1987). In humans, testosterone is available from the fetal testes beginning at approximately 9 weeks post-conception, rising to a peak around 16 weeks and then falling (Williams, 1981). This period encompasses the critical period for differentiation of the internal and external genitalia. Theoretically, this is also the period (perhaps the only period) during which the brain undergoes at least the first important phase of sexual differentiation.

Normal levels of testosterone in males appear to produce asymptotic (fully masculinized) morphological genital differentiation, since the only overlap between the sexes occurs under pathological conditions (e.g., adrenogenital syndrome, androgen insensitivity syndrome, 5α-reductase deficiency). Extra testosterone *in utero* does not lead to hypermasculinization of the genitalia at birth. In the expression of behavioral sex differences, however, there is considerable overlap between males and females, as shown in Figure 3A. In fact, human behavioral sex differences are relatively small. Our data on the effect of prenatal exposure to androgenic synthetic progestins suggest that if more testosterone were normally available to males during the period of central nervous system differentiation, the magnitude of behavioral differences between the sexes would be significantly greater. Thus, human males normally may be exposed to subasymptotic prenatal androgen stimulation, which in turn has consequences for the sexual differentiation of the brain and behavior. If this assumption is correct, the levels of androgen normally secreted during prenatal development are not high enough to produce the fullest possible expression of the male behavioral genotype but suffice to stimulate full expression of the genotype for morphology. This may reflect a difference between neural and somatic cells with

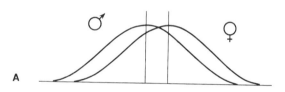

Figure 3. Phenotypic expression of quantitative sex differences. (A) Overlapping normal distributions representing the result of normal prenatal hormonal environmental conditions. B and C represent the possible effects of elevated levels of androgen on the male distribution of behavior. (B) The range remains the same, but the distribution is skewed. (C) The distribution is normal, but the kurtosis is affected. The result of the alterations in the distributions reflected in B and C is an increase of the difference between the means for males and females. (Reprinted from Reinisch & Sanders, 1984.)

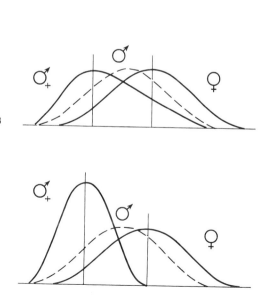

respect to (1) the amount of testosterone they require for masculinization or (2) differences in the ease with which testosterone infiltrates the tissue.

A question arises as to ways in which alteration of the hormonal milieu (elevation of androgen levels) could change the distribution of male behavior so as to augment the differences between the sexes. We assume that the full range of the male behavioral genotype is expressed in the general population of human males. Theoretically, there are at least two ways in which additional androgen could affect the distribution of masculine behavior in the male population so that the distributions of male and female behavior would be further apart. One possibility is that an elevation of prenatal androgens might increase the frequencies of highly masculine behaviors while the full range of behaviors would still be expressed in the population (see Figure 3B). In such a distribution the endpoints would remain the same but the distribution would be skewed toward the masculine pole, with more individuals exhibiting higher levels of masculine response. Alternatively, the addition of androgen might decrease variability among males, altering the range of male behavior and moving the mean of the distribution toward the masculine pole (see Figure 3C). In this case, the distribution would remain symmetrical, but the kurtosis would be altered.

Whatever the mechanism, we propose the following hypothesis. Most human males are naturally exposed to androgen levels insufficient to establish the full expression of the male behavioral genotype, resulting in smaller behavioral differences between the sexes than would exist if more androgen were available. It has been demonstrated that there are large differences among individual human male fetuses in the levels of androgen present prenatally (Warne, Faiman, Reyes, & Winter, 1977). Thus, variations among individuals in androgen exposure during prenatal development may contribute to differences in the degree of masculinization of the brain, leading to behavioral variability within each sex as well as between males and females.

MALE VULNERABILITY

A number of studies have shown that males may be more likely to be affected, or more severely affected, by prenatal exposure to drugs and hormones than are females (see Reinisch & Sanders, 1984, 1987). Further, the frequency of mental retardation, stuttering, dyslexia, learning disabilities, delayed speech development, and left-handedness is higher among males than females (Vandenberg, 1987). Gualtieri and Hicks (1985) have suggested that males may be more vulnerable to environmental insults, perhaps because a male fetus presents greater antigenicity (due to H-Y antigen) to the pregnant mother than does a female fetus. This antigenicity induces a state of maternal immunoreactivity, which can lead directly or indirectly to fetal damage and greater male susceptibility to environmental insult. Alternatively, it is possible that the presence of the Y chromosome results in hypercytoresponsivity to steroid hormones and perhaps other substances as well.

Additional data from our laboratory suggesting greater male susceptibility to the effects of exogenous substances during ontogeny derive from a preliminary study of subjects exposed prenatally to barbiturates. (Barbiturate-containing compounds were widely used for the treatment of morning sickness, anxiety, and insomnia as well as a wide range of medical conditions that can occur during pregnancy. We estimate that 22,346,543 children were born of barbiturate-treated pregnancies between 1950 and 1977 in the United States alone.) Data from laboratory animals indicate that barbiturates can affect the developing brain either directly by influencing neurological development or indirectly by affecting liver metabolism of other substances such as steroid hormones, which in turn may affect neurogenesis (Reinisch & Sanders, 1982). Eighteen exposed males and 20 exposed females were compared to unexposed same-sex controls, who were members of the birth cohort and had been matched on 15 variables. Compared to controls, barbiturate-exposed males were found to have higher scores on the M–F scale of the Extended Personal Attributes Questionnaire (Helmreich, Spence, & Holahan, 1979), had higher physical aggression scores on the Leifer–Roberts Response Hierarchy—Reinisch Revision (Reinisch & Sanders, 1986), and self-reported more antisocial and violent acts during adolescence. These are all measures that could be considered to reflect either hypermasculinity or an increase in social deviance. Although females appeared to be affected in similar ways by prenatal exposure to barbiturates, the effects were significantly larger in exposed males and encompassed more dimensions. Confirmations of these findings await final analysis of the full sample of 100 matched pairs.

INFANT SEX DIFFERENCES: POSSIBLE HORMONAL COMPONENTS

The postnatal environment, of course, exerts a considerable influence on human development, including psychosexual differentiation. To differentiate prenatal from postnatal influences, it is relevant to ask if behavioral sex differences have been identified during early infancy. In an evaluation of data on the development of behavioral milestones from the 1959–1961 Copenhagen birth cohort, we discovered subtle but highly significant sex differences during the first year of life (Reinisch, Rosenblum, Rubin, & Schulsinger, 1991). Sex differences in the age of attainment of the following developmental milestones were examined: lifts head while lying on stomach (M1); smiles (M2); holds head up (M3); reaches for objects (M4); sits without support (M5); stands with support (M6); crawls independently (M7); walks with support (M8); stands without support (M9); walks without support (M10). Although the sequence of behavioral development was identical for males and females, potentially important differences in timing were apparent. Boys reached three milestones significantly earlier than girls: lifts head while lying on stomach (M1); stands with support (M6); and crawls independently (M7).

Differences in the duration of the intervals between milestones also revealed significant differences between the sexes. Of the 45 intervals among the 10 milestones, 19 showed significant sex differences. Although males and females reached the last milestone, "walks without support" (M10), at the same time (just prior to the 13th month), females spent significantly more time between the milestones, "sits without support" (M5) and "stands with support" (M6), whereas boys spent more time between "crawls independently" (M7) and "walks with support" (M8). The greatest difference between the sexes was found in these two intervals.

One possible interpretation of the interval sex differences listed above is that girls move more quickly to attain milestones that facilitate *en face* interactions between the infant and the caretaker (Korner, 1969; Lewis, Kagan, & Kalafat, 1966). Having attained them, girls spend more time in this stage than do boys before moving on to activities that foster independence. In contrast, after attaining milestones that foster social interaction, boys appear to proceed rapidly to milestones that have the potential for facilitating independent, nonsocial action and separation from the caretaker. Boys then spend longer intervals than girls involved in these activities. Thus, boys may have relatively less opportunity for face-to-face contact with caretakers. Data from both the field and laboratory indicate that males of most nonhuman primate species move more rapidly toward independent activity than do females (Meaney & Stewart, 1985; Rosenblum, 1974). This developmental pattern also may reflect earlier development of physical strength and spatial skills in males (Beatty, 1984; Bell & Darling, 1965; Hobson, 1947; Korner, 1969).

It is possible that data showing sex differences in the age at which various milestones are attained may reflect a systematic reporting bias between mothers of male and female children. More likely, sex differences are the result, at least in part, of differences between the sexes with respect to (1) genetic and/or prenatal environmental factors (De Vries *et al.*, 1984; Glucksman, 1981; Hall, 1982; Reinisch, 1983), (2) maturational rates (Garai & Scheinfeld, 1968), and (3) specific expectations or the general pattern of interaction between the infant and caretaker (Lewis, 1972; Meaney & Stewart, 1985; Olley, 1973). It is our opinion that these differences are most reasonably interpreted as being a product of a series of biological–environmental interactions.

In support of a prenatal biological contribution to infant sex differences are data on newborns demonstrating significant sex differences in (1) sensory thresholds, with females showing more responsivity than males to haptic (touch) stimulation (Bell & Costello, 1964; Wolff, 1969), electrical stimulation (Lipsett & Levy, 1959), the taste of sweetness (Nisbett & Gurwitz, 1970), and photic stimulation (Engel, Crowell, & Nishijima, 1968); (2) several types of oral behavior, which females exhibit more frequently than males (Balint, 1948; Korner, 1969, 1973); and (3) greater physical strength and vigor of males (Korner, 1969; Rosenblum, 1974). In addition to the behavioral sex differences that have been identified in infants (Kagan, 1969; Lewis *et al.*, 1966; Silverman, 1970; Watson, 1969), sex differences in the functional development of the central nervous system have been reported as early as 3 months of age (Shucard *et al.*, 1981). These data clearly demonstrate that human males and females exhibit a range of differences in the expression of behavior beginning in the first year of life. We believe that these early sex differences reflect, at least in part, the divergent prenatal hormonal milieu of males and females.

MULTIPLIER EFFECT

Our view of the role of prenatal hormones, whether from an endogenous or exogenous source, in the development of human behavioral sex differences is illustrated in the schematic diagram presented in Figure 4. This model, derived from the "multiplier effect" concept of E. O. Wilson (1975), represents our conceptualization of the transactions between the organism and the environment that contribute to the development of sex differences in behavior.

As described earlier, at the time of conception the only difference between the sexes is chromosomal/genetic. As the testes of a normal male differentiate, androgens are produced, leading to divergent prenatal hormone environments for male and female fetuses. Beginning during the second month of gestation, the presence or absence of virilizing or masculinizing hormones (primarily testosterone in normally developing males) produces morphological (somatic and central nervous sys-

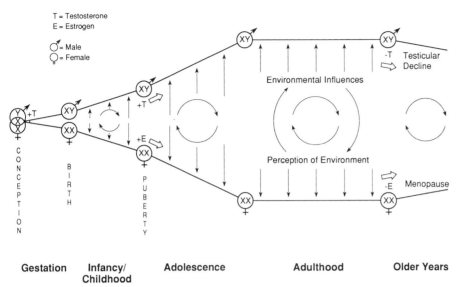

Figure 4. Multiplier effect. Role of the interaction between biological factors and the environment in the development of sex differences. Prior to birth, genetic and hormonal differences lead to sexual differentiation of the genitalia and brain. The relatively small behavioral differences between the sexes that exist at birth are augmented by successive interactions between the individual and the social environment. Pubertal hormones lead to further physical differentiation, which in turn magnifies the differences in societal expectations and social interactions experienced by males and females, thus further enhancing sex differences in behavior. (Modified from Reinisch and Sanders, 1987.)

tem) differences between the sexes, which contribute to the development of behavioral differences. At birth, sex differences are small but relevant to (1) how the organism responds to the environment and (2) how the social environment responds to the baby. First, the sexes differ with respect to their perception of environmental stimuli (Gandelman, 1983; Reinisch et al., 1979). These perceptual differences could be the result of somatic sex differences in the CNS or in peripheral (sensory) receptors. It is likely that perceptual, cognitive, and temperamental differences lead to different interpretations of, and responses to, the environment. Secondly, most societies treat infants, children, and adults differently, based on their external genitalia.

The results of interactions between the organism and the environment over time are probably responsible for behavioral sex differences. We believe that small sex differences in appearance, behavior, and responsiveness at birth combine with cultural expectations about appropriate behavior for each sex and lead to augmentation of the behavioral sex differences. Each successive interaction alters the organism slightly. The modified organism then perceives and responds to the environment a little differently and, in turn, is perceived and treated in a slightly different fashion. Through this dynamic transaction between the organism and its environment involving biological, psychological, and social factors, major sex differences in behavior emerge over time. (Additional supporting evidence can be found in Chapter 5 by I. L. Ward in this volume.) At puberty, the development of secondary sexual characteristics prompted by the release of adult levels of reproductive hormones dramatically increases the physical and perceptual differences between the sexes. These changes elicit increasingly different societal expectations of how the

JUNE MACHOVER
REINISCH AND
STEPHANIE A.
SANDERS

young man or woman should behave. Overall sex differences are thus maximized during the reproductive years.

As the individual continues to age, hormone levels drop, and reproductive capacity diminishes. Some evidence suggests that behavioral androgyny increases in parallel, and, thus, sex differences decrease with age following the reproductive years (Maracek, 1979).

In summary, gonadal hormones and the resultant sexual dimorphism of somatotype and central nervous system, make essential biological contributions to the development of human behavioral sex differences. Hormones present during fetal development can be thought of as influencing later behavior by either imposing organization on the neural substrate (Goy, 1968; Goy & McEwen, 1980; Harris, 1964; Phoenix, Goy, Gerall, & Young, 1959; Young, Goy, & Phoenix, 1964) or sensitizing the organism to later hormonal stimulation (Beach, 1945, 1971) or environmental input. The more divergent the sex stereotypes within a culture are, the more one would expect biologically influenced sexual dimorphisms in behavior to be magnified by postnatal socioenvironmental influences. In such societies, sex differences that have no basis in biology also may emerge as a result of highly differential sex role expectations. However, even if differential treatment of the sexes did not occur, behavioral differences would remain, nevertheless, between males and females.

SUMMARY

This chapter presents concepts relevant to understanding the interaction between biological and socioenvironmental variables contributing to the development of human behavioral sex differences. These include (1) models of masculinity and femininity, (2) the proposition of human subasymptotic levels of prenatal androgen stimulation, (3) the male vulnerability principle, and (4) the multiplier effect.

Data from human clinical endocrine syndromes and studies of subjects exposed to exogenous hormones during gestation (Reinisch, 1974, 1977, 1983; Reinisch & Sanders, 1984, 1987; Reinisch *et al.*, 1991; Sanders & Reinisch, 1985) provide strong evidence that the relative presence or absence of androgenic substances *in utero* influences cognitive and personality development. Differences in the prenatal hormone environment may contribute not only to the development of sex differences but also to the naturally occurring variation among individuals within each sex (Maccoby & Jacklin, 1974; Reinisch & Sanders, 1984). Additionally, researchers from a number of disciplines have suggested that sex differences in cognitive abilities are related to differences in brain hemispheric laterality that are initially established by the influence of the prenatal hormonal milieu. This suggestion has gained empirical support from our evaluation of the abilities of individuals prenatally exposed to abnormal hormonal environments.

It is important to note that such biological contributions to human behavioral development are not understood as being deterministic. Rather, the prenatal hormonal environment may affect behavior by biasing the neural system in such a way as to create behavioral predispositions. It is then that culture adds its influence by rewarding, punishing, and/or ignoring behaviors and thereby can exaggerate, diminish, distort, or permit the expression of biologically based sex differences.

Acknowledgments

239

HUMAN
COGNITIVE AND
PERSONALITY
DEVELOPMENT

This work was supported, in part, by Public Health Service grants HD20263 and HD17655 to J. M. Reinisch, The Kinsey Institute, and Indiana University. We gratefully acknowledge the assistance of Carolyn Kaufman and Mary Ziemba-Davis in the preparation of this manuscript and the editors of this volume for their constructive commentary.

REFERENCES

Baker, S. W., & Ehrhardt, A. A. (1974). Prenatal androgen, intelligence and cognitive sex differences. In R. C. Friedman, R. M. Richart, & R. L. Vande Wiele (Eds.), *Sex differences in behavior* (pp. 53–76). New York: John Wiley & Sons.

Balint, M. (1948). Individual differences of behavior in early infancy and an objective way of recording them. *Journal of Genetic Psychology, 73,* 57–117.

Beach, F. A. (1945). Bisexual mating behavior in the male rat: Effects of castration and hormone administration. *Physiological Zoology, 18,* 390–402.

Beach, F. A. (1971). Hormonal factors controlling the differentiation, development, and display of copulatory behavior in the ramstergig and related species. In E. Tobach, L. R. Aronson, & E. Shaw (Eds.), *Biopsychology of development* (pp. 249–296). New York: Academic Press.

Beatty, W. W. (1984) Hormonal organization of sex differences in play fighting and spatial behavior. In G. J. DeVries, J. P. C. De Bruin, H. B. M. Uylings, & M. A. Corner (Eds.), *Progress in brain research, Vol. 61: Sex differences in the brain* (pp. 315–330). Amsterdam: Elsevier.

Bell, R. Q., & Costello, N. (1964). Three tests for sex differences in tactile sensitivity in the newborn. *Biologica Neonatorum, 7,* 335–347.

Bell, R. Q., & Darling, J. F. (1965). The prone head reaction in the human neonate: Relation with sex and tactile sensitivity. *Child Development, 36,* 943–949.

Bem, S. L. (1974). The measurement of psychological androgyny. *Journal of Consulting and Clinical Psychology, 42,* 155–162.

Benbow, C. P., & Benbow, R. M. (1984). Biological correlates of high mathematical reasoning ability. In G. J. De Vries, J. P. C. De Bruin, H. B. M. Uylings, & M. A. Corner (Eds.), *Progress in brain research, Vol. 61: Sex differences in the brain* (pp. 469–490). Amsterdam: Elsevier.

Benbow, C. P., & Stanley, J. C. (1980). Sex differences in mathematical ability: Fact or artifact? *Science, 210,* 1262–1264.

Benbow, C. P., & Stanley, J. C. (1981). Mathematical ability: Is sex a factor? *Science, 212,* 118–121.

Benbow, C. P., & Stanley, J. C. (1983). Sex differences in mathematical reasoning ability: More facts. *Science, 222,* 1029–1031.

Berzins, J. I., Welling, M. A., & Wetter, R. E. (1978). A new measure of psychological androgyny based on the Personality Research Form. *Journal of Consulting and Clinical Psychology, 46,* 126–138.

Broverman, D. M., Vogel, W., Klaiber, E. L., Majcher, D., Shea, D., & Paul, V. (1981). Changes in cognitive task performance across the menstrual cycle. *Journal of Comparative and Physiological Psychology, 95,* 646–654.

Bruyer, R., & Craps, V. (1985). Facial asymmetry: Perceptual awareness and lateral differences. *Canadian Journal of Psychology, 39,* 54–69.

Burstein, B., Bank, L., & Jarvik, L. F. (1980). Sex differences in cognitive functioning: Evidence, determinants, implications. *Human Development, 23,* 289–313.

Constantinople, A. (1973). Masculinity–femininity: An exception to a famous dictum? *Psychological Bulletin, 80,* 389–407.

de Lacoste-Utamsing, C., & Holloway, R. L. (1982). Sexual dimorphism in the human corpus callosum. *Science, 216,* 1431–1432.

De Vries, G. J., De Bruin, J. P. C., Uylings, H. B. M., & Corner, M. A. (Eds.). (1984). *Progress in brain research, Vol. 61: Sex differences in the brain.* Amsterdam: Elsevier.

Diamond, M., Llacuna, A., & Wong, C. L. (1973). Sex behavior after neonatal progesterone, testosterone, estrogen, or antiandrogens. *Hormones and Behavior, 4,* 73–88.

Dörner, G., Döcke, F., & Hinz, G. (1971). Paradoxical effects of estrogen on brain differentiation. *Neuroendocrinology, 7,* 146–155.

Ehrhardt, A. A., & Baker, S. W. (1974). Fetal androgens, human central nervous system differentiation and behavior sex differences. In R. C. Friedman, R. M. Richart, & R. L. Vande Wiele (Eds.), *Sex differences in behavior* (pp. 33–51). New York: John Wiley & Sons.

Ehrhardt, A. A., & Money, J. (1967). Progestin-induced hermaphroditism: IQ and psychosexual identity in a study of ten girls. *Journal of Sex Research, 3,* 83–100.

Ehrhardt, A. A., Epstein, R., & Money, J. (1968). Fetal androgens and female gender identity in the early-treated adrenogenital syndrome. *Johns Hopkins Medical Journal, 122,* 160–167.

Ehrhardt, A. A., Evers, K., & Money, J. (1968). Influence of androgen and some aspects of sexually dimorphic behavior in women with the late-treated adrenogenital syndrome. *Johns Hopkins Medical Journal, 123,* 115–122.

Ehrhardt, A. A., Meyer-Bahlburg, H. F. L., Rosen, L. R., Feldman, J. F., Verdiano, N. P., Zimmerman, I., & McEwen, B. S. (1985). Sexual orientation after prenatal exposure to exogenous estrogen. *Archives of Sexual Behavior, 14*(1), 57–75.

Ellis, L. (1982). Developmental androgen fluctuations and the five dimensions of mammalian sex (with emphasis upon the behavioral dimension and the human species). *Ethology and Sociobiology, 3,* 171–197.

Engel, R., Crowell, D., & Nishijima, S. (1968). Visual and auditory response latencies in neonates. In B. N. D. Fernando (Ed.), *Felicitation Volume in Honour of C. C. De Silva* (pp. 31–40) Ceylon: Kularatne & Company.

Galaburda, A. M., LeMay, M., Kemper, T. L., & Geschwind, N. (1978). Right–left asymmetries in the brain. *Science, 199,* 852–856.

Gandelman, R. (1983). Gonadal hormones and sensory function. *Neuroscience and Biobehavioral Reviews, 7,* 1–17.

Garai, J. E., & Scheinfeld, A. (1968). Sex differences in mental and behavioral traits. *Genetic Psychological Monographs, 77,* 169–299.

Gazzaniga, M. (1972). One brain—two minds. *American Scientist, 60,* 311–317.

Geschwind, N. (1970). The organization of language and the brain. *Science, 170,* 940–944.

Geschwind, N., & Behan, P. (1982). Left-handedness: Association with immune disease, migraine, and developmental learning disorder. *Proceedings of the National Academy of Science, U.S.A., 79,* 5097–5100.

Geschwind, N., & Galaburda, A. M. (1985). Cerebral lateralization. *Archives of Neurology, 42,* 428–654.

Glucksman, A. (1981). *Sexual dimorphism in human and mammalian biology and pathology.* New York: Academic Press.

Gorski, R. A. (1987). Sex differences in the rodent brain: Their nature and origin. In J. M. Reinisch, L. A. Rosenblum, & S. A. Sanders (Eds.), *Masculinity/femininity: Basic perspectives.* New York: Oxford University Press.

Gorski, R. A., Gordon, J. H., Shryne, J. E., & Southam, A. M. (1978). Evidence for a morphological sex difference within the medial preoptic area of the rat brain. *Brain Research, 148,* 333–346.

Goy, R. W. (1968). Organizing effects of androgens on the behaviour of rhesus monkeys. In R. P. Michael (Ed.), *Endocrinology and human behavior* (pp. 12–31). London: Oxford University Press.

Goy, R. W., & McEwen, B. S. (1980). *Sexual differentiation of the brain.* Cambridge, MA: MIT Press.

Goy, R. W., Wolf, J. E., & Eisele, S. G. (1977). Experimental female hermaphroditism in rhesus monkeys: Anatomical and psychological characteristics. In J. Money & H. Musaph (Eds.), *Handbook of sexology* (pp. 139–156). Amsterdam: Elsevier/North-Holland Biomedical Press.

Gray, J. S., Levine, S., & Broadhurst, P. L. (1965). Gonadal hormone injections in infancy and adult emotional behaviour. *Animal Behaviour, 13,* 33–45.

Gualtieri, T., & Hicks, R. E. (1985). An immunoreactive theory of selective male affliction. *Behavioral and Brain Sciences, 8*(3), 427–477.

Hall, R. L. (1982). *Sexual dimorphism in Homo sapiens: A question of size.* New York: Praeger.

Halpern, D. F. (1986). *Sex differences in cognitive ability.* Hillsdale, NJ: Lawrence Erlbaum Associates.

Harris, G. W. (1964). Sex hormones, brain development and brain function. *Endocrinology, 75,* 627–648.

Heilbrun, A. B. (1976). Measurement of masculine and feminine sex role identities as independent dimensions. *Journal of Consulting and Clinical Psychology, 44,* 183–190.

Helmreich, R. L., Spence, J. T., & Holahan, C. K. (1979). Psychological androgyny and sex-role flexibility: A test of two hypotheses. *Journal of Personality and Social Psychology, 37,* 1631–1644.

Hines, M. (1981). *Prenatal diethylstilbestrol (DES) exposure, human sexually dimorphic behavior and cerebral lateralization.* Doctoral Thesis: University of California–Los Angeles. *Dissertation Abstracts, 42,* 423B.

Hines, M. (1982). Prenatal gonadal hormones and sex differences in human behavior. *Psychological Bulletin, 92,* 56–80.

Hines, M., & Shipley, C. (1984). Prenatal exposure to diethylstilbestrol (DES) and the development of sexually dimorphic cognitive abilities and cerebral lateralization. *Developmental Psychology, 20,* 81–94.

Hines, M., Alsum, P., Gorski, R. A., & Goy, R. W. (1982). Prenatal exposure to estrogen masculinizes and defeminizes behavior in the guinea pig. *Society for Neuroscience Abstracts, 8,* 196.

Hobson, J. R. (1947). Sex differences in primary mental abilities. *Journal of Educational Research, 41,* 126–132.

Inglis, J., Ruckman, M., Lawson, J. S., MacLean, A. W., & Monga, T. N. (1983). Sex differences in the cognitive effects of unilateral brain damage: Comparison of stroke patients and normal control subjects. *Cortex, 19,* 551–555.

Jacobsen, B. D. (1962). Hazards of norethindrone therapy during pregnancy. *American Journal of Obstetrics and Gynecology, 84,* 962–968.

Jost, A. (1972). Becoming a male. In G. Raspe and S. Bernhard (Eds.), *Advances in the biosciences, Vol. 10* (pp. 3–13). New York: Pergamon Press.

Kagan, J. (1969). On the meaning of behavior: Illustrations from the infant. *Child Development, 40,* 1121–1134.

Kester, P., Green, R., Finch, S. J., & Williams, K. (1980). Prenatal female hormone administration and psychosexual development in human males. *Psychoneuroendocrinology, 5,* 269–285.

Kimura, D. (1961). Cerebral dominance and the perception of verbal stimuli. *Canadian Journal of Psychology, 15,* 166–171.

Kimura, D. (1983). Sex differences in cerebral organization for speech and praxic function. *Canadian Journal of Psychology, 37,* 19–35.

Korner, A. F. (1969). Neonatal startles, smiles, erections, and reflex sucks as related to state, sex, and individuality. *Child Development, 40,* 1039–1054.

Korner, A. F. (1973). Sex differences in newborns with special reference to differences in the organization of oral behavior. *Journal of Child Psychology & Psychiatry, 14,* 19–29.

Levy, J. (1972). Lateral specialization of the human brain: Behavioral manifestations and possible evolutionary basis. In J. A. Kliger (Ed.), *The biology of behavior.* Corvallis, OR: Oregon State University Press.

Levy, J., & Levy, J. M. (1978). Human lateralization from head to foot: Sex-related factors. *Science, 200,* 1291–1292.

Lewis, M. (1972). State as an infant-environment interaction: An analysis of mother–infant interaction as a function of sex. *Merrill-Palmer Quarterly, 18,* 95–121.

Lewis, M., Kagan, J., & Kalafat, J. (1966). Patterns of fixation in the young infant. *Child Development, 37,* 331–341.

Lipsett, L. P., & Levy, N. (1959). Electrotactual threshold in the human neonate. *Child Development, 30,* 547–554.

Maccoby, E. E., & Jacklin, C. N. (1974). *The psychology of sex differences.* Palo Alto: Stanford University Press.

MacLusky, N. J., & Naftolin, F. (1981). Sexual differentiation of the central nervous system. *Science, 211,* 1294–1302.

Maracek, J. (1979). Social change, positive mental health, and psychological androgyny. *Psychology of Women Quarterly, 3,* 241–247.

McEwen, B. S. (1987). Observations on brain sexual differentiation: A biochemist's view. In J. M. Reinisch, L. A. Rosenblum, & S. A. Sanders (Eds.), *Masculinity/femininity: Basic perspectives.* New York: Oxford University Press.

McGee, M. G. (1979). Human spatial abilities: Psychometric studies and environmental, genetic, hormonal, and neurological influences. *Psychological Bulletin, 86,* 889–918.

McGlone, J. (1980). Sex differences in human brain asymmetry: A critical survey. *The Behavioral and Brain Sciences, 3,* 215–263.

Meaney, M. J., & Stewart, J. (1985). In J. S. Rosenblatt, C. Beer, M. C. Busnel, & P. J. B. Slater (Eds.), *Advances in the study of behavior* (pp. 2–58). New York: Academic Press.

Money, J., & Ehrhardt, A. A. (1972) *Man and woman, boy and girl.* Baltimore: Johns Hopkins University Press.

Money, J., & Ogunro, C. (1974). Behavioral sexology: Ten cases of genetic male intersexuality with impaired prenatal and pubertal androgenization. *Archives of Sexual Behavior, 3,* 181–205.

Money, J., & Schwartz, M. (1977). Dating, romantic and nonromantic friendships, and sexuality in 17 early-treated adrenogenital females, aged 16–25. In P. A. Lee, L. P. Plotnick, A. A. Kowarski, and C. J. Migeon (Eds.), *Congenital adrenal hyperplasia* (pp. 419–431). Baltimore: University Park Press.

Money, J., Ehrhardt, A. A., & Masica, D. N. (1968). Fetal feminization induced by androgen insensitivity in the testicular feminizing syndrome: Effect on marriage and maternalism. *Johns Hopkins Medical Journal, 123,* 105–114.

Money, J., Schwartz, M., & Lewis, V. G. (1984). Adult erotosexual status and fetal hormonal masculinization and demasculinization: 46,XX congenital virilizing adrenal hyperplasia (CVAH) and 46,XY androgen insensitivity syndrome (AIS) compared. *Psychoneuroendocrinology, 9,* 405–414.

Nisbett, R. E., & Gurwitz, S. B. (1970). Weight, sex and the eating behavior of human newborns. *Journal of Comparative and Physiological Psychology, 73,* 245–253.

Ojemann, G. A. (1983). The intrahemispheric organization of human language, derived with electrical stimulation techniques. *Trends in Neurosciences, 16,* 184–189.

Olley, J. G. (1973). *Sex differences in human behavior in the first year of life.* Major Area Paper, Department of Psychology, George Peabody College for Teachers, Nashville, Tennessee. Cited in Korner (1973)

Phoenix, C. H., Goy, R. W., Gerall, A. A., & Young, W. C. (1959). Organizing action of prenatally administered testosterone propionate on the tissues mediating mating behavior in the female guinea pig. *Endocrinology, 65,* 369–382.

Reinisch, J. M. (1974). Fetal hormones, the brain, and human sex differences: A heuristic integrative review of the recent literature. *Archives of Sexual Behavior, 3,* 51–90.

Reinisch, J. M. (1976). Effects of prenatal hormone exposure on physical and psychological development in humans and animals: With a note on the state of the field. In E. J. Sacher (Ed.), *Hormones, behavior and psychopathology* (pp. 69–94). New York: Raven Press.

Reinisch, J. M. (1977). Prenatal exposure of human foetuses to synthetic progestin and oestrogen affects personality. *Nature, 266,* 561–562.

Reinisch, J. M. (1981). Prenatal exposure to synthetic progestins increases potential for aggression in humans. *Science, 211,* 1171–1173.

Reinisch, J. M. (1983). Influence of early exposure to steroid hormones on behavioral development. In W. Everaerd, C. B. Hindley, A. Bot, & J. J. van der Werff ten Bosch (Eds.), *Development in adolescence. Psychological, social and biological aspects* (pp. 63–113). Boston: Martinus Nijhoff.

Reinisch, J. M., & Gandelman, R. (1978). Human research in behavioral endocrinology: Methodological and theoretical considerations. In G. Dörner & M. Kawakami (Eds.), *Hormones and brain development* (pp. 77–86). Amsterdam: Elsevier/North-Holland Biomedical Press.

Reinisch, J. M., & Karow, W. G. (1977). Prenatal exposure to synthetic progestins and estrogens: Effects on human development. *Archives of Sexual Behavior, 6,* 257–288.

Reinisch, J. M., & Sanders, S. A. (1982). Early barbiturate exposure: The brain, sexually dimorphic behavior, and learning. *Neuroscience and Biobehavioral Reviews, 6,* 311–319.

Reinisch, J. M., & Sanders, S. A. (1984). Prenatal gonadal steroidal influences on gender-related behavior. In J. G. De Vries, J. P. C. De Bruin, H. B. M. Uylings, & M. A. Corner (Eds.), *Progress in brain research, Vol. 61: Sex differences in the brain* (pp. 407–416). Amsterdam: Elsevier.

Reinisch, J. M., & Sanders, S. A. (1986). A test of sex differences in aggressive response to hypothetical conflict situations. *Journal of Personality and Social Psychology, 50*(5), 1045–1049.

Reinisch, J. M., & Sanders, S. A. (1987). Behavioral influences of prenatal hormones. In C. B. Nemeroff & P. T. Loosen (Eds.), *Handbook of clinical psychoneuroendocrinology* (pp. 431–448). New York: The Guilford Press.

Reinisch, J. M., & Sanders, S. A. (in press). Effects of prenatal exposure to diethylstilbestrol (DES) on hemispheric laterality and spatial ability in human males. *Hormones and Behavior.*

Reinisch, J. M., Gandelman, R., & Spiegel, F. S. (1979). Prenatal influences on cognitive abilities: Data from experimental animals and human genetic and endocrine syndromes. In M. A. Wittig & A. C. Petersen (Eds.), *Sex-related differences in cognitive functioning* (pp. 215–239). New York: Academic Press.

Reinisch, J. M., Rosenblum, L. A., Rubin, D. B., & Schulsinger, M. F. (1991). Sex differences in behavioral milestones during the first year of life. *Journal of Psychology and Human Sexuality, 4*(2), 19–36.

Reinisch, J. M., Ziemba-Davis, M., & Sanders, S. A. (1991). Hormonal contributions to sexually dimorphic behavioral development in humans. *Psychoneuroendocrinology, 16,* 213–278.

Remillard, G. M., Andermann, F., Testa, G. F., Gloor, P., Aubé, M., Martin, J. B., Feindel, W., Gubermann, A., & Simpson, C. (1983). Sexual ictal manifestations predominate in women with temporal lobe epilepsy: A finding suggesting sexual dimorphism in the human brain. *Neurology, 33,* 323–330.

Rosenblum, L. A. (1974). Sex differences in mother–infant attachment in monkeys. In R. C. Friedman, R. M. Richart, & R. L. Vande Wiele (Eds.), *Sex differences in behavior* (pp. 123–145). New York: John Wiley & Sons.

Sanders, S. A., & Reinisch, J. M. (1985). Behavioral effects on humans of progesterone-related compounds during development and in the adult. In D. Ganten & D. Pfaff (Eds.), *Current topics in neuroendocrinology, Vol. 5: Actions of progesterone on the brain* (pp. 175–205). Berlin, Heidelberg: Springer-Verlag.

Shucard, J. L., Shucard, D. W., Cummins, K. R., & Campos, J. J. (1981). Auditory evoked potentials and sex-related differences in brain development. *Brain and Language, 13,* 91–102.

Silverman, J. (1970). Attentional styles and the study of sex differences. In D. Mostofsky (Ed.), *Attention: Contemporary theory and analysis* (pp. 61–98). New York: Appleton-Century-Crofts.

Spence, J. T., & Helmreich, R. L. (1978). *Masculinity and femininity: Their psychological dimensions, correlates and antecedents.* Austin, TX: University of Texas Press.

Spence, J. T., Helmreich, R. L., & Stapp, J. (1974). The PAQ: A measure of sex-role stereotypes and M–F. *JSAS Catalogue of Selected Documents in Psychology, 4,* 127.

Swaab, D. F., & Fliers, E. (1985). A sexually dimorphic nucleus in the human brain. *Science, 228,* 1112–1115.

Swaab, D. F., & Hofman, M. A. (1984). Sexual differentiation of the human brain: A historical perspective. In G. J. De Vries, J. P. C. De Bruin, H. B. M. Uylings, & M. A. Corner (Eds.), *Progress in brain research, Vol. 61: Sex differences in the brain* (pp. 361–374). Amsterdam: Elsevier.

Vandenberg, S. G. (1987). Sex differences in mental retardation and their implications for sex differences in ability. In J. M. Reinisch, L. A. Rosenblum, & S. A. Sanders (Eds.), *Masculinity/femininity: Basic perspectives* (pp. 157–171). New York: Oxford University Press.

Waber, D. P. (1979). Cognitive abilities and sex-related variations in the maturation of cerebral cortical functions. In M. A. Wittig & A. C. Petersen (Eds.), *Sex-related differences in cognitive functioning: Developmental issues* (pp. 161–186). New York: Academic Press.

Warne, G. L., Faiman, C., Reyes, F. I., & Winter, J. S. D. (1977). Studies on human sexual development. V. Concentrations of testosterone, 17-hydroxyprogesterone and progesterone in human amniotic fluid throughout gestation. *Journal of Clinical Endocrinology and Metabolism, 44,* 934–938.

Watson, J. S. (1969). Operant conditioning of visual fixation in infants under visual and auditory reinforcement. *Developmental Psychology, 1,* 508–516.

Whalen, R. E. (1974). Sexual differentiation: Models, methods and mechanisms. In R. C. Friedman, R. M. Richart, & R. L. Vande Wiele (Eds.), *Sex differences in behavior* (pp. 467–481). New York: John Wiley & Sons.

Williams, R. H. (1981). *Textbook of endocrinology.* Philadelphia: W. B. Saunders.

Wilson, E. O. (1975). *Sociobiology: The new synthesis* (pp. 11–13). Cambridge, MA: The Belknap Press, Harvard University.

Wilson, J. R. (1987). The fragile X chromosome anomaly: A comment. In J. M. Reinisch, L. A. Rosenblum, & S. A. Sanders (Eds.), *Masculinity/femininity: Basic perspectives* (pp. 172–175). New York: Oxford University Press.

Wilson, J. R., & Vandenberg, S. G. (1979). Sex differences in cognition: Evidence from the Hawaii Family Study. In T. E. McGill, D. A. Dewsbury, & B. D. Sachs (Eds.), *Sex and behavior* (pp. 317–335). New York: Plenum Press.

Witelson, S. F. (1976). Hemispheric specialization for linguistic and nonlinguistic tactual perception using a dichotomous stimulation technique. *Cortex, 10,* 3–17.

Witelson, S. F. (1985). Sex and the single hemisphere: Specialization of the right hemisphere for spatial processing. *Science, 194,* 425–427.

Wolff, P. H. (1969). The natural history of crying and other vocalizations in early infancy. In B. M. Foss (Ed.), *Determinants of infant behaviour,* Vol. 3 (pp. 113–138). London: Methuen.

Yalom, I., Green, R., & Fisk, N. (1973). Prenatal exposure to female hormones: Effect on psychosexual development in boys. *Archives of General Psychiatry, 28,* 554–561.

Young, W. C., Goy, R. W., & Phoenix, C. H. (1964). Hormones and sexual behavior. *Science, 142,* 212–218.

8

Gender Differences in Human Neuropsychological Function

JERRE LEVY AND WENDY HELLER

INTRODUCTION

Scientific studies of sex differences in cognitive abilities have been stimulated by a long history of popular interest in psychological differences between men and women. According to Anastasi, when Samuel Johnson was asked whether men or women were more intelligent, he replied by asking, "Which man, which woman?" (p. 453, Anastasi, 1958). Johnson's reply highlights the fact that for a very large number of cognitive capacities (e.g., vocabulary, verbal reasoning, the ability to recognize an object from incomplete parts), no consistent sex differences exist. For other cognitive characteristics, the average difference between sexes is very small compared to within-gender variations. As Anastasi (1958) says, "Owing to the large extent of individual differences within any one group as contrasted to the relatively small difference between group averages, an individual's membership in a given group furnishes little or no information about his status in most traits" (p. 453).

We would add that if predictions are derived about an individual's probable success in a particular educational program or occupation based on regressed true scores of psychometric evaluations, the addition of gender to the prediction equation does not improve its accuracy. Claims that feminist aims would be invalidated were science to establish that cognitive sex differences are caused partly or totally by biological factors (e.g., Levin, 1980; see Levy, 1981, for rebuttal) reflect a failure to understand the scientific literature. The scientific findings strongly support the conclusion that an effective use of human talent compels sex-blind evaluations in meeting occupational requirements and the demands of educational programs.

JERRE LEVY   Department of Psychology, University of Chicago, Chicago, Illinois 60637.   WENDY HELLER   Department of Psychology, University of Illinois, Urbana, Illinois 61820.

Sexual Differentiation, Volume 11 of *Handbook of Behavioral Neurobiology*, edited by Arnold A. Gerall, Howard Moltz, and Ingeborg L. Ward, Plenum Press, New York, 1992.

JERRE LEVY AND
WENDY HELLER

Many psychological investigations have found average differences between h
man males and females in a variety of cognitive abilities (see Wittig & Peterse
1979, for review; see Table 1). Females perform at a higher average level on tests
verbal fluency, spelling, reading speed, and reading comprehension. Speech ar
language fluency develop earlier in girls than boys, and reading disorders are abo
four or five times more common in boys than girls (see Maccoby & Jacklin, 1974, f
developmental studies). Female superiority in the communicative aspects of la
guage does not, however, generalize to the formal manipulation of symbols as cha
acterized by mathematical reasoning, in which males obtain average higher score
The males' better performance on tests of mathematical reasoning (as distinct fro
arithmetical calculation) even appears among youngsters whose mathematical ta
ents are at the upper end of the range and among boys and girls who have had equ
training and express equal interest in mathematics (Benbow & Benbow, 1984).

The sexes differ not only in aspects of natural and formal language but also
nonverbal processes. Males, on average, perform better than females in unde
standing and mentally manipulating spatial relations on paper-and-pencil maze tes
(Porteus, 1965). Higher performance on spatial tests is correlated with a great
ability of males than females to perceive underlying visuospatial relations in a co
fusing and conflicting context, an ability that has been called "field independence
(Witkin, Dyk, Faterson, Goodenough, & Karp, 1962/1974). Thus, there is an ave
age male superiority in identifying a simple figure embedded in a more complex o
(Embedded Figures Test), setting a rod to the absolute vertical when the rod
surrounded by a tilted frame (Rod-and-Frame Test), and correctly specifying th
horizontal water line in a tilted bottle (Morris, 1971; Thomas, Jamison,
Hummel, 1973).

Field independence, though positively correlated with other spatial abilities,
negatively correlated with many personal attributes involved in social interaction.
comparison to field-independent people, field-dependent individuals are more a
tentive to social sources of information, more open in expressing thoughts ar
feelings, and more effective in social interactions (see Witkin, Goodenough, & Ol

TABLE 1. AVERAGE PERFORMANCE OF FEMALES RELATIVE
TO MALES FOR VARIOUS COGNITIVE ABILITIES

Cognitive ability	Average performance of females (F) relative to males (M)
Verbal fluency	F > M
Spelling	F > M
Reading speed and comprehension	F > M
Social comprehension, understanding emotional information	F > M
Field independence	M > F
Mathematical reasoning	M > F
Comprehension of spatial relations	M > F
Verbal reasoning	F = M
Verbal IQ	F = M
Performance IQ	F = M

man, 1979, for review). These observations suggest that an increase in field dependence, such as typically seen in women, may reflect a greater use of contextual information. Such sensitivity may be important for social understanding. Females, cross-culturally, are better than men at interpreting the meaning of facial expression and other social, emotional information (Rosenthal, Hall, DiMatteo, Rogers, & Archer, 1979). Women also discriminate fine visual details better than men (Schneider & Patterson, 1942).

Sex Differences in Human Neuropsychological Organization

In the large majority of right-handers, language functions, including speech, speech comprehension, reading, writing, mathematics, and arithmetic abilities, are mediated predominately by the left cerebral hemisphere. The right hemisphere is specialized for a variety of nonverbal processes, including interpretation of facial expression and emotional information, understanding of spatial relations, and nonverbal representation of sensory experience (see Bradshaw & Nettleton, 1983, for review).

Neither gender is consistently superior to the other in the functions of either hemisphere. The left hemisphere is specialized for verbal fluency, in which females are superior, for mathematical reasoning, in which males are superior, and for verbal reasoning, in which the genders are equal. Similarly, the right hemisphere is specialized for understanding spatial relations, in which males are superior, for interpreting emotional information, in which females are superior, and for imagistic representations of sensory events (in which the genders are equal). Additionally, the genders are equal in Performance IQ (which measures nonverbal intelligence) and in Verbal IQ (which measures verbal intelligence) on the Wechsler Adult Intelligence Scale (Matarazzo, 1972). Brain metabolic activity, as assessed from positron emission tomography (PET), is more strongly correlated with Performance IQ (PIQ) in the right than in the left hemisphere and with Verbal IQ (VIQ) in the left than right hemisphere (Chase *et al.*, 1984). We can conclude, therefore, that the overall level of intelligence of each hemisphere is no different in men and women. Nonetheless, specific processes associated with each hemisphere may not be equally expressed by the sexes.

Sex Differences in the Degree of Functional Brain Asymmetry

In patients with unilateral brain damage, the association between verbal disorders and left hemisphere damage and between nonverbal disorders and right hemisphere damage is stronger in men than in women (Basso, Cupitani, & Monaschini, 1982; Edwards, Ellams, & Thompson, 1976; Inglis, Ruckner, Lawson, MacLean, & Monga, 1982; Kimura & Harshman, 1984; McGlone, 1977, 1980; Messerli, Tissot, & Rodriguez, 1976). The fact that the side of brain damage is more predictive of the nature (verbal and nonverbal) and severity of behavioral deficits in men than it is in women suggests a more symmetrical functional organization in women.

The greater similarity of the female's cerebral hemispheres in cognitive functioning appears to reflect both increases in the functional capacity of the nonspecialized hemisphere and decreases in the functional capacity of the specialized hemisphere. Kimura and Harshman (1984) examined groups of nonaphasic patients with unilateral hemispheric lesions to determine whether cognitive deficits differed for the two sexes. Digit span was significantly depressed in men but not in women with left-side damage, which suggests that the right hemisphere of women may be more involved in digit-span processing than the right hemisphere of men. In both sexes, vocabulary knowledge was deficient following left hemisphere lesions, but only women showed a vocabulary deficit with right hemisphere lesions. Quite possibly, word representation in women is more dependent on contributions from both hemispheres.

Both male and female patients with right hemisphere damage (Kimura & Harshman, 1984) were deficient in their ability to detect a missing portion of a picture [Wechsler Adult Intelligence Scale (WAIS); Picture Completion subtest], but only men were deficient in copying designs with colored blocks (WAIS; Block Design subtest), in arranging cartoon pictures to tell a story (WAIS; Picture Arrangement), and in putting together simple jigsaw puzzles (WAIS; Object Assembly). The infrequency of deficits in women suggests that the left hemisphere represents and processes nonverbal information to a greater degree than it does in men.

Although these observations suggest that verbal and nonverbal functions are more bilaterally represented in the female than male brain, no evidence indicates that the right hemisphere of women mediates speech. In both genders, speech is entirely dependent on the left hemisphere in about 98% of right-handers.

Electrophysiological studies of normal people also suggest that cognitive functions are more symmetrically organized in women than in men. When engaging in laterally specialized tasks, men's α rhythm is suppressed more over the specialized side than women's (see Butler, 1984).

Investigations of normal right-handers responding to lateralized tachistoscopic input have often reported that perceptual asymmetries are smaller in women than in men. Typically, right-handers are faster or more accurate at identifying verbal stimuli in the right visual field/left hemisphere and certain nonverbal stimuli in the left visual field/right hemisphere. Many studies using verbal tasks, however, have revealed that the magnitude of the left hemisphere superiority is smaller in women than men (Bradshaw & Gates, 1978; Bradshaw, Gates, & Nettleton, 1977; Bryden, 1979; Hannay & Malone, 1976; Kail & Siegal, 1978), whereas on nonemotional visuospatial tasks, the right hemisphere superiority is smaller in women than men (Rizzolatti & Buchtel, 1977; Davidoff, 1977; Levy & Reid, 1978; McGlone & Davidson, 1973). It should be noted that many investigations of perceptual asymmetries fail to find gender differences on nonemotional tasks (e.g., Kershner & Jeng, 1972; Hannay & Boyer, 1978; Leehey, Carey, Diamond, & Cahn, 1978; Piazza, 1980). Investigators who have found them, however, consistently report greater asymmetries for men than women (e.g., Bradshaw & Gates, 1978; Rizzolatti & Buchtel, 1977).

The results of the foregoing studies on sex differences in perceptual asymmetries are consistent with those from electroencephalographic studies and from investigations of brain-damaged patients. Two quite distinct explanations can be made for the smaller perceptual asymmetries observed in women. The greater similarity between hemispheres can result, in some cases, from a reduced functional competency of the specialized hemisphere. In other cases, it can be caused by an increased representation of function in the unspecialized hemisphere (see Figure 1). Both causes for relative hemispheric similarity have been observed in women. First, in Levy and Reid's (1978) study, a right hemisphere superiority that was observed for both genders on a spatial task was smaller for women than men. Since the performance for the two genders was equal for the less specialized left hemisphere, the

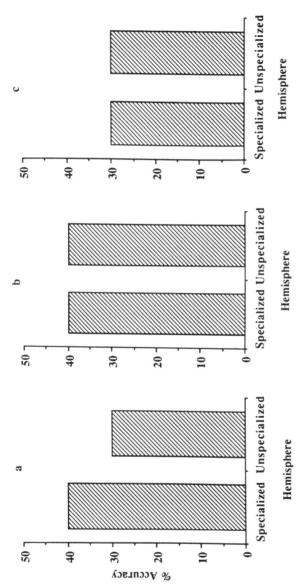

Figure 1. Two possible explanations for a reduced asymmetry in performance. (a) Suppose the typical asymmetry in favor of the specialized hemisphere is 10%. A reduced asymmetry can result from (b) an increase in the level of performance by the unspecialized hemisphere or (c) a decrease in the level of performance by the specialized hemisphere.

reduced asymmetry in women had to reflect a relatively poorer performance by the more specialized right hemisphere. The inference is that the right hemisphere of women is less specialized for spatial functions than the right hemisphere of men, resulting in a greater similarity of the left and right hemispheres for females than males. In other words, spatial functions in this case are not more bilaterally represented in women; rather, the right hemisphere of females is only slightly better than the left hemisphere in spatial functions. Failures of women to show spatial disorders with right-hemisphere damage, therefore, may reflect the fact that the right hemisphere was little better than the left hemisphere premorbidly.

For verbal functions, however, the reduced perceptual asymmetry of women compared to men may arise from a much better performance by the less specialized right hemisphere. Thus, in a study by Bradshaw *et al.* (1977), right and left hemisphere performance was similar in women because language functions were well represented in the right side of the brain. In this case, the reduced asymmetry of women compared to men reflects a greater bilateral representation of language, not a reduced competency of the specialized hemisphere. The relatively good language capacities of the right hemisphere could explain cases where women fail to show verbal disorders following left-side damage or do manifest verbal disorders with right-side damage.

A greater hemispheric similarity in women for certain verbal and spatial functions does not imply that the female brain is more symmetrical than the male brain with respect to *all* functions. In fact, Ladavas, Umilta, and Ricci-Bitti (1980) found that women, but not men, had a right hemisphere superiority in identifying facial expressions. The greater symmetry in male performance resulted from a depression of right hemisphere accuracy compared to female performance, not from an enhancement of left hemisphere performance. These results suggest that the male's right hemisphere lacks the emotional specialization of the female's right hemisphere, just as the female's right hemisphere appears to lack the spatial specialization of the male's right hemisphere. On another task requiring interpretation of facial expression, Strauss and Moscovitch (1981) confirmed the results of Ladavas *et al.* (1980) with respect to both the lack of asymmetry in men and the depression of right hemisphere performance in men as compared to women. These data are concordant with findings from dichotic-listening tasks, where two different auditory inputs are simultaneously presented to the two ears. When melodic patterns or environmental sounds are presented, both of which have an emotional impact, discrimination is better for the left ear/right hemisphere than for the right ear/left hemisphere. However, the superiority of the right hemisphere over the left is larger for women than men (Piazza, 1980).

We conclude that the female brain is more functionally symmetric than the male brain in verbal and spatial processes and less functionally symmetric in emotional processes. The male's right hemisphere, although differing greatly from the left in its spatial and verbal abilities, is only a little better in interpreting emotional information. In contrast, the female's right hemisphere is much better than the left hemisphere at interpreting emotional information but is only slightly superior in spatial ability. Although the female's right hemisphere, like that of the male, lacks expressive speech capacities, certain receptive language functions may have a relatively high level of representation.

The fact that the genders differ in the degree of similarity of the two hemispheres, in the nature of specializations, and in cognition does not, of course, establish that psychological and neurological sex differences are functionally related. Males, on the average, are taller than females and have shorter hair. Hair length and height are two variables that differentiate males and females, yet, within genders, height is not related to hair length. In contrast, height and weight differentiate the genders, and they also covary within genders. The within-gender covariation establishes that the between-gender difference in one variable can partially explain (depending on the within-gender correlation) the between-gender difference in the other variable. To infer that neurological gender differences are functionally related to psychological gender differences, it is necessary to show that within-gender relations between the two variables correspond to gender relations.

WITHIN-GENDER ASSOCIATION BETWEEN COGNITIVE ABILITIES AND PERCEPTUAL ASYMMETRIES

If the greater similarity of the two hemispheres of female brains for verbal and spatial processes is functionally related to their cognitive abilities, then *within* genders, greater similarity between hemispheres should be associated with decreased spatial abilities and increased field dependence. In line with this expectation are data provided by Oltman, Ehrlichman, and Cox (1977) and Zoccolotti and Oltman (1978), who reported that, within genders, the greater the field dependency the smaller the perceptual asymmetry on lateralized tasks. Further, O'Connor and Shaw (1978) and Oltman, Semple, and Goldstein (1979) found that the electroencephalographic patterns of the two hemispheres were more similar when field dependency scores increased, suggesting that the two hemispheres were more in phase for field-dependent than for field-independent people.

In Zoccolotti and Oltman's (1978) study of males, a tachistoscopically presented letter-identification task yielded better right visual field/left hemisphere than left visual field/right hemisphere performance in field-independent people and no asymmetry in field-dependent people. The lack of asymmetry in field-dependent men was related to unusually high performance by the right hemisphere. This outcome is consistent with the conclusion that field dependency is associated with increased language representation in the right hemisphere. On a nonemotional face recognition task, field-independent people had better right hemisphere than left hemisphere performance, and field-dependent people had no asymmetry. However, in this case, the lack of asymmetry in field-dependent people resulted from unusually poor performance by the right hemisphere. These results suggest that for nonemotional visuospatial functions, the specialized right hemisphere of field-dependent people is inferior to that of field-independent people.

Rapaczynski and Ehrlichman (1979) examined a group of women on a tachistoscopically presented face-recognition task similar to that used in men by Zoccolatti and Oltman (1978). As was the case for males, field-independent women had better right hemisphere performance than left. However, field-dependent women, rather than having no asymmetry, had better left hemisphere performance than right. The performance of the left hemisphere did not differ significantly for the two groups, suggesting that visuospatial function in the left hemisphere was not elevated in field-dependent subjects. Rather, the performance of the right hemisphere was much lower for field-dependent than for field-independent subjects. The field-dependency scores of women in Rapaczynski and Ehrlichman's study were considerably higher than the scores of the field-dependent men in Zoccolatti and Oltman's (1978) study. This may explain why field-dependent women performed better with the left than right hemisphere, whereas field-dependent men showed no asymmetry.

Extreme degrees of field dependency may be associated with a relative failure of right hemisphere specialization for spatial function as a result of an unusually high degree of right hemisphere specialization for emotional or language processes. In such cases, a verbal coding of visuospatial stimuli by the left hemisphere may provide a better representation of information than an imagistic coding by the right hemisphere. Thus, the observations from Rapaczynski and Ehrlichman's (1979) study are not inconsistent with those of Zoccolatti and Oltman (1978), since the field-dependent women in the former were more field dependent than the men in the latter.

Sheehan and Smith (1986) examined a group of boys (ages 11–13) on lateralized verbal and spatial tachistoscopic tests and on verbal and spatial ability tests. For each subject, they derived an index of lateralization, based on the two laterality tasks, and examined the relationship between the degree of hemispheric asymmetry and cognitive ability. The verbal factor score (based on verbal fluency, vocabulary, and verbal reasoning) was unrelated to the degree of lateralization, but the spatial factor score, which included a measure of field independence, was strongly predicted by lateralization. The more similar the hemispheres, the lower the spatial ability. Unfortunately, separate performance scores for each visual field were not presented.

The fact that verbal performance was not related to the degree of hemispheric asymmetry in Sheehan and Smith's (1986) study is not surprising. There is no evidence that the normal female superiority in verbal fluency and the quite minor female superiority in vocabulary [0.12 standard deviation relative to within-gender variation (Matarazzo 1972)] are related to the greater bilateral representation of language functions in females than males. Further, even if there were a relationship, the small female superiority in vocabulary is only detectable with very large samples, and the genders do not differ in verbal reasoning. Briefly, had females been included in Sheehan and Smith's (1986) sample, it is unlikely that they would have differed significantly from males on the verbal ability measure.

In contrast, the smaller asymmetry scores for spatial tasks observed in females compared to males result from poorer right hemisphere performance. Females are considerably more field dependent than males, and Sheehan and Smith's (1986) spatial ability measure included an index of field independence as well as measures of other spatial processes in which females are inferior to males. Their finding that hemispheric asymmetry declined as spatial ability decreased within males is to be expected if within-gender relationships are the same as between-gender relationships.

Weak lateralization in association with field dependence has also been demonstrated by Dawson (1977), Pizzamiglio (1974), and Pizzamiglio and Cecchini (1971) on verbal dichotic-listening tasks. These latter studies suggest that as the verbal capacities of the right hemisphere increase (which reduce verbal asymmetry), the spatial specialization of the right hemisphere decreases. Such a relationship may be explained by presuming that as the organizational patterns of the right hemisphere become progressively specialized for verbal processing, they would become progressively less specialized for spatial processing.

The various investigations show that the degree of spatial ability and field independence is positively correlated with the magnitude of performance asymmetries between the hemispheres. When studies reported separate accuracy and speed scores for each hemisphere, small asymmetries on verbal tasks were associated with better right hemisphere performance, whereas small asymmetries on spatial tasks were associated with poorer right hemisphere performance.

We can infer that developmental factors that lead to small asymmetries between hemispheres result in reduced spatial abilities and reduced field independence. If reduced spatial specialization of the right hemisphere is associated with increased emotional specialization, then small hemispheric asymmetries should also be associated with enhanced right hemisphere specialization for interpreting emotional information. Not only does this relationship occur between genders, but as discussed, field dependency is associated with social interest and social competence (Witkin *et al.*, 1979). Unfortunately, no studies to date have investigated the relationship between hemispheric asymmetry within genders and emotional and social understanding.

In summary, the within-gender relationships between hemisphere asymmetry and cognitive patterns appear to be the same as the between-gender relationships. We can therefore conclude that, at least to some degree, cognitive differences between men and women are functionally related to gender differences in hemispheric specialization.

SEX DIFFERENCES IN ANTERIOR AND POSTERIOR CORTICAL ORGANIZATION

Differences in cortical organization of function in normal people can be inferred from patterns of perceptual asymmetries on behavioral tasks. Such differences have also been found in studies of men and women with brain damage. Kimura (1983) found that aphasic disorders were equally common in men with either anterior or posterior left-side lesions. In women, however, aphasia was rarely observed after posterior damage but was as common as in men after anterior damage. In the few female patients who manifested aphasia with damage to the posterior regions, receptive language function was considerably better than in women with anterior lesions. In contrast, males with posterior lesions had poorer receptive functions than those with anterior lesions.

Mateer, Polen, and Ojeman (1982) electrically stimulated regions of the left hemisphere cortex to determine the regions of stimulation that disrupted speech. In 100% of both men and women, stimulation of the frontal speech zone blocked the ability of the patient to name an object. Stimulation in a relatively confined area around this zone also disrupted speech in men but never in women. In men, speech was also disrupted with stimulation in any postcentral region of either the middle or superior temporal gyrus or the inferior parietal cortex. In women, however, the only postcentral stimulation that ever interfered with speech was in a highly restricted zone in the posterior portion of the superior temporal gyrus. These studies show that speech functions are more widely distributed in the left hemisphere of men than women.

The Mateer *et al.* (1982) findings are consistent with those of Kimura (1983), since posterior damage in women would be expected to disrupt speech only if the lesion invaded the critical portion identified by Mateer *et al.* In contrast, for men, speech disorders would be expected with damage to any region of the posterior left hemisphere that included the inferior parietal cortex or any portion of either the middle or superior temporal gyrus.

It is of interest that the large perisylvian cortex, both temporal and parietal, which is important for speech functions in men but not women, includes regions critical for reading. Females are superior to males in reading comprehension and speed (Hitchcock & Pinder, 1974; Schaie & Roberts, 1970), and reading disabilities are much more common in boys than girls (Biller, 1974; Garai & Scheinfeld, 1968). Possibly, much of the female postcentral cortex of the left hemisphere is specialized for skills involved in reading. The comparable regions of males may be specialized more for language in general, including speech and mathematics, but less for reading in particular. Neither reading nor mathematical ability has yet been compared for men and women with posterior left hemisphere damage.

Gender differences in the functional organization of anterior and posterior regions are also present for the right hemisphere. Kimura (1983) found that in male patients, performance on the WAIS Block Design and Object Assembly tasks was equally depressed for those with anterior and posterior damage of the right side of the brain. In female patients, however, performance on these tasks was essentially normal or close to normal with posterior right hemisphere damage and significantly depressed with anterior right hemisphere damage. Clearly, the postcentral regions of the female brain are not without function. As suggested, the postcentral left hemisphere of women may be more highly specialized than the comparable area of men for skills involved in reading. Perhaps the postcentral right hemisphere of females is specialized for interpreting social information, and to a greater extent than is the case for males, just as the posterior right hemisphere of males may be more specialized than that of females for spatial processing.

Although a relationship is present between sex differences in cognition and in the organization of the anterior and posterior cortical regions, its magnitude is relatively small. The sex difference in anterior and posterior cortical organization is profound and larger than the within-gender variation. This relationship stands in dramatic contrast to the small between-gender difference in cognition and hemispheric asymmetry, compared to the large within-gender variance. Factors that generate a massive difference between genders in functional localization in the anterior and posterior portions of the brain do not generate a similarly large gender difference in either cognition or hemispheric asymmetry. The implication is that the structures of neural programs in the two hemispheres of males and females may be quite different yet have quite similar, though not identical, capacities to solve various cognitive problems.

Sex Differences in the Maturation of the Two Hemispheres

Basic patterns of lateral specialization are known to be present even in infancy (Best, Hoffman, & Glanville, 1982; Entus, 1977; Glanville, Best, & Levenson, 1977; Molfese, 1977; Molfese & Hess, 1978). In both male and female infants, the left hemisphere is more responsive to and better able to discriminate verbal stimuli than the right hemisphere, and the reverse holds for musical stimuli. The functional asymmetries present at birth represent different potentialities of the two hemispheres for the development of verbal and nonverbal skills. The differing potentialities, however, do not define either the nature of the verbal and nonverbal processes that will be developed or the degree of functional hemisphere asymmetry that will be present in the adult.

In infantile hemiplegics, in whom one hemisphere is dysfunctional from birth and hemispherectomy is performed in infancy or childhood, the intact hemisphere develops a notable level of bifunctional competence in both the verbal and nonverbal domains. When infantile hemiplegics with left hemispherectomy reach adulthood, speech and other language functions are remarkably normal, and disorders of language can only be detected by rather subtle linguistic tests (Dennis & Kohn, 1975). In infantile hemiplegics with right hemispherectomy, spatial functioning is as good as in normal children up to 10 years of age, and the minor disorders that are present are restricted to spatial processes that emerge after age 10 (Kohn & Dennis, 1974). Thus, even though at birth the two hemispheres have differing potentialities for the development of verbal and nonverbal abilities, neither hemisphere is committed to any fixed level of adult capacities. Each hemisphere can develop either a highly restricted set of abilities very different from those of the other hemisphere or a broader, more general set of abilities similar to those of the other hemisphere.

As each hemisphere matures, new representational strategies develop, which, in turn, affect its organization and capacities. As Levine (1985) has shown, children under the age of 10 fail to show the right hemisphere superiority present in older children and adults for recognizing unfamiliar faces flashed to the lateral visual fields. Performance by the left hemisphere is as good for younger as older children, but performance by the right hemisphere is significantly poorer in younger children. Prior to age 10, the right hemisphere's superior potential for representing unfamiliar faces is not yet established.

Levine (1985) has also shown, however, that even for the youngest children tested (age 7), the right hemisphere is superior, as in older children and adults, for recognizing highly familiar faces of classmates. With sufficient experience, hemispheric potentials can apparently be reached at an earlier age, which illustrates the interactive effects between hemispheric maturity and environmental influences. Thus, the maturity of each hemisphere during development, in interaction with environmental influences, governs both the nature and level of capacities that each hemisphere can display. This conclusion is extremely important in understanding the emergence of adult gender differences in cognition, because evidence is accumulating that the right hemisphere matures faster in males and the left hemisphere matures faster in females.

By 3 months of age, and possibly earlier, boys and girls differ in the functional reactivity of the two sides of the brain. Shucard, Shucard, Cummins, and Campos (1981) examined evoked responses over the two hemispheres to a pair of clicks superimposed over either a music or speech background in 3-month-old infants. Regardless of background, girls showed larger responses over the left than right hemisphere, whereas boys had larger responses over the right than left hemisphere. By 6 months of age (Shucard, Shucard, & Thomas, 1984), male infants continued to show larger right hemisphere evoked responses for either a music-background or a speech-background condition, whereas females showed the adult pattern; i.e., evoked responses to clicks were larger over the left hemisphere with a speech background but larger over the right hemisphere with a music background (Shucard, Shucard, & Thomas, 1977).

The gender differences in evoked response patterns are specific for neutral stimuli (clicks) that are superimposed over a background noise of either speech or music. As stated earlier, newborn boys and girls show the same asymmetry of evoked responses to direct speech or musical stimuli (Molfese, 1977; Molfese & Hess, 1978) and, by a few months of age, the same asymmetry in discriminating speech or musical stimuli (Best *et al.*, 1982; Entus, 1977; Glanville *et al.*, 1977). The basic verbal and nonverbal characteristics of the two hemispheres are present in both genders, as reflected in the electrophysiological response to and discrimination of speech and musical stimuli. However, the click-evoked responses studied by Shucard *et al.* (1981, 1984) in infants index a different aspect of hemispheric function.

When these clicks are superimposed over a complex background, the magnitude of the evoked response will depend on how well the clicks can be discriminated from background complexity. In infants 3 months of age, possible differences between hemispheres in differentiating clicks from speech or clicks from music are minor compared to hemispheric differences in the ability to discriminate the click signal from *any* complex auditory background. Such differences depend on the functional maturity of the two sides of the brain. Only after a sufficient level of maturity has been achieved by both sides of the brain can the more subtle distinction of signal versus speech or signal versus music affect the asymmetry of evoked responses to the click signal.

The evoked-response studies of Shucard *et al.* (1981, 1984) suggest that the female's left hemisphere and the male's right hemisphere predominate in sensory analysis at 3 months of age. By 6 months of age, right hemisphere predominance is still present in males, suggesting a rather pronounced maturational lag of the left hemisphere. The fact that 6-month-old female infants show the normal adult pattern of evoked-response asymmetries for speech and music backgrounds demonstrates that at this age, the female's left hemisphere is evidently far advanced in development compared to that of the male. During the first year of life, it appears that the maturational asymmetry in favor of the right hemisphere in males is considerably more pronounced than the maturational asymmetry in favor of the left hemisphere in females.

In support of this inference are findings of Taylor (1969) on adult patients with temporal lobe epilepsy in association with mesial temporal sclerotic (MTS) lesions induced by febrile convulsions. Retrospective histories show that the majority of such patients have their first seizure prior to the first birthday and that the proportion with later onset of the first seizure declines monotonically with age. Thus, the longer an infant and child is seizure-free, the less likely it becomes that an MTS disorder will develop.

Taylor (1969) found that males with left hemisphere lesions were significantly overrepresented among patients with a seizure onset prior to the first year of age. As the age of onset of the first seizure increased, the overrepresentation of males with left hemisphere lesions declined. These observations indicate that during the first year of life, the male's left hemisphere is less mature in its resistance to seizures than the male right hemisphere or either hemisphere of females. No significant difference was found in the frequency of left and right hemisphere lesions in female patients with a seizure onset prior to the first birthday, suggesting that the maturity levels of the two female hemispheres were considerably more similar than those of males. These data support the suggestion that the observations of Shucard *et al.* (1981, 1984) on evoked response asymmetries in 3- and 6-month-old infants imply a larger discrepancy between the hemispheres of boys (in favor of the right) than of girls.

Maturational asymmetries of the two hemispheres established during infancy continue during the childhood period and can be expected to influence the development of communicative language, symbolic representation, spatial skills, and emotional and social abilities. The kinds of activities in which children engage, the degree of competence they display at various ages for different skills, and the amount of reinforcement they receive are in part determined by how mature the hemispheres are at various stages of development. Environmental factors, in turn, affect the subsequent development and specializations of each hemisphere. Thus, sexual differentiation of cognitive function and hemispheric specialization continues during childhood and is probably heavily influenced by an interaction between the rate of hemispheric maturation and various environmental influences. Studies of children support our suggestion that gender differences in hemispheric maturation, established during the first year of life and present throughout development, condition how boys and girls process incoming information, as we discuss in the next section.

HEMISPHERIC ASYMMETRIES DURING CHILDHOOD

A number of behavioral studies indicate that maturational differences are present in the two hemispheres of school-age boys and girls, especially as reflected in tactile perception by the left and right hands. These gender differences may relate to Conel's finding (cited by Landsdell, 1964) that at 4 years of age, myelinization of axons for the hand-projection region of the cortex is more complete in the right hemisphere of boys and in the left hemisphere of girls.

Based on Hermelin and O'Connor's (1971) observation that blind readers are more proficient in reading Braille with the left than right hand, Rudel, Denkla, and Spalten (1974) examined the two hands of normal children to determine which was more proficient at learning to associate letters with Braille. In boys, there was no hand asymmetry in the youngest group of 7- and 8-year-olds, but, thereafter, a left-hand (right hemisphere) superiority was evident. For girls, a left-hand superiority only emerged at age 11, and prior to that age, either no asymmetry or a right-hand (left hemisphere) superiority was detected.

In interpreting these data, it is essential to keep in mind that in adults, Braille configurations are better represented in the right than left hemisphere. This adult right hemisphere superiority depends on the maturity of its specialized representational systems. Prior to such maturation, the representational strategies of the right hemisphere would be as primitive as those of the left hemisphere. The observations of Rudel *et al.* (1974) therefore imply that an adequate maturity of spatial specialization of the right hemisphere for the Braille-letter task was present by age 9 in boys but was delayed until age 11 in girls. Thus, the maturity level of the right hemisphere of girls is still behind that of boys during school years.

The fact that boys had a left-hand superiority by age 9 does not, however, imply that the male right hemisphere was fully mature by this age. It only implies that, for the particular task given, with its particular level of difficulty, the male right hemisphere was *sufficiently* mature to utilize its specialized capabilities.

Rudel, Denkla, and Hirsch (1977) examined children on a more difficult Braille task than that of Rudel *et al.* (1974), which required the comparing of *two* Braille configurations and deciding whether they were the same. Thus, demands on the specialized spatial abilities of the right hemisphere were greater for this task than for that of Rudel *et al.* (1974). A left-hand superiority on the two-configuration task did not appear until age 11 in boys and age 13 in girls. Thus, with sufficient task difficulty, it was possible to demonstrate that right hemisphere maturation was still discrepant between boys and girls late in development.

Cioffi and Kandel (1979) compared the left and right hands of boys and girls on cross-modal tactovisual tasks. Tactile stimuli were presented to each hand for simultaneous palpation, and children were then asked to select the matching stimuli from a visual display. In one task, the stimuli were two-letter words formed from plastic letters. At all ages tested (from age 7), performance was superior for the right hand (left hemisphere) in both boys and girls. The absence of a gender difference on the word task is not surprising. Most children learn to read at age 6, and at birth both male and female infants show larger evoked responses over the left than right hemisphere to speech stimuli (Molfese, 1977). Even if the male's left hemisphere is less mature than the right, its word-representation capacities would still be expected to exceed those of the right hemisphere by age 7.

On a second task, stimuli consisted of consonant bigrams formed from two plastic letters. This task differs from the first in that the letters did not form words. Nonetheless, they were highly familiar stimuli that could be represented easily either as letter names or as familiar visual images. At all ages, girls showed a right-hand (left hemisphere) superiority, and boys showed a left-hand (right hemisphere) superiority. Overall performance did not differ between genders. Thus, for highly familiar stimuli that could be equally well represented as letter names or as visual images, girls were biased to rely on left hemisphere representation, and boys on right hemisphere representation. These biases may reflect a better linguistic encoding for girls and a better imagistic–spatial encoding for boys.

On a third task, which was developed by Witelson (1976), stimuli consisted of unfamiliar nonsense shapes. On this task, Cioffi and Kandel (1979) found a left-hand (right hemisphere) superiority in both boys and girls from 7 years of age on. Although the task was basically spatial in nature, it was evidently much simpler than even the single-configuration Braille task of Rudel et al. (1974), since on the Braille task, a left-hand superiority emerged for boys only at age 9, and for girls at age 11. Apparently, the spatial specialization of the right hemisphere at age 7 in both genders was sufficiently mature to provide a better representation of the nonsense stimuli than was possible for the left hemisphere.

In contrast to Cioffi and Kandel (1979), Witelson gave children very extensive practice with the shapes prior to the experimental trials. In principle, at least, such practice would allow children to develop verbal descriptions or even names (e.g., "the house shape") for the nonsense stimuli. In Witelson's study, boys had a left-hand (right hemisphere) superiority at the youngest age tested (age 6), whereas girls did not. However, close examination of Witelson's data shows that between ages 10 and 12, the left hand of girls showed an improvement, whereas the right hand did not. As a result, 12-year-old girls had a left-hand advantage very similar to that of boys. Apparently, by age 12, the right hemisphere of girls was superior to the left in representing the nonsense shapes, even though extensive practice probably allowed the left hemisphere to develop verbal codes for the nonsense shapes.

We do not believe that Cioffi and Kandel's (1979) observations for girls conflict with those of Witelson (1976). Rather, we suggest that the extensive practice provided by Witelson, but not by Cioffi and Kandel, allowed the left hemisphere of girls to utilize verbal encoding strategies that could be applied to the nonsense shapes. In consequence, the verbal representations of the left hemisphere were as adequate as the spatial representations of the right hemisphere up through age 10. With continued maturation of the right hemisphere, however, its spatial representations by age 12 surpassed the verbal representations of the left hemisphere in their accuracy in stimulus encoding. The lack of any right-hand improvement for girls between ages 10 and 12 suggests that by age 10, verbal representation of the inherently spatial stimuli had reached its asymptote.

To summarize, a left-hand (right hemisphere) superiority in discriminating Braille configurations emerges later in girls than boys, and later for both genders as spatial complexity increases. A right-hand (left hemisphere) superiority is present for tactile word discrimination in children as young as 7, and the asymmetry does not differ between genders. However, for consonant bigrams, which are equally amenable to verbal and spatial representation, girls have a right-hand (left hemisphere) and boys have a left-hand (right hemisphere) superiority. Thus, boys are biased to represent stimuli spatially, and girls are biased to represent stimuli verbally.

For tactile nonsense shapes, boys as young as 6 or 7 have a left-hand (right hemisphere) superiority regardless of whether they have been given the opportunity to learn verbal encoding strategies or not. Girls, however, manifest a left-hand (right hemisphere) superiority from age 7 onward only if there has been no opportunity to learn verbal representation strategies. When such an opportunity is available, the two hemispheres of girls perform at equal levels up through age 10, although by age 12 the right hemisphere surpasses the left.

The various studies indicate that up to the beginning of adolescence, the right hemisphere is more functionally mature in males than females. They further suggest that when encoding stimuli that are equally well represented by either strategy, girls have a bias to rely on verbal processes of the left hemisphere, and boys a bias to rely on spatial processes of the right hemisphere. In addition, given the opportunity, girls can and do learn to encode inherently spatial stimuli with a verbal strategy. The verbal encoding produces left hemisphere performance as good as right hemisphere performance until the right hemisphere has achieved complete or almost complete maturity. The differing biases of prepubertal boys and girls show that gender differences in hemispheric maturity throughout the developmental period have major effects on how information is processed and represented.

Such biases suggest that during infant and toddler development, verbal communications and interactions between young children and parents would be highly effective in stimulating a left hemisphere development of verbal communicative skills in girls but would be much less effective in boys, since their left hemisphere might not be sufficiently mature to process and encode such information. Since the right hemisphere of girls lags only slightly behind the left, the emphasis on verbal activities might also tend to stimulate the right hemisphere to develop some level of language competency. Simultaneously, skill in understanding nonverbal aspects of social communication and emotion, which go hand in hand with verbal communication, would develop a strong representation in the female right hemisphere. These effects, taking place in preschool years, might influence the subsequent maturation of the female hemispheres and cognitive profiles.

Since the verbal environment is relatively inaccessible to the immature left hemisphere of males at the time when the right hemisphere is maturationally advanced, boys would be expected to center their interest on the physical environment and its mechanics and spatial relations as spatial schemas develop. Their activities would be self-reinforcing and would foster a strong right hemisphere specialization for spatial skills that does not depend on understanding social communications. By the time the left hemisphere becomes sufficiently mature to process verbal communicative information effectively, the right hemisphere would already be committed to a spatial specialization. This might limit the right hemisphere's level of specialization for interpreting social communication, either verbal or emotional.

As it matures, the male's left hemisphere would, of course, develop skill in the communicative aspects of language. However, this communicative understanding might not be reinforced as strongly for boys as for girls, since it was not temporally linked and thereby integrated with the developing specialization in the right hemisphere for emotional information. Further, a high level of spatial specialization in the right hemisphere might encourage the development of a left hemisphere specialization for representing symbolic relations that can be mapped onto the spatial representations of the right hemisphere.

The effects of learning, social interaction, and general environmental stimulation on hemispheric development are dramatically highlighted in the case of Genie, who from the age of 18 months until 13 years was confined to a room and had neither social interaction nor the opportunity to learn language (Curtis, 1977). Objects in the room, as well as environmental sights and sounds through an open window, provided sensory stimulation. As an adult, Genie developed a highly defective language that, among other limitations, lacks syntax and function words and is echolalic. Laterality testing shows a deficiency in left hemisphere processing of all stimuli, and electroencephalographic studies reveal severe dysfunctions in left hemisphere activity. On cognitive tests, Genie is profoundly retarded on all left hemisphere processes. In contrast, Genie scores at the 98th percentile or better on many right hemisphere spatial functions.

Genie's development (Curtis, 1977) reveals not only that her left hemisphere failed to develop in the absence of verbal and social stimulation but that her right hemisphere developed hypernormal spatial abilities. Presumably, Genie's right hemisphere could develop neither a typical female language representation nor a specialization for interpreting nonverbal social and emotional information. The only stimulation available induced a very high level of spatial specialization in the right side of the brain.

In the discussions to follow of the relationships between gonadal hormones in early development and cognition and hemispheric specialization and asymmetry, the foregoing developmental model should be kept in mind. Studies of children and adults with abnormal early hormonal environments strongly suggest that fetal or infant gonadal androgen and estrogen affect the development of the two sides of the brain and, consequently, the development of cognitive abilities. Optimal levels of both androgens and estrogens appear to be necessary for normal neuropsychological development, and departures from the normal range are associated with a variety of cognitive deficits. However, we can relate the effects of hormones to hemispheric maturation rates only by considering the normal gender differences in left and right hemisphere maturation, in the early hormonal environment, in adult cognitive patterns, and in adult specializations of the two hemispheres.

GENETIC AND HORMONAL EFFECTS ON HUMAN NEUROPSYCHOLOGICAL GENDER DIFFERENCES

Studies of cognitive patterns in individuals with genetic or hormonal anomalies provide otherwise unattainable information on the effects of abnormal karyotypes and hormonal environments in humans. Two genetic anomalies that have been studied are men with Klinefelter syndrome (XXY) and women with Turner syndrome (XO).

NEUROPSYCHOLOGICAL PROFILES IN ABNORMAL KARYOTYPES

KLINEFELTER SYNDROME. In Klinefelter syndrome (KS), there is an XXY karyotype, but androgen production and testicular development are normal during infancy (Ratcliffe & Tierney, 1982). At puberty, however, the testes begin to degenerate, androgens fail to reach the normal levels, and some physical feminization occurs.

Compared to normal men, KS men have a normal performance IQ but significant and severe disabilities in a wide spectrum of verbal abilities (Netley & Rovet, 1982a, 1987; Nielsen, Sorensen, Theilgaard, Froland, & Johnsen, 1969; Ratcliffe, Bancroft, Axworthy, & McLaren, 1982; Walzer, Graham, Bashir, & Sibert, 1982). Furthermore, in tests of hemispheric lateralization, KS men show depressed performance for the left hemisphere but normal or better than normal performance for the right hemisphere (Netley & Rovet, 1984, 1987).

Netley and Rovet (1987) hypothesized that abnormal functioning of the left hemisphere in KS men could be related to developmental slowing in the rate of cell division prenatally. Developmental slowing in the KS fetus has been inferred from studies of fingerprint patterns, which show an abnormally low total ridge count (Hreczko & Sigmon, 1980; Netley & Rovet, 1987; Penrose, 1967; Valentine, 1969) typically associated with a slow rate of cell division prenatally (Barlow, 1973; Mittwoch, 1973).

Using total ridge count as an index of developmental slowing, Netley and Rovet (1987) found that KS boys with the lowest ridge counts (slowest early development) manifested the lowest verbal IQ and the most abnormal laterality patterns, whereas those with the highest ridge counts manifested the highest verbal IQ and the most normal laterality patterns.

An association between a slowing in the rate of cell division and abnormal left hemisphere development has also been suggested by Geschwind and Galaburda (1987), who argue that excessive androgenic activity is responsible. Although KS infants do not differ from normal infants in plasma androgen levels, it might be the case that KS individuals manifest higher sensitivity to androgen, at least in some tissues, since they have more autoimmune disorders than normal males (Segami & Alarcon-Segovia, 1978). Increased susceptibility to autoimmune disorders has been linked to increased action of androgens, which suppress the development of the "self-versus-other" discrimination of the immune system (see Geschwind & Galaburda, 1987, for review).

Since infant androgen levels are normal in the KS male (Ratcliffe & Tierney, 1982), some other mechanism, such as an increase in the density of androgen receptors, may lead to excessive androgenic action in early development. Further research is clearly needed, however, to identify the hormonal and genetic mechanisms that underlie the neuropsychological findings. Furthermore, although it is evident that left hemisphere functions are severely affected in KS individuals, there are also data suggesting mild impairments in right hemisphere functions (Nyborg & Nielsen, 1981; Serra, Pizzamiglio, Boari, & Spera, 1978). These results must be reconciled with other findings of normal or superior right hemisphere performance on lateralized tasks.

TURNER SYNDROME. Turner syndrome (TS) is a congenital abnormality in phenotypic females in which one of the two X chromosomes is missing or its short arm is deleted or translocated. The ovaries are typically represented by two dysfunctional streaks of ovarian tissue. In addition to gonadal dysgenesis, these women display a variety of somatic stigmata such as short stature, webbed neck, and skeletal abnormalities. Turner syndrome women can vary greatly in the degree to which they manifest the classical stigmata (Turner, 1960; Simpson, 1975) as well as in the degree to which the gonads are dysfunctional (Lippe, 1982; Grumbach & Conte,

1985). All develop the external genitalia of normal women and are, therefore, reared as such, and some appear in clinics only when they fail to undergo puberty.

As a group of subjects in whom gonadal hormones are essentially absent, TS women have been extensively studied. Originally, TS women were thought to be retarded, based on overall IQ scores (Turner, 1960; Bekker & Van Gemund, 1968). Further research demonstrated, however, that TS women have a normal verbal IQ but very significantly depressed abilities on a wide spectrum of visuospatial tasks, including those in which there is a normal male superiority (e.g., understanding spatial relations) and those in which there is a normal gender equality (e.g., performance IQ, which is standardized to gender equality; see Alexander, Walker, & Money, 1964; Garron, 1977; Money & Alexander, 1966; Shaffer, 1962).

Although verbal IQ is normal in TS women, they show minor and selective disorders of certain verbal functions. First, TS women score lower than normal women in verbal fluency (Heller & Levy, 1986; Money & Alexander, 1966; Waber, 1979). Second, on lateralized verbal tachistoscopic tasks, performance for the right hemisphere does not differ between TS and normal women, but performance for the left hemisphere is significantly lower (Heller & Levy, unpublished manuscript; McGlone, 1985). Heller and Levy (1986) also showed, through an analysis of the pattern of errors, that the depression of left hemisphere performance for TS as compared to normal women was caused by a reduction in linguistic encoding of information (see Levy, Heller, Banich, & Burton, 1983). Heller and Levy (1986) found that verbal fluency was predicted by the level of linguistic encoding of left hemisphere stimuli: as linguistic encoding increased, verbal fluency improved. Verbal reasoning, however, was unrelated to the linguistic encoding of stimuli by the left hemisphere, which indicates that the left hemisphere disorders of TS women may be specific for those verbal processes for which there is a normal female superiority.

On verbal dichotic-listening tests, TS women also have a reduced left hemisphere advantage compared to controls (Gordon & Galatzer, 1980; Netley & Rovet, 1982b; Rovet & Netley, 1983), and in the one study where separate ear performances could be derived (Gordon & Galatzer, 1980), the decreased left hemisphere advantage for TS women resulted from normal right hemisphere performance and depressed left hemisphere performance. The normal right hemisphere performance of TS women on verbal laterality tasks shows that their right hemisphere has no more verbal capacity than the right hemisphere of normal women. The spatial disorders in TS women are therefore not a consequence of an unusual degree of verbal specialization in the right hemisphere.

In brief, TS women appear to have a disorder of right hemisphere spatial function compared to either normal men or women and, in addition, a more selective disorder of certain left hemisphere verbal processes as compared to normal women. Their verbal reasoning does not differ from that of normal men and women, but they have a deficit in verbal fluency compared to normal women and decreased linguistic encoding of verbal stimuli by the left hemisphere. The level of linguistic encoding by the left hemisphere, furthermore, is predictive of verbal fluency.

Some authors have argued that the cognitive differences between TS and normal women are unlikely to be caused by the effects of hormones. They claim that the normal fetal and infant ovaries are inactive and do not contribute to normal female differentiation (see Döhler, 1978; Rovet & Netley, 1982). This view has been challenged, however, by many studies (see Bidlingmaier, Strom, Dorr, Eisenmenger, &

Knorr, 1987). The evidence is strong that significant amounts of gonadal hormones (androgens and estrogens) circulate in both sexes during early postnatal life (see MacLusky & Naftolin, 1981), and the first 6 months of life are associated with fluctuating, and often high, plasma levels of estradiol in girls (Winter, Hughes, Reyes, & Faiman, 1976). Furthermore, the theca interna of fetal ovarian follicles yields the same strong alkaline phosphatase reaction as shown by these cells in ovaries of mature women, a reaction that has been related directly to gonadotropic stimulation and steroidogenesis (Ross & Vande Wiele, 1974). In addition, Bidlingmaier *et al.* (1987) report that gonadal estrogen concentrations in female infants paralleled changes in gonadal morphology. In particular, ovarian weights varied as a function of estrogen concentration, and the biggest ovaries contained multiple macroscopic cysts.

In normal women, a small number of genes are thought to specify gonadal differentiation, with consequent phenotypic sexual dimorphism organized almost exclusively by gonadal products (i.e., androgens, estrogens, Müllerian duct inhibitor; Haseltine & Ohno, 1981). Somatic gene activity in TS women and in normal women is virtually identical, since in normal women a large portion of one of the X chromosomes is inactivated early in embryogenesis (Gordon & Ruddle, 1981). One might argue that if the second X chromosome in normal females is truly inactive, TS women ought to be phenotypically normal. Although most regions are inactivated, two loci that mediate gonadal differentiation escape inactivation (Gordon & Ruddle, 1981).

Therefore, although it is possible that the loss of genetic activity of the second X chromosome acts through a nonhormonal mechanism to produce the cognitive abnormalities of TS women, it seems equally probable that they reflect the lack of gonadal hormones in early development. Evidence that the cognitive differences between TS and normal women may be hormonally mediated is strengthened by studies of individuals with other hormonal disorders, which indicate that cognitive abilities are affected by gonadal hormone status, even in the presence of a normal karyotype.

Neuropsychological Profiles in Abnormal Hormonal Environments

Androgen-Insensitivity Syndrome. Women with the androgen-insensitivity (AI) syndrome have an XY male karyotype and normal androgen production by the testes, but cells are deficient in androgen receptors. Androgen itself, therefore, cannot regulate cell differentiation. The fetus develops female external genitalia, and testicular estrogens induce a feminizing puberty. At puberty the testes are typically removed, and AI women are maintained on estrogen therapy. Their juvenile and adult behaviors are stereotypically feminine (Money, Schwartz, & Lewis, 1984).

Masica, Money, Ehrhardt, and Lewis (1969) assessed verbal and performance IQ in 15 girls and women from 12 families with a complete testicular feminizing syndrome. With IQs averaged within sets of siblings to provide a single performance and verbal IQ score for each of the 12 families, mean performance IQ was 99.6 and mean verbal IQ was 111.7. The 12.1-IQ-point offset in verbal and performance IQ was significant, and in 10 of the 12 cases, verbal IQ exceeded performance IQ. In one case, this was reversed, and in one case, verbal and performance IQs were equal. In a single case report, Spellacy, Bernstein, and Cohen (1965) describe an AI girl whose verbal IQ was 128 and performance IQ was 105.

Unfortunately, neither Masica *et al.* (1969) nor Spellacy *et al.* (1965) had control groups, and their data therefore can only be compared to standardizations for the population. These standardizations show that for both males and females, the average discrepancy in verbal and performance IQ is zero. The discrepancy shown by AI women, therefore, deviates equally from those of normal females and normal males. Without a matched control group, it is not possible to know whether the verbal–performance discrepancy in AI women was caused by an increased verbal IQ, a reduced performance IQ, or both.

The discrepancy between verbal and performance IQ does indicate, however, that visuospatial and visuomotor abilities are inferior to verbal abilities in AI women. Since these are abilities typically associated with the right hemisphere, these data suggest that right hemisphere function may be impaired relative to left when androgen is depleted or when its actions are prevented.

IDIOPATHIC HYPOGONADOTROPIC HYPOGONADISM. Hier and Crowley (1982) examined verbal and spatial abilities in 19 men with idiopathic hypogonadotropic hypogonadism (IHH) as compared to 19 normal men and five men who developed acquired hypogonadotropic hypogonadism following a normal virilizing puberty (postpubertal AHH). These patients have a normal 46,XY karyotype and undergo apparently normal masculinization *in utero,* a process that is presumably mediated by maternal gonadotropins. Puberty does not occur, however, because of a deficiency in gonadotropin-releasing factor. Depending on the severity of the disorder, the testes vary in size and in androgen production. Thus, in the IHH male, prenatal androgen levels appear relatively normal because of maternal gonadotropins, but androgen levels during infancy are extremely low.

The IHH men in Hier and Crowley's (1982) study all had a normal XY karyotype, a failure of pubertal development by age 18, a normal pituitary as inferred from x-ray analysis of the sella turcica, small testes, and defective virilization prior to androgen therapy. They had been withdrawn from androgen therapy for 3 months prior to psychometric assessment. These men showed a marked decrement in spatial skills compared to both normal men and males who developed the androgen deficiency after puberty. On the Block Design subtest of the Wechsler performance scale, IHH men were a full standard deviation below control subjects. On a test of field independence (Embedded Figures), they required almost twice as long (69.8 seconds) to complete the task as normal men (37.4 seconds) or AHH men (37.8 seconds). Furthermore, in IHH men, testicular volume (which is indicative of androgen production) was uncorrelated with verbal ability but was correlated with spatial ability such that the more severe the androgen deficiency, the greater the spatial deficit. Even when the androgen deficiency was corrected through replacement therapy, spatial ability did not improve.

Equivalent standardized scores were available for the Block Design subtest and the three verbal tests in Hier and Crowley's (1982) study. For the 19 control and five AHH men, the average verbal ability score was 0.19 standard deviations above the Block Design score (equivalent to 2.8 IQ points). However, for IHH men, the average verbal ability score was 0.84 standard deviations above the Block Design score (equivalent to 12.6 IQ points), which is very similar to the observations of Masica *et al.* (1969) in AI women. Further, the verbal and performance IQs of IHH men (based on a prorated score) were practically identical to those of AI women in the Masica *et al.* (1969) study.

It is relevant to note that in the standardization population, the difference for males between the three verbal tasks that Hier and Crowley (1982) administered and the Block Design subtest is equivalent to 1.28 IQ points in favor of the verbal tasks, and the difference for females is equivalent to 3.15 IQ points in favor of the verbal tasks (Matarazzo, 1972). Thus, the difference for IHH males in favor of verbal performance is not only greater than for the male control subjects but also greater than for females in the standardization population.

The studies reviewed above suggest that right hemisphere spatial functions are more poorly developed in both AI women and IHH men than in normal women. Normal women simply do not show anything like a 12-point IQ discrepancy between verbal and performance IQ or between performance on the three verbal subtests that Hier and Crowley (1982) examined and the Block Design subtest.

THE CONGENITAL ADRENAL HYPERPLASIA SYNDROME.   A common androgen disorder in early development is congenital adrenal hyperplasia (CAH), in which the fetal adrenals secrete abnormally high levels of androgen. When detected, usually at birth, medical intervention is typically initiated to normalize adrenal corticoid output. In CAH children, therefore, postnatal androgen status can be within normal limits. Although CAH children have been reported to be masculinized in various emotional and personality variables as compared to control children (see Hines, 1982, for review), investigations of cognitive function have generally found few differences between CAH girls and others.

Early studies that claimed that CAH female subjects possessed higher than normal intellectual function (Dalton, 1968, 1976; Ehrhardt & Money, 1967; Money & Lewis, 1966) were not confirmed when appropriate control samples were used (Baker & Ehrhardt, 1974; McGuire, Ryan, & Omenn, 1975; Perlman, 1973). With respect to specific abilities, Lewis, Money, and Epstein (1968) report that verbal and performance IQs were equal in CAH subjects, which indicates no selective enhancement or depression of left or right hemisphere functions.

The CAH males and CAH females could not be distinguished from control subjects on any of a set of verbal and spatial cognitive tests (McGuire *et al.*, 1975). In a small sample, Perlman (1973) found that, compared to normal girls, CAH girls performed less well on verbal IQ, on arithmetic tasks, and on Block Design tasks but obtained higher scores in a picture completion task, whereas CAH boys did not differ from normal boys. Perlman's (1973) observations for girls reveal no interpretable pattern and, in addition, conflict with negative outcomes in other studies that used identical tasks.

Two studies report meaningful differences between CAH and normal subjects. In Baker and Ehrhardt's (1974) research, both CAH males and CAH females manifest lower mathematics ability than their unaffected siblings. Resnick and Berenbaum's (1982) research found that CAH girls and women were superior to normal women in spatial tasks for which men are generally superior to women. Thus, Resnick and Berenbaum's female CAH subjects manifested a cognitive masculinization. The reliability of Baker and Ehrhardt's and Resnick and Berenbaum's observations can only be determined from future investigations. However, the inferences drawn from the two studies are in opposite directions. Spatial and mathematical abilities are generally positively correlated, which makes it surprising that CAH children were simultaneously deficient in mathematics ability and superior in spatial ability. Conceivably, CAH children are highly variable in cognitive profiles, depending on the severity and timing of the androgen disorder and on their postnatal hormonal status. It should also be noted that the androgen environment to which CAH children are exposed is not that typical of normal males, which consists predominantly of the testicular hormone, testosterone. Adrenal androgens may not be comparable in their effects.

PRENATAL EXPOSURE TO DIETHYLSTILBESTROL. Excessive estrogen levels in early development are present in two conditions. Prenatally, fetuses are exposed to an abnormal estrogen concentration when mothers have been treated with diethylstilbestrol (DES) in an attempt to control a threatened abortion. Postnatally, excessive estrogens accumulate as a result of a disorder of liver metabolism typically, but not always, in association with protein starvation during infancy (kwashiorkor).

Before it was discovered to have many adverse effects on development, DES, a synthetic estrogenic compound, was prescribed to a large number of pregnant women at risk for abortion. Although studies have shown that prenatal exposure of human fetuses to synthetic progestin or estrogen affects various emotional and personality characteristics (see Hines, 1982, for review), most studies have found no effects on cognitive abilities. Reinisch and Karow (1977) differentiated between subjects who had been exposed predominantly to prenatal estrogen versus prenatal progestin and compared the two groups to their siblings on verbal IQ, performance IQ, and verbal, spatial, and attentional cognitive factors. The two groups of affected individuals did not differ, and neither differed from sibling controls on any of the measures. No cognitive effects of abnormal prenatal exposure to estrogen or progestin were reported by Yalom, Green, and Fisk (1973), Kester, Green, Finch, and Williams (1980), or Hines (1981). Hines (1981) reported, however, that DES-exposed women showed a larger asymmetry in favor of the left hemisphere on a verbal dichotic-listening test than did control women, even though she found no cognitive differences between the two groups.

KWASHIORKOR. In kwashiorkor excessive estrogens accumulate as a consequence of a disorder in liver metabolism. Even with subsequent diet correction, symptoms of kwashiorkor are resistant to cure, and gynecomastia (breast enlargement) as well as other signs are frequent in the adult man (Davies, 1947).

We are aware of only one cognitive study of men with symptoms of kwashiorkor. Dawson (1966) studied cognitive characteristics of a group of men with gynecomastia and a large control group. Both gynecomastic ($N = 10$) and control ($N = 149$) men were apprentices working at an iron-ore mine in Sierra Leone. The two groups did not differ in general intelligence, but men with gynecomastia were significantly superior on a measure of verbal ability (by 0.40 S.D. relative to within-group variance) and significantly inferior in spatial ability, as indexed by a block design task (by 0.65 S.D. relative to within-group variance) and a test of spatial relations (by 0.75 S.D. relative to within-group variance). The nature of the verbal ability measure is unknown, since Dawson (1966) only says that it is the verbal portion of an intelligence test constructed in terms of the Sierra Leone indigenous culture. The offset in standard units between groups cannot be compared to the typical gender offset, since no data are available for the Sierra Leone culture.

CONCLUSIONS AND SPECULATIONS

The studies reviewed above suggest that in both normal males and females, right hemisphere maturation and the consequent development of spatial abilities may be dependent on stimulating effects of gonadal hormones, particularly androgen but also estrogen, in early development. When gonadal hormones are absent (TS women), depleted (IHH males), or otherwise reduced in potency (AI women), spatial abilities suffer. Furthermore, the results from IHH males suggest that infancy may be a crucial period for these hormonal influences, since prenatal androgens are apparently normal, and androgen replacement therapy in later years did not enhance their spatial abilities. This inference is supported by observations of relatively high levels of plasma gonadal hormones during the first 6 months of life in normal infants.

Whether androgenic and estrogenic activity is totally absent in infancy or reduced compared to that in normal babies, general verbal intelligence appears to be unaffected. Since men and women do not differ in verbal intelligence, the lack of effects would be expected. However, since TS women are impaired in both verbal fluency and left hemisphere linguistic development, which favor females, the question is raised regarding the possibility that estrogen may have a selective feminizing effect on the development of the left hemisphere. Infant ovarian hormones may have a restricted effect on the development of those left hemisphere specializations that are better organized in women than men. A pattern of better verbal abilities in association with abnormally high estrogen levels during early development in kwashiorkor suggests that estrogen may stimulate left hemisphere development, possibly, in these cases, by inhibiting effects of androgen.

More research is needed to clarify the effects of estrogen versus androgen on development. However, the hypothesis that estrogens act to feminize certain aspects of cognitive function is not without precedent in animal studies, which have shown a number of nonreproductive behaviors to be feminized by estrogen in infancy. Male rats are superior to female rats in complex maze learning, which in part is related to greater open-field activity in females and in part to superior spatial memory in males (see Beatty, 1984). Open-field exploration is more dependent on the left than the right cerebral hemisphere in rats, since it is dramatically reduced after ligation of the left but not the right middle cerebral artery (Robinson, 1979).

Estrogens administered during infancy impair maze learning in male rats (Dawson, Cheung, & Law, 1975), and neonatal ovariectomy reduces open-field activity in females (Blizard & Denef, 1973; Steward & Cygan, 1980) unless replacement estrogens are given during infancy (Stewart & Cygan, 1980). Steward and Cygan demonstrated that feminine open-field behavior is preserved in rats ovariectomized during the first days of life if low doses of exogenous estrogen are supplied between the 10th and 20th postnatal days or either low or high doses are supplied between the 20th and 30th days.

The foregoing data are not consistent with the traditional concept that brain differentiation in mammals follows a feminine pattern unless sufficient levels of androgens are present during initial periods in early development (see Dohler *et al.*, 1984; Toran-Allerand, 1984, for reviews). Similarly, the evidence reviewed from humans suggests that physiological levels of estrogen in infancy may function to feminize certain neurological structures. In particular, the data suggest the possibility that under some circumstances, an absence of estrogen may lead to a lack of feminine development of the left hemisphere.

To summarize our reading of this literature, the various studies are consistent with a model of cortical development in which the right hemisphere is highly dependent on and sensitive to the stimulating effects of gonadal steroids in infancy, and in which the effects of androgen are more potent than the effects of estrogen.

In contrast to the strong steroid dependence of right hemisphere spatial development, we suggest that left hemisphere cortical maturation proceeds in the absence of gonadal steroids in infancy but may be slowed by androgen (as suggested by Geschwind & Galabunda, 1987) and may be stimulated by estrogen. Within the normal range of male and female infant gonadal steroid levels, or when these hormones are depleted, variations in the rate of left hemisphere maturation have no effects on general verbal intelligence, which does not differ between sexes. The normal slowing effects of androgen, under this model, in conjunction with a stimulating effect on the right hemisphere, lead to faster right than left hemisphere development in normal males. This, in turn, results, through interactions with environmental influences, in the typical male pattern of left hemisphere specialization for symbol manipulation in mathematics and reduced verbal communicative skills as compared to females, and to right hemisphere specialization for spatial skills and reduced social–emotional abilities as compared to females.

The foregoing considerations help to explain why within-gender variations in cognition and hemispheric asymmetry are so large compared to between-gender differences in spite of the very large gender difference in infant gonadal steroids. Even minor variations in infant gonadal steroids could lead to major changes in hemispheric maturation rates, so that a fairly large fraction of males have a maturation pattern as feminine as that of the average female, and a similar fraction of females have a maturation pattern as masculine as that of the average male. If, as we argue, hemispheric maturational rates govern adult cognitive profiles, hemispheric specialization, and hemispheric asymmetry, then the average gender difference in these rates underlies cognitive gender differences, and within-gender variations in these rates underlie within-gender cognitive variations. The fact that hemispheric asymmetry and specialization have the same relation to cognition within and between genders then becomes completely comprehensible.

Finally, in spite of the relative gender similarity in hemispheric specialization, hemispheric asymmetry, and cognition, the gender difference in posterior and anterior cortical organization is extremely large, larger than the within-gender variation. Selection pressures demand gender differences in behaviors directly related to reproductive success, but a secondary result is that neural organization in general is regulated by ovarian versus testicular hormones in females and males. The fact that the between-gender difference in cognition is quite small compared to within-gender variations or to the neurological gender difference means, however, that cognitive demands on men and women throughout human evolution differed only to a minor degree. The central problem of evolution was not, therefore, to generate this minor difference in cognition, given an obligatory major difference in male and female neural programs, but was rather to mold the different male and female programs to serve similar cognitive functions. In brief, the observations strongly indicate that similar selective forces acted on the cognitive evolution of men and women to produce a functional convergence of two very different structures of neural programs in the two genders. Although selective pressures could not produce similar patterns of functional localization within the hemispheres in men and women, they could and did produce similar, but not identical, hemispheric maturation rates and functional cognitive capacities.

Thus, we conclude that early gonadal hormones influence the development of gender differences in cognition. However, in spite of such hormonal influences, and the consequent major sex differences in certain aspects of the organization of the

cortex, the cognitive functions controlled by the different underlying neural programs are highly similar for men and women.

Acknowledgments

We thank the Spencer Foundation for a grant to J. Levy.

REFERENCES

Alexander, D., Walker, H. T., Jr., & Money, J. (1964). Studies in direction sense. Turner's syndrome. *Archives of General Psychiatry, 10,* 337–339.

Anastasi, A. (1958). *Differential psychology.* New York: Macmillan.

Baker, S. W., & Ehrhardt, A. A. (1974). Prenatal androgen, intelligence and cognitive sex differences. In R. C. Friedman, R. N. Richart, & R. L. Vande Wiele (Eds.), *Sex differences in behavior* (pp. 33–51). New York: John Wiley & Sons.

Barlow, P. (1973). The influence of inactive chromosomes on human development. *Humangenetik, 17,* 105–136.

Basso, A., Cupitani, E., & Monaschini, S. (1982). Sex differences in recovery from aphasia. *Cortex. 18,* 469–475.

Beatty, W. W. (1984). Hormonal organization of sex differences in play fighting and spatial behavior. In G. J. De Vries, J. P. C. De Bruin, H. B. M. Uylings, & M. A. Corner (Eds.), *Sex differences in the brain, The relation between structure and function, Vol. 61, Progress in brain research* (pp. 315–330). Amsterdam: Elsevier.

Bekker, F. J., & Van Gemund, J. J. (1968). Mental retardation and cognitive deficits in XO Turner's syndrome. *Maandschrift voor Kindergeneeskunde, 36,* 148–156.

Benbow, C. M., & Benbow, R. M. (1984). Biological correlates of high mathematical reasoning ability. In G. J. De Vries, J. P. C. De Bruin, H. B. M. Uylings, & M. A. Corner (Eds.), *Sex differences in the brain, The relation between structure and function, Vol. 61, Progress in brain research* (pp. 469–490). Amsterdam: Elsevier.

Best, C. T., Hoffman, H., & Glanville, B. B. (1982). Development of infant ear asymmetries for speech and music. *Perception and Psychophysics, 31,* 75–85.

Bidlingmaier, F., Strom, T. M., Dorr, H. G., Eisenmenger, W., & Knorr, D. (1987). Estrone and estradiol concentrations in human ovaries, testes, and adrenals during the first two years of life. *Journal of Clinical Endocrinology and Metabolism, 65,* 862–867.

Biller, H. B. (1974). Paternal deprivation, cognitive functioning, and the feminized classroom. In A. Davids (Ed.), *Child personality and psychopathology: Current topics* (pp. 11–52). New York: John Wiley & Sons.

Blizard, D. A., & Denef, C. (1973). Neonatal androgen effects on open-field activity and sexual behavior in the female rat: The modifying influence of ovarian secretions during development. *Physiology and Behavior, 11,* 65–69.

Bradshaw, J. L., & Gates, A. (1978). Visual field differences in verbal tasks: Effects of task familiarity and sex of subject. *Brain and Language, 5,* 166–187.

Bradshaw, J. L., & Nettleton, N. C. (1983). *Human cerebral asymmetry.* Englewood Cliffs, NJ: Prentice-Hall.

Bradshaw, J. L., Gates, A., & Nettleton, N. (1977). Bihemispheric involvement in lexical decisions: Handedness and a possible sex difference. *Neuropsychologia, 15,* 277–286.

Bryden, M. (1979). Evidence for sex differences in cerebral organization. In M. Wittig & A. Petersen (Eds.), *Sex-related differences in cognitive functioning: Developmental issues* (pp. 121–139). New York: Academic Press.

Butler, S. (1984). Sex differences in cerebral function. In G. J. De Vries, J. P. C. De Bruin, H. B. M. Uylings, & M. A. Corner (Eds.), *Sex differences in the brain. The relation between structure and function, Vol. 61, Progress in brain research* (pp. 443–455). Amsterdam: Elsevier.

Chase, T. N., Fedio, P., Foster, N. L., Brooks, R., Di Chiro, G., & Mansi, L. (1984). Wechsler adult intelligence scale performance: Cortical localization by fluorodeoxyglucose F[18]-positron emission tomography. *Archives of Neurology, 41,* 1244–1247.

Cioffi, J., & Kandel, G. (1979). Laterality of stereognostic accuracy of children for words, shapes, and bigrams: A sex difference for bigrams. *Science, 204,* 1432–1434.

Curtis, S. (1977). *Genie: A psycholinguistic study of a modern-day "wild child."* New York: Academic Press

Dalton, K. (1968). Ante-natal progesterone and intelligence. *British Journal of Psychiatry, 114,* 1377–1382.

Dalton, K. (1976). Prenatal progesterone and educational attainments. *British Journal of Psychiatry, 129,* 438–442.

Davidoff, J. (1977). Hemispheric differences in dot detection. *Cortex, 13,* 434–444.

Davies, J. N. P. (1947). Pathology of Central African natives; Mulago Hospital post-mortem studies. *East African Medical Journal, 24,* 180.

Dawson, J. L. M. (1977). An anthropological perspective on the evolution and lateralization of the brain. In S. J. Dimond and D. A. Blizard (Eds.), *Evolution and lateralization of the brain, Vol. 299, Annals of the New York Academy of Sciences* (pp. 424–447). New York: The New York Academy of Sciences.

Dawson, J. L. M. (1966). Kwashiorkor, gynaecomastia, and feminization processes. *The Journal of Tropical Medicine and Hygiene, 69,* 175–179.

Dawson, J. L. M., Cheung, Y. M., & Law, R. T. S. (1975). Developmental effects of neonatal sex hormones on spatial and activity skills in the white rat. *Biological Psychology, 3,* 213–229.

Dennis, M., & Kohn, B. (1975). Comprehension of syntax in infantile hemiplegics after cerebral hemidecortication: Left hemisphere superiority. *Brain and Language, 2,* 472–482.

Döhler, K.-D. (1978). Is female sexual differentiation hormone mediated? *Trends in Neuroscience, 1,* 138–140.

Döhler, K.-D., Hancke, J. L., Srivastava, S. S., Hoffman, C., Shryne, J. E., & Gorski, R. A. (1984). Participation of estrogens in female sexual differentiation of the brain: Neuroanatomical, neuroendocrine, and behavioral evidence. In G. J. De Vries, J. P. C. De Bruin, H. B. M. Uylings, & M. A. Corner (Eds.), *Sex differences in the brain. The relation between structure and function, Vol. 61, Progress in brain research* (pp. 99–117). Amsterdam: Elsevier.

Edwards, S., Ellams, J., & Thompson, J. (1976). Language and intelligence in dysphasia: Are they related? *British Journal of Disorders of Communication, 11,* 83–114.

Ehrhardt, A. A., & Money, J. (1967). Progestin-induced hermaphroditism: IQ and psychosexual identity in a sample of 10 girls. *Journal of Sex Research, 2,* 83–100.

Entus, A. (1977). Hemispheric asymmetry in processing of dichotically presented speech and nonspeech stimuli by infants. In S. Segalowitz & F. Gruber (Eds.), *Language development and neurological theory* (pp. 63–73). New York: Academic Press.

Garai, J. E., & Scheinfield, A. (1968). Sex differences in mental and behavioral traits. *Genetic Psychology Monographs, 77,* 169–299.

Garron, D. C. (1977). Intelligence among persons with Turner's syndrome. *Behavioral Genetics, 7,* 105–127.

Geschwind, N., & Galaburda, A. M. (1987). *Cerebral lateralization: Biological mechanisms, associations, and pathology.* Cambridge, MA: The MIT Press.

Glanville, B. B., Best, C. T., & Levenson, R. (1977). A cardiac measure of cerebral asymmetries in infant auditory perception. *Developmental Psychology, 13,* 55–59.

Gordon, H. W., & Galatzer, A. (1980). Cerebral organization in patients with gonadal dysgenesis. *Psychoneuroendocrinology, 5,* 235–244.

Gordon, J. W., & Ruddle, F. H. (1981). Mammalian gonadal determination and gametogenesis. *Science, 211,* 1265–1272.

Grumbach, M. M., & Conte, F. A. (1985). Disorders of sexual differentiation. In J. D. Wilson & D. W. Foster (Eds.), *Williams' textbook of endocrinology* (7th ed.). Philadelphia: W. B. Saunders.

Hannay, H., & Boyer, C. (1978). Sex differences in hemispheric asymmetry revisited. *Perceptual and Motor Skills, 47,* 317–321.

Hannay, H., & Malone, D. (1976). Visual field effects and short-term memory for verbal material. *Neuropsychologia, 14,* 203–209.

Haseltine, F. P., & Ohno, S. (1981). Mechanisms of gonadal differentiation. *Science, 211,* 1272–1278.

Heller, W., & Levy, J. (1986). Deficits in left-hemisphere linguistic function in Turner's syndrome. *Society for Neuroscience Abstracts 12,* 1442.

Hermelin, B., & O'Connor, N. (1971). Functional asymmetry in the reading of Braille. *Neuropsychologia 9,* 431–435.

Hier, D. B., & Crowley, W. F., Jr. (1982). Spatial ability in androgen-deficient men. *The New England Journal of Medicine, 306,* 1202–1205.

Hines, M. (1981). *Prenatal diethylstilbestrol (DES) exposure, human sexually dimophic behavior and cerebral lateralization.* Doctoral dissertation, University of California, Los Angeles. *Dissertation Abstracts International, 42,* 423B.

Hines, M. (1982). Prenatal gonadal hormones and sex differences in human behavior. *Psychological Bulletin, 92,* 56–80.

Hitchcock, D. C., & Pinder, G. D. (1974). *Reading and arithmetic achievement among youths 12–17 years as measured by the Wide Range Achievement Test: Vital and health statistics—Series 11—No. 136.* Washington, DC: U. S. Government Printing Office.

Hreczko, T., & Sigmon, B. (1980). The dermatoglyphics of a Toronto sample of children with XXY, XYY and XXX aneuploides. *American Journal of Physical Anthropology, 52,* 33–42.

Inglis, J., Ruckner, M., Lawson, J. S., MacLean, A. W., & Monga, T. N. (1982). Sex differences in the cognitive effects of unilateral brain damage. *Cortex, 18,* 277–286.

Johnson, M., & Everitt, B. (1980). *Essential reproduction.* Boston: Blackwell Scientific Publications.

Kail, R., & Siegel, A. (1978). Sex and hemispheric differences in the recall of verbal and spatial information. *Cortex, 14,* 557–563.

Kershner, J. R., & Jeng, A. G. (1972). Dual functional hemispheric asymmetry in visual perception: Effects of ocular dominance and postexposural processes. *Neuropsychologia, 10,* 437–445.

Kester, P., Green, R., Finch, S. J., & Williams, K. (1980). Prenatal "female hormone" administration and psychosexual development in human males. *Psychoneuroendocrinology, 5,* 269–285.

Kimura, D. (1983). Sex differences in cerebral organization for speech and praxic function. *Canadian Journal of Psychology, 37,* 19–35.

Kimura, D., & Harshman, R. A. (1984). Sex differences in brain organization for verbal and nonverbal functions. In G. J. De Vries, J. P. C. De Bruin, H. B. M. Uylings & M. A. Corner (Eds.), *Sex differences in the brain. The relation between structure and function, Vol. 61, Progress in brain research* (pp. 423–441). Amsterdam: Elsevier.

Kohn, B., & Dennis, M. (1974). Selective impairments of visuospatial abilities in infantile hemiplegics after right cerebral hemidecortication. *Neuropsychologia, 12,* 505–512.

Lavadas, E., Umilta, C., & Ricci-Bitti, P. E. (1980). Evidence for sex differences in right-hemisphere dominance for emotions. *Neuropsychologia, 18,* 361–366.

Lansdell, H. (1964). Sex differences in hemispheric asymmetries of the human brain. *Nature, 203,* 550.

Leehey, S. C., Carey, S., Diamond, R., & Cahn, A. (1978). Upright and inverted faces: The right hemisphere knows the difference. *Cortex, 14,* 411–419.

Levin, M. (1980). The feminist mystique. *Commentary, 70,* 25–30.

Levine, S. C. (1985). Developmental changes in right-hemisphere involvement in face recognition. In C. T. Best (Ed.), *Hemispheric function and collaboration in the child* (pp. 157–191). New York: Academic Press.

Levy, J. (1981). Letter. *Commentary, 71,* 10–11.

Levy, J., & Reid, M. (1978). Variations in cerebral organization as a function of handeness, hand posture in writing, and sex. *Journal of Experimental Psychology: General, 107,* 119–144.

Levy, J., Heller, W., Banich, M. T., & Burton, L. A. (1983). Are variations among right-handed individuals in perceptual asymmetries caused by characteristic arousal differences between hemispheres? *Journal of Experimental Psychology: Human Perception and Performance, 9,* 329–359.

Lewis, V. G., Money, J., & Epstein, R. (1968). Concordance of verbal and nonverbal ability in the adrenogenital syndrome. *Johns Hopkins Medical Journal, 122,* 192–195.

Lippe, B. (1982). Primary ovarian failure. In S. A. Kaplan (Ed.), *Clinical pediatric and adolescent endocrinology* (pp. 269–299). Philadelphia: W. B. Saunders.

Maccoby, E. E., & Jacklin, C. N. (1974). *The psychology of sex differences.* Stanford: Stanford University Press.

MacLusky, N. S., & Naftolin, F. (1981). Sexual differentiation of the nervous system. *Science, 211,* 1294.

Masica, D. N., Money, J., Ehrhadt, A. A., & Lewis, V. G. (1969). I.Q., fetal sex hormones and cognitive patterns: Studies in the testicular feminizing syndrome of androgen insensitivity. *Johns Hopkins Medical Journal, 124,* 34–43.

Matarazzo, J. D. (1972). *Wechsler's measurement and appraisal of adult intelligence* (5th ed.). Baltimore: Williams & Wilkins.

Mateer, C. A., Polen, S. B., & Ojemann, G. A. (1982). Sexual variation in cortical localization of naming as determined by stimulation mapping. *The Behavioral and Brain Sciences, 5,* 310–311.

McGlone, J. (1977). Sex differences in human brain asymmetry: A critical survey. *The Behavioral and Brain Sciences, 3,* 215–263.

McGlone, J. (1980). Sex differences in human brain asymmetry: A critical review. *The Behavioral and Brain Sciences, 3,* 214–263.

McGlone, J. (1985). Can spatial deficits in Turner's syndrome be explained by focal CNS dysfunction of atypical speech laterization? *Journal of Clinical and Experimental Neuropsychology, 7,* 375–394.

McGlone, J., & Davidson, W. (1973). The relationship between cerebral speech laterality and spatial ability with special reference to sex and hand preference. *Neuropsychologia, 11,* 105–111.

McGuire, L. S., Ryan, K. O., & Omenn, G. S. (1975). Congenital adrenal hyperplasia: II. Cognitive and behavioral studies. *Behavior Genetics, 5,* 175–188.

Messerli, P., Tissot, A., & Rodriguez, J. (1976). Recovery from aphasia: Some factors and prognosis. Y Lebrun & R. Hoops (Eds.), *Recovery in aphasics* (pp. 124–135). Amsterdam: Swets & Zeitlinger.

Mittwoch, U. (1973). *Genetics of sex differentiation.* New York: Academic Press.

Molfese, D. (1977). Infant cerebral asymmetry. In S. J. Segalowitz & F. F. Gruber (Eds.), *Language development and neurological theory* (pp. 21–35). New York: Academic Press.

Molfese, D., & Hess, T. M. (1978). Hemispheric specialization for VOT perception in the preschool child. *Journal of Experimental Psychology, 26,* 71–84.

Money, J., & Alexander, D. (1966). Turner's syndrome: Further demonstration of the presence of specific cognitional deficiencies. *Journal of Medical Genetics, 3,* 47–48.

Money, J., & Lewis, V. (1966). IQ, genetics and accelerated growth: Adrenogenital syndrome. *Johns Hopkins Hospital Bulletin, 118,* 365–373.

Money, J., Schwartz, M., & Lewis, V. G. (1984). Adult erotosexual status and fetal hormonal masculinization and demasculinization: 46,XX congenital virilizing adrenal hyperplasia (CVAH) and 46,XY androgen insensitivity syndrome (AIS) compared. *Psychoneuroendocrinology, 9,* 405–414.

Morris, B. B. (1971). Effects of angle, sex, and cue on adults' perception of the horizontal. *Perceptual and Motor Skills, 32,* 827–830.

Netley, C., & Rovet, J. (1982a). Verbal deficits in children with 47,XXY and 47,XXX karyotypes: A descriptive and experimental study. *Brain and Language, 17,* 58–72.

Netley, C., & Rovet, J. (1982b). Atypical hemisphere lateralization in Turner syndrome subjects. *Cortex, 18,* 377–384.

Netley, C., & Rovet, J. (1984). Hemispheric lateralization in 47,XXY Klinefelter's boys. *Brain and Cognition, 3,* 10–18.

Netley, C., & Rovet, J. (1987). Relations between a dermatoglyphic measure, hemispheric specialization, and intellectual abilities in 47,XXY males. *Brain and Cognition, 6,* 153–160.

Nielsen, J., Sorensen, A., Theilgaard, A., Froland, A., & Johnsen, S. G. (1969). A psychiatric–psychological study of 50 severely hypogonadal male patients, including 34 with Klinefelter's syndrome, 47,XYY. *Acta Jutlandica, 41,* 3.

Nyborg, H., & Nielsen, J. (1981). Spatial ability of men with karyotype 47,XXY or normal controls. In W. Schmid & J. Nielsen (Eds.), *Human behavior and genetics* (pp. 85–106). Amsterdam: Elsevier/North-Holland.

O'Connor, K. P., & Shaw, J. C. (1978). Field dependence, laterality, and the EEG. *Biological Psychology, 6,* 93–109.

Oltman, P. K., Ehrlichman, H., & Cox, P. W. (1977). Field independence and laterality in the perception of faces. *Perceptual and Motor Skills, 45,* 255–260.

Oltman, P. K., Semple, C., & Goldstein, L. (1979). Cognitive style and interhemispheric differentiation in the EEG. *Neuropsychologia, 17,* 699–702.

Penrose, L. (1967). Finger print patterns and the sex chromosome. *The Lancet, 1,* 298–300.

Perlman, S. M. (1973). Cognitive abilities of children with hormone abnormalities: Screening by psychoeducational tests. *Journal of Learning Disabilities, 6,* 21–29.

Piazza, D. M. (1980). The influence of sex and handedness in the hemispheric specialization of verbal and nonverbal tasks. *Neuropsychologia, 18,* 163–176.

Pizzamiglio, L. (1974). Handedness, ear preference and field dependence. *Perceptual and Motor Skills, 38,* 700–702.

Pizzamiglio, L., & Cecchini, M. (1971). Development of the hemispheric dominance in children five to ten years of age and their relations with development of cognitive processes. *Brain Research, 31,* 361–378.

Porteus, S. D. (1965). *Porteus maze test: Fifty years' application.* Palo Alto, CA: Pacific Books.

Rapaczynski, W., & Ehrlichman, H. (1979). Opposite visual hemifield superiorities in face recognition as a function of cognitive style. *Neuropsychologia, 17,* 645–652.

Ratcliffe, S. G., & Tierney, I. R. (1982). 47,XXY males and handedness. *The Lancet, 2,* 716.

Ratcliffe, S. G., Bancroft, J., Axworthy, D., & McLaren, W. (1982). Klinefelter's syndrome in adolescence. *Archives of Diseases of Childhood, 57,* 6–12.

Reinisch, J. M., & Karow, W. G. (1977). Prenatal exposure to synthetic progestins and estrogens: Effects on human development. *Archives of Sexual Behavior, 6,* 257–288.

Resnick, S., & Berenbaum, S. A. (1982). Cognitive functioning in individuals with congenital adrenal hyperplasia. *Behavior Genetics, 12,* 594–595.

Rizzolatti, G., & Buchtel, H. (1977). Hemispheric superiority in reaction time to faces: A sex difference. *Cortex, 13,* 300–305.

Robinson, R. G. (1979). Differential behavioral and biochemical effects hemispheric infarction in the rat. *Science, 205,* 707–710.

Rosenthal, R., Hall, J. A., DiMatteo, M. R., Rogers, P. L., and Archer, D. (1979). *Sensitivity to nonverbal communication.* Baltimore: The Johns Hopkins University Press.

Ross, G. T., & Vande Wiele, R. L. (1974). The ovaries. R. H. Williams (Ed.), *Textbook of Endocrinology* (pp. 368–422). Philadelphia: W. B. Saunders.

Rovet, J., & Netley, C. (1982). Processing deficits in Turner's syndrome. *Developmental Psychology, 18,* 77–94.

Rovet, J., & Netley, C. (1983). Hemisphere specialization in Turner's syndrome. *The INS Bulletin,* October, p. 35. (as cited by McGlone, 1985).

Rudel, R., Denckla, M., & Spalten, E. (1974). The functional asymmetry of Braille letter learning in normal sighted children. *Neurology, 24,* 733–738.

Rudel, R., Denckla, M., & Hirsch, S. (1977). The development of left-hand superiority for discriminating Braille configurations. *Neurology, 27,* 160–164.

Schaie, K. W., & Roberts, J. (1970). *School achievement of children as measured by the reading and arithmetic subtests of the Wide Range Achievement Test. Vital and Health Statistics—Series 11—No. 103.* Washington, DC: U. S. Government Printing Office.

Schneidler, G. G., & Paterson, D. G. (1942). Sex differences in clerical aptitude. *Journal of Experimental Psychology, 33,* 303–309.

Segami, M. I., & Alarcon-Segovia, U. S. A. (1978). Systemic lupus erythematosis and Klinefelter's syndrome. *Archives of Rheumatology, 20,* 1565–1567.

Serra, A., Pizzamiglio, L., Boari, A., & Spera, S. (1978). A comparative study of cognitive traits in human sex chromosome aneuploids and sterile and fertile euploids. *Behavior Genetics, 8,* 143–154.

Shaffer, J. (1962). A specific cognitive deficit observed in gonadal aplasia (Turner's syndrome). *Journal of Clinical Psychology, 18,* 403–406.

Sheehan, E. P., & Smith, H. V. (1986). Cerebral lateralization and handedness and their effects on verbal and spatial reasoning. *Neuropsychologia, 24,* 531–540.

Shucard, D. W., Shucard, J. L., & Thomas, D. G. (1977). Auditory evoked potentials of probes of hemispheric differences in cognitive processing. *Science, 197,* 1295–1298.

Shucard, J. L., Shucard, D. W., Cummins, K. R., & Campos, J. J. (1981). Auditory evoked potentials and sex-related differences in brain development. *Brain and Language, 13,* 91–102.

Shucard, D. W., Shucard, J. L., & Thomas, D. G. (1984). The development of cerebral specialization in infants. R. J. Emde & J. Harmon (Eds.), *Continuities and discontinuities in development.* New York: Plenum Press.

Simpson, J. L. (1975). Gonadal dysgenesis and abnormalities of the human sex chromosomes: Current status of phenotypic–karyotypic correlations. *Birth Defects, 11,* 23–59.

Spellacy, W. N., Bernstein, I. C., & Cohen, W. H. (1965). Complete form of testicular feminization syndrome. Report of a case with biochemical psychiatric studies. *Obstetrics and Gynecology, 26,* 499–503.

Stewart, J., & Cygan, D. (1980). Ovarian hormones act early in development to feminize open-field behavior in the rat. *Hormones and Behavior, 14,* 20–32.

Strauss, E., & Moscovitch, M. (1981). Perception of facial expressions. *Brain and Language, 13,* 308–332.

Taylor, D. C. (1969). Differential rates of cerebral maturation between sexes and between hemispheres. *The Lancet, 2,* 140–142.

Thomas, H., Jamison, W., & Hummel, D. D. (1973). Observation is insufficient for discovering that the surface of still water is invariantly horizontal. *Science, 181,* 173–174.

Toran-Allerand, C. D. (1984). On the genesis of sexual differentiation of the central nervous system. Morphogenetic consequences of steroidal exposure and possible role of alpha-fetoprotein. In G. J. De Vries, J. P. C. De Bruin, H. B. M. Uylings, & M. A. Corner (Eds.), *Sex difference in the brain. The relation between structure and function, Vol. 61: Progress in brain research* (pp. 63–98). Amsterdam: Elsevier.

Turner, H. H. (1960). Ovarian dwarfism and rudimentary ovaries. E. B. Astwood (Ed.), *Clinical endocri* nology, *Vol. 1* (pp. 455–467). New York: Grune & Stratton.

Valentine, G. H. (1969). *The chromosome disorders: An introduction for clinicians,* Philadelphia: Lippincott

Waber, D. (1979). Neuropsychological aspects of Turner's syndrome. *Developmental Medicine and Chile* Neurology, 21, *58–70.*

Walzer, S., Graham, J. M., Bashir, A. S., & Sibert, A. R. (1982). Preliminary observations on language and learning in XXY boys. *Birth Defects: Original Article Series, 18,* 185–192.

Winter, J. S. D., Hughes, I. A., Reyes, F. I., & Faiman, C. (1976). Pituitary–gonadal relations in infancy 2. Patterns of serum gonadal steroid concentrations in men from birth to two years of age. *Journal o* Clinical Endocrinology and Metabolism, 42, *679–686.*

Witelson, S. (1976). Sex and the single hemisphere: Right hemisphere specialization for spatial process ing. *Science, 193,* 425–427.

Witkin, H. A., Dyk, R. B., Faterson, H. F., Goodenough, D. R., & Karp, S. A. (1974). *Psychologica* differentiation. *Potomac, MD: Lawrence Erlbaum. (Original work published 1962)*

Witkin, H. A., Goodenough, D. R., & Oltman, P. K. (1979). Psychological differentiation: Current status *Journal of Personality and Social Psychology, 37,* 1127–1145.

Wittig, M., & Petersen, A. (Eds.). (1979). *Sex-related differences in cognitive functioning: Developmenta* issues. *New York: Academic Press.*

Yalom, I. D., Green, R., & Fisk, N. (1973). Prenatal exposure to female hormones: Effect on psychosex ual development in boys. *Archives of General Psychiatry, 28,* 554–561.

Zoccolotti, P., & Oltman, P. K. (1978). Field independence and lateralization of verbal and configura tional processing. *Cortex, 14,* 155–168.

9

Factors Determining the Onset of Puberty

MARK E. WILSON

INTRODUCTION

Puberty represents a stage in a mammal's life span during which dramatic behavioral and physiological changes occur. Although significant progress has been made describing the complex cascade of events that characterize puberty, nonetheless, the factors that initiate and maintain the process and thus set the timing of the maturational process (Tanner, 1962) are poorly understood. The purpose of this chapter is to examine those variables that influence the timing of puberty. The "timing of puberty" refers not only to the age at which somatic and neuroendocrinological changes in the reproductive system are initiated but also to the rate at which these are completed. The analysis focuses on female primates, but attention is given to data from nonprimate mammals that help elucidate possible mechanisms controlling puberty onset and the ontogeny of reproductive changes leading to sexual maturity. This chapter thus does not describe the status of our current understanding of mammalian puberty, as this has been done quite thoroughly in several recent comprehensive reviews (Reiter & Grumbach, 1982; Foster, 1988; Plant, 1988a, 1988b; Ojeda & Urbanski, 1988), but rather attempts to define the mechanisms that initiate sexual maturation and attainment of adult reproductive functioning in female primates.

Historically, puberty was assessed by changes in the behavior or physical appearance of the female, such as vaginal opening for mice, estrus for domestic animals, and menarche for Old World monkeys and apes. These overt signs, though,

MARK E. WILSON   Yerkes Regional Primate Research Center, Field Station, Emory University, Lawrenceville, Georgia 30243.

Sexual Differentiation, Volume 11 of *Handbook of Behavioral Neurobiology,* edited by Arnold A. Gerall, Howard Moltz, and Ingeborg L. Ward, Plenum Press, New York, 1992.

are the products of changes in hypothalamic–pituitary–ovarian function. Indeed, the critical factor in the initiation of adult reproductive cycles is the appropriate pituitary secretion of luteinizing hormone (LH) and follicle-stimulating hormone (FSH) in response to the hypothalamic release of gonadotropin-releasing hormone (GnRH) (Reiter & Grumbach, 1982). The "rate-limiting step" then for the onset of adult reproductive cycles is the appearance of a particular pattern of GnRH secretion, which drives the pituitary and, in turn, stimulates ovarian activity (Wildt, Marshall, & Knobil, 1980). The ovaries are capable of responding to gonadotropic stimulation prior to puberty (Weiss, Rifkin, & Atkinson, 1976; Foster, Ryan, & Papkoff, 1984), underscoring the importance of the appropriate pattern of GnRH release for the initiation of puberty.

The onset of puberty is best characterized then by an increase in basal gonadotropin secretion. Furthermore, the development of the "positive-feedback mode" of gonadotropin release in primates, characteristic of periovulatory hormone dynamics (Knobil, 1974), is directly related to the basal secretion of gonadotropin (Terasawa, 1985; Wilson & Gordon, 1986). Indeed, during prepuberty, when gonadotropin secretion is low or absent, estradiol (E_2) can only elicit an LH surge if administered concomitantly with GnRH (Wildt *et al.*, 1980). This is in contrast to other mammals, e.g., sheep (Foster, 1984), in which an LH surge can be induced by exogenous E_2 long before tonic LH levels increase. Thus, for primates, the capacity to induce an ovulatory-like surge in LH occurs late in puberty (Dierschke, Weiss, & Knobil, 1974) and is directly related to the capacity of the hypothalamic–pituitary unit to release GnRH and, consequently, gonadotropins. Barring overt pathology in other systems resulting in abnormal development (Ducharme & Collu, 1982), factors that influence the developmental secretion and effectiveness of GnRH regulate the timing of the onset of puberty. Puberty can be induced with pulsatile administration of GnRH in monkeys (Wildt *et al.*, 1980; Sopelak, Collins, & Hodgen, 1985) and boys (Delemarre, van de Waal, & Schoemaker, 1983) and can be arrested with the use of GnRH analogues that block the activity of endogenous GnRH (Mansfield *et al.*, 1983).

CHARACTERISTICS OF PUBERTY

Although the importance of GnRH secretion in the initiation of adult reproductive function is well accepted, the developmental time course of GnRH release is not fully understood. Before we proceed with an assessment of potential factors that may initiate the appropriate secretory GnRH patterns, a brief review of the ontogeny of gonadotropin secretion is presented (see Reiter & Grumbach, 1982; Ryan, 1986; Plant, 1988a, 1988b, for further information). Illustrated in Figure 1 is a diagrammatic timetable of the maturation of gonadotropin secretion in a female monkey or ape. Following birth, the hypothalamic–pituitary–ovarian system is functional for several months during the neonatal period, as characterized by the hypersecretion of gonadotropins following the removal of the ovaries (Plant, 1986). Although the GnRH secretory mechanism is active in females during this neonatal period, a greater increase in gonadotropin occurs following gonadectomy in male monkeys (Plant, 1986).

Approximately 2 months from birth for monkeys and 8 months from birth for humans, even in the absence of the ovaries, serum gonadotropin levels in peripheral circulation decline to unmeasurable levels. The ensuing juvenile period is charac-

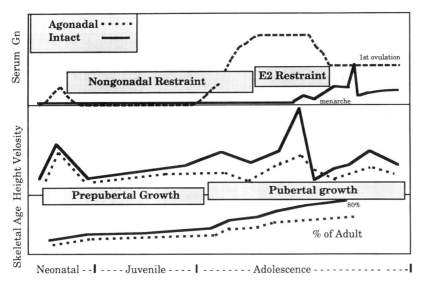

Figure 1. Schematic representation of maturation of gonadotropin (Gn) secretion in agonadal and gonadally intact female primates.

terized by low serum levels of gonadotropins, presumedly reflecting an absence of appropriate GnRH stimulation. However, blood samples collected at frequent intervals throughout an extended portion of the day in juvenile girls reveal evidence of pulsatile LH release, albeit at very low levels (Penny, Olambiwonnu, & Frasier, 1977; Jakacki et al., 1982). Additional evidence from humans (Boyar et al., 1974) and monkeys (Terasawa, Bridson, Nass, Noonan, & Dierschke, 1984) reveals that, during the juvenile stage, nighttime samples yield higher concentrations of LH than daytime samples, but all are lower than those observed in sexually mature females. Neuroanatomical evidence also indicates that GnRH cell bodies, although smaller than those in adult animals, are nevertheless present in the hypothalami of juvenile monkeys and that treatment with GnRH antisera further reduces the low serum levels of LH in immature animals (Cameron, McNeill, et al., 1985). Furthermore, studies directly measuring GnRH in the stalk–median eminence vasculature reveal that pulses of GnRH are present in immature female monkeys, albeit at a lower frequency and amplitude than in sexually mature animals, and that maturation is best characterized by an increase in pulse amplitude (Watanabe & Terasawa, 1989). Furthermore, the GnRH gene is fully expressed during the juvenile phase (Adams, Vician, Clifton, & Steiner, 1987), and the capacity to synthesize (Cameron, McNeill, et al., 1985) as well as the hypothalamic content of GnRH (Fraser, Pohl, & Plant, 1989) are similar between juvenile and adult monkeys. Episodic administration of an analogue of the excitatory amino acid aspartate stimulates GnRH secretion and results in an immediate and robust release in LH (Gay & Plant, 1987, 1988), which within 11 weeks produces an adult-like pattern of LH and testosterone secretion (Plant, Gay, Marshall, & Arslan, 1989).

Taken together, these data indicate that GnRH secretory activity is present during the juvenile period. If so, the onset of gonadotropin secretion marking the transition from the juvenile to the peripubertal stage is most likely caused by an increase in the amplitude of GnRH secretion. It is not known whether this low level of GnRH secretion during the juvenile phase results from an active inhibition of

GnRH-secreting neurons or an absence of neuronal input into GnRH-secreting neurons (Kelch, Marshall, Sauder, Hopwood, & Reame, 1983). Although data from the male suggest a "reawakening" of the GnRH pulse generator and thus a loss of active inhibition, data from the female suggest a gradual awakening (Plant, 1986). The critical question, then, is what factors initiate the change in GnRH secretion. Available evidence indicates that the maturational change in GnRH release is caused by an alteration in both a steroid-independent and a steroid-dependent influence.

As the juvenile phase ends, basal levels of serum LH increase spontaneously in the absence of the gonads (Terasawa *et al.*, 1984; Plant, 1983; Wilson, Gordon, & Collins, 1986; Winter & Faiman, 1972; Conte, Grumbach, & Kaplan, 1975), indicating that the hiatus in gonadotropin secretion during the juvenile phase is caused by some "intrinsic central nervous system inhibitory mechanism" on the GnRH system and not by the influence of the gonad (Reiter & Grumbach, 1982). Once this nongonadal restraint of GnRH release has decreased, an apparent hypersensitivity to steroid hormone negative feedback regulates the timing of the final stages of puberty (see Figure 1; Kulin, Grumbach, & Kaplan, 1969; Wilson *et al.*, 1986; Rapisarda, Bergman, Steiner, & Foster, 1983; Winter *et al.*, 1987). Thus, a modified version of the original gonadostat hypothesis, which stated that the maturation of gonadotropin secretion is caused by a diminution in the steroid negative feedback (Holweg & Dohrn, 1932; Ramirez & McCann, 1963), is useful in explaining the onset of gonadotropin secretion in intact females.

This is illustrated by examining the developmental time course of LH and FSH secretion in gonadally intact and ovariectomized E_2-treated and untreated females (Figure 2). Exposure to low levels of estradiol delays the maturational increases in LH compared to agonadal, ovariectomized females, but in time these low E_2 levels lose the capacity to suppress LH release (Rapisarda *et al.*, 1983; Wilson, 1989; Wilson *et al.*, 1986). This release from an apparent hypersensitivity to E_2 negative feedback occurs in rhesus monkeys prior to the expected age at menarche and first ovulation and some 5 months following the release from nongonadal restraint (Wilson *et al.*, 1986; Wilson, in press). Indeed, the effectiveness of E_2 in inhibiting LH release continues to decline during the interval from menarche to the establishment of normal ovulatory cycles and reflects the period of adolescent sterility characteristic of postpubertal primate females (Rapisarda *et al.*, 1983).

If one accepts the notion that a hypersensitivity to gonadal steroid negative feedback regulates the development of gonadotropin secretion during the late pubertal period, several criteria must be met: (1) once an adult-like pattern of episodic gonadotropin secretion is attained in an ovariectomized adolescent female, replacement therapy with E_2 must abolish gonadotropin release, and (2) increasing doses of E_2, which would be ineffective in suppressing gonadotropin secretion in adult ovariectomized females, would be required to suppress LH and FSH secretion as the female nears the expected age when first ovulation would occur. If, on the other hand, E_2 is ineffective in suppressing episodic gonadotropin release once maximal levels are attained in an ovariectomized adolescent, the apparent hypersensitivity to E_2 negative feedback (see Figure 2) could simply reflect a normal E_2 negative feedback mechanism operating on an immature GnRH pulse generator (Plant, 1988a). The distinction between a two-stage and a single-stage sequence is important in attempting to determine the variables that initiate the entire process and thus set the timing of pubertal changes. If a two-step change in GnRH secretion occurs, one must identify those factors that initiate the change in nongonadal restraint and, later, the change in hypersensitivity to E_2 negative feedback. Conversely, if GnRH

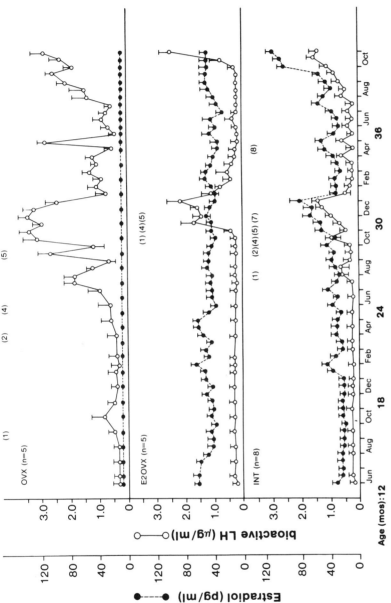

Figure 2. Maturational changes in bioactive LH and E_2 concentrations in outdoor-housed female rhesus monkeys that were either gonadally intact (INT), ovariectomized-treated with E_2 (E2OVX), or untreated ovariectomized (OVX). Numbers in parentheses above each panel represent the cumulative number of animals in each group exhibiting a significant elevation in LH at the time indicated. (From Wilson et al., 1986.)

secretion simply increases throughout maturation, then the focus would be on the variables that initiate and maintain this change.

Brain manipulation studies in primates demonstrate both an inhibitory and an active excitatory influence on the development of gonadotropin release. Unlike rodent species (Donovan & van der Werff ten Bosch, 1956; Ojeda, Aguado, & White, 1983), anterior hypothalamic lesions in female rhesus monkeys do not affect the timing of sexual maturation (van der Werff ten Bosch, Dierschke, Terasawa, & Slob, 1983). Posterior hypothalamic lesions, however, do advance the age at menarche and first ovulation (Terasawa, Noonan, Nass, & Loose, 1984) and the loss of nongonadal restraint of LH release (Schultz & Terasawa, 1988), suggesting that neural input to this hypothalamic locus restrains the onset of puberty. These results are congruent with clinical studies regarding hypothalamic tumors and sexual precocity in humans (Weinberger & Grant, 1941). In contrast, lesions in the medial basal hypothalamus including the posterior median eminence significantly delay menarche and first ovulation in rhesus monkeys (Schultz, Terasawa, & Goy, 1986). These data suggest one of two possibilities: (1) this area provides tonic excitatory input to the control of GnRH release, and ablation of this area delays puberty, or (2) the lesions simply disrupt the normal pathways from the hypothalamus to the pituitary. More definitive studies are needed to determine whether this area of the basal hypothalamus exhibits an ontogenetic change in excitatory stimulation.

Available data on humans (Corley, Valk, Kelch, & Marshall, 1981; Ross, Loriaux, & Cutler, 1983) and monkeys (Terasawa, Nass, Yeoman, Loose, & Schultz, 1983; Bercu et al., 1983) indicate that pulse frequency is not altered throughout maturation, although an increase in pulse amplitude does occur. Indeed, analysis of hypothalamic release of GnRH in female monkeys indicates that pulse frequency increases slightly prior to the onset of puberty but that, once initiated, developmental increases in LH during the adolescent period are sustained by an increase in GnRH pulse amplitude (Watanabe & Terasawa, 1989).

Several methodological problems make the assessment of ontogenetic changes in pulsatile gonadotropin release difficult to interpret. First, the sensitivity of assays for gonadotropins has been limited, although bioassays for LH (e.g., Steiner & Bremner, 1981) and FSH (e.g., Padmanabham, Chappel, & Beitins, 1987) have alleviated some of these problems. Second, a physiologically relevant sampling frequency necessary for confidently identifying and characterizing pulsatile gonadotropin secretion has not been consistently employed (Veldhuis, Rogol, & Johnson, 1985). Third, the use of LH pulse characteristics as an index of pulsatile GnRH is problematic but, given the difficulty in measuring GnRH, is currently unavoidable. These difficulties are particularly troublesome when attempting to distinguish between a nongonadal restraint mechanism and a hypersensitivity to E_2 negative feedback and between hypothalamic and pituitary sites of action. Indeed, current data from the adult are equivocal as to whether E_2 acts at the pituitary to suppress LH (Nakai, Plant, Hess, Keogh, & Knobil, 1978) or at the hypothalamus to suppress GnRH (Plant & Dubey, 1984) or both (Chappel, Resko, Norman, & Spies, 1981). Pituitary LH responsiveness to GnRH does increase throughout puberty in female primates (Grumbach, Roth, Kaplan, & Kelch, 1974; Monroe, Yamamoto, & Jaffe, 1983), likely because of a priming effect on the pituitary of increasing endogenous GnRH secretion and not a change in pituitary responsiveness (Kelch et al., 1983; Delemarre-van de Waal, Van den Brande, & Schoemaker, 1985). The recent development of techniques to allow the measurement of GnRH in portal circulation in primates will certainly help resolve this problem (Watanabe & Terasawa, 1989).

The discussion thus far has implied that the ontogeny of FSH secretion is similar to that of LH. Although GnRH is certainly a releasing hormone for FSH, other factors may selectively influence the development of FSH secretion. Although maturational increases occur coincidentally with LH (Winter *et al.*, 1987), the ratio of FSH to LH has been reported to decrease throughout maturation in females (Kelch *et al.*, 1983).

A similar situation is obtained in the adult when the frequency of pulsatile GnRH administration is increased (Knobil, 1980), suggesting that puberty may be characterized by a gradual increase in GnRH release. Other studies have shown, however, that as puberty is initiated in children, FSH levels are elevated in a fashion similar to those of LH (Oerter, Uriarte, Rose, Barnes, & Cutler, 1990). In addition to this neuroendocrine influence, the ovary may regulate the development of FSH secretion. Although the age of the initial increase in basal FSH secretion was similar in E_2-treated ovariectomized and gonadally intact female rhesus monkeys, the absolute levels of FSH in serum were four times higher in the E_2-treated ovariectomized group than in the intact animals despite similar serum concentrations of E_2 (M. E. Wilson, unpublished observations). This comparison suggests that an ovarian factor in addition to E_2 may modulate the development of FSH secretion.

Inhibin, a gonadal peptide that selectively inhibits FSH release (Channing, Anderson, Hoover, Gagliano, & Hodgen, 1981; Ying, 1988), may play a role in the development of FSH. FSH stimulates ovarian inhibin production (Ying, 1988), which could account for the observation that inhibin level varies throughout the menstrual cycle in women (McLachlan, Robertson, Healy, Burger, & de Krester, 1987) and rises throughout puberty in girls (Burger *et al.*, 1988). The possibility exists then that, in addition to E_2 negative feedback, rising FSH levels during puberty are held in check by rising inhibin release. The importance of inhibin for the initiation of adult-like FSH release remains to be determined.

Attempts to identify factors that initiate and maintain the progression of puberty should thus focus on manipulations that alter the developmental time course of LH and FSH secretion (as an index of GnRH release) and should include an assessment of the developmental changes in steroid-independent and steroid-dependent influences on episodic gonadotropin release. This fine-grained analysis should ultimately identify factors that alter the timing of the first ovulation, the basic dependent variable in the assessment of puberty.

Variables Affecting the Timing of Puberty

Ponderal Growth and Nutrition

It has long been thought that a critical amount of body weight or, more precisely, a critical amount of body fat is necessary for the initiation of reproductive competency. Retrospective analyses of puberty in humans suggests that the occurrence of menarche is more closely related to body weight or a body weight-to-height ratio than to chronological age (Frisch & Revelle, 1969, 1970, 1971; Frisch, 1974). Further studies reveal that menarche is delayed in malnourished children (Kulin, Bwibo, Mutie, & Santner, 1982) or young athletic females whose body weights are reduced compared to sedentary agemates (Frisch, Wyshak, & Vincent, 1980). Although this view has a high degree of intuitive appeal, empirical confirmation has not been obtained.

The hypothesis that the onset of puberty is related to a critical level of body fat could be tested in one of two ways: by obtaining additional correlational data ör by directly manipulating body weight. Correlational data from nonhuman primate species (Wilen & Naftolin, 1976; Resko, Goy, Robinson, & Norman, 1982; Wilson *et al.*, 1986) as well as further analysis of human data (Ellison, 1981) suggest that as chronological age at menarche or first ovulation increases, individuals are bigger and heavier. If a critical percentage of body fat were necessary, no correlation should be present between body weight or fat and age at maturity. That is, within any given population, body weights at menarche would be constant at increasing chronological ages. But this is not the case, as illustrated in Figure 3. A significant positive relationship between weight and age at menarche and at first ovulation is observed for rhesus monkeys (M. E. Wilson, unpublished observations). Older animals experience these reproductive events at larger body weights.

Analyses of alterations in body weight, either from experiments of nature or by design, have provided insights into alternative mechanisms that influence the onset of puberty. A cross-sectional analysis of both girls and boys from prepuberty through young adult status indicates that changes in secretion of gonadotropins and subsequent gonadal steroids precede changes in body weight and fat rather than the converse (Penny, Goldstein, & Frasier, 1978). Studies of children with precocious puberty also suggest a dissociation between body composition and menarche (Crawford & Osler, 1975). Additionally, although malnourishment slows both adolescent growth and puberty, once menarche occurs, females are smaller than would be expected by a critical body weight hypothesis (Kulin *et al.*, 1982). Indeed, significant

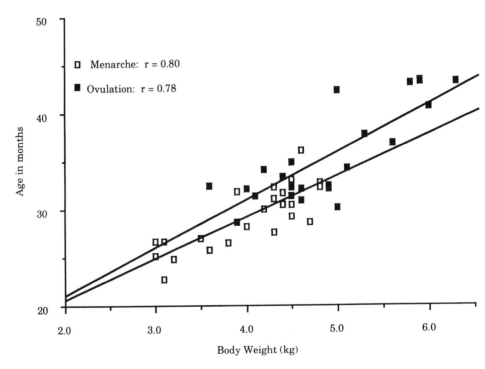

Figure 3. Relationship between body weight and age at menarche and age at first ovulation in rhesus monkeys (*n* = 31). First ovulation was inferred from sustained increases in serum progesterone.

weight gains occur following menarche in girls, suggesting that body size, *per se,* has little relation to the timing of maturation (Garn, LaVelle, Rosenberg, & Hawthorne, 1986).

Studies of food restriction or dietary manipulations with animals confirm that malnutrition does delay puberty but that the delay is not likely caused by alterations in body composition. Vaginal opening in mice is significantly delayed by restriction of diet. However, this delay is unrelated to differences in body fat (Hansen, Schillo, Hinshelwood, & Hauser, 1983) because it occurs at a lower percentage body fat in diet-restricted animals. Similarly, in female rats, the degree of sexual maturation is not predicted from body fat status (Glass & Swerdloff, 1980), and underfed male rats are more sexually mature than controls of the same body weight, since these animals are older when puberty is reached (Glass, Anderson, Herbert, & Vigersky, 1984).

Systematic evaluation of male rats indicates (Figure 4) that the delay in the age at puberty accompanying dietary restriction is associated with reduced secretion of LH, FSH, and growth hormone (GH) (Sisk & Bronson, 1986). Interestingly, when food-restricted rats are given *ad libitum* access to food, gonadotropin and GH secretion recover rapidly to control levels before normal body weight is restored. Similar results are observed in sheep (Foster *et al.,* 1989). Acute food restriction in adult male monkeys also diminishes the frequency of pulsatile LH secretion, which is reversed by refeeding (Cameron & Nosbisch, 1991). Again, these data indicate that changes in the reproductive system occur before significant increments in weight

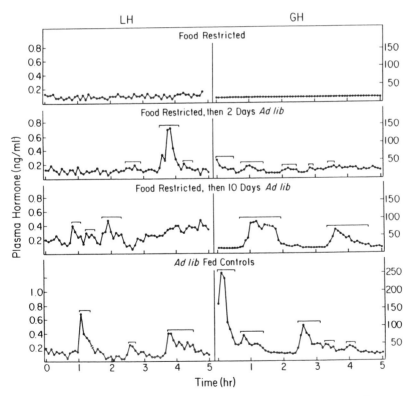

Figure 4. Pulsatile secretion of serum LH and GH in representative rats exposed to a particular feeding regimen. Brackets indicate significant hormone pulses. (From Sisk & Bronson, 1986.)

and that a blood-borne signal may initiate gonadotropin release. Also, disruptions in menstrual cyclicity from dieting in women precede any significant loss of weight (Pirke, Schweiger, Lemmel, Krieg, & Berger, 1985), and refeeding results in an immediate onset of LH secretion followed in time by an accumulation of body weight (Warren, 1983). These data are consonant with the observation that female rats that are required to exercise to obtain food have vaginal opening at a significantly younger age and lower body weight than either diet-restricted or *ad-lib*-fed animals (Bronson, 1987), suggesting that increased metabolic efficiency rather than weight *per se* may influence reproductive onset.

Similar studies in lambs reveal that food restriction delays first ovulation (Fitzgerald, Michael, & Butler, 1982), and refeeding initiates ovulatory cycles but at a lower body weight than that of controls at their first ovulation (Foster & Olster, 1985). Further analyses indicate that the delayed first ovulation in the diet-restricted animals is associated with enhanced E_2 negative feedback suppression of LH secretion (Foster & Olster, 1985). It is not known whether this represents a continuation of development or a diet effect on the regulation of gonadotropin secretion.

Experimental diet restriction studies have not been done with immature primates. Alternatively, diet enhancement has been used. Immature female rhesus monkeys fed a high-fat diet exhibit menarche and first ovulation at an earlier age than control females (Figure 5; Schwartz, Wilson, Walker, & Collins, 1988). These precocious reproductive events are not associated with higher body weights or fat but with earlier elevations in serum insulin, GH, and insulin-like growth factor-1 (IGF-1). These data support the conclusion that the reproductive deficits and delays induced by dietary restriction are not secondary to changes in body weight or fat.

A problem with the body-weight hypothesis has been the specification of how a critical amount of body fat could affect neuroendocrine processes involved in the reproductive system. The presence of fat tissue as a modulator of E_2 bioavailability has been hypothesized as a mechanism affecting maturation (Frisch, 1984). Since body fat acts as a site for the aromatization of androgens to estrogens, lower

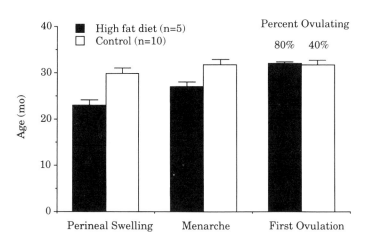

Figure 5. Comparison of timing of reproductive events in female rhesus monkeys fed a normal, control diet (CND) or a high fat diet. Age at first ovulation is calculated for only those females within each group which ovulated by 35 months of age. (Redrawn from Schwartz *et al.*, 1985.)

amounts of body fat during prepuberty or delayed puberty results in less availability of estrogen. As maturation proceeds and more body fat is accumulated, the availability of E_2 would increase. Unfortunately, this explanation does not fit with the existing understanding of puberty onset. If, indeed, changes in hypersensitivity to E_2 negative feedback on LH secretion characterizes the final stages of puberty, then increased E_2 release from accumulated body fat would actually result in a further delay in the initiation of LH secretion. Estradiol is not the key to the onset of puberty, but rather the initiation of gonadotropin secretion. In support of this, treatment of prepubertal monkeys with E_2 does not advance sexual maturation (Dierschke et al., 1974).

It also has been argued (Kennedy & Mitra, 1963; Frisch, 1974) that a critical amount of body weight or fat results in a change in metabolic rate, which subsequently affects E_2 negative-feedback suppression of LH. As an alternative to a critical body-weight hypothesis, a metabolic hypothesis has been proffered to account for the initiation of reproductive cycles (Steiner, Cameron, McNeill, Clifton, & Bremner, 1983). In adult females, severe weight loss is followed by anovulation because of a suppression of gonadotropin secretion (Vigersky, Anderson, Thompson, & Loriaux, 1977; Wentz, 1980; Warren & Vande Wiele, 1973). The analogy of reduced gonadotropin secretion and an anovulatory condition in undernourished adults to an immature reproductive condition suggests that a common metabolic cue rather than body weight is the critical factor (Cameron, Hansen, et al., 1985).

The acceleration in the timing of menarche and first ovulation observed in monkeys on a high-fat diet (Schwartz et al., 1988) could be explained by a "metabolic hypothesis." The lack of certain essential metabolic factors, which could act either directly on the system governing gonadotropin secretion or as precursors for the neurotransmitters involved in GnRH secretion, could result in an anovulatory state. For example, metabolic factors that would increase the availability of precursors for serotonin, acetylcholine, norepinephrine, and dopamine may vary greatly following a meal and thus influence the synthesis of these neurotransmitters (Wurtman, 1979). One candidate that could serve as a cue for gonadotropin release is insulin, whose concentration in serum varies with body weight (Cameron, Hansen, et al., 1985). Insulin infusion elevates LH levels in nonprimate mammals (McCann, 1984), suggesting that increases in insulin may be tied to maturational increases in LH. Although fasting levels of serum insulin are lower in prepubertal monkeys of both sexes compared to adults (Steiner et al., 1983), episodic insulin secretion in response to fasting is similar in juvenile and adult male monkeys (Plant, 1988b). Basal levels of circulating glucose increase with advancing age (Chaussain, Georges, Calzada, & Job, 1977), suggesting that maturational increases in another regulator of glucose metabolism is important for the initiation of gonadotropin secretion. An analysis of changes in metabolism following a meal also reveals that juvenile male monkeys have a more rapid transition to a fasted state in terms of lower levels of insulin, glucose, and all large neutral amino acids with the exception of tryptophan than adult males (Cameron, Koerker, & Steiner, 1985). The study failed to find an age difference in ratios of either tryptophan or tyrosine to other amino acids—important as precursors for serotonin and norepinephrine, respectively. On the other hand, infusion of dextrose and a complex of neutral amino acids to juvenile male monkeys resulted in an increase in LH secretion in three of six animals (Cameron, Hansen, et al., 1985). Changes in LH secretion were not related to alterations in body weight or daily caloric intake.

The utilization of available nutrients such as glucose and specific amino acids may be important for the support of gonadotropin secretion. The brain may be sensitive to caloric balance such as a disproportionate use of available calories by the brain relative to the body, which may occur during postnatal development when brain size and cognitive capacity are growing exponentially (Winterer, Cútler, & Loriaux, 1984). Developmentally, once a balance in metabolic fuel utilization is reached, GnRH activity may be initiated and sustained. For example, patients who lack the capacity to store glycogen, and thus produce glucose, have delayed puberty, which is reversed following chronic nocturnal glucose administration (Winterer *et al.*, 1984). Furthermore, recent *in vitro* studies reveal that the pulsatile release of GnRH from the hypothalamus is dependent on glucose availability (Bourguignon, Gerard, Debougnoux, Rose, & Franchimont, 1987). Additional support for the importance of metabolic fuels is evident in the induction of anestrus in hamsters when fatty acid and glucose utilization are both blocked (Schneider & Wade, 1989).

The suppression of gonadotropin secretion by diet restriction can be reversed by pulsatile infusion of GnRH, suggesting that the metabolic alteration may directly affect GnRH release independent of steroid negative feedback (Dubey, Cameron, Steiner, & Plant, 1986). The only metabolic substrate that is consistently suppressed during food restriction is glutamate (Dubey *et al.*, 1986), but glutamate levels do not vary between adults and juveniles following a meal (Cameron, Koerker, *et al.*, 1985). As mentioned above, episodic administration of an analogue to this excitatory neurotransmitter results in precocious puberty in male monkeys by the commencement of endogenous release of GnRH (Plant *et al.*, 1989). Whether the induction of this neurotransmitter system is responsible for the initiation of puberty is not known. What is clear is that a diet-restricted adult is not necessarily equivalent to a normal, unfasted juvenile (Dubey *et al.*, 1986). But unlike the dynamics of insulin secretion, it appears that the secretion of IGF-1 is similar in a juvenile and an adult fasted state. As illustrated in Figure 6, serum concentrations of IGF-1 rise throughout development in ovariectomized rhesus monkeys. Although maturational increases in E_2 facilitate IGF-1 secretion (Wilson, Schwartz, Walker, & Gordon, 1984), these data support observations in the rat that there is a nongonadal drive to IGF-1 secretion during maturation (Handelsman, Spaliviero, Scott, & Baxter, 1987): IGF-1 is under the primary regulation of GH and consequently mediates GH effects on bone (Froesch, Schmid, Schwander, & Zapf, 1985). In addition, IGF-1 also has metabolic properties, as it name implies: it enhances glucose uptake without suppressing glucose production or reducing protein breakdown (Jacob, Barrett, Plewe, Fagin, & Sherwin, 1989). Interestingly, fasting reduces serum IGF-1 concentrations (Underwood, Clemmons, Maes, D'Ercole, & Ketelsleger, 1986) and hepatic mRNA (Straus & Takemoto, 1989), effects that are reversed by refeeding. These data suggest that

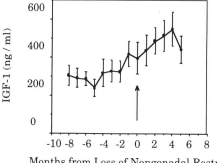

Figure 6. Developmental changes in serum concentrations of insulin-like growth factor-1 (IGF-1) in ovariectomized female rhesus (*n* = 5) aligned to the initial increase in serum LH, representing the loss of nongonadal restraint.

changes in the peripheral production of IGF-1 are similar between juvenile animals and a fasted and refed adult. Whether maturational increases in IGF-1 provide the signal for an increase in the availability of metabolic fuels for the induction of GnRH and consequently gonadotropin secretion is not known at this time.

A metabolic hypothesis provides a testable framework in which to identify the factors involved in the initiation of sexual maturation. As is the case for any potential "initiator," the isolation of any one metabolic cue merely necessitates the search for the initiation of that change. Assuming a metabolic factor is identified, the mechanism for *its* age-dependent change must, in turn, be identified. For example, if a change in IGF-1 secretion instigates adult-like gonadotropin secretion, two questions arise: (1) what drives the maturational increase in IGF-1 and (2) how, at the cellular level, does IGF-1 affect the developmental drive to GnRH secretion? Although nocturnal serum concentrations of GH do increase during development in ovariectomized rhesus monkeys (Wilson, 1989), which could account for the increase in IGF-1, some neuroendocrinological change must account for this change in GH secretion. Furthermore, studies must assess whether IGF-1 directly affects neurons impacting on GnRH-secreting cells or whether these effects are mediated via changes in metabolic blood-borne signals.

A second issue that must be addressed is the use of diet restriction to study sexual maturation. Food restriction at any age imposes a dramatic biological insult. It is not clear whether malnourished or diet-restricted individuals provide an adequate model of puberty onset (see discussions in Ellison, 1981; Foster, Karsch, Olster, Ryan, & Yellon, 1986). Luteinizing hormone release may be suppressed during fasting by the activation of an endogenous opioid inhibitory tone (Dyer, Mansfield, Corbet, & Dean, 1985). Since some evidence suggests that a change in the inhibitory tone of these peptides may account for increases in LH during late puberty (see below), the influence of metabolism on the development of gonadotropin secretion may be mediated by endogenous opioid activity within the hypothalamus. Obviously, in order to determine what regulates the timing of the onset of puberty, manipulations must be made to speed up or delay the process. In addition to delaying puberty by diet restriction, the use of biologicals that block specific metabolic processes (cf. Schneider & Wade, 1989) or the use of selective metabolic regulators (e.g., IGF-1) to alter essential nutrients may prove to be as efficacious and physiologically relevant for the identification of factors that alter the timing of puberty.

SKELETAL GROWTH

As with changes in body weight, skeletal maturation varies with sexual maturation. Scales of physical development based on characteristics of bones have been developed that coincide with specific stages of reproductive development (Tanner & Whitehouse, 1976). Indeed, menarche is more closely related to skeletal age, than to body weight in humans (Sizonenko, Burr, Kaplan, & Grumbach, 1970; Marshall, 1974) and monkeys (Wilson, Gordon, Rudman, & Tanner, 1988). Skeletal age, based on the ossification of bones, is a robust predictor of the degree of sexual maturation. Although the precise timing of changes in bone maturation and sexual development is unknown for female primates in general because of a paucity of longitudinal data (Watts, 1985), menarche in humans follows the growth spurt in skeletal maturation. Compared to the development of secondary sexual characteristics, menarche thus occurs relatively late in the pubertal process in humans (Tanner,

1972). Nevertheless, the *initiation* of the height spurt in humans precedes, in general, the appearance of secondary sexual characteristics (Marshall & Tanner, 1970), suggesting that the occurrence of these reproductive landmarks may be the consequence of endocrine activities regulating the growth spurt in height (Zacharias & Rand, 1983). This relationship in humans is quite similar to that in rhesus monkeys (Wilson, Gordon, Rudman, & Tanner, 1989), in which menarche consistently follows an acceleration in height (Figure 7). Also, bone age or skeletal maturity is highly correlated with the occurrence of first ovulation in rhesus monkeys (Krey *et al.,* 1981; Wilson *et al.,* 1988, 1989). These data raise the question as to whether the timing of puberty is causally tied to the regulation of skeletal maturation.

Bone maturation is controlled by GH secretion from the pituitary (Tanner, 1972). Although it is assumed that GH secretion increases during development, supporting data on a variety of species are equivocal (cf. Parker, Rossman, Kripke, Gibson, & Wilson, 1979). Maturational increases in gonadal steroids may enhance GH release (Harris *et al.,* 1985), but an increase in GH independent of the gonads does occur as evidenced by the fact that daytime (Figure 8) and nighttime GH concentrations increase throughout maturation in ovariectomized monkeys (Wilson, 1989). Regardless of whether there is a CNS-dependent elevation in GH secretion, gonadal steroids promote growth by an augmentation of GH secretion. Females with gonadal dysgenesis (Kirkland, Lin, & Kirkland, 1987; Copeland, 1988) or ovariectomized monkeys grow less overall than estradiol-treated agemates (Loy, Loy, Kiefer, & Conaway, 1984; Wilson, Schwartz, *et al.,* 1984). Although the height spurt occurs at a similar age for gonadally intact, ovariectomized E_2-treated, and ovariectomized juvenile rhesus monkeys, the magnitude of the peak in height velocity is significantly lower in agonadal females not treated with steroids (Wilson, 1989). The important point is that agonadal individuals with a functional pituitary grow, but more slowly and to a lesser degree than gonadally intact agemates.

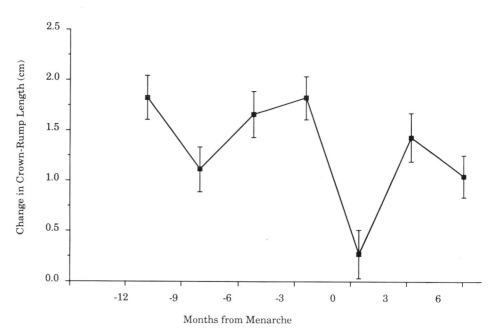

Figure 7.   Developmental changes in crown–rump length at 3-month intervals aligned to menarche for rhesus monkeys ($n = 14$; redrawn from Wilson *et al.,* 1988).

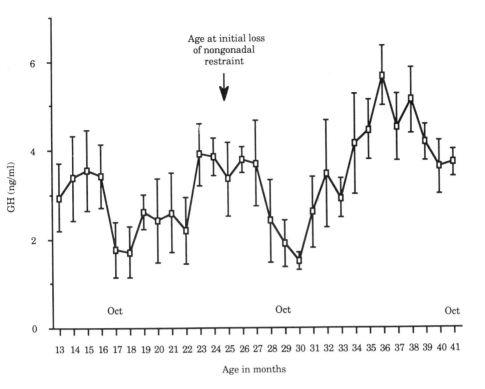

Figure 8.   Maturational changes in daytime serum GH concentrations in ovariectomized female rhesus monkeys ($n = 5$). Females were ovariectomized at 12 months of age; GH levels were determined by previously validated radioimmunoassay. (From Wilson *et al.*, 1986.)

To address the question of whether factors that regulate skeletal maturation are causally related to sexual maturation, one must determine whether changes in growth factors result in an alteration in the time course of puberty. Children with isolated GH deficiency show a slower rate of bone maturation (Tanner, 1972) and may exhibit a deficiency in gonadotropin secretion at a chronological age when indices of sexual maturation would otherwise be expected (Tanner & Whitehouse, 1975). Nevertheless, these children exhibit indices of puberty at a delayed chronological age but at a bone age similar to that of normal children. As mentioned above, malnutrition slows body weight gain and skeletal growth and delays sexual maturation (Driezen, Stone, & Spies, 1961; Driezen, Spivakis, & Stone, 1967). The eventual occurrence of menarche in malnourished children is at the same bone age as in well-nourished girls despite a difference in chronological age of approximately 2 years (Driezen *et al.*, 1967). This tight coupling of bone age to indices of sexual maturation suggests that a common variable may be regulating each.

Several sets of data indicate that GH may be important for maturational increases in gonadotropin secretion in addition to its effects on growth. Suppression of endogenous GH by treatment of rats with the tapeworm *Spirometra mansonoides*, which secretes a variant GH, does not adversely affect growth rates (Ramaley & Phares, 1980, 1983) but does delay first ovulation in females (Ramaley & Phares, 1980, 1982) and balanopreputial separation in males (Ramaley & Phares, 1983). Concurrent treatment with bovine GH normalizes the occurrence of vaginal opening and first ovulation. Furthermore, GH-suppressed animals were more sensitive to the negative feedback effects of E_2 following prepubertal ovariectomy (Ramaley & Phares, 1982). These data suggest that GH may be directly involved in the maturation of gonadotropin secretion.

An analysis of the ontogeny of LH release in female rhesus monkeys reveals that elevations in daytime serum GH precede maturational increases in LH in gonadally intact and E_2-treated ovariectomized animals (Figure 9; Wilson et al., 1986). Furthermore, GH levels, but not body weight, clearly discriminate precocious females from agemates that ovulate later, with early-ovulating females having significantly higher GH concentrations (Wilson et al., 1986). These data suggest that maturational increases in GH may be involved in the decrease in hypersensitivity to E_2 negative feedback of LH that characterizes the final stages of sexual maturation. Data are equivocal whether children with GH deficiency show a blunted response in gonadotropin secretion to GnRH (Kelch, Markovs, & Hess, 1976; Cohen, Kaplan, & Grumbach, 1983). Furthermore, 6 to 12 months of human GH therapy to GH-deficient children did not enhance the responsiveness to GnRH (Kelch et al., 1976), suggesting that the GH and gonadotropin deficiencies may be independent.

Although these descriptive data suggest a role for GH in the initiation of gonadotropin secretion, administration of human GH to adolescent female rhesus monkeys did not consistently advance the timing of the increases in LH: the age at the initial increase in LH was advanced only in gonadally intact, GH-treated females and not in E_2-treated ovariectomized females receiving GH (Wilson et al., 1989). Although menarche was not advanced in comparison to untreated control females, first ovulation occurred significantly earlier in three of five hGH-treated females. Although the timing of LH release was not consistently affected, GH treatment did augment E_2 secretion once puberty had begun (Figure 10), suggesting that GH may affect reproductive development by augmenting ovarian responsivity to gonadotropins. This hypothesis is supported by studies indicating that GH-induced increases in ovarian IGF-1 (Davoren & Hsueh, 1986) may synergize with a developing pattern of gonadotropin secretion to facilitate granulosa cell steroid production (Adashi, Resnick, D'Ercole, Svoboda, & van Wyk, 1985). The delay of puberty in GH-deficient rats results from a reduction in ovarian responsivity to gonadotropin (Advis, Smith-White, & Ojeda, 1981). This paracrine role of IGF-1 may influence granulosa cell growth and enhance the action of a still-developing secretory pattern of gonadotropins. Data from humans, however, indicate that GH administration to GH-deficient prepubertal children does not augment steroid secretion in response to acute hCG administration (Kulin, Samojlik, Demers, & Santner, 1985). Nevertheless, a GH-dependent facilitation of ovarian maturation may occur during puberty (Adashi et al., 1985) and may be one way in which this growth peptide shares a common link to both reproductive and skeletal maturation.

In addition to its possible ovarian effects, the peripheral metabolic activities of GH also may induce changes that, in turn, result in alterations in GnRH activity. For example, GH deficiency in children is associated with symptomatic hypoglycemia (Goodman, Grumbach, & Kaplan, 1968; Bougneres, Artavia-Loria, Ferre, Chaussain, & Job, 1985) and may be associated with low postprandial levels of insulin (Hopwood, Forsman, Kenny, & Drash, 1975). Chronic administration of hGH to these hypopituitary children does normalize postprandial blood glucose levels (Gold, Spector, & Samaan, 1968; Bougneres et al., 1985). The hypoglycemia during prepuberty is the result of a decreased production of glucose by the liver (Boug-

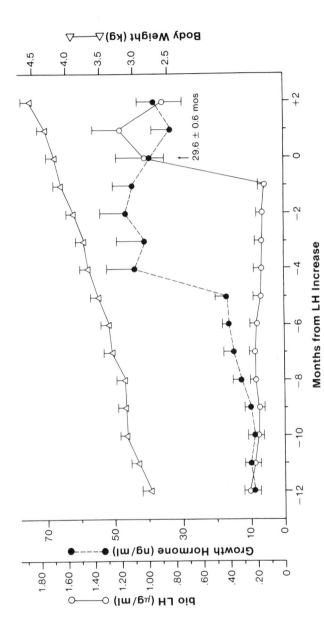

Figure 9. Maturational changes in serum GH, bioactive LH, and body weight aligned to the increase in LH for gonadally intact and E₂-treated ovariectomized females combined (*n* = 13). (From Wilson *et al.*, 1986.)

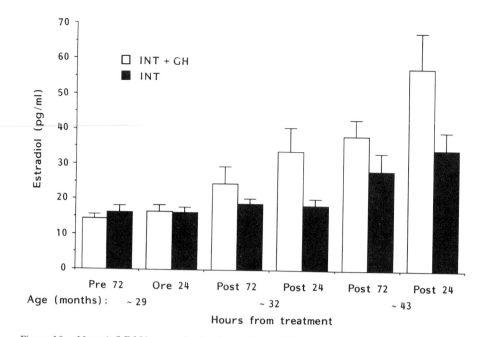

Figure 10. Mean (±S.E.M.) serum levels of estradiol at different ages in female rhesus monkeys that received GH three times weekly (250 μg/kg, SC, M, W, F; *n* = 5) from 20 to 44 months of age and in untreated control females (*n* = 6). "Pre" and "post" values refer to the number of hours samples were taken following GH administration. Maturational increases in LH for both groups occurred by ~31 months of age. (From Wilson *et al.*, 1989.)

neres *et al.*, 1985). Rather than a decrease in glucogenic precursors, the decrease in glucose production may be the result of diminished hepatic glycogen.

These data are consistent with the predictions generated from the metabolic hypothesis. Recall that normal pituitary juvenile males had lower glucose levels following a meal than adult males (Cameron, Koerker, *et al.*, 1985) and that infusion of glucose in combination with a complex of neutral amino acids resulted in LH release (Cameron, Hansen, *et al.*, 1985). Although this suggests that maturational increases in GH could increase the availability of glucose for utilization by the brain, GH treatment did not consistently affect the onset of puberty in monkeys (Wilson *et al.*, 1989). The possibility remains that GH-dependent changes in IGF-1 may be the signal for the initiation of gonadotropin secretion and the onset of sexual maturation. Studies must first separate the metabolic actions of GH from those of IGF-1 and then determine whether these metabolic effects initiate gonadotropin release.

It is possible that the relationship between GH physiology and sexual maturation may not be causal but simply correlational. Evidence from the rat has suggested that pulsatile administration of GnRH to pituitary cells *in vitro* results not only in gonadotropin but also GH release (Badger, Crowley, Millard, Mak, & Babu, 1987). Although this suggests that increased GH release is the result of GnRH stimulation, the physiological relevance of these data is not clear, since it is known that GH release is governed by the hypothalamic release of GH-releasing factor and somatostatin (Martin, 1986). In contrast, the infusion of the putative excitatory neurotransmitter aspartate, which initiates LH secretion in juvenile male monkeys (Gay & Plant, 1987), also increases serum concentrations of GH and prolactin. Thus, an increase in excitatory neuronal firing within the hypothalamus may cause a concomitant increase in both GH and LH. Although this would argue against a causal role of GH in the initiation of puberty, the temporal sequence of events with skeletal maturation preceding sexual maturation must be considered. Therefore, at this time, one cannot determine whether GH itself is involved or is itself a response to yet another unidentified factor that first initiates GH release and then GnRH-dependent gonadotropin release.

MELATONIN

Tumors of the pineal gland result in precocious puberty (Holweg & Dohrn, 1932). The hormonal secretions of the pineal gland are regulated by the prevailing photoperiod (Reiter, 1980). Melatonin, the most studied pineal hormone, is secreted maximally during the nighttime and minimally during the daylight hours. Attempts to determine whether changes in pineal melatonin secretion are related to sexual maturation have produced mixed results, primarily because of the use of different sampling regimens and an absence of longitudinal data. Analysis of children at different stages of puberty reveals a small maturational increase in daytime serum samples (Penny, 1985) and overnight urinary excretion of melatonin (Penny, 1982). On the other hand, no differences were observed in either 24-hour melatonin excretion (Tetsuo, Poth, & Markey, 1982) or daytime blood samples from children at different stages of puberty or in those exhibiting precocious puberty (Ehrenkranz et al., 1982; Lenko, Lang, Aubert, Paunier, & Sizonenko, 1982). Reversal of precocious puberty by treatment with a GnRH agonist fails to alter nocturnal melatonin secretion (Berga, Jones, & Yen, 1987), suggesting that neither GnRH nor gonadal steroids alter melatonin release. In contrast, with advancing sexual maturation, daytime melatonin levels are decreased in boys but not girls (Silman, Leone, Hooper, & Preece, 1979). Furthermore, a distinct rhythm is evident in children from infancy through adolescence with nocturnal levels of melatonin significantly higher at earlier stages of puberty (Attanasio, Borrelli, & Gupta, 1985; Gupta, Riedel, Frick, Attanasio, & Ranke, 1983).

In support of the cross-sectional studies in children, longitudinal data from female rhesus monkeys indicate that nocturnal melatonin concentrations decrease significantly with advancing sexual maturity (Figure 11; Wilson & Gordon, 1989a). Female monkeys that exhibit first ovulation earlier than agemates show a more rapid decline in nocturnal melatonin secretion. In contrast, no change in diurnal melatonin secretion is observed with advancing age in lambs housed under controlled photoperiods (Foster et al., 1986). Also, since young sexually mature hamsters secrete more melatonin than older animals, the decrease in melatonin may be more related to age than to puberty, per se (Reiter, Richardson, Johnson, Ferguson, & Dihn, 1980). In addition, pineal tumors may secrete an LH-like substance that would result in an earlier onset of puberty (Neuwelt, Mickey, & Lewy, 1986).

The idea that the pineal gland is the clock setting the timing for sexual maturation is intriguing, given its functional physiological relationship to other anatomic areas, primarily the suprachiasmatic nucleus and the superior cervical ganglion. These anatomic loci control the circadian rhythm of a variety of biological and behavioral functions, regardless of whether they are entrained by external cues (Moore, 1979). Ablation of the suprachiasmatic nucleus abolishes almost every diurnal rhythm (Moore & Eichler, 1972). Furthermore, the suprachiasmatic nucleus has an inherent endogenous diurnal rhythmic activity itself (Green & Gillette, 1982). Since circadian rhythms may simply be a component of circannual rhythms

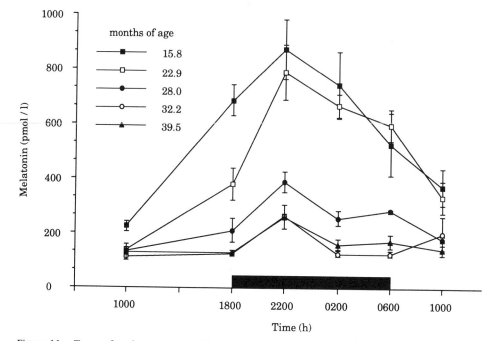

Figure 11. Twenty-four-hour patterns of serum melatonin (mean ± S.D.) in gonadally intact female rhesus monkeys (*n* = 4) throughout development. Menarche occurred at ~26 months of age. (From Wilson & Gordon, 1989a.)

(Gwinner, 1981), one could speculate that daily rhythms may influence the timing of events with a period greater than 1 year. Nevertheless, the notion that melatonin regulates sexual maturation also implies extrahypothalamic control of GnRH release and thus of the ontogeny of sexual maturation. Melatonin does bind to receptors in the median eminence and arcuate nucleus area in hamsters (Weaver, Namboodiri, & Reppert, 1988) and to the suprachiasmatic nucleus in humans (Reppert, Weaver, Rivkees, & Stopa, 1988). Isolation of the medial basal hypothalamus resulting in complete medial basal hypothalamic disconnection advanced first ovulation relative to control rhesus monkeys in one study (Norman & Spies, 1981) but not in another (Krey *et al.*, 1981). Differences between these two studies have not been reconciled. If the pineal gland modulates the timing of puberty, one would hypothesize that removal of pineal input would accelerate maturation, or, conversely, that maintenance of high nocturnal melatonin concentrations would delay puberty.

The experimental analysis of the role of the pineal and melatonin secretion in sexual maturation is complicated by the fact that most animals observed in experimental situations are seasonal breeders under natural conditions and that this annual rhythm in reproduction is photoperiodically entrained via melatonin secretion (Lincoln & Short, 1980). One must resolve the issue of how developmental changes in melatonin secretion interact with seasonal changes in the melatonin rhythm, particularly when the species under study is a seasonal breeder susceptible to environmental control. Furthermore, the effect of melatonin secretion may not be uniform: i.e., hamsters breed exclusively under long days, when the duration of nocturnal melatonin is short (Turek & Campbell, 1979), and sheep breed under short days, when the duration of nocturnal melatonin is long (Karsch *et al.*, 1984).

Rather than being simply pro- or antigonadal, the pineal gland acts as the neuroendocrine transducer converting light to a hormonal signal (Reiter, 1980). The duration and amplitude of the nocturnal increases in melatonin may alter the neuroendocrine–gonadal system by interacting with an endogenous diurnal rhythm in photosensitivity (Hoffman, 1981). Data on both long-day breeders (Reiter, Richardson, & King, 1983) and short-day breeders (Karsch *et al.*, 1984) indicate that annual changes in the diurnal secretion of melatonin are critical for seasonal shifts in gonadotropin secretion.

For a seasonal breeder, then, photoperiodically driven changes in melatonin secretion may determine the initiation of sexually maturity by overriding endogenous changes in gonadotropin secretion that occur independently of season (Foster, 1983; Sisk & Turek, 1983; Wilson & Gordon, 1989b; Smith, Matt, Prestowitz, & Stetson, 1982). For example, if a female lamb's GnRH pulse generator, and consequently gonadotropin secretion, matures at an age coincident with the long days of spring, first ovulation will be delayed until the fall, the typical season of ovulation for this species (Foster *et al.*, 1986). If female lambs are exposed to continuous long days from birth, first ovulation is delayed for more than 20 weeks beyond that observed in unmanipulated females (Yellon & Foster, 1985). Continuous exposure to short days advances first ovulation in some females, although these cycles are characterized by insufficient luteal function, and delays first ovulation in others for more than a year (Yellon & Foster, 1985; Foster, 1983). A normal onset of first ovulation is observed if lambs are exposed to as little as five long days at 20 weeks of age followed by continuous short days. These data suggest that in lambs photoperiodic information, mediated via diurnal melatonin secretion (Yellon & Foster, 1986), is crucial to timing of the onset of puberty. Since lambs on constant long or short days eventually ovulate, photoperiodic changes are not entirely necessary for initiating puberty; rather, these changes establish its appropriate timing in this species.

Although it is not yet clear whether the seasonal pattern of reproduction in rhesus monkeys is photoperiodically regulated as in sheep, females that reach the typical age of sexual maturity in the spring have a delay of first ovulation until fall (Wilson, Gordon, Blank, & Collins, 1984; Wilson & Gordon, 1989b). Based on the available data, different housing environments may have a less pronounced effect on the age of first ovulation in rhesus monkeys than in sheep. Monkeys housed under controlled environmental conditions (generally 12L:12D at 22–26°C) exhibit menarche and first ovulation at the same age range (Resko *et al.*, 1982; Terasawa *et al.*, 1984; Foster, 1977; Lotz, Saxton, & Hess, 1987; Wilson *et al.*, 1988) as animals in outdoor environments (Wilson, Gordon, *et al.*, 1984, 1986). Indeed, a seasonal distribution in the occurrence of first ovulation is also observed in animals housed under constant conditions (Terasawa *et al.*, 1983; Lotz *et al.*, 1987), although it is more distinct in animals housed outdoors. Unfortunately, season of birth is often not reported, so it is not known if animals housed indoors are simply maturing at a particular chronological age with season having no effect.

In animals under controlled laboratory conditions, any reference to season is irrelevant; the animals never experience changes in the environment. If first ovulation is synchronized among rhesus monkeys housed indoors, several possibilities to account for the pattern exist: (1) these animals are simply maturing within the same age range, (2) they are being exposed inadvertently to some external cue, or (3) there is an endogenous clock that times puberty to occur within several separate age ranges. Thus, as in sheep, seasonal cues are not required for the initiation of puberty in rhesus monkeys, but they do set the time frame within which these events occur. Indeed, maturational increases in LH release are observed in female monkeys during the spring and summer months, but further increases leading to first ovulation do not occur until the subsequent autumn (Wilson *et al.*, 1986). Systematic study of rhesus monkeys exposed to selected photoperiodic changes is needed to answer this question fully. In support of the importance of photoperiod for the seasonal timing of first ovulation, administration of a short-day melatonin pattern to adolescent rhesus monkeys advances the age of menarche and first ovulation compared to age-matched controls (Wilson & Gordon, 1989c). These data suggest that these rhesus females are neuroendocrinologically mature enough for the initiation of puberty during the spring and summer months of their second year but that the long-day melatonin pattern operating at this time delays maturation to the fall months. Removal of this long-day pattern by superimposing a short-day pattern advances puberty.

There is no evidence that sexual maturation in humans is influenced by season. As in monkeys (Wilson, Schwartz, *et al.*, 1984) and sheep (Schanbacher & Crouse, 1981), growth rates in humans are enhanced during the long, warm days of summer (Marshall & Swan, 1971). Interestingly, growth rates of hamsters, which are long-day breeders, are enhanced during short days (Zucker & Boshes, 1982), in part because of melatonin (Bartness & Wade, 1984). These comparisons suggest that, in a given species, a melatonin rhythm that inhibits reproduction may enhance growth. If melatonin is having an effect on the timing of sexual maturation in humans, it must be independent of changes in the particular 24-hour pattern of secretion, i.e., duration of nighttime levels, and dependent on other parameters of its secretory profile. Thus, two effects of melatonin may be hypothesized: (1) a decrease in the nocturnal amplitude of melatonin may provide the signal to the GnRH pulse generator for the initiation of puberty, and (2) the duration of the daily melatonin rhythm may signal the appropriate time of year for the onset of puberty. If a decreased nocturnal pattern of melatonin secretion is characteristic of maturation, higher levels of melatonin somehow restrain the GnRH pulse generator. This hypothesis is not supported by the observation that pinealectomy fails to elevate gonadotropin levels in gonadectomized infantile male rhesus monkeys (Plant & Zorub, 1986), suggesting that pineal secretions are not involved in nongonadal restraint of LH during this stage of development. This analysis, however, was limited to the first 36 weeks of life, some 70 weeks prior to the occurrence of normal maturational increases in LH (Plant, 1986). Thus, it is not known whether pinealectomy advances onset of the peripubertal stage. Melatonin implants in ewes, which elevate the overall diurnal pattern of endogenous melatonin, do delay puberty (Kennaway, Peek, Gilmore, & Royles, 1986). Daily administration of melatonin also delays vaginal opening in rats (Rivest, Lang, Aubert, & Sizonenko, 1985). In contrast, abnormally high levels of melatonin during the night do not adversely affect the timing of first ovulation (Yellon & Foster, 1986), suggesting that a change in the overall 24-hour pattern may not be important.

One interesting hypothesis stems from the observation that melatonin suppresses GH secretion in humans (Smythe & Lazarus, 1974) and rats (Smythe & Lazarus, 1973). A maturationally dependent decrease in melatonin output could, in turn, affect GH secretion, which, as described above, could influence the initiation of puberty. It was proposed (Smythe & Lazarus, 1973) that melatonin acts as a receptor antagonist of serotonin at hypothalamic sites to account for the opposing effects of melatonin and serotonin on GH release. A second possibility is that maturational changes in somatostatin derived from the pineal gland alter both GH and GnRH. In support of this, somatostatin-secreting neurons have been identified in the pineal gland (Patel & Reichlin, 1978). Maturational changes in diurnal melato-nin secretion may merely reflect a maturational change in the activity of the pineal gland itself, with its secretory products showing similar developmental changes. Manipulation of melatonin secretion during puberty would be of interest in terms of its influence not only on sexual maturation but also on bone maturation.

ENDOGENOUS OPIOID PEPTIDES

In monkeys, endogenous opioid peptides (EOP), including the enkephalins and β-endorphin, arise in cell bodies concentrated in the arcuate nucleus of the medial basal hypothalamus (Ferin, van Vugt, & Wardlaw, 1984), an area involved in the regulation of GnRH (Knobil & Hotchkiss, 1988). Treatment of immature rats with naloxone, an antagonist of β-endorphin, increases LH secretion prior to first ovula-tion (Blank, Panerai, & Friesen, 1979; Bruni, van Vugt, Marshall, & Meites, 1977), and long-term treatment with naloxone advances first ovulation (Sirinathsinghji, Motta, & Martini, 1985). A significant response in LH release to acute naloxone treatment in children occurs only after the appearance of secondary sexual charac-teristics, once puberty has begun (Mauras, Veldhuis, & Rogol, 1986; Petraglia *et al.*, 1986; Sauder, Case, Hopwood, Kelch, & Marshall, 1984; Fraioli *et al.*, 1984). This temporal association between the effect of EOP antagonists and stage of puberty has led to the hypothesis that a decrease in EOP inhibitory tone may mediate the age-dependent decrease in hypersensitivity to gonadal steroid inhibition of gonado-tropin secretion that occurs during late puberty (Bhanot & Wilkinson, 1983a). The EOP, however, are ineffective in suppressing LH in the absence of gonadal steroids in rats (Bhanot & Wilkinson, 1983b, 1984). Furthermore, EOP or their antagonists do not affect GnRH-stimulated release of gonadotropins in rats (Bhanot & Wilkin-son, 1983a) or monkeys (Ferin, Wehrenberg, Lam, Alston, & Vande Wiele, 1982), suggesting that the effect is localized in the hypothalamic control of GnRH itself. In contrast, morphine administration attenuates the pituitary response to GnRH in rats (Kalra, Saha, & Kalra, 1988), indicating that the effects of EOP on gonadotro-pin secretion are at both the hypothalamic and hypophyseal level.

These data indicate that a decrease in the inhibitory tone of EOP may mediate the decrease in E_2 negative feedback suppression of LH characterizing late puberty. No data are available on the effect of EOP on E_2 negative-feedback suppression of LH during primate puberty. Data from adult female rhesus monkeys indicate that the administration of EOP antagonists augments LH release primarily during the periovulatory (Rossmanith, Wirth, Sterzik, & Yen, 1989) luteal phase (van Vugt, Bakst, Dyrenfurth, & Ferin, 1983; Orstead, Hess, & Spies, 1987). If EOP are in-volved in the maturation of LH release, these data suggest that the particular ste-roidal mediation of EOP may change during the transition from puberty to adulthood.

These data suggest that a decrease in EOP inhibitory tone may characterize late puberty. Rather than a decrease, hypothalamic mRNA content for proopiomelano-cortin (POMC), the precursor for β-endorphin, is higher in adult rats (Wiemann, Clifton, & Steiner, 1989) and monkeys (Vician, Adams, & Clifton, 1989) than in juvenile animals. Interestingly, castration decreases POMC mRNA content in the arcuate nucleus of male rat, an effect that is reversed by testosterone treatment (Chowen-Breed *et al.*, 1989). This augmentation of POMC by gonadal steroids would explain why EOP antagonists do not stimulate gonadotropin in the absence of steroids: there is no endogenous EOP activity to block. Thus, the maturational difference in mRNA content of POMC between juvenile and adult animals could result from the pubertal rise in gonadal steroid secretion and thus could represent the molecular mechanism for the hypersensitivity to gonadal negative feedback. Although gonadal steroids would stimulate EOP activity, the subsequent decrease in hypersensitivity to E_2 negative feedback may be mediated by a change in the inhibitory tone of EOP independent of absolute changes in gene expression. On the other hand, pubertal increases in some other factor may eventually antagonize the effect of E_2 on POMC gene expression, thereby reducing EOP inhibitory tone. Data are not available from adolescent individuals on whether POMC content and EOP activity are higher at this time than in adulthood.

In addition to this proposed role in the maturation of gonadotropin secretion, EOP may also influence GH release (Rivier, Vale, Ling, Brown, & Guillemin, 1977; Bruhn, Tresco, Mueller, & Jackson, 1989) by acting at the hypothalamus to reduce somatostatin and increase GH secretion (Drouva, Epelbaum, Tapia-Arancibia, Laplante, & Kordon, 1981). The significance of the effect of EOP on GH release during puberty has not been investigated, but EOP may antagonize the action of somatostatin (Drouva, Epelbaum, Tapia-Arancibia, Laplante, & Kordon, 1981), thereby increasing GH secretion. The hypothesis must be entertained that EOP activity during the juvenile phase may mediate GH release and the subsequent acceleration in skeletal growth. In the presence of low levels of E_2, this activity on EOP maximally suppresses LH. As the animal matures, the inhibitory tone of EOP diminishes, resulting in increased LH secretion and the onset of puberty. Although the decrease in EOP activity at this time might result in a concomitant decrease in GH release, the peripubertal increases in E_2 may compensate and maintain elevated GH levels through this developmental phase (Harris *et al.*, 1985). Although this scheme is intuitively attractive, each step needs systematic evaluation.

SOCIAL FACTORS

Psychosocial stresses delay the onset of puberty and retard growth (Patton & Gardner, 1962; Powell, Brasel, Raiti, & Blizzard, 1967), and these phenomena have a generalized effect on the reproductive system in adults as well as pubertal individuals. Nevertheless, it is interesting to note that one manifestation of emotional deprivation in children is depressed GH secretion and altered glucose metabolism (Powell *et al.*, 1967).

The timing of puberty in a variety of species of mice is affected by the presence or absence of particular conspecifics (Vandenbergh, 1983). The social effect is related to exposure to urine such that the onset of female puberty, as indicated by vaginal estrus, is accelerated by exposure to urine from adult males (Vandenbergh, 1969; Drickamer, 1983), estrous females (Drickamer, 1986), and pregnant or lactating females (Drickamer & Hoover, 1979). In contrast, puberty is delayed if females are exposed to urine collected from group-housed females (Drickamer, 1977; Massey & Vandenbergh, 1980). The effectiveness with which these chemosignals modify the timing of puberty varies seasonally (Drickamer, 1984, 1987; Garcia & Whitsett, 1983) despite the fact that mice housed under a 12L:12D light cycle breed throughout the year. Since time of year at which the urine is collected does not influence the timing of puberty, the responsivity of the females to the chemosignals must vary seasonally (Drickamer, 1987).

These social effects may provide yet another cue, in addition to photoperiod, by which the timing of puberty is altered to coincide with an optimal time to reproduce.

The analysis of chemosignals on puberty has focused primarily on age at vaginal estrus. In general the effect is small, e.g., differences of 32.3 ± 0.4 days versus 30.6 ± 0.5 days (Drickamer, 1987), but may nevertheless be biologically significant to a species whose life span is limited to approximately 2 years. Furthermore, it is not known how the hormonal cascade of puberty is affected by the socially derived chemosignals. One assumes that gonadotropin secretion is affected, but future studies must bear this out.

Although chemosignals may synchronize menstruation in human females (McClintock, 1971), no systematic evidence is available indicating that such chemosignals affect either the onset of sexual maturation in primates or their reproduction in general (Rogel, 1978). In select species of monkeys, phenomena analogous to those in mice have been observed. First ovulation in a majority of offspring in family units of marmoset monkeys is delayed (Abbott & Hearn, 1978), as are reproductive cycles in adult, subordinate females (Abbott, McNeilly, Lunn, Hulme, & Burden, 1981). There is no evidence of elevated prolactin levels, as a measure of social stress, in her anovulatory offspring. Young females typically exhibit first ovulation when removed from their family group (Abbott, 1984). Similar observations are reported from closely related tamarins (French, Abbott, & Snowdon, 1984; Katz & Epple, 1979). The suppression of ovarian function in offspring does not appear to be related to social dominance, as few agonistic interactions are observed between mothers and daughters. It is not known whether this ovarian suppression is mediated by chemosignals (Abbott, 1984), but in any event, single-male–single-female breeding units are maintained in these monogamous species.

Since the age at first ovulation is similar in rhesus monkeys whether housed indoors singly (Terasawa et al., 1983; Foster, 1977), housed indoors in peer groups (Wilson et al., 1988), outdoor compound housed socially (Wilson et al., 1986), or free ranging (Drickamer, 1974), one might conclude that cues from conspecifics do not alter the timing of puberty. Nonetheless, social variables, manifested in terms of individual social dominance rank, can modify the timing of puberty in rhesus monkeys. Monkeys housed socially outdoors exhibit first ovulation at two discrete ages: approximately 20% ovulated at 32 months, whereas the majority ovulate at about 43 months (Wilson, Gordon, et al., 1984; Wilson et al., 1986). The bimodal distribution in outdoor-housed monkeys results from the fact that first ovulation is seasonally restricted to the fall (Wilson et al., 1986). Higher-ranking females are most often the ones that have a successful first pregnancy at an earlier age than later-maturing agemates (Wilson, Walker, & Gordon, 1983).

An analysis of the age at first ovulation, indicated by changes in serum progesterone levels, of 31 rhesus monkeys studied longitudinally in the past 5 years at the Yerkes colony supports the hypothesis that rank affects the timing of sexual maturation: early ovulators are significantly more often socially dominant (Figure 12). Although monkeys housed in nonsocial settings exhibited first ovulation in the same age range (e.g., Terasawa et al., 1983; Krey et al., 1981), data from socially housed monkeys help explain some of the variance in this distribution. From the data on nonsocially housed animals, it is not known what percentage ovulate by a particular age. The important point is that first ovulation is likely accelerated in the higher-ranking females rather than delayed in the lower-ranking females, since these latter females—which account for about 80% of maturing females—ovulate in the same age range as animals housed individually (Terasawa et al., 1983; Foster, 1977; Resko et al., 1982).

Manipulation of social rank during development supports this hypothesis

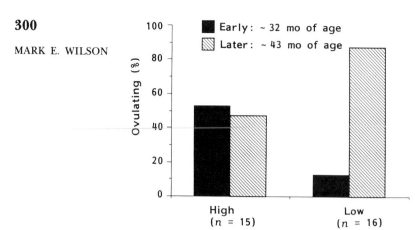

Figure 12.   Distribution of the age at first ovulation by social rank in outdoor-housed female rhesus. First ovulation was inferred from sustained elevations in serum progesterone. Social ranks were assigned on the basis of the outcome of agonistic interactions. (From Bernstein, 1970.)

(Schwartz, Wilson, Walker, & Collins, 1985). Formation of a new social group of rhesus monkeys resulted in some pubertal animals maintaining their previous social rank while others either fell or rose in rank. The subsequent occurrence of first ovulation was predicted more accurately from current rather than previous rank. The early-ovulating females more often were higher ranking. These data do not imply that genetic factors have no role in the pubertal timing of maturational events, but rather differences in social dominance can account for some of the variance in the age at first ovulation and thus the timing of puberty.

The intriguing question is how social rank influences the timing of maturation. A previous analysis indicates that the markers of puberty are advanced in early-ovulating females (Wilson et al., 1986), including maturational increases in LH, menarche, and maturational increases in daytime GH. In contrast, body weights are not different between early-ovulating females and later-maturing agemates. Although these descriptive data indicate that high social status may advance the timing of puberty, it is not known how this is accomplished. Again, since body weight does not predict early versus later-ovulating females, it is unlikely that amount of food ingested plays a role (Schwartz et al., 1985). Furthermore, in all situations in which these data were generated, food was available ad libitum. Preliminary data do suggest, however, that the pattern of feeding is different between high- and low-ranking animals, with the former able to gain access undisturbed by other animals (S. M. Schwartz, M. L. Walker, & M. E. Wilson, unpublished observation). In contrast, lower-ranking animals must be more opportunistic in feeding. These rank differences in feeding order and pattern of food intake are not surprising, since access to food resources may be dictated, in part, by social dominance rank (Bernstein, 1970). One might hypothesize that the pattern of food ingestion may alter the utilization of metabolic substrates and thus affect gonadotropin secretion. This indeed may be the case, since food ingestion results in elevations in serum insulin (Mayhew, Wright, & Ashmore, 1969), which, in turn, may influence the availability of precursors for neurotransmitter synthesis in the brain (Fernstrom, 1979; Wurtman, 1979).

A second possibility is that the rank-related differences in the timing of first ovulation are related to sleeping patterns. Growth hormone secretion is increased at night and is augmented by sleep in humans (Parker et al., 1979; Martin, 1986) but may not be so in adult monkeys (Quabbe, Gregor, Bumke-Vogt, Eckhof, & Witt, 1981). The 24-hour GH secretory profile in young monkeys is not known, nor is it known how sleep disruption affects the pattern. Elevations in serum concentrations of gonadotropins are observed initially in samples collected at night (Boyar *et al.*, 1974; Terasawa *et al.*, 1984). Furthermore, at least in humans, this nocturnal release of LH is tied to sleep onset (Boyar *et al.*, 1974). The significance of this early diurnal difference in LH secretion is not known, other than that it disappears with the onset of adult reproductive cycles (Terasawa *et al.*, 1984a). Perhaps higher-ranking monkeys have more undisturbed sleeping patterns such that the mechanisms regulating the peripubertal increase in LH release that occurs during the night are undisturbed. Rank-related differences in sleeping patterns have not been documented. Nevertheless, one could hypothesize that lower-ranking animals could have a more disturbed sleeping pattern. Systematic studies are needed to address this issue.

One point that must be emphasized is that not all high-ranking juveniles ovulate early. As illustrated in Figure 1, 47% of the high-ranking females ovulate at the typical age of 43 months. Physiologically, the high-ranking females that ovulate later are obviously distinct from their "rankmates" that ovulate earlier. If social rank is playing a role in the timing of maturation, differences in puberty onset must be caused by differences in behavior. This has not been systematically examined. Since some females classified as "high rank" do not ovulate early but some low-ranking females do, the behavioral differences between early- and later-ovulating monkeys are not likely related to dominance rank in a simple fashion.

SUMMARY

Our understanding of the neuroendocrine events associated with the onset of ovarian function is quite advanced. By simply modulating the GnRH signal, we can either induce puberty precociously (Wildt *et al.*, 1980) or arrest it (Crowley *et al.*, 1981). The unknown, of course, is how the endogenous GnRH signal is spontaneously initiated and maintained during maturation. Is there a specific change in the system regulating GnRH release, or is the maturational release in GnRH simply caused by a generalized increase in neuronal activity (cf. Gay & Plant, 1987). Furthermore, manipulations effective in acutely inducing LH and FSH release prepubertally must also be shown to be effective in maintaining gonadotropin secretion resulting in first ovulation. Otherwise, the manipulation may simply be either a pharmacological or transitory experimental effect.

Our understanding of the endocrinology of growth is increasing rapidly (Styne *et al.*, 1985; Harris *et al.*, 1985), but little is known concerning the causal relationship between physical and sexual maturation. A potential problem in the search for the initiator of puberty is the particular animal model chosen. Obviously, an understanding of puberty onset is important in two regards: to improve our basic understanding of the regulation of reproduction in the species under study and to provide information to alleviate the distress caused by disorders of growth and reproductive maturation. Although mammalian species exhibit many similarities in the neuroendocrine regulation of puberty, basic differences in the ontogeny of gonadotropin secretion nevertheless exist (Foster, 1980). For example, an ovulatory-like LH surge can be induced in immature sheep (Foster *et al.*, 1984), whereas this is a late pubertal event in primates (Dierschke *et al.*, 1974). Although it is generally assumed that the regulation of GnRH release is controlled by the same general mechanism in mammals and that perturbations in the timing of puberty—from photoperiodic influences and social variables—are regulated by common mechanism, these assumptions are basically untested. Again, a comparison of sheep and monkeys is illustrative: sheep require specific photoperiodic information to exhibit normal ovulatory cycles at the appropriate age (Foster *et al.*, 1986), whereas ovulations can occur within the same age range regardless of housing conditions in rhesus monkeys (Terasawa *et al.*, 1983; Wilson *et al.*, 1986, 1988).

An appreciation of the environmental situations in which animals mature is critical for assessing the factors that initiate puberty. As Reiter (1980) noted, "For the sake of their own convenience, scientists have created completely abnormal environmental conditions for many experimental rodents . . . which in their natural habitats are exposed to the vicissitudes of temperature and constantly changing photoperiod" (p. 109). This statement should not be limited to rodent species. Consideration must be given to the ecological pressures that shaped the natural biology of the species in order to understand fully how the environment, nutrition and food intake, and resultant metabolic changes affect reproductive processes. There indeed may be a common mammalian mechanism regulating GnRH release, but careful comparative study is necessary before we fully understand the initiation of puberty.

Acknowledgments

Special thanks for technical assistance are given to J. Wilson, E. Osterud, K. Chikazawa, S. Glasco, and E. Seres. I appreciate the editorial assistance of D. Vinson and L. Wright. I appreciate the valuable insight from discussions with C. G. Rudman of Genentech, Inc., and J. M. Tanner of the Institute of Child Health, London, and helpful comments on earlier versions of this manuscript from T. P. Gordon and S. M. Schwartz. Preparation of this chapter was supported by NIH grants HD16305 and, in part, RR00165. The Yerkes Primate Research Center is fully accredited by the American Association for Accreditation of Laboratory Animal Care.

REFERENCES

Abbott, D. H. (1984). Behavioral and physiological suppression of fertility in subordinate marmoset monkeys, *American Journal of Primatology, 6,* 169–186.

Abbott, D. H., & Hearn, J. P. (1978). Physical, hormonal and behavioral aspects of sexual development in the marmoset monkey, *Callithrix jacchus. Journal of Reproduction and Fertility, 53,* 155–166.

Abbott, D. H., McNeilly, A. S., Lunn, S. F., Hulme, M. J., & Burden, F. J. (1981). Inhibition of ovarian function in subordinate female marmoset monkeys (*Callithrix jacchus jacchus*). *Journal of Reproduction and Fertility, 53,* 335–345.

Adams, L. A., Vician, L., Clifton, D. K., & Steiner, R. A. (1987). Gonadotrophin releasing hormone (GnRH) mRNA content in GnRH neurons is similar in the brains of juvenile and adult male monkeys, *Macaca fascicularis.* In *Annual Meeting of the Endocrine Society* (Abstract 64). Baltimore: Williams & Wilkins.

Adashi, E. Y., Resnick, C. E., D'Ercole, A. J., Svoboda, M. E., & van Wyk, J. J. (1985). Insulin-like growth factors as intraovarian regulators of granulosa cell growth and function. *Endocrine Reviews, 6,* 400–420.

Advis, J. P., Smith-White, S., & Ojeda, S. R. (1981). Activation of growth hormone short loop negative feedback delays puberty in the female rat. *Endocrinology, 108,* 1343–1348.

Attanasio, A., Borrelli, P., & Gupta, D. (1985). Circadian rhythms in serum melatonin from infancy to adolescence. *Journal of Clinical Endocrinology and Metabolism, 61,* 388–390.

Badger, T. M., Crowley, W. F., Millard, W. J., Mak, P., & Babu, G. N. (1987). Growth hormone secretion following gonadotropin releasing hormone or bombesin administration in perifused rat pituitary cells. In *Annual Meeting of the Endocrine Society* (abstract 109). Baltimore: Williams & Wilkins.

Bartness, T. J., & Wade, G. N. (1984). Photoperiodic control of body weight and energy metabolism in

Syrian hamsters (*Mesocricetus auratus*): Role of pineal gland, melatonin, gonads and diet, *Endocrinology, 114*, 492–498.

Bercu, B. B., Lee, B. C., Spiliotis, B. E., Pineda, J. L., Denman, D. W., Hoffman, H. J., & Brown, T. J. (1983). Male sexual development in the monkey. II. Cross-sectional analysis of pulsatile pituitary secretion in castrated males. *Journal of Clinical Endocrinology and Metabolism, 56*, 1227–1235.

Berga, S. L., Jones, K. L., & Yen, S. S. C. (1987). Nocturnal melatonin levels in girls with idiopathic precocious puberty. In *Annual Meeting of the Endocrine Society* (abstract 887). Baltimore: Williams & Wilkins.

Bernstein, I. S. (1970). Primate status hierarchies. In L. A. Rosenblum (Ed.), *Primate behavior: Developments in field and laboratory research*, Vol. 1. (pp. 71–109). New York: Academic Press.

Bhanot, R., & Wilkinson, M. (1983a). Opiatergic control of gonadotropin secretion during puberty in the rat: A neurochemical basis for the hypothalamic gonadostat. *Endocrinology, 113*, 596–603.

Bhanot, R., & Wilkinson, M. (1983b). Opiatergic control of LH secretion is eliminated by gonadectomy. *Endocrinology, 112*, 399–401.

Bhanot, R., & Wilkinson, M. (1984). The inhibitory effect of opiates on gonadotropin secretion is dependent upon gonadal steroids. *Journal of Endocrinology, 102*, 133–141.

Blank, M. S., Panerai, A. E., & Friesen, H. G. (1979). Opioid peptides modulate luteinizing hormone secretion during sexual maturation. *Science, 203*, 1129–1131.

Bougneres, P. F., Artavia-Loria, E., Ferre, P., Chaussain, J. L., & Job, J. C. (1985). Effects of hypopituitarism and growth hormone replacement therapy on the production and utilization of glucose in childhood. *Journal of Clinical Endocrinology and Metabolism, 61*, 1152–1157.

Bourguignon, P. B., Gerard, A., Debougnoux, G., Rose, J., & Franchimont, P. (1987). Pulsatile release of gonadotropin-releasing hormone (GnRH) from the rat hypothalamus *in vitro*: Calcium and glucose dependency and inhibition by superactive GnRH analogs. *Endocrinology, 121*, 993–999.

Boyar, R. M., Rosenfeld, R. S., Kaplan, S., Finkelstein, J. W., Roffwarg, H. P., Weitzman, E. D., & Hellman, L. (1974). Human puberty—simultaneous augmented secretion of luteinizing hormone and testosterone during sleep. *Journal of Clinical Investigation, 54*, 609–618.

Bronson, F. H. (1987). Puberty in female rats: Relative effect of exercise and food restriction. *American Journal of Physiology, 252*, R140–R184.

Bruhn, T. O., Tresco, P. A., Mueller, G. P., & Jackson, I. M. D. (1989). Beta-endorphin mediates clonidine stimulated growth hormone release. *Neuroendocrinology, 50*, 460–463.

Bruni, J. F., van Vugt, D., Marshall, S., & Meites, J. (1977). Effects of naloxone, morphine, methionine enkephalin on serum prolactin, luteinizing hormone, follicle stimulating hormone, thyroid hormone, and growth hormone. *Life Sciences, 21*, 461–466.

Burger, H. G., McLachlan, R. I., Bangah, M., Quigg, H., Findlay, J. K., Robertson, D. M., de Krester, D. M., Warne, G. L., Werther, G. A., Hudson, I. L., Cook, J. J., Fiedler, R., Greco, S., Yong, G. B. W., & Smith, P. (1988). Serum inhibin concentrations rise throughout normal male and female puberty. *Journal of Clinical Endocrinology and Metabolism, 67*, 689–694.

Cameron, J. L., & Nosbisch, C. (1991). Suppression of pulsatile luteinizing hormone and testosterone secretion during short term food restriction in the adult male rhesus monkey (*Macaca mulatta*). *Endocrinology, 128*, 1532–1540.

Cameron, J. L., Hansen, P. D., McNeill, T. H., Koerker, D. J., Clifton, D. K., Rogers, K. V., Bremner, W. J., & Steiner, R. A. (1985). Metabolic cues for the onset of puberty in primate species. In S. Venturoli, C. Flamigni, & J. R. Givens (Eds.), *Adolescence in females* (pp. 59–78). New York: Year Book Medical Publishers.

Cameron, J. L., Koerker, D. J., & Steiner, R. A. (1985). Metabolic changes during maturation of male monkeys: Possible signals for onset of puberty. *American Journal of Physiology, 249*, E385–E391.

Cameron, J. L., McNeill, T. H., Fraser, H. M., Bremner, W. J., Clifton, D. K., & Steiner, R. A. (1985). The role of endogenous gonadotropin-releasing hormone in the control of luteinizing hormone and testosterone secretion in the juvenile male monkey, *Macaca fascicularis*. *Biology of Reproduction, 33*, 147–156.

Channing, C. P., Anderson, L. D., Hoover, D. J., Gagliano, P., & Hodgen, G. D. (1981). Inhibitory effects of porcine follicular fluid on monkey serum FSH levels and follicular maturation. *Biology of Reproduction, 25*, 885–903.

Chappel, S. C., Resko, J. A., Norman, R. L., & Spies, H. A. (1981). Studies in the rhesus monkey on the site where estrogen inhibits gonadotropins: Delivery of pituitary gland. *Journal of Clinical Endocrinology and Metabolism, 52*, 1–8.

Chaussain, J. L., Georges, P., Calzada, L., & Job, J. C. (1977). Glycemic response to 24-h fast in normal children. III. Influence of age. *Journal of Pediatrics, 91*, 711–714.

Chowen-Breed, J., Fraser, H. M., Vician, L., Damassa, D. A., Clifton, D. K., & Steiner, R. A. (1989). Testosterone regulation of proopiomelanocortin ribonucleic acid in the arcuate nucleus of the male rat. *Endocrinology, 124,* 1697–1702.

Cohen, H. N., Kaplan, S. L., & Grumbach, M. M. (1983). Serum luteinizing hormone and follicle-stimulating hormone and the response to luteinizing hormone-releasing factor in children and adolescents with isolated growth hormone deficiency. *Journal of Clinical Endocrinology and Metabolism, 57,* 855–858.

Conte, F. A., Grumbach, M. M., & Kaplan, S. L. (1975). A diphasic pattern of gonadotropin secretion in patients with the syndrome of gonadal dysgenesis. *Journal of Clinical Endocrinology and Metabolism, 40,* 670–674.

Copeland, K. C. (1988). Effects of acute high dose and chronic low dose estrogen on plasma somatomedin-C and growth in patients with Turner's syndrome. *Journal of Clinical Endocrinology and Metabolism, 66,* 1278–1282.

Corley, K. P., Valk, T. W., Kelch, R. P., & Marshall, J. C. (1981). Estimation of GnRH pulse amplitude during pubertal development. *Pediatric Research, 15,* 157–162.

Crawford, J. D., & Osler, D. C. (1975). Body composition at menarche: The Frisch–Revelle hypothesis revisited. *Pediatrics, 56,* 449–458.

Crowley, W. F., Comite, F., Vale, W., Rivier, J., Loriaux, D. L., & Cutler, G. B. (1981). Therapeutic use of pituitary desensitization with a long-acting LHRH agonist: A potential new treatment for idiopathic precocious puberty. *Journal of Clinical Endocrinology and Metabolism, 52,* 370–372.

Davoren, J. B., & Hsueh, J. W. (1986). Growth hormone increases ovarian a levels of immunoreactive somatomedin-C/insulin-like growth factor *in vivo. Journal of Clinical Endocrinology and Metabolism, 118,* 888–892.

Delemarre-van de Waal, H. A., & Schoemaker, J. (1983). Induction of puberty by prolonged pulsatile LRH administration. *Acta Endocrinologica, 102,* 603–609.

Delemarre-van de Waal, H. A., Van den Brande, J. L., & Schoemaker, J. (1985). Prolonged pulsatile administration of luteinizing hormone-releasing hormone in prepubertal children: Diagnostic and physiologic aspects, *Journal of Clinical Endocrinology and Metabolism, 61,* 859–867.

Dierschke, D. J., Weiss, G., & Knobil, E. (1974). Sexual maturation in the female rhesus monkey and the development of estrogen-induced gonadotropic hormone release. *Endocrinology, 94,* 198–206.

Donovan, B. T., & van der Werff ten Bosch, J. J. (1956). Precocious puberty in rats with hypothalamic lesions. *Nature, 178,* 745.

Drickamer, L. C. (1974). A ten year summary of reproductive data for free ranging *Macaca mulatta. Folia Primatologica, 21,* 61–80.

Drickamer, L. C. (1977). Delay of sexual maturation in female house mice by exposure to grouped females or urine from grouped females. *Journal of Reproduction and Fertility, 51,* 77–81.

Drickamer, L. C. (1983). Male acceleration of puberty in female mice (*Mus musculus*). *Journal of Comparative Psychology, 97,* 191–200.

Drickamer, L. C. (1984). Seasonal variation in acceleration and delay of sexual maturation in female mice by urinary chemosignals. *Journal of Reproduction and Fertility, 72,* 55–58.

Drickamer, L. C. (1986). Effects of urine from females in oestrus on puberty in female mice. *Journal of Reproduction and Fertility, 77,* 613–622.

Drickamer, L. C. (1987). Seasonal variations in the effectiveness of urinary chemosignals influencing puberty in female house mice. *Journal of Reproduction and Fertility, 80,* 295–300.

Drickamer, L. C., & Hoover, J. F. (1979). Effects of urine from pregnant and lactating female house mice on sexual maturation of juvenile females. *Developmental Psychobiology, 12,* 545–551.

Driezen, S., Stone, R. E., & Spies, T. D. (1961). The influence of chronic undernutrition on bone growth in children. *Postgraduate Medicine, 29,* 182.

Driezen, S., Spivakis, C. N., & Stone, R. E. (1967). A comparison of skeletal growth and maturation in undernourished and well-nourished girls before and after menarche. *The Journal of Pediatrics, 70,* 256–263.

Drouva, S. V., Epelbaum, J., Tapia-Arancibia, L., Laplante, E., & Kordon, C. (1981). Opiate receptors modulate LHRH and SRIF release from mediobasal hypothalamic neurons. *Neuroendocrinology, 32,* 163–167.

Dubey, A. K., Cameron, J. L., Steiner, R. A., & Plant, T. M. (1986). Inhibition of gonadotropin secretion in castrated male rhesus monkeys (*Macaca mulatta*) induced by dietary restriction: Analogy with the prepubertal hiatus of gonadotropin release. *Endocrinology, 118,* 518–525.

Ducharme, J. R., & Collu, R. (1982). Pubertal development: Normal, precocious and delayed. *Clinics in Endocrinology and Metabolism, 11,* 57–87.

Dyer, R. G., Mansfield, S., Corbet, H., & Dean, A. D. P. (1985). Fasting impairs LH secretion in female rats by activating an inhibitory opioid pathway. *Journal of Endocrinology, 105,* 91–97.

Ehrenkranz, J. R. L., Tamarkin, L., Comite, F., Johnsonbaugh, R. E., Bybee, D. E., Loriaux, L., & Cutler, G. B., Jr. (1982). Daily rhythm of plasma melatonin in normal and precocious puberty. *Journal of Clinical Endocrinology and Metabolism, 55,* 307–310.

Ellison, P. T. (1981). Threshold hypotheses, developmental age, and menstrual function. *American Journal of Physical Anthropology, 54,* 337–340.

Ferin, M., Wehrenberg, W. B., Lam, N. Y., Alston, E. J., and van de Wiele, R. L. (1982). Effects and site of action of morphine on gonadotropin secretion in the female rhesus monkey. *Endocrinology, 111,* 1652–1656.

Ferin, M., van Vugt, D., & Wardlaw, S. (1984). The hypothalamic control of the menstrual cycle and the role of endogenous opioid peptides. *Recent Progress in Hormone Research, 40,* 441–485.

Fernstrom, J. D. (1979). The influence of circadian variations in plasma amino acid concentrations of monoamine synthesis in the brain. In D. T. Krieger (Ed.) *Endocrine rhythms* (pp. 89–122). New York: Raven Press.

Fitzgerald, J., Michael, F., & Butler, W. R. (1982). Growth and sexual maturation in ewes: Dietary and seasonal effects modulating luteinizing hormone secretion and first ovulation. *Biology of Reproduction, 27,* 864–870.

Foster, D. L. (1977). Luteinizing hormone and progesterone secretion during sexual maturation of the rhesus monkey: Short luteal phases during the initial menstrual cycles. *Biology of Reproduction, 17,* 584–590.

Foster, D. L. (1980). Comparative development of mammalian females: Proposed analogies among patterns of LH secretion in various species. In L. LaCauza & A. W. Root (Eds.), *Problems in pediatric endocrinology* (pp. 193–210). New York: Academic Press.

Foster, D. L. (1983). Photoperiod and sexual maturation of the female lamb: Early exposure to short days perturbs estradiol feedback inhibition of luteinizing hormone secretion and produces abnormal ovarian cycles. *Endocrinology, 112,* 11–17.

Foster, D. L. (1984). Preovulatory gonadotropin surge systems of prepubertal female sheep is exquisitely sensitive to the stimulatory feedback action of estradiol. *Endocrinology, 115,* 1186–1189.

Foster, D. L. (1988). Puberty in female sheep. In E. Knobil and J. Neill (Eds.), *The physiology of reproduction* (pp. 1739–1762). New York: Raven Press.

Foster, D. L., & Olster, D. H. (1985). Effect of restricted nutrition on puberty in the lamb: Patterns of tonic luteinizing hormone (LH) secretion and competency of the LH surge system. *Endocrinology, 116,* 375–381.

Foster, D. L., Ryan, K. D., & Papkoff, H. (1984). Hourly administration of luteinizing hormone induces ovulation in prepubertal female sheep. *Endocrinology, 115,* 1179–1185.

Foster, D. L., Karsch, F. J., Olster, D. H., Ryan, K. D., & Yellon, S. M. (1986). Determinants of puberty in a seasonal breeder. *Recent Progress in Hormone Research, 42,* 331–384.

Foster, D. L., Ebling, F. J. P., Micka, A. F., Vannerson, L. A., Bucholtz, D. C., Wood, R. I., Suttie, J. M., & Fenner, D. E. (1989). Metabolic interfaces between growth and reproduction. I. Nutritional modulation of gonadotropin, prolactin, and growth hormone secretion in the growth-limited lamb. *Endocrinology, 125,* 342–350.

Fraioli, F., Cappa, M., Fabbri, A., Gnessi, L., Moretti, C., Borrelli, P., & Isidori, A. (1984). Lack of endogenous opioid inhibitory tone on LH secretion in early puberty. *Clinical Endocrinology, 20,* 299–305.

Fraser, M. O., Pohl, C. R., & Plant, T. M. (1989). The hypogonadotropic state of the prepubertal male rhesus monkey (*Macaca mulatta*) is not associated with a decrease in hypothalamic gonadotropin-releasing hormone content. *Biology of Reproduction, 40,* 972–980.

French, J. A., Abbott, D. H., & Snowdon, C. T. (1984). The effect of social environment on estrogen excretion, scent marking, and sociosexual behavior in tamarins (*Saguinus oedipus*). *American Journal of Primatology, 6,* 155–167.

Frisch, R. E. (1974). Critical weight at menarche, initiation of the adolescent growth spurt and control of puberty. In M. M. Grumbach, G. D. Grave, & F. E. Meyer (Eds.), *Control of the onset of puberty* (pp. 403–423). New York: John Wiley & Sons.

Frisch, R. E. (1984). Body fat, puberty and fertility, *Biological Review, 59,* 161–188.

Frisch, R. E., & Revelle, R. (1969). The height and weight of adolescent boys and girls at the time of peak velocity of growth in height and weight: Longitudinal data. *Human Biology, 41,* 536–559.

Frisch, R. E., & Revelle, R. (1970). Height and weight at menarche and a hypothesis of critical body weights and adolescent events. *Science, 169,* 397–399.

Frisch, R. E., & Revelle, R. (1971). Height and weight at menarche and a hypothesis of menarche. *Archive of Disease in Childhood, 46,* 695–701.

Frisch, R. E., Wyshak, G., & Vincent, L. (1980). Delayed menarche and amenorrhea of ballet dancers. *New England Journal of Medicine, 303,* 1125–1126.

Froesch, E. R., Schmid, C., Schwander, J., & Zapf, J. (1985). Actions of insulin like growth factors. *Annual Review of Physiology, 47,* 443–467.

Garcia, I. M. P. S., & Whitsett, J. M. (1983). Influence of photoperiod and social environment on sexual maturation in female deer mice (*Peromyscus maniculatus bairdii*). *Journal of Comparative Psychology, 97,* 127–134.

Garn, S. M., LaVelle, M., Rosenberg, K. R., & Hawthorne, V. M. (1986). Maturational timing as a factor in female fatness and obesity. *American Journal of Clinical Nutrition, 43,* 879–883.

Gay, V. L., & Plant, T. M. (1987). N-Methyl-D,L-aspartate elicits hypothalamic gonadotropin-releasing hormone release in prepubertal male rhesus monkeys (*Macaca mulatta*). *Endocrinology, 120,* 2289–2296.

Gay, V. L., & Plant, T. M. (1988). Sustained intermittent release of gonadotropin-releasing hormone in prepubertal male rhesus monkeys induced by N-methyl-DL-aspartic acid. *Neuroendocrinology, 48,* 147–152.

Glass, A. R., & Swerdloff, R. S. (1980). Nutritional influences on sexual maturation in the rat. *Federation Proceedings, 39,* 2360.

Glass, A. R., Anderson, J., Herbert, D., & Vigersky, R. A. (1984). Relationship between pubertal timing and body size in underfed male rats. *Endocrinology, 115,* 19–24.

Gold, H., Spector, S., & Samaan, N. A. (1968). Effects of growth hormone on the carbohydrate metabolism in hypopituitary dwarfs. *Metabolism, 17,* 74.

Goodman, H. G., Grumbach, M. M., & Kaplan, S. L. (1968). Growth and growth hormone II. A comparison of isolated hormone deficiency and multiple pituitary hormone deficiencies in 35 patients with idiopathic hypopituitary dwarfism. *New England Journal of Medicine, 278,* 57.

Green, D. J., & Gillette, R. (1982). Circadian rhythm of firing rate recorded from single cells in the rat suprachiasmatic brain slice. *Brain Research, 245,* 198–200.

Grumbach, M. M., Roth, J. C., Kaplan, S. L., & Kelch, R. P. (1974). Hypothalamic–pituitary regulation of puberty in men: Evidence and concepts derived from clinical research. In M. M. Grumbach, G. D. Grave, & F. E. Mayer (Eds.), *Control of the onset of puberty* (pp. 115–137). New York: John Wiley & Sons.

Gupta, D., Riedel, L., Frick, H. J., Attanasio, A., & Ranke, M. B. (1983). Circulating melatonin in children: in relation to puberty endocrine disorders functional tests and racial origins, *Neuroendocrinology Letters, 5,* 63.

Gwinner, E. (1981). Circannual systems. In J. Aschoff (Ed.), *Handbook of behavioral neurobiology, Vol. 4: Biological rhythms* (pp. 391–410). New York: Plenum Press.

Handelsman, D. J., Spaliviero, J. A., Scott, C. D., & Baxter, R. C. (1987). Hormonal regulation of the peripubertal surge of insulin-like growth factor in the rat. *Endocrinology, 120,* 491–496.

Hansen, P. J., Schillo, K. K., Hinshelwood, M. M., & Hauser, E. R. (1983). Body composition at vaginal opening in mice as influenced by food intake and photoperiod: Tests of critical body weight and composition hypothesis for puberty onset. *Biology of Reproduction, 29,* 924–931.

Harris, D. A., Van Vliet, G., Egli, C. A., Grumbach, M. M., Kaplan, S. L., Styne, D. M., & Vainsal, M. (1985). Somatomedin-C in normal puberty and in time precocious puberty before and after treatment with a potent luteinizing hormone-releasing hormone agonist. *Journal of Clinical Endocrinology and Metabolism, 61,* 152–159.

Hoffman, K. (1981). Photoperiodism in vertebrates. In J. Aschoff (Ed.), *Handbook of behavioral neurobiology, Vol. 4: Biological rhythms* (pp. 449–473). New York: Plenum Press.

Holweg, W., & Dohrn, M. (1932). Uber die Bezeihungen zwischen Hypophysenborderlappen und Keimdrusen. *Klinische Wochenschrift, 6,* 233–235.

Hopwood, N. J., Forsman, P. J., Kenny, F. M., & Drash, A. L. (1975). Hypoglycemia in hypopituitary children. *American Journal Diseases of Children, 129,* 918.

Jacob, R., Barrett, E., Plewe, G., Fagin, K. D., & Sherwin, R. S. (1989). Acute effects of insulin-like growth factor 1 on glucose and amino acid metabolism in the awake fasted rat. *Journal of Clinical Investigation, 83,* 1717–1723.

Jakacki, R. I., Kelch, R. P., Sander, S., Lloyd, J. S., Hopwood, N. J., & Marshall, J. C. (1982). Pulsatile secretion of luteinizing hormone in children. *Journal of Clinical Endocrinology and Metabolism, 55,* 453–458.

Kalra, P. S., Sahu, A., & Kalra, S. P. (1988). Opiate-induced hypersensitivity to testosterone feedback: Pituitary involvement. *Endocrinology, 122,* 997–1003.

Karsch, F. J., Bittman, E. L., Foster, D. L., Goodman, R. L., Legan, S. J., & Robinson, J. E. (1984). Neuroendocrine basis of seasonal reproduction. *Recent Progress in Hormonal Research, 40,* 185–232.

Katz, Y., & Epple, G. (1979). The coming of age in female *Saguinus*. *Endocrinology, 104,* 259.

Kelch, R. P., Markovs, M., & Hess, J. (1976). LH and FSH responsiveness to intravenous gonadotropin-releasing hormone (GnRH) in children with hypothalamic or pituitary disorders: Lack of effect of replacement therapy with human growth hormone. *Journal of Clinical Endocrinology and Metabolism, 42,* 1104–1113.

Kelch, R. P., Marshall, J. C., Sauder, S., Hopwood, N. J., & Reame, N. E. (1983). Gonadotropin regulation during human puberty. In R. L. Norman (Ed.), *Neuroendocrine aspects of reproduction* (pp. 229–253). New York: Academic Press.

Kennaway, D. J., Peek, J. C., Gilmore, T. A., & Royles, P. (1986). Pituitary response to LHRH, LH pulsatility and plasma melatonin and prolactin changes in ewe lambs treated with melatonin implants to delay puberty. *Journal of Reproduction and Fertility, 78,* 137–148.

Kennedy, G. C., & Mitra, J. (1963). Body weight and food intake as initiation factors for puberty in the rat. *Journal of Physiology, 166,* 408–418.

Kirkland, R. T., Lin, T. H., & Kirkland, J. L. (1987). Growth hormone studies and therapeutic trails in Turner syndrome. In *Annual Meeting of the Endocrine Society* (abstract 230). Baltimore: Williams & Wilkins.

Knobil, E. (1974). On the control of gonadotropin secretion in the rhesus monkey. *Recent Progress in Hormone Research, 30,* 1–32.

Knobil, E. (1980). The neuroendocrine control of the menstrual cycle. *Recent Progress in Hormone Research, 36,* 53–88.

Knobil, E., & Hotchkiss, J. (1988). The menstrual cycle and its neuroendocrine control. In E. Knobil and J. Neill (Eds.), *Physiology of reproduction* (pp. 1971–1994). New York: Raven Press.

Krey, L. C., Hess, D. L., Butler, W. R., Espinosa-Campos, J., Lu, K. H., Piva, F., Plant, T. M., & Knobil, E. (1981). Medial basal hypothalamic disconnection and the onset of puberty in the female rhesus monkey. *Endocrinology, 108,* 1944–1948.

Kulin, H. E., Grumbach, M. M., & Kaplan, S. L. (1969). Changing sensitivity of the pubertal gonadal hypothalamic feedback mechanism in man. *Science, 166,* 1012–1013.

Kulin, H. E., Bwibo, N., Mutie, D., & Santner, S. (1982). The effect of chronic childhood malnutrition on pubertal growth and development. *American Journal of Clinical Nutrition, 36,* 527–536.

Kulin, H. E., Samojlik, E., Demers, L. M., & Santner, S. J. (1985). Response of the prepubertal ovary to acute chorionic gonadotropin administration: Absence of modulation by growth hormone. *Journal of Clinical Endocrinology and Metabolism, 60,* 208–211.

Lenko, H. L., Lang, U., Aubert, M. L., Paunier, L., & Sizonenko, P. C. (1982). Hormonal changes in puberty. VII, lack of variation of daytime plasma melatonin. *Journal of Clinical Endocrinology and Metabolism, 54,* 1056.

Lincoln, G. A., & Short, R. V. (1980). Seasonal breeding: Nature's contraceptive. *Recent Progress in Hormone Research, 36,* 1–52.

Lotz, W. G., Saxton, J. L., & Hess, D. L. (1987). Seasonal patterns of menarche and ovulation in female rhesus monkeys raised under controlled environmental conditions. In *Annual Meeting of the American Society of Experimental Biologists* (abstract 1229). Lancaster, Pennsylvania: Lancaster Press.

Loy, J., Loy, K., Kiefer, G., & Conaway, C. (1984). *The behavior of gonadectomized rhesus monkeys.* New York: S. Karger.

Mansfield, M. J., Bearsworth, D. E., Loughlin, J. S., Crawford, J. D., Bode, H. H., Rivier, J., Vale, W., Kushner, D. C., Crigler, J. F., & Crowley, W. F. (1983). Long-term treatment of central precocious puberty with a long-acting analogue of luteinizing hormone-releasing hormone. *New England Journal of Medicine, 309,* 1286–1290.

Marshall, W. A. (1974). Interrelationships of skeletal maturation sexual development and somatic growth in man. *Annals of Human Biology, 1,* 29–40.

Marshall, W. A., & Swan, A. V. (1971). Seasonal variation in growth rates of normal and blind children. *Journal of the Society for the Study of Human Biology, 43,* 502–516.

Marshall, W. A., & Tanner, J. M. (1970). Variation in the pattern of pubertal changes in boys. *Archives of Disease in Childhood, 45,* 3.

Martin, J. B. (1986). Regulation of growth hormone secretion. In S. Raiti & R. A. Tolman, (Eds.), *Human growth hormone* (pp. 303–323). New York: Plenum Press.

Massey, A., & Vandenbergh, J. G. (1980). Puberty delay by a urinary cue from female house mice in feral populations. *Science, 209,* 821–822.

Mauras, N., Veldhuis, J. D., & Rogol, A. D. (1986). Role of endogenous opiates in pubertal maturation: Opposing actions of naltrexone in prepubertal and late pubertal boys. *Journal of Clinical Endocrinology and Metabolism, 62,* 1256–1263.

Mayhew, D. A., Wright, P. H., & Ashmore, J. (1969). Regulation of insulin secretion. *Pharmacology Review, 21,* 184–212.

McCann, J. P. (1984). Effects of acute changes in insulin and glucose metabolism on LH and progesterone production. *Biology of Reproduction, 30,* 90.

McClintock, M. K. (1971). Menstrual synchrony and suppression. *Nature, 229,* 244–245.

McLachlan, R. I., Robertson, D. M., Healy, D. L., Burger, H. G., & de Krester, D. M. (1987). Circulating immunoreactive inhibin levels during the normal human menstrual cycle. *Journal of Clinical Endocrinology and Metabolism, 65,* 654–670.

Monroe, S. E., Yamamoto, M., & Jaffe, R. B. (1983). Changes in gonadotrope responsivity to gonadotropin releasing hormone during development of the rhesus monkey. *Biology of Reproduction, 29,* 422–431.

Moore, R. Y. (1979). The anatomy of central neural mechanisms regulating endocrine rhythms. In D. T. Krieger (Ed.), *Endocrine rhythms* (pp. 63–87). New York: Raven Press.

Moore, R. Y., & Eichler, V. B. (1972). Loss of circadian adrenal corticosterone rhythm following suprachiasmatic lesions in the rat. *Brain Research, 42,* 201–206.

Nakai, Y., Plant, T. M., Hess, D. L., Keogh, E. J., & Knobil, E. (1978). On the sites of the negative and positive feedback actions of estradiol in the control of gonadotropin secretion in the rhesus monkey. *Endocrinology, 102,* 1008–1014.

Neuwelt, E. A., Mickey, B., & Lewy, A. J. (1986). The importance of melatonin and tumor markers in pineal tumors. *Journal of Neural Transmission, Supplement 21,* 397–414.

Norman, R. L., & Spies, H. G. (1981). Brain lesions in infant female rhesus monkeys: Effects on menarche and first ovulation and on diurnal rhythms of prolactin and cortisol. *Endocrinology, 108,* 1723–1729.

Oerter, K. E., Uriarte, M. M., Rose, S. R., Barnes, K. M., & Cutler, G. B. (1990). Gonadotropin secretory dynamics during puberty in normal girls and boys. *Journal of Clinical Endocrinology and Metabolism, 71,* 1251–1258.

Ojeda, S. R., & Urbanski, H. F. (1988). Puberty in the rat. In E. Knobil & J. Neill (Eds.), *The physiology of reproduction* (pp. 1699–1737). Raven Press, New York.

Ojeda, S. R., Aguado, L. I., & White, S. S. (1983). Neuroendocrine mechanisms controlling the onset of female puberty: The rat as a model. *Progress in Neuroendocrinology, 37,* 306–313.

Orstead, K. M., Hess, D. L., & Spies, H. G. (1987). Opiatergic inhibition of pulsatile luteinizing hormone release during the menstrual cycle of rhesus macaques. *Proceedings of the Society for Experimental Biology and Medicine, 184,* 312–319.

Padmanabham, V., Chappel, S. C., & Beitins, I. Z. (1987). An improved *in vitro* bioassay of FSH in unextracted serum. *Endocrinology, 121,* 1089–1095.

Parker, D. C., Rossman, L. G., Kripke, D. F., Gibson, W., & Wilson, K. (1979). Rhythmicities in human growth hormone concentrations in plasma. In D. T. Krieger (Ed.), *Endocrine rhythms* (pp. 143–173). New York: Raven Press.

Patel, Y. C., & Reichlin, S. (1978). Somatostatin in hypothalamus, extrahypothalamic brain and peripheral tissue of the rat. *Endocrinology, 102,* 523–530.

Patton, R. G., & Gardner, L. I. (1962). Influence of family environment on growth: The syndrome of "maternal deprivation." *Pediatrics, 30,* 957–962.

Penny, R. (1982). Melatonin excretion in normal males and females: Increase during puberty. *Metabolism, 31,* 816–823.

Penny, R. (1985). Episodic secretion of melatonin in pre- and postpubertal girls and boys. *Journal of Clinical Endocrinology and Metabolism, 60,* 751–756.

Penny, R., Olambiwonnu, N. O., & Frasier, S. D. (1977). Episodic fluctuations of serum gonadotropins in pre- and postpubertal girls and boys. *Journal of Clinical Endocrinology and Metabolism, 45,* 307–311.

Penny, R., Goldstein, I. P., & Frasier, S. D. (1978). Gonadotropin excretion and body composition. *Pediatrics, 61,* 294–300.

Petraglia, F., Bernasconi, S., Iughetti, L., Loche, S., Romanini, F., Facchinetti, F., Marcellini, C., & Genazzani, A. R. (1986). Naloxone-induced luteinizing hormone secretion in normal, precocious, and delayed puberty. *Journal of Clinical Endocrinology and Metabolism, 63,* 1112–1116.

Pirke, K. M., Schweiger, V., Lemmel, W., Krieg, J. C., & Berger, M. (1985). The influence of dieting on the menstrual cycle of healthy young women. *Journal of Clinical Endocrinology and Metabolism, 60,* 1174–1179.

Plant, T. M. (1983). Ontogeny of gonadotropin secretion in the rhesus macaque (*Macaca mulatta*). In R. L. Norman (Ed.), *Neuroendocrine aspects of reproduction* (pp. 133–147). New York: Academic Press.

Plant, T. M. (1986). A striking sex difference in the gonadotropin response to gonadectomy during infantile development in the rhesus monkey (*Macaca mulatta*). *Endocrinology, 119,* 539–545.

Plant, T. M. (1988a). Puberty in primates. In: E. Knobil & J. Neill (Eds.), *The physiology of reproduction* (pp. 1763–1788). New York: Raven Press.

Plant, T. M. (1988b). Neuroendocrine basis of puberty in the rhesus monkey (*Macaca mulatta*). In L. Martini & W. F. Ganong (Eds.), *Frontiers in neuroendocrinology* (pp. 215–238). New York: Raven Press.

Plant, T. M., & Dubey, A. K. (1984). Evidence from the rhesus monkey (*Macaca mulatta*) for the view that negative feedback control of luteinizing hormone secretion by the testis is mediated by a deceleration of hypothalamic gonadotropic releasing hormone pulse frequency. *Endocrinology, 115*, 2145–2153.

Plant, T. M., & Zorub, D. S. (1986). Pinealectomy in agonadal infantile male rhesus monkeys (*Macaca mulatta*) does not interrupt initiation of the prepubertal hiatus in gonadotropin secretion. *Endocrinology, 118*, 227–232.

Plant, T. M., Gay, V. L., Marshall, G. R., & Arslan, M. (1989). Puberty in monkeys is triggered by chemical stimulation of the hypothalamus. *Proceedings of the National Academy of Sciences U.S.A., 86*, 2506–2510.

Powell, G. F., Brasel, J. A., Raiti, S., & Blizzard, R. M. (1967). Emotional deprivation and growth retardation simulating hypopituitarism. II. Endocrinologic evaluation of the syndrome. *New England Journal of Medicine, 276*, 1279–1285.

Quabbe, H. J., Gregor, M., Bumke-Vogt, C., Eckhof, A., & Witt, I. (1981). Twenty four hour pattern of growth hormone secretion in the rhesus monkey: Studies including alterations of the sleep/wake and sleep stage cycles. *Endocrinology, 109*, 513–522.

Ramaley, J. A., & Phares, C. K. (1980). Delay of puberty onset in females due to suppression of growth hormone. *Endocrinology, 106*, 1989–1993.

Ramaley, J. A., & Phares, C. K. (1982). Regulation of gonadotropin secretion in the prepubertal period. *Neuroendocrinology, 35*, 439–448.

Ramaley, J. A., & Phares, C. K. (1983). Delay of puberty onset in males due to suppression of growth hormone. *Neuroendocrinology, 36*, 321–329.

Ramirez, V. D., & McCann, S. M. (1963). Comparison of the regulation of luteinizing hormone (LH) secretion in immature and adult rats. *Endocrinology, 72*, 452–465.

Rapisarda, J. J., Bergman, K. S., Steiner, R. A., & Foster, D. L. (1983). Response to estradiol inhibition of tonic luteinizing hormone secretion decreases during the final stage of puberty in the rhesus monkey. *Endocrinology, 112*, 1172–1179.

Reiter, E. O., & Grumbach, M. M. (1982). Neuroendocrine control mechanisms and the onset of puberty. *Annual Review of Physiology, 44*, 595–613.

Reiter, R. J. (1980). The pineal and its hormones in the control of reproduction in mammals. *Endocrine Reviews, 1*, 109–131.

Reiter, R. J., Richardson, B. A., Johnson, L. Y., Ferguson, B. N., & Dihn, D. T. (1980). Pineal melatonin rhythm: Reduction with age in Syrian hamsters. *Science, 210*, 1372–1373.

Reiter, R. J., Richardson, B. A., & King, T. S. (1983). The pineal gland and its indole products: Their importance in the control of reproduction in mammals. In R. Relkin (Ed.), *The pineal gland* (pp. 151–199). New York: Elsevier.

Reppert, S. M., Weaver, D. R., Rivkees, S. A., & Stopa, E. G. (1988). Putative melatonin receptors in a human biological clock. *Science, 242*, 78–81.

Resko, J. A., Goy, R. W., Robinson, J. A., & Norman, R. C. (1982). The pubescent rhesus monkey: Some characteristics of the menstrual cycle. *Biology of Reproduction, 27*, 354–361.

Rivest, R. W., Lang, U., Aubert, M. L., & Sizonenko, P. C. (1985). Daily administration of melatonin delays rat vaginal opening and disrupts the first estrous cycles: Evidence that these effects are synchronized by the onset of light. *Endocrinology, 116*, 779–787.

Rivier, C., Vale, W., Ling, N., Brown, M., & Guillemin, R. (1977). Stimulation *in vivo* of the secretion of prolactin and growth hormone by β-endorphin. *Endocrinology, 100*, 238–241.

Rogel, M. J. (1978). A critical evaluation of the possibility of higher primate reproductive and sexual pheromones. *Psychological Bulletin, 85*, 810–830.

Ross, J. L., Loriaux, D. L., & Cutler, G. B. (1983). Developmental changes in neuroendocrine regulation of gonadotropin secretion in gonadal dysgenesis. *Journal of Clinical Endocrinology and Metabolism, 57*, 288–293.

Rossmanith, W. G., Wirth, U., Sterzik, K., & Yen, S. C. C. (1989). The effects of prolonged opiodergic blockade on LH pulsatile secretion during the menstrual cycle. *Journal of Clinical Investigation, 12*, 245–252.

Ryan, K. D. (1986). Maturation of the hypothalamic–pituitary axis regulating gonadotropin secretion in the primate. *Seminars in Reproductive Endocrinology, 4*, 223–231.

Sauder, S. E., Case, G. D., Hopwood, N. J., Kelch, R. P., & Marshall, J. C. (1984). The effects of opiate antagonism on gonadotropin secretion in children and in women with hypothalamic amenorrhea. *Pediatric Research, 18,* 322–328.

Schanbacher, B. D., & Crouse, J. D. (1981). Photoperiodic regulation of growth: A photosensitive phase during light–dark cycle. *American Journal of Physiology, 241,* E1–E5.

Schneider, J. E., & Wade, G. N. (1989). Availability of metabolic fuels controls estrous cyclicity of Syrian hamsters. *Science, 244,* 1326–1328.

Schultz, N. J., & Terasawa, E. (1988). Posterior hypothalamic lesions advance the time of pubertal changes in luteinizing hormone release in ovariectomized female rhesus monkeys. *Endocrinology, 123,* 445–455.

Schultz, N. J., Terasawa, E., & Goy, R. W. (1986). Delayed puberty induced by lesions of the posterior median eminence in female rhesus monkeys. In *Annual Meeting of the Endocrine Society* (abstract 914). Baltimore: Williams & Wilkins.

Schwartz, S. M., Wilson, M. E., Walker, M. L., & Collins, D. C. (1985). Social and growth correlates of puberty onset in female rhesus monkeys. *Nutrition and Behavior, 2,* 225–232.

Schwartz, S. M., Wilson, M. E., Walker, M. L., & Collins, D. C. (1988). Dietary influences on growth and sexual maturation in premenarchial rhesus monkeys. *Hormones and Behavior, 22,* 231–251.

Silman, R. E., Leone, R. M., Hooper, J. L., & Preece, M. A. (1979). Melatonin, the pineal gland and human puberty. *Nature, 282,* 301–303.

Sirinathsinghji, D. J. S., Motta, M., & Martini, L. (1985). Induction of precocious puberty in the female rat after chronic naloxone administration during the neonatal period: The opiate "break" on prepubertal gonadotropin secretion. *Journal of Endocrinology, 104,* 299–307.

Sisk, C. L., & Bronson, F. H. (1986). Effects of food restriction and restoration on gonadotropin and growth hormone secretion in immature male rats. *Biology of Reproduction, 35,* 554–561.

Sisk, C. L., & Turek, F. W. (1983). Developmental time course of pubertal and photoperiodic changes in testosterone negative feedback on gonadotropin secretion in the golden hamster. *Endocrinology, 112,* 1208–1215.

Sizonenko, P. C., Burr, I. M., Kaplan, S. L., & Grumbach, M. M. (1970). Hormonal changes in puberty. II. Correlation of serum luteinizing hormone and follicle stimulating hormone with stages of puberty and bone age in normal girls. *Pediatric Research, 4,* 36–45.

Smith, S. G. III, Matt, K. S., Prestowitz, W. F., & Stetson, M. H. (1982). Regulation of tonic gonadotropin release in prepubertal female hamsters. *Endocrinology, 110,* 1262–1267.

Smythe, G. A., & Lazarus, L. (1973). Growth hormone regulation by melatonin and serotonin. *Nature, 244,* 230–231.

Smythe, G. A., & Lazarus, L. (1974). Suppression of human growth hormone secretion by melatonin and cyproheptadine. *The Journal of Clinical Investigation, 54,* 116–121.

Sopelak, V. M., Collins, R. L., & Hodgen, G. D. (1985). Follicular stimulation versus ovulation induction in juvenile primates: Importance of gonadotropin-releasing hormone dose. *Journal of Clinical Endocrinology and Metabolism, 62,* 557–562.

Steiner, R. A., & Bremner, W. J. (1981). Endocrine correlates of sexual development in the male monkey, *Macaca fascicularis. Endocrinology, 109,* 914–919.

Steiner, E. A., Cameron, J. L., McNeill, T. H., Clifton, D. K., & Bremner, W. J. (1983). Metabolic signals for the onset of puberty. In R. C. Norman (Ed.), *Neuroendocrine aspects of reproduction* (pp. 183–225). New York: Academic Press.

Straus, D. S., & Takemoto, C. D. (1989). *Effect of fasting and refeeding on IGF-1 mRNA levels and gene transcription in young male rats.* Paper presented at the Annual Meeting of the Endocrine Society, Seattle, WA. Baltimore: Williams & Wilkins.

Styne, D. M., Harris, D. A., Egli, C. A., Conte, F. A., Kaplan, S. L., Rivier, J., Vale, W., & Grumbach, M. M. (1985). Treatment of true precocious puberty with a potent luteinizing hormone-releasing factor agonist: Effect on growth, sexual maturation, pelvic sinography and the hypothalamic–pituitary–gonadal axis. *Journal of Clinical Endocrinology and Metabolism, 61,* 142–151.

Tanner, J. M. (1962). *Growth at adolescence* (2nd ed.). Oxford: Blackwell Scientific Publishers.

Tanner, J. M. (1972). Human growth hormone. *Nature, 237,* 433–439.

Tanner, J. M., & Whitehouse, R. H. (1975). A note on the bone age at which patients with true isolated growth hormone deficiency enter puberty. *Journal of Clinical Endocrinology and Metabolism, 41,* 788–790.

Tanner, J. M., & Whitehouse, R. H. (1976). Clinical longitudinal standards for height, weight, height velocity, weight velocity and stages of puberty. *Archives of Diseases of Childhood, 51,* 170.

Terasawa, E. (1985). Developmental changes in the positive feedback effect of estrogen on luteinizing hormone release in ovariectomized female rhesus monkeys. *Endocrinology, 117,* 2490–2497.

Terasawa, E., Nass, T. E., Yeoman, R. R., Loose, M. D., & Schultz, N. J. (1983). Hypothalamic control of puberty in the female rhesus macaque. In R. L. Norman (Ed.), *Neuroendocrine aspects of reproduction* (pp. 149–182). New York: Academic Press.

Terasawa, E., Bridson, W. E., Nass, T. E., Noonan, J. J., & Dierschke, D. J. (1984a). Developmental changes in luteinizing hormone secretory pattern in peripubertal female rhesus monkeys: Comparisons between gonadally intact and ovariectomized animals. *Endocrinology, 115,* 2233–2240.

Terasawa, E., Noonan, J. J., Nass, T. E., & Loose, M. D. (1984b). Posterior hypothalamic lesions advance the onset of puberty in the female rhesus monkey. *Endocrinology, 115,* 2241–2250.

Tetsuo, M., Poth, M., & Markey, S. P. (1982). Melatonin metabolite excretion during childhood and puberty. *Journal of Clinical Endocrinology and Metabolism, 55,* 311–313.

Turek, F. W., & Campbell, C. S. (1979). Photoperiodic regulation of neuroendocrine–gonadal activity. *Biology of Reproduction, 20,* 32–50.

Underwood, L. E., Clemmons, D. R., Maes, M., D'Ercole, A. J., & Ketelsleger, J. M. (1986). Regulation of somatomedin-C/insulin-like growth factor 1 by nutrients. *Hormone Research, 24,* 166–176.

Vandenbergh, J. G. (1969). Male order accelerates female sexual maturation in mice. *Endocrinology, 84,* 658–660.

Vandenbergh, J. C. (1983). Pheromonal regulation of puberty. In J. G. Vandenbergh (Ed.), *Pheromones and reproduction in mammals* (pp. 95–112). New York: Academic Press.

van der Werff Ten Bosch, J. J., Dierschke, D. J., Terasawa, E., & Slob, A. K. (1983). Anterior hypothalamic lesions and pubertal development in female rhesus monkeys. *Behavioral Brain Research, 7,* 321–330.

van Vugt, D. A., Bakst, G., Dyrenfurth, I., & Ferin, M. (1983). Naloxone stimulation of luteinizing hormone secretion in the female monkey: Influence of endocrine and experimental conditions. *Endocrinology, 113,* 1858–1864.

Veldhuis, J. D., Rogol, A. D., & Johnson, M. L. (1985). Minimizing false-positive errors in hormone pulse detection. *American Journal of Physiology, 248,* E475–E481.

Vician, L., Adams, L. A., & Clifton, D. K. (1989). *Pro-opiomelanocortin messenger RNA in the arcuate nucleus increases with puberty in the male macaque.* Paper presented at the Annual Meeting of the Endocrine Society, Seattle, WA.

Vigersky, R. A., Anderson, A. E., Thompson, R. H., & Loriaux, D. L. (1977). Hypothalamic dysfunction in secondary amenorrhea associated with simple weight loss. *New England Journal of Medicine, 297,* 1141–1148.

Warren, M. D. (1983). Effects of under nutrition on reproductive function in the human. *Endocrine Reviews, 4,* 363–377.

Warren, M. P., & Vande Wiele, R. L. (1973). Clinical and metabolic features of anorexia nervosa. *American Journal of Obstetrics and Gynecology, 117,* 435–449.

Watanabe, G., & Terasawa, E. (1989). *In vivo* release in luteinizing hormone releasing hormone increases with puberty in female rhesus monkeys. *Endocrinology, 125,* 92–99.

Watts, E. S. (1985). Adolescent growth and development of monkeys, apes and humans. In E. S. Watts (Ed.), *Nonhuman primate models for human growth and development* (pp. 41–65). New York: Alan R. Liss.

Weaver, D. R., Namboodiri, M. A. A., & Reppert, S. M. (1988). Iodinated melatonin mimics melatonin action and reveals discrete finding sites fetal brain. *FEBS Letters, 228,* 123–127.

Weinberger, L. M., & Grant, F. C. (1941). Precocious puberty and tumors of the hypothalamus. *Archives of Internal medicine, 67,* 762–792.

Weiss, G., Rifkin, I., & Atkinson, L. E. (1976). Induction of ovulation in premenarchial rhesus monkeys with human gonadotropins. *Biology of Reproduction, 14,* 401–404.

Wentz, A. C. (1980). Body weight and amenorrhea. *Obstetrics and Gynecology, 56,* 482–487.

Wiemann, J. N., Clifton, D. K., & Steiner, R. A. (1989). Pubertal changes in gonadotropin-releasing hormone and proopiomelanocortin gene expression in the brain of the male rat. *Endocrinology, 124,* 1760–1767.

Wildt, L., Marshall, G., & Knobil, E. (1980). Experimental induction of puberty in the infantile female rhesus monkey. *Science, 107,* 1373–1375.

Wilen, R., & Naftolin, F. (1976). Age, weight, and weight gain in the individual pubertal female rhesus monkey (*Macaca mulatta*). *Biology of Reproduction, 15,* 356–360.

Wilson, M. E. (1989). Relationship between growth and puberty in the rhesus monkey. In H. A. Delemarre, T. M. Plant, G. P. van Rees, & J. Schoemaker (Eds.), *Control of the Onset of Puberty III* (pp. 137–149). Amsterdam: Elsevier.

Wilson, M. E., & Gordon, T. P. (1986). Ontogeny of the positive feedback response of luteinizing hormone in seasonal breeding rhesus monkeys. In *Annual Meeting of the Endocrine Society* (abstract 913). Baltimore: Williams & Wilkins.

Wilson, M. E., & Gordon, T. P. (1989a). Nocturnal changes in serum melatonin during female puberty in rhesus monkeys: A longitudinal study. *Journal of Endocrinology, 121,* 553–562.

Wilson, M. E., & Gordon, T. P. (1989b). Season determines the timing of first ovulation in outdoor housed rhesus monkeys. *Journal of Reproduction and Fertility, 85,* 583–591.

Wilson, M. E., & Gordon, T. P. (1989c). Short day melatonin pattern advances puberty in seasonal breeding rhesus monkeys. *Journal of Reproduction and Fertility, 86,* 435–444.

Wilson, M. E., Walker, M. L., & Gordon, T. P. (1983). Consequences of first pregnancy in rhesus monkeys. *American Journal of Physical Anthropology, 61,* 103–110.

Wilson, M. E., Gordon, T. P., Blank, M. S., & Collins, D. C. (1984). Timing of sexual maturity in female rhesus monkeys (*Macaca mulatta*) housed outdoors. *Journal of Reproduction and Fertility, 70,* 625–633.

Wilson, M. E., Schwartz, S. M., Walker, M. L., & Gordon, T. P. (1984). Oestradiol and somatomedin-C influence body weight patterns in premenarchial rhesus monkeys housed outdoors. *Journal of Endocrinology, 102,* 311–317.

Wilson, M. E., Gordon, T. P., & Collins, D. C. (1986). Ontogeny of luteinizing hormone secretion and first ovulation in seasonal breeding rhesus monkeys. *Endocrinology, 118,* 293–301.

Wilson, M. E., Gordon, T. P., Rudman, C. G., & Tanner, J. M. (1988). Effect of a natural versus artificial environment on the tempo of maturation in female rhesus monkeys. *Endocrinology, 123,* 2653–2661.

Wilson, M. E., Gordon, T. P., Rudman, C. G., & Tanner, J. M. (1989). Effects of growth hormone on the tempo of sexual maturation in female rhesus monkeys. *Journal of Clinical Endocrinology and Metabolism, 68,* 29–38.

Winter, J. S. D., & Faiman, C. (1972). Serum gonadotropin concentrations in agonadal children and adults. *Journal of Clinical Endocrinology and Metabolism, 35,* 561–564.

Winter, J. S. D., Ellsworth, L., Fuller, G., Hobson, W. C., Reyes, F. I., & Faiman, C. (1987). The role of gonadal steroids in feedback regulation of gonadotropin secretion at different stages of primate development. *Acta Endocrinologica, 114,* 257–268.

Winterer, J., Cutler, G. B., & Loriaux, D. L. (1984). Caloric balance, brain to body ratio, and the timing of menarche. *Medical Hypothesis, 15,* 87–91.

Wurtman, R. J. (1979). When—and why—should nutritional state control neurotransmitter synthesis? *Journal of Neural Transmission, 15,* (Supplement), 69–79.

Yellon, S. M., & Foster, D. L. (1985). Alternate photoperiods time puberty in the female lamb. *Endocrinology, 116,* 2090–2097.

Yellon, S. M., & Foster, D. L. (1986). Melatonin rhythms time photoperiod-induced puberty in the female lamb. *Endocrinology, 119,* 44–49.

Yen, S. S. C., & Vicic, W. J. (1970). Serum follicle-stimulating hormone levels in puberty. *American Journal of Obstetrics and Gynecology, 106,* 134–137.

Ying, S. Y. (1988). Inhibins, activins and follistatins: Gonadal proteins modulating the secretion of follicle-stimulating hormone. *Endocrine Reviews, 9,* 267–293.

Zacharias, L., & Rand, W. M. (1983). Adolescent growth height and its relation to menarche in contemporary American girls. *Annals of Human Biology, 10,* 209–222.

Zucker, I., & Boshes, M. (1982). Circannual body weight rhythms of ground squirrels: Role of gonadal hormones. *American Journal of Physiology, 243,* R546–R551.

10

Early Androgen and Age-Related Modifications in Female Rat Reproduction

Arnold A. Gerall and Lior Givon

Introduction

Organizational Hypothesis

BASIC ASSUMPTIONS. The organizational hypothesis of sexual differentiation proposed by Phoenix, Goy, Gerall, and Young (1959) set the stage for research establishing that gonadal hormones permanently alter the expression of reproductive physiological and behavioral processes. It distinguished between two modes of steroid hormone action: activational or transient and organizational or permanent consequences on brain action. Organizational effects of hormones were restricted to fetal and neonatal developmental periods, whereas activational effects occur at any time after birth.

According to the organizational hypothesis, the intrinsic tendency of the mammalian brain is to develop female structural and functional characteristics. During a restricted developmental phase the brain is assumed to possess bisexual potentiality, and, if androgen is present at this time, neural differentiation proceeds toward the masculine and away from the feminine direction. Thus, for mammals, circulating androgen or its intracellular metabolite, estradiol-17β (E_2), is considered to be the critical epigenetic factor guiding differentiation of sexually dimorphic neural structure and function (Naftolin *et al.*, 1975).

Arnold A. Gerall and Lior Givon   Department of Psychology, Tulane University, New Orleans, Louisiana 70112.

Sexual Differentiation, Volume 11 in *Handbook of Behavioral Neurobiology*, edited by Arnold A. Gerall, Howard Moltz, and Ingeborg L. Ward, Plenum Press, New York, 1992.

In exploring organizational effects in laboratory rodents, most investigators viewed the late prenatal and early postnatal periods as encompassing the time when androgens have their most potent organizing consequences on reproductive processes. For both male and female sexual processes governed by the brain, this period was empirically determined to be approximately from 5 days before to 10 days after birth in the rat (Flerkó, Petrusz, & Tima, 1967; Gerall & Ward, 1966; Gerall, Hendricks, Johnson, & Bounds, 1967) and between 30 and 65 days before birth in the guinea pig (Goy, Bridson, & Young, 1964). Recent evidence has suggested that the time when hormones can have organizational influences on neurally mediated behavioral processes should be extended to other periods including some around the time of puberty in rats (Arnold & Breedlove, 1985).

EXPERIMENTAL SUPPORT.   Most of the research performed to test the implications of the organizational hypothesis focused on the manipulation of male and female reproductive characteristics. Androgen administered during appropriate perinatal periods was shown to defeminize and masculinize female laboratory rodents, particularly rats, hamsters, and guinea pigs. Defeminized female animals neither display receptive sexual behaviors in the presence of activating hormones nor cyclically release pituitary hormones (Goy & McEwen, 1980). Masculinized female rats can exhibit complete male copulatory behaviors, including those accompanying ejaculation (Ward, 1969). Furthermore, depriving genetic male rodents of perinatal androgen by surgical or chemical methods demasculinizes and feminizes them. Males rats castrated on the day of birth and given adequate androgen replacement in adulthood infrequently mount estrous females and rarely ejaculate (Grady, Phoenix, & Young, 1965). However, when given E_2 and progesterone (P) in adulthood, these males readily exhibit female receptive behaviors and phasically release pituitary gonadotropic hormones, as revealed by the presence of corpora lutea in implanted ovaries (Harris, 1964). As reviewed by W. W. Beatty in Chapter 3, sex differences in nonreproductive behaviors also have been permanently modified by varying perinatal hormone status.

Of significance to the understanding of sexual differentiation are the demonstrations that manipulation of perinatal steroids can modify sexually dimorphic neuronal characteristics in the central nervous system in directions predicted by the organizational hypothesis (Dörner & Staudt, 1968; Gorski, Harlan, Jacobson, Shryne, & Southam, 1980; Raisman & Field, 1973). For many, the fact that perinatal hormones alter neuronal morphology lent credence to the organizational hypothesis not provided by studies using either behavioral or physiological endpoints. In reviewing the investigations carried out in his laboratories on the effects of early hormone exposure on the brain, Gorski (1984) suggested that the sexually dimorphic nucleus of the medial preoptic area in the rat ". . . is a morphological signature of the organizational effects of gonadal hormones on the developing brain (p. 135)." The results of investigations on sexual dimorphisms in the brain are reviewed by S. A. Tobet and T. O. Fox in Chapter 2.

EXTENSION OF HYPOTHESIS TO TIME-DEPENDENT MODIFICATIONS

A logical extension of the organizational hypothesis is that perinatal hormones not only determine masculine–feminine or qualitative but also quantitative aspects of reproductive processes in each animal. Quantitative variables can include amount and timing of secretion of pituitary hormones or intensity and duration of sexual behavior. In influencing reproductive capacity, early hormonal exposure could determine reproductive competence throughout the animal's life span. This chapter reviews results from studies restricted to female reproductive processes in the rat that might help determine whether exposure to neonatal androgen alters the display and underlying mechanisms of reproduction in a manner that relates to age-dependent modifications. It presents one hypothesis, based on the organizational view of how neonatal androgen alters reproductive capacity, that has been examined only at the behavioral level. Also, an overview is presented of neural structures and a limited number of neuroactive substances involved in female rat reproduction. These are viewed in terms of how they might participate in processes modified over time and by neonatal androgen.

MAXIMUM, RATE, AND THRESHOLD PARAMETERS OF RESPONSIVENESS. The basic assumption made in extending the organizational hypothesis is that androgens circulating during the critical period of sexual differentiation establish the maximum level at which the reproductive system can respond to activating hormones later in life (Gerall, Dunlap, & Sonntag, 1980). Specifically, in the laboratory rat, circulating perinatal testosterone and androstenedione or one of their metabolites set the maximum value of neural tissue responsiveness to the activating action of E_2. Furthermore, this maximum value is lowered in direct proportion to both the amount of circulating androgens and the duration of exposure to them during the perinatal sensitive period.

The second assumption made is that the rate of decline in female reproductive processes over time from the maximum value is dependent on both intrinsic and extrinsic factors, similar to those suggested by Finch (1987). Intrinsic factors are localized at the cellular level and could involve changes in DNA or protein synthesis mechanisms that can cause tissues to be less able to utilize or respond to circulating substances. The intrinsic rate of decline is viewed as being independent of exposure to neonatal hormone and therefore as prevailing regardless of reproductive and gonadal history. Being dependent on cellular factors, the rate parameter can have a distinctive value for each organ and brain region. When examining a single component of the reproductive system, one would anticipate that the rate of decline from different maxima will be the same, and hence, curves depicting performance over time would be parallel. Predictions of rate of decline in performance of systems with heterogeneous components are more complicated than those for any one of its components and may vary for animals exposed to different levels of neonatal androgen.

Extrinsic factors superimpose their effects on the basic intrinsic rate of decline. Current research suggests that one of the most likely extrinsic factors that might influence rate of decline of female reproductive function is E_2 (Arai, 1971; LaPolt, Matt, Judd, & Lu, 1986). If E_2 is found to affect the rate of aging, it would be necessary to determine if it is a transitory or permanent effect. If the latter is the case, this effect should be separated from the usual organizational category in that it is restricted to decremental actions on existing cellular structures or on the function of a system and does not involve traditional forms of cellular differentiation, rearrangement, or growth.

A third assumption is that activational thresholds exist that must be exceeded before a physiological or behavioral function can be manifested. Figure 1 summarizes how various levels of perinatal steroids might determine the age at which female rats manifest reduced fertility and, eventually, sterility and loss of receptive

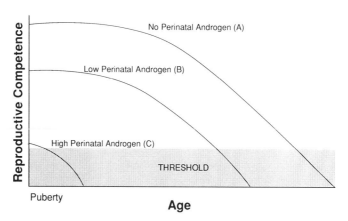

Figure 1. Hypothetical representation of degrees of reproductive competence associated with three levels of neonatal androgen as a function of aging. Reproductive activity is not manifested when values fall below the indicated threshold. Neonatal androgen is represented as determining maximum starting level and not rate parameters.

behavior. Female rats exposed to high levels of exogenous androgen (curve C) will not be reproductively competent from the onset of puberty because their maximum value of responsiveness to physiological amounts of activating hormone has been set below threshold. This curve also applies to male rats, which are intact neonatally, with regard to their ability to display feminine reproductive behaviors and neuroendocrine patterns. With progressively lower levels of perinatal androgen exposure (curves B and A, respectively), spontaneous ovulation and feminine reproductive behaviors would be exhibited after puberty for the period of time it takes for the intrinsically determined decline in responsiveness to pass below threshold.

The last assumption is that time-dependent changes in certain structures and physiological systems might be predicted from a knowledge of the nature of their sexually dimorphic representation in the brain. In some systems, masculine and feminine characteristics can coexist, such as male and female sexual behaviors, and in others they are mutually exclusive, such as the absence or presence of E_2 positive feedback necessary for ovulation. In the former case, it is assumed that the brain areas underlying each of the characteristics are independent so that alterations in one do not directly affect activity in the other. In the latter case, specific brain areas involved in the ovulatory mechanism should be developmentally sexually dimorphic, and neonatal androgens should alter their structure and/or function, rendering them incapable of orchestrating ovulatory processes. It is hypothesized that decreases in ovulatory capability of the female rat over time would be mediated by changes in these underlying neural systems. Furthermore, these changes might involve modifications that make feminine neural morphology and neurochemical systems more similar to those in the male.

SEXUAL DIMORPHISM.   In extending the organizational hypothesis to age-related modifications in reproductive capacity, the analysis of its usefulness must cut across several levels of behavioral and physiological observations. One value of this reductionist approach is that, in general, observed phenomena at the higher levels, such as sexual behaviors or ovulation, are more complex and variable than those at lower levels to which they are being related. For example, female proceptive, receptive, and maternal behaviors are more diverse than the actions of the mediating steroid hormones and their receptors. Thus, it might be possible to identify simpler sets of variables to account for some of the diversity of higher-order activities. Also, if the approach is to be useful and valid, it must be consistent with what is known about participating structural and chemical events.

Pregnancy and Fertility: Age-Related and Neonatal Androgen Modifications

Age-related modifications in the female laboratory rat's reproductive behavior and physiology as she passes from her sexually active and fertile period into an extended phase of sterility and sexual behavior refractoriness have been described in detail by several investigators (Aschheim, 1976, 1983; Finch & Langfield, 1985; Meites, Huang, & Simpkins, 1978). The female rat displays a 4- to 5-day cycle of spontaneous ovulation and sexual receptivity. During the dark phase of the illumination cycle ovulation occurs, and the female exhibits estrous behaviors that, in the presence of a male, eventuate in copulation and conception. After a gestation period of 22–23 days, the young female rat delivers between 4 and 18 offspring.

As the female rat ages, her fertility and likelihood of pregnancy and of exhibiting receptive behaviors gradually decline (Asdell, Bogart, & Sperling, 1941; Ingram, Mandl, & Zuckerman, 1958). The consequences of exposure to a low level of neonatal androgen on these three indices of reproductive capability are shown in Figure 2 (Gerall *et al.*, 1980). In this study, the control animals were injected with oil and experimental littermates with 0.5 μg testosterone propionate (TP) at 3 days of age. In each of the seven mating opportunities provided over 650 days, each female was paired with a vigorous male and, if receptive, received five ejaculations.

At 90 days of age the control and neonatally androgenized females had similar numbers of successful pregnancies and offspring. Starting at 270 days of age the neonatally androgenized females became more variable in their reproductive performance, and thereafter, their measures of fecundity declined more rapidly than those of control animals. Thus, at the biological level, complex reproductive events

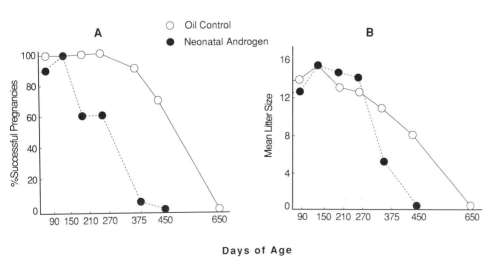

Figure 2. Percentage of successful pregnancies and mean litter size of oil-injected and neonatally androgenized rats as a function of age. (Modified from Gerall *et al.*, 1980).

in the rat quantitatively change over time, and the rate of this change can be influenced by relatively small modifications in the neonatal hormonal milieu.

The rate of decline in the biological measures of reproductive success manifested in the neonatally androgenized animals is the end product of many interacting physiological components. Specifically, successful pregnancy and fertility depend on the neuroendocrine systems regulating ovarian secretions and ovulation as well those mediating receptive behavior, pregnancy, and parturition. Because of dissimilar periods of sensitivity to neonatal androgen, the maximum responsiveness of each component to circulating hormones in adulthood could be altered differentially. This would result in changing the slope of decline in the overall system from that manifested by each of its components. To determine whether neonatal androgen alters only the maximum of responsiveness and not its rate of decline would require studying a single component of the reproductive process, such as estrous behavior induced in gonadectomized animals by exogenous hormones.

RECEPTIVE BEHAVIOR: AGE-RELATED AND NEONATAL ANDROGEN MODIFICATIONS

The intensity of feminine receptive behavior as measured by evaluating lordosis and proceptivity can be shown to be inversely related to age and degree of exposure to neonatal androgen. One example of age-related decline in receptivity is depicted in Figure 3. Female rats ovariectomized (ovx) at 35 days of age had the first of a series of mating tests initiated at either 40, 80, or 120 days of age. Each female received 3.3 μg of estradiol benzoate (EB) 48 hours and 0.5 mg P 4 hours before being tested with a vigorous male. Degree of lordosis intensity and expression of proceptive behaviors were measured in determining receptivity. Receptivity was

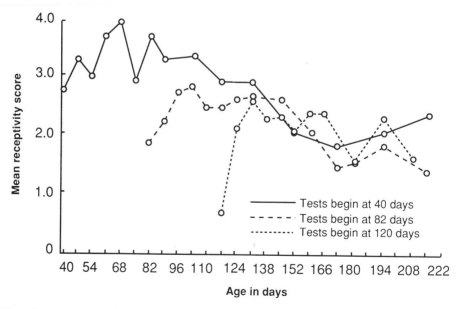

Figure 3.   Mean receptivity scores of female rats ovariectomized at 35 days of age and having their first of a series of mating tests started at 40, 82, or 120 days of age. All animals received the same doses of estrogen and progesterone before each test (Dunlap *et al.*, 1978).

highest at the earlier ages and declined gradually over the 180-day testing period. Following three to five hormone exposures and tests, animals that received their initial tests at older ages reached the same level of receptivity manifested by those that started earlier, and thereafter they declined at a rate related to age (Dunlap, Gerall, & Carlton, 1978).

Using the same activating hormone and testing procedures described above, Dunlap, Gerall, & Hendricks (1972) obtained the same relationship between age and receptivity with males castrated within 6 hours after birth and tested beginning at either 35, 60, 90, or 120 days of age, as recorded with females. As shown in Figure 4, after the commonly observed period of adjustment to the hormones and testing procedure, mean receptivity reached a level displayed by similar-aged animals that started testing earlier and then declined at a rate related to age.

Research has consistently found that expression of receptive behaviors is reduced in direct relationship to the amount and duration of perinatal testosterone exposure. The following two studies support the hypothesis that the maximum level of responsiveness to activating hormone is set by neonatal androgen. The mean level of receptivity manifested by female rats injected with either oil or 10, 50, or 1,250 μg TP at 5 days of age is illustrated in Figure 5 (Gerall & Kenney, 1970). In this study, the inverse relationship between degree of receptivity and level of exposure to neonatal androgen was maintained when dose and timing of activating hormones were varied. Receptivity exhibited by males castrated at successively later days after birth and receiving EB and P is provided in Figure 6. The highest degree of receptivity displayed at each testing age was inversely related to the duration of exposure to neonatal hormone (unpublished data).

In summary, gradual time-dependent declines in fertility, probability of preg-

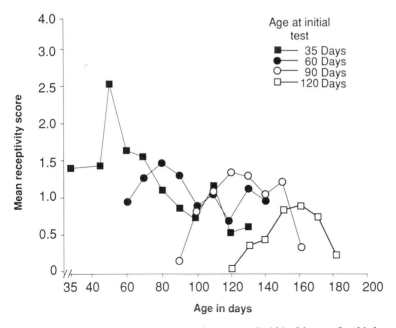

Figure 4.   Mean female receptivity exhibited by males castrated within 6 hours after birth over seven mating tests. Males were injected with 3.3 μg estradiol benzoate and 0.5 mg progesterone before each test. The first test in the series was given at the age of 35, 60, 90, or 120 days. (Adapted from Dunlap *et al.*, 1972.)

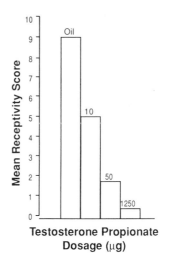

Figure 5. Mean receptivity scores of female rats injected when 5 days old with different doses of testosterone propionate.

nancy, and degree of female receptivity in the rat are systematically altered by androgen circulating during perinatal periods. The data can be interpreted to indicate that the female's maximum reproductive capability or responsiveness is reduced in proportion to the degree of androgen exposure. High levels of neonatal androgen render the female immediately sterile and nonreceptive at the time of

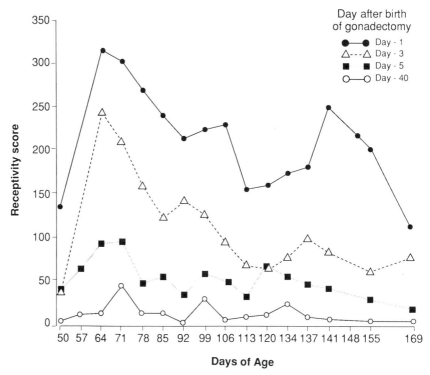

Figure 6. Mean receptivity scores of males neonatally castrated when either 1, 3, 5, or 40 days old as a function of age. All animals received 3.3 μg estradiol benzoate and 0.5 mg progesterone before each test.

puberty. At lower androgen levels, the loss is gradual and deviates from that occurring naturally according to the amount of androgen and the complexity of reproductive activity being observed.

Estrous Cycles and Vaginal Smear Patterns

Normal daily variation in and deviations from the rat's reproductive neuroendocrine cycle can be conveniently and noninvasively detected by evaluating cell types in vaginal smears. The proportion of cell types in the smears is related to the amount of E_2 and P secreted by the ovaries. In the 4- to 5-day cycling rat, diestrus is indicated by smears that contain predominantly leukocytes, proestrus by nucleated epithelial cells, estrus by cornified cells, and metestrus by a mixture of the three cell types. One of the first signs of age-related changes in the reproductive system is irregularity introduced into the daily sequence of vaginal cytology. In most strains of rats, the number of days yielding predominantly cornified cells increases. With the passing of time successively more, and eventually all, of the daily vaginal samples contain only cornified cells. This condition, labeled persistent vaginal estrus or constant estrus (CE), is the most frequently occurring of three types of reproductive senescence in the rat.

A small proportion of rats enter into a repetitive pseudopregnancy condition in which sequences of 9–15 days of predominantly leukocytic smears are interrupted by 1–3 days of cornified smears, indicating not only that corpora lutea are being maintained but also that ovulation is occurring periodically. Lastly, some aging rats have as their initial acyclic state persistent diestrus, which is indicated by unvarying daily leukocytic smears. With increasing age, animals in CE or repetitive pseudopregnancy might develop persistent diestrus (Huang & Meites, 1975).

Of the three senescent states, CE has received the most experimental attention. In the CE condition the ovaries contain many unruptured follicles and no corpora lutea, resulting in an uninterrupted daily secretion of E_2. Progesterone, synthesized by newly formed corpora lutea, is unavailable to alter the vaginal epithelium or behavior. Both the ovary and the pituitary of the CE rat are capable of normal functioning, as shown by their restored cyclic release when transplanted into younger rats (Peng & Huang, 1972). In the rat, experimental evidence strongly indicates that the major neuroendocrine function compromised with the passing of time is associated with the central nervous system, in particular, hypothalamic regulation of the preovulatory luteinizing hormone (LH) surge (Meites, Huang, & Riegle, 1976).

Serum LH

Normal Pattern

Luteinizing hormone released from anterior pituitary gonadotropes promotes gonadal steroidogenesis and, in addition, serves as the primary trigger for follicular rupture in ovulation. In gonadally intact females during the luteal and early follicular phases of the cycle, LH release is under negative feedback control by E_2. The LH is released in a pulsatile, episodic pattern with a species-dependent frequency and amplitude that is demonstrated by removing negative feedback exerted by gonadal vals (Gallo, 1980). Estradiol administered to either gonadectomized female or male rats diminishes LH pulse amplitude. Although LH pulse amplitude of females has been reported to be several times higher than that of males, interpulse latencies do not differ between the sexes (Watts & Fink, 1984).

The primary endocrine factor differentiating the sexes is the phasic preovulatory LH surge. As E_2 produced by developing follicles reaches a critical serum level, negative feedback is turned off, and serum LH levels continue to rise and finally peak between 1400 and 1800 hours on the afternoon of proestrus when rats are maintained in illumination conditions of 14 hours of light and 10 hours of dark with light onset at 0500. Both LH pulse amplitude and frequency increase during the preovulatory surge (Gallo, 1981). Female rats ovx in adulthood and given sufficient E_2 replacement will also exhibit afternoon surges of LH superimposed on the episodic pulsatile pattern of release. Male rats castrated in adulthood and treated with E_2 do not display this afternoon LH surge. Thus, the characteristically feminine phasic release of LH is tied to photoperiod and is dependent on the strength and duration of endogenous or exogenous E_2 exposure. In intact female rats, the preovulatory surge of LH reflects a shift from an inhibitory to a facilitatory or positive-feedback relationship between E_2 and LH.

AGE-RELATED MODIFICATIONS

Characteristics of LH release have been compared in young, middle-aged, and old female rats. In most studies, the young animals are 3–6 months old, middle-aged are 8–14 months old, and old or senescent animals are over 18 months old. Middle-aged animals, designated here as TCE, are in transition between regular cycles and the CE state; TCE animals represent a heterogeneous group with reference to vaginal smear profiles but not necessarily in terms of other indices of hypothalamic–pituitary function.

Senescent, in comparison to young, female rats have a lower average LH content in the pituitary gland (Clemens & Meites, 1971) and, after ovx, two to four times less serum LH (Huang, Marshall, & Meites, 1976). Middle-aged animals had significantly lower mean LH plasma concentrations than young females on the afternoon of proestrus (Gray, Tennent, Smith, & Davidson, 1980). The rise in plasma LH levels following ovx was significantly lower in middle-aged than young animals (Gray et al., 1980). Furthermore, lower amounts of LH were released by exogenous E_2 in both middle-aged cycling and aged noncycling animals than in young cycling animals (Gray et al., 1980; Huang, Steger, Sonntag, & Meites, 1980).

In addition to its concentration, the timing of LH release was modified by age (van der Schoot, 1976). Cooper, Conn, and Walker (1980) found that the extent of the LH rise in middle-aged animals was equal to that of young animals but that it occurred later on the afternoon of proestrus. Recent methodology has enabled researchers to measure more precisely temporal modifications of LH release in middle-aged and old animals. Using indwelling cannulae, Wise, Dueker, and Wuttke (1988) and Estes and Simpkins (1982) sampled LH concentrations in ovx female rats at 10-minute intervals over 3 hours. All of the young females, but only about 75% and 40% of the middle-aged and old animals, respectively, exhibited pulsatile LH release. Mean LH serum concentration as well as pulse frequency was lower in direct relation to age.

Many similarities between the CE condition in naturally aging and neonatally androgenized younger rats are now recognized. Historically, the initial demonstration that perinatal steroids induce reproductive acyclicity and sterility in rodents was provided by Pfeiffer (1936), who found that testes transplanted into newborn female rats caused their ovaries to be polyfollicular and devoid of corpora lutea when examined at 90 days of age. Wilson (1943) reported that estradiol dipropionate administered at either 1, 5, or 10 days of age but not at later ages induced the same anovulatory condition. Barraclough and co-workers (Barraclough, 1955, 1961, 1973; Barraclough & Leathem, 1954), along with Gorski and his colleagues (Gorski, 1973, 1979), have primarily used single injections of TP to explore systematically the action of perinatal steroids on brain–ovarian function. The temporal and dosage parameters for inducing the CE condition and altering hypothalamic function have been defined. Dosages of TP greater than 50 μg injected before 8 days of age uniformly cause the rat to be permanently anovulatory from the onset of puberty.

Investigators were also aware that when TP was injected in amounts lower than 50 μg or after the fifth day of life, a proportion of the experimental animals were not CE at puberty (Gorski & Barraclough, 1963). Swanson and van der Werff ten Bosch (1964) found that female rats receiving low doses of neonatal androgen passed more quickly than oil-injected control animals from regular to irregular vaginal cyclicity before entering the CE state and suggested that these animals were exhibiting premature aging. In describing this phenomenon, Gorski coined the commonly used designation, the delayed anovulatory syndrome, and demonstrated that the extent of the delay between puberty and the onset of CE is inversely related to the neonatal age at which androgen is administered (Gorski, 1968). The inverse relationship between dosage of neonatal androgen and the delay between the onset of puberty and the cessation of vaginal cycles has been confirmed by Napoli and Gerall (1970).

AGE-RELATED AND NEONATAL ANDROGEN MODIFICATIONS

The parallelism between neonatal androgenization and premature aging was shown with reference to the LH system by Harlan and Gorski (1977), who demonstrated that the degree to which serum LH is elevated by exogenous steroids is related to age and dose of neonatal TP. They examined the LH response of three groups of females to graded doses of EB supplemented or not by P at different ages. The control group received no exogenous steroid, a second group (TCE) received a low dose, and a third (CE) a high dose, of neonatal androgen. Intact animals from the control and low- but not from the high-dose androgen groups were exhibiting vaginal cycles at 30, 60, and 120 days of age when the LH tests were performed. At 30 days of age, LH responses to EB were similar in the control and TCE groups but absent in the CE group. At 60 days of age, the LH response of the TCE group to exogenous steroids was significantly lower than that of the control group. At later ages, all groups showed diminished LH responsiveness to exogenous steroids.

A comprehensive series of studies elucidating modifications in the female reproductive system associated with exposure to neonatal androgen has been performed by Barraclough and his colleagues. Lookingland and Barraclough (1982) steroids. Gonadectomy increases LH pulse amplitude and shortens interpulse inter-

and Lookingland, Wise, and Barraclough (1982) compared the profile of pituitary hormones in neonatally androgenized animals that reached the anovulatory state with that in animals still exhibiting vaginal cycles at the time of obtaining blood samples. The experimental subjects were injected with 50 μg TP on day 5 and at 50–55 days of age were separated into two groups according to whether their vaginal smears were persistently cornified (CE animals) or showed vaginal cyclicity (TCE animals). Control animals received oil on day 5. Circulating hormone levels were evaluated in blood sampled hourly from indwelling jugular cannulae and, in some groups, in trunk blood to avoid stress factors associated with cannula implantation and sequential blood sampling.

Informative data were obtained from animals ovx and implanted with E_2-containing Silastic capsules or a sequence of E_2 and P capsules. In this preparation, both the amplitude and the temporal course of LH release initiated by E_2 or P can be determined in the animals without confounding effects from differential ovarian secretions. Those CE animals implanted with E_2 on day 1 exhibited LH surges on days 3 and 4 that were 80% lower than the amount of LH released by similarly implanted control animals. Progesterone provided at 0900 on day 3 elicited a mean LH peak during the afternoon that in CE animals was about 80% of the amplitude of the preovulatory LH peak recorded in intact untreated animals. Identical E_2 and P treatment did not elevate LH above base line in the CE animals. Similar results were obtained in groups from which trunk blood was assayed.

Because the TCE animals in this study exhibited vaginal cycles at 50–55 days of age, their spontaneous release of LH could be compared with that of control animals. Between 1200 and 1600 hours on proestrus, LH increased about 2700% over base line in control and only 68% in TCE animals. The LH levels did not differ between the two groups on other days of the estrous cycle.

After being ovx and implanted with E_2 for 3 days, all control animals displayed LH increases on days 2–4. Only 36% of the TCE animals exhibited elevated LH on day 2, which was about 25% of the magnitude of that measured in control animals. No increase in LH occurred in the CE animals. Trunk blood LH concentrations from ovx animals provided with E_2 or E_2 and P on various days were always on a continuum, with the greatest amount in the control and least amount in the CE animals. When measured after receiving E_2 and P capsules, the control, TCE, and CE animals' LH values at 1700 hours were 2,239, 1,314, and 89 ng/mL, respectively.

In summary, studies by Harlan and Gorski (1977), Lookingland and Barraclough (1982), Lookingland et al. (1982), and other researchers show that time-dependent modifications in LH release include reduced amplitude and delayed pulsatile release in response to E_2. Neonatal androgen has the same consequences on LH as the passing of time. Furthermore, increasing the level of exposure to neonatal androgen increases the similarity of serum LH characteristics to those of older animals. Animals that receive low doses of neonatal androgen and still manifest vaginal and other signs of estrous periodicity share serum LH characteristics of untreated middle-aged animals before they enter into the CE condition. Overall, the preceding data sufficiently support the view that aging and perinatal exposure to androgen have similar effects on the magnitude and timing of LH release to encourage exploration of the extent to which they involve identical components of the underlying neuroendocrine system.

HYPOTHALAMIC DISTRIBUTION

The resting level, ultradian pulse amplitude and frequency, and preovulatory surge of LH are controlled by the hypothalamic gonadotropin-releasing hormone (GnRH). On the day of proestrus, endogenous E_2 triggers the release of sufficient GnRH to initiate the preovulatory LH surge. In the rat, the majority of GnRH-synthesizing cell bodies are located in the diagonal band of Broca (DBB), the medial septum, and the medial preoptic area (MPOA). Other cell bodies are found more caudally in the anterior hypothalamic area (AHA) and medial basal hypothalamus (MBH), mostly in the vicinity of the periventricular region (Barry, Hoffman, & Wray, 1985; Ibata, Watanabe, Kinoshita, Kubo, & Sano, 1979; King, Tobet, Snavely, & Arimura, 1982). Emanating from GnRH perikarya are one laterally and two medially coursing networks of fibers, which are diagrammatically depicted in Figure 7. In general, fibers originating in the DBB and medial septal and rostral MPOA that project to the median eminence (ME) aggregate in a medial pathway that courses along the walls and floor of the third ventricle and under the optic chiasm (Barry *et al.*, 1985; Hoffman & Gibbs, 1982; Merchenthaler, Kovács, Lovász, & Sétáló, 1980). Rostral GnRH neurons also form a middle pathway, running lateral to the periventricular region. Fibers from GnRH neurons located in the more

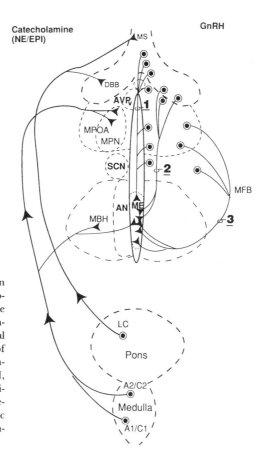

Figure 7. A schematic horizontal brain section illustrating GnRH and NE/EPI cell bodies, processes, and terminal fields. On the right three GnRH pathways to the ME are shown. Abbreviations: AN, arcuate nucleus; AVPv, anteroventral periventricular nucleus; DBB, diagonal band of Broca; EPI, epinephrine; GnRH, gonadotropin-releasing hormone; LC, locus coeruleus; MBH, medial basal hypothalamus; ME, median eminence; MFB, medial forebrain bundle; MPN, medial preoptic nucleus; MPOA, medial preoptic area; MS, medial septal nucleus; NE, norepinephrine; SCN, suprachiasmatic nucleus.

caudal POA and MBH regions project to the ME through a lateral pathway that merges into the medial forebrain bundle (Barry *et al.*, 1985; Merchenthaler, Kovács, *et al.*, 1980). The GnRH neurons terminate in the external palisade layer of the ME, where the GnRH they discharge enters fenestrated capillaries of the hypothalamic–hypophyseal portal system and is transported to anterior pituitary LH gonadotropes.

ABSENCE OF ALTERATIONS BY AGE AND NEONATAL ANDROGEN

Data from diverse experimental approaches indicate that neither the age-related nor the neonatal androgenized CE condition can be attributed to changes in the number or functional capacity of GnRH neurons (Barraclough, Wise, & Selmanoff, 1984). No differences in GnRH fiber density as revealed by immunocytochemistry, or in morphology as analyzed by electron microscopy, have been detected between normal cycling and neonatally androgenized CE females throughout the hypothalamus (King, Tobet, Snavely, & Arimura, 1980; Leranth, Palkovitz, MacLusky, Shanabrough, & Naftolin, 1986). Using immunocytochemical identification of GnRH, Merchenthaler, Lengvári, Horváth, and Sétáló (1980) have reported identical distributions of GnRH in young and middle-aged females. The higher intensity of GnRH-immunoreactive staining in older animals that has been obtained in some studies can be interpreted as indicating an age-related change in release rather than in the synthesis, processing, or axonal transport of the decapeptide. In support of this conclusion, Rubin, King, and Bridges (1984) found no difference between young and older animals either in the immunoreactive characteristics or in the subcellular compartmentalization of the GnRH prohormone and the smaller forms into which it is cleaved. An age-related difference in two types of synaptosome-like particles containing GnRH, fragile and rigid types, has been reported by Barnea, Cho, and Porter (1980). However, the relative amounts of these two forms differ primarily between immature and adult animals; no difference was present between 10- and 24-month-old animals.

Several studies report that GnRH concentrations in the ME and other brain sites do not differ between young and middle-aged animals as well as between androgenized CE animals and their young or cycling controls (Lookingland *et al.*, 1982; Wise, 1982a). No difference in ME GnRH concentration was found between young and old animals before and after ovx and after E_2 replacement (Wilkes & Yen, 1981). Although some studies report lower amounts of GnRH (Wise, 1982a; Wise & Ratner, 1980) and others, higher amounts of GnRH in the hypothalamus of older animals (Barnea *et al.*, 1980), most of the reports conclude that sufficient releasable GnRH is present in the ME to evoke ovulation. Indeed, the ease with which many different pharmacological and natural agents induce ovulation in aged and androgenized CE animals attests to the preservation of the GnRH system (Handa, Condon, Whitmoyer, & Gorski, 1986; Huang *et al.*, 1976; Linnoila & Cooper, 1976; Quadri, Kledzik, & Meites, 1973).

AREAS OF THE BRAIN REGULATING GnRH RELEASE

IDENTIFICATION OF RELEVANT REGIONS

Congruent with the conclusion that modifications in reproductive processes are not directly traceable to alterations in GnRH neurons is the finding by Shivers,

Harlan, Morrell, and Pfaff (1983) that with few possible exceptions, GnRH neurons do not concentrate $[^3H]E_2$ and therefore are incapable of responding directly to E_2 with genomic alterations. Rather, the GnRH regulation of tonic and phasic LH release involves intermediary steroid-receptor-containing neurons that release neuroactive substances in the vicinity of GnRH perikarya, axons, or terminals. The location of brain areas containing E_2-responsive neurons will be considered along with the GnRH system.

A consensus has not been reached as to the relative contributions made by the rostral septal MPOA and the more caudal MBH to the phasic GnRH and LH release characteristics. On the basis of existing evidence at the time, Barraclough and Gorski (1961) proposed that the MPOA mediates the phasic and the MBH the tonic release of LH in the rat. Several lines of evidence encourage refinement of this conception with regard to the locus of control of phasic LH release. Female rats with large lesions in the MPOA manifest repetitive pseudopregnancy, a condition made possible because ovulation recurs periodically (Clemens, Smalstig, & Sawyer, 1976). Furthermore, surgically isolating rostral areas from the MBH does not eliminate the LH pulses that occur approximately every 30 minutes in ovx animals (Soper & Weick, 1980).

Detailed cytoarchitectonic parcellation of the MPOA along with the use of small lesions throughout the extent of the preoptic–hypothalamic continuum has identified specific nuclear areas involved in the regulation of LH release. A brief description of the location of some of these areas within the MPOA and MBH along with results of discrete lesion studies follows. Schematic illustrations of the relative positions of these areas are provided in Figures 8 and 9.

The MPOA and Its Subdivisions

Anteroventral Periventricular Nucleus. The first nuclear grouping encountered at the rostral border of the MPOA is the anteroventral periventricular

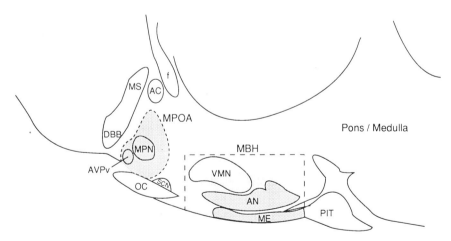

Figure 8. A composite sketch of a midsagittal and an adjacent parasagittal section of the hypothalamus of the rat. Stippled areas encompass neural structures generally included in the MPOA and MBH regions. Abbreviations: AC, anterior commissure; AN, arcuate nucleus; AVPv, anteroventral periventricular nucleus; DBB, diagonal band of Broca; f, fornix; MBH, medial basal hypothalamus; ME, median eminence; MPN, medial preoptic nucleus; MPOA, medial preoptic area; MS, medial septal nucleus; OC, optic chiasm; PIT, pituitary gland; SCN, suprachiasmatic nucleus; VMN, ventromedial nucleus.

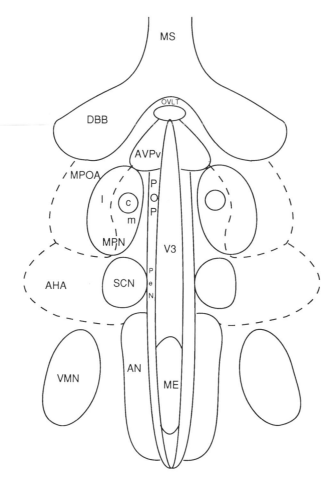

Figure 9. A horizontal section of the brain telencephalic and diencephalic nuclei, some of which are associated with GnRH neurons and the regulation of LH release. Abbreviations: AHA, anterior hypothalamic area; AN, arcuate nucleus; AVPv, anteroventral periventricular nucleus; DBB, diagonal band of Broca; ME, median eminence; MPN, medial preoptic nucleus; MPOA, medial preoptic area; MS, medial septal nucleus; OVLT, organum vasculosa of the lamina terminalis; PeN, hypothalamic periventricular nucleus; POP, preoptic periventricular nucleus; SCN, suprachiasmatic nucleus; V3, third ventricle; VMN, ventromedial nucleus.

nucleus (AVPv). The AVPv lies on either side of the third ventricle, extending 200–300 μm from immediately caudal to the organum vasculosa of the lamina terminalis (OVLT) to the rostromedial border of the medial preoptic nucleus (MPN) (Simerly, Swanson, & Gorski, 1984). The AVPv is one site whose destruction reliably eliminates spontaneous ovulation, induces the CE condition, and eliminates the ability of P to induce an LH surge (Popolow, King, & Gerall, 1981; Wiegand & Terasawa, 1982; Wiegand, Terasawa, & Bridson, 1978). Although the AVPv itself has been reported to be devoid of GnRH-containing cell bodies and to contain few GnRH fibers, GnRH neurons are found in close proximity to this nucleus (Simerly & Swanson, 1987). In addition, the volume of the AVPv is greater in the adult female rat than in the male (Bleier, Byne, & Siggelkow, 1982). Thus, the AVPv encompasses an area of potential influence on GnRH neuronal activity.

PREOPTIC PERIVENTRICULAR AND HYPOTHALAMIC PERIVENTRICULAR NUCLEI. Beginning at the caudal border of the AVPv and continuing throughout the extent of the hypothalamus, the periventricular nucleus borders the third ventricle laterally. In the region of the MPOA, this nucleus is referred to as the preoptic periventricular nucleus (POP), and throughout the hypothalamus proper as the hypothalamic periventricular nucleus (PeN). Some GnRH-containing cell bodies are distributed throughout this nucleus, and GnRH fibers in the medial pathways pass through the POP–PeN on their way to the ME. Thus, the POP and PeN provide other areas of potential influence on transmission in GnRH fibers.

MEDIAL PREOPTIC NUCLEUS.   The medial preoptic nucleus (MPN) begins at the caudal level of the AVPv, just lateral to the POP, and continues to the caudal boundary of the MPOA. It is the largest of the nuclear groups found within the MPOA. Simerly and co-workers (Simerly & Swanson, 1986; Simerly, Gorski, & Swanson, 1986; Simerly *et al.*, 1984) have divided the MPN into three distinct areas based on cyto- and chemoarchitectural characteristics. Each of the subdivisions is sexually dimorphic in the rat, and the dimorphism can be reversed by neonatal steroidal manipulations. The lateral portion of the MPN (MPNl), like the AVPv, is larger in the female than in the male rat. The MPNl does not directly modulate the phasic release of LH. The volume of the medial subdivision of the MPN (MPNm) is larger in the male than in the female rat. Lesions aimed at the AVPv often damage part of the MPNm, suggesting the possibility that this area may be involved in the LH surge. Finally, the central subdivision of the MPN (MPNc) is also larger in the male rat and corresponds to what Gorski *et al.* (1980) identified as the sexually dimorphic nucleus (SDN) of the MPOA. The MPN does not project directly to the ME. However, connections between the MPNm and AVPv (Simerly & Swanson, 1988) may provide for influence of MPNm activity on the LH surge.

THE MBH

ANTERIOR HYPOTHALAMIC AREA AND SUPRACHIASMATIC NUCLEUS.   Immediately caudal to the MPOA is the anterior hypothalamic area (AHA), which is often included as part of the MBH and contains the suprachiasmatic nucleus (SCN). In rodents, the SCN is the center of the endogenous circadian pacemaker (Moore, 1979; Rusak & Zucker, 1975). Lesions placed in the SCN disrupt spontaneous ovulation and induce the CE condition (Brown-Grant & Raisman, 1977). However, unlike the relatively permanent disruption of the LH surge resulting from lesions in the AVPv, the WH surge can be induced in SCN-lesioned females by injections of P (Wiegand *et al.*, 1978). According to Watts, Swanson, and Sanchez-Watts (1987), only a small number of SCN fibers extend directly to and terminate in the MPOA. The largest number of efferents from the SCN terminate in a subparaventricular zone that is bordered dorsally by the paraventricular nucleus, laterally by the AHA, medially by the PeN, and ventrally by the SCN. This zone may serve as a relay point for SCN fibers that are distributed to other parts of the hypothalamus.

Sexual dimorphisms in volume and shape of the SCN have been reported in normal intact male and female rats by Robinson, Fox, Dikkes, and Pearlstein (1986). LeBlond, Morris, Karakiulakis, Powell, and Thomas (1982) reported that the number and length of synaptic contacts are greater in male than in female rats. Neonatal castration of males reduces these measures to values found in females, but neonatal androgenization of females does not raise the values to the level of normal males.

ARCUATE NUCLEUS.   The arcuate nucleus (AN) is located along the ventrolateral third ventricle beginning in the retrochiasmatic area and extending to the caudal border of the hypothalamus (Palkovits & Brownstein, 1988). Many of the neurons within the AN terminate in the external layer of the ME. Lesions placed in the AN can lead to gonadal atrophy (Bogdanove, 1954), indicating its importance in the regulation of tonic gonadotropin release. However, in the rat, most research indicates that the AN does not contain GnRH-synthesizing cell bodies, so that lesions in this area must interrupt GnRH fibers en route to the ME or neurons that interact with GnRH fibers in the ME. The AN is sexually dimorphic in that the female has a larger proportion of axodendritic spine synapses than the male rat, and this difference is reversible with neonatal manipulations (Matsumoto & Arai, 1981).

MEDIAN EMINENCE.   The ME extends throughout the middle half of the AN from the ventral surface of the brain along the floor of the third ventricle to the infundibular stalk. Because of its function as the vascular link between the brain and the pituitary gland, damage to the ME results in numerous endocrine pathologies including gonadal atrophy. Because GnRH neurons terminate here, the ME has been studied as a possible locus of regulation of GnRH release by neurotransmitters.

It appears from studies placing lesions in the hypothalamus that areas of the brain involved in the regulation of LH release do not necessarily contain GnRH neurons. Instead, these areas, such as the AVPv, are in close proximity to GnRH neuronal structures and thus could influence their activity. In addition, as proposed earlier, the brain areas shown to be involved in LH release do display sexual dimorphisms that are determined perinatally by androgens. The AVPv and AN are larger in their respective dimorphisms in the female than in the male, arguably implying that perinatal androgen serves to diminish certain functions controlled by these areas, quite possibly the LH surge. However, in order to act as intermediaries in the regulation of GnRH, neurons in these areas need to be able to respond to E_2. Examination of brain areas that concentrate $[^3H]E_2$ supports the contention that the AVPv and AN may in fact be the areas underlying the ovulatory mechanism in the female rat.

ESTROGEN RECEPTOR LOCATION

The E_2-concentrating neurons in the brain localized by autoradiographic $[^3H]E_2$-labeling methods have been confirmed by immunocytochemical methods using monoclonal antibodies generated against various portions of the E_2 receptor. Based on the detailed autoradiographic studies of Pfaff and Keiner (1973) and Stumpf, Sar, and Keefer (1975), E_2-concentrating neurons that might be involved in regulating GnRH release can be identified in the following brain areas. The AVPv, MPNm, MPNc, and AN are rich in E_2 receptors (Cintra *et al.,* 1986; Simerly, Chang, Muramatsu, Swanson, 1990) whereas the MPNl contains a moderately low concentration of E_2 receptors. In addition, the periventricular nucleus contains E_2 receptors, with the concentration being highest in the POP and gradually declining through the caudal extent of the PeN. The SCN itself has been reported to contain only occasional E_2 receptors; however, the region of its maximal projection, the subparaventricular zone, concentrates E_2.

Thus, nuclei demonstrated through lesion and stimulation studies to be involved in the regulation of GnRH release are in fact capable of genomically responding to E_2 based on the presence of high concentrations of $[^3H]E_2$ binding. Because a substantial portion of the subparaventricular zone contains E_2 receptors, it represents a potential site where E_2 could influence the SCN output and possibly the periodicity of the estrous cycle or other time-related reproductive processes.

Norepinephrine and Epinephrine Regulation
of GnRH and LH Release

331

Androgens and
Female
Reproduction

The MPOA's contributory role in the phasic release of LH is supported by many studies showing that electrical and chemical stimulation of the MPOA induces the LH surge and that intrinsic neurochemical processes are correlated with the occurrence or absence of the LH surge. Barraclough *et al.* (1984) have developed a neural model of ovulation that states that the neuroactive substance providing stimulatory control of GnRH neuronal activity is the catecholamine, norepinephrine (NE). They posit that the endogenous opiate peptides (EOP) provide an indirect inhibitory influence on GnRH activity through direct inhibition of the NE system. Certain groups of NE-containing and EOP-containing cell bodies concentrate $[^3H]E_2$, and thus E_2 can directly alter activity in both of these systems. This model is used here to examine the effects of aging and neonatal androgenization on the neurochemical control of phasic release of LH.

In general, E_2 can act on NE/epinephrine (EPI) cell bodies located in the pons and medulla and on EOP neurons in the AN, AVPv, and brainstem regions. The NE excitatory influence can be mediated anywhere in the GnRH system: NE released from axonal varicosities can act on GnRH cell bodies, their processes, and their terminals in the ME; EOPs can alter NE neuronal activity presynaptically at NE terminals or at the cell body itself. Furthermore, EOP neurons can also directly affect GnRH neurons (Hoffman, Fitzsimmons, & Watson, 1989).

EVIDENCE FOR CONTROL OF LH

Norepinephrine has received a considerable amount of empirical support for having a critical role in determining the occurrence and amplitude of the LH surge. As early as 1949, Everett, Sawyer, and Markee attributed to NE a function in the ovulatory process. When delivered centrally or systemically, NE agonists facilitate and its antagonists inhibit ovulation. A variety of pharmacological agents affect the NE system and thus influence the occurrence of the LH surge and ovulation (Barraclough & Wise, 1982). Barbiturates, such as pentobarbital and phenobarbital, given to proestrous rats block the LH surge and ovulation. Reserpine, which depletes catecholamines from synaptic vesicles, and synthetic enzyme inhibitors, such as α-methyl-*para*-tyrosine (α-MPT), block the surge and ovulation. Also, α-MPT prevents the P-induced LH surge. In addition, antagonists to the α-adrenergic receptor, which is stimulatory to the surge (Leung, Arendash, Whitmoyer, Gorski, & Sawyer, 1982), such as dibenamine, phenoxybenzamine, and chlorpromazine impede ovulation.

Norepinephrine fibers and terminals, as illustrated in Figure 7, are found throughout the septal–preoptic and MBH regions, where they could interact with GnRH cell bodies and processes. Only a few NE fibers appear near GnRH processes in the OVLT and ME (Jennes, Beckman, Stumpf, & Grzanna, 1982). The majority of NE fibers projecting to the forebrain originate from cell bodies located bilaterally in the pons and medulla, designated by Dahlström and Fuxe (1964) as A1, A2, A5, A6, and A7 NE groups. Fibers emanating from the A1 and A2 (nucleus of the solitary tract) medullary NE neuronal groups project to the MPOA, SCN, AN, and ME via the medial forebrain bundle (Moore & Card, 1984). Additionally, NE fibers originating in A6 (locus coeruleus) project to limbic system structures such as the hippocampus and amygdala as well as to the cerebral cortex, cerebellum, and areas involved in olfaction.

Electrical stimulation of A1 following electrochemical stimulation of the contralateral MPN significantly increases serum LH above base line (Gitler & Barraclough, 1988). About one-fifth of the MPOA–medial septal neurons (putative GnRH neurons) that project to the AN/ME respond to A1 stimulation (Kim, Dudley, & Moss, 1987). Finally, the NE-synthesizing cell bodies in A1, A2, and A6 concentrate $[^3H]E_2$ (Sar & Stumpf, 1981). Thus, a structural basis exists for regulation of GnRH neuronal activity by E_2 via the NE system.

The MBH is also innervated by EPI fibers whose cell bodies are located in the brainstem C1 and C2 groups. When administered to proestrous rats, EPI facilitates ovulation (Rubinstein & Sawyer, 1970). Centrally acting selective inhibitors of the enzyme required for synthesis of EPI block LH surges induced by E_2 followed by P in ovx animals (Crowley & Terry, 1981; Crowley, Terry, & Johnson, 1982). In hormonally untreated ovx rats, blocking EPI synthesis does not impair pulsatile LH secretion (Crowley *et al.*, 1982). Thus, EPI could act in concert with NE in mediating E_2 induction of the preovulatory LH surge.

AGE-RELATED MODIFICATIONS

The supposition that release mechanisms are similarly altered by age and neonatal androgenization is supported by findings that both are associated with changes in NE content and activity in parts of the hypothalamus that influence phasic LH release (Lookingland & Barraclough, 1982; Lookingland *et al.*, 1982; Rance, Wise, Selmanoff, & Barraclough, 1981; Wise, 1982b, 1984b). In general, NE content and turnover rate, the latter representing an index of neuronal activity, were measured in the ME, AN, MPN, and SCN at various morning and afternoon intervals of proestrus. Middle-aged and androgenized CE animals, whether intact or ovx and implanted with Silastic E_2 capsules, have lower resting or basal NE levels in the ME and MPN than their younger or cycling control peers (Lookingland *et al.*, 1982; Wise, 1982b). Also, middle-aged cycling animals exhibit an attenuated LH surge and a reduced NE turnover rate in the MPN and ME on the afternoon of proestrus. In general, spontaneous LH surges in these older animals, particularly ovx animals with E_2 Silastic implants, occur later and reach lower peak values than those measured in younger animals.

Norepinephrine turnover rates have been examined in two strains of aging rats by Estes and Simpkins (1984a, 1984b). In the Long–Evans strain, NE turnover rates were lower in old than in young ovx rats in the ME and part of the MPOA. In ovx Fischer 334 rats, NE turnover rates in the MPOA and ME regions were increased rather than decreased. As noted by Estes and Simpkins (1984b), the difference between the results of their study and those obtained by Wise (1982a) could be related to strain differences. As they age, Fischer 334, in contrast to Sprague–Dawley and Long–Evans, rats tend to enter into repetitive pseudopregnancy rather than CE. In addition, animals studied by Estes and Simpkins (1984a, 1984b) were ovx 14 days before the assays were performed and not given steroid replacement. The age-related increase in NE turnover rates in the Fischer 334 rats, therefore, may reflect their response to the ovx or involve neuroendocrine activity associated with the low-E_2 pseudopregnancy condition.

In androgenized CE animals, whether intact or ovx with E_2 implants, NE turnover rates did not increase from their morning level during the afternoon LH surge (Lookingland *et al.*, 1982). In intact androgenized TCE cycling animals, LH significantly increased on proestrous afternoon, but its amplitude was 1/20 of that in control animals. Similarly, ovx TCE animals exhibited spontaneous LH bursts on the second and third days after implantation of E_2 capsules, but their amplitude, particularly on the second day, was significantly lower than that of controls. Correspondingly, afternoon NE turnover rates increased in the ME and MPN of the ovx TCE animals, but the amplitude in the MPN was 42% lower than that in the controls. Thus, an increase in NE turnover rates in the MPN and ME that occurred around the time of LH surges was present in a reduced form in the TCE animals.

Although differing in degree, incremental changes in NE turnover rates in the MPN, and perhaps to a lesser extent in the ME, associated with spontaneous LH surges are similarly reduced in the cycling middle-aged and androgenized animals below that in control animals.

ENDOGENOUS OPIOID REGULATION OF GnRH AND LH RELEASE

CLASSIFICATION AND LOCATION

Evidence from pharmacological and physiological studies has led to the hypothesis that E_2 controls the LH surge not by directly activating NE neurons but by modulating intermediary EOP neurons, which tonically exert a restraining influence on NE release. Other views are emerging, based on anatomic evidence, that the EOP neurons might affect release of LH by acting directly on GnRH neurons (Chen, Witkin, & Silverman, 1989; Hoffman *et al.*, 1989; Morrell, McGinty, & Pfaff, 1985).

Three families of EOPs, proopiomelanocortin (POMC), proenkephalins, and prodynorphins, each derived from separate genes, are widely distributed throughout the central nervous system. Comparatively little work has been published on the role of the prodynorphin family of EOPs in reproductive neuroendocrine function, so it is not discussed here. The POMC family includes β-endorphin, and the proenkephalin family includes both Leu- and Met-enkephalin. Each type of EOP binds preferentially to a specific subtype of EOP receptor. β-Endorphin binds with the highest affinity to the ϵ receptor; however, it also readily binds the μ receptor. Met-enkephalin and the potent EOP agonist morphine preferentially bind the μ receptor subtype. Leu-enkephalin almost always binds only the δ receptor (Morley, Levine, Yim, & Lowy, 1983). However, in some circumstances, such as when high pharmacological doses are used, the specificity of receptor subtype is lost. The most frequently used EOP antagonist, naloxone, binds to all receptor subtypes.

β-Endorphin-synthesizing cell bodies in the hypothalamus are restricted to the AN and the region immediately lateral to it. Their processes project to the ME, POP, PeN, MPOA, SCN, and AHA, among other areas (Barden & Dupont, 1982; Finley, Lindstrom, & Petrusz, 1981). The highest hypothalamic concentrations of β-endorphin, as determined by radioimmunoassay, are found in the AN–ME areas (Barden & Dupont, 1982). Enkephalin-containing perikarya are located throughout the MPOA, SCN, ventromedial nucleus (VMN), and AN, and their processes are found in the POA and ME (Finley, Maderdut, & Petrusz, 1981; Sar, Stumpf, Miller, Chang, & Cuatrecasas, 1978).

RELATIONSHIP TO NE/EPI ROLE IN LH RELEASE

With the possible exception of Leu-enkephalin, all of the EOPs have been demonstrated to be capable of inhibiting the tonic and phasic release of LH (Crowley, 1988; Kalra, 1983; Kalra & Kalra, 1983; Leadem & Kalra, 1985). The EOP agonists reduce or block the preovulatory LH surge (Köves, Marton, Molnár, & Halász, 1981; Pang, Zimmermann, & Sawyer, 1977), decrease basal LH serum levels in ovx animals (Van Vugt, Sylvester, Aylsworth, & Meites, 1982), and reduce the magnitude and frequency of pulsatile LH bursts in ovx animals (Kinoshita *et al.*, 1980; Leadem & Kalra, 1985). The EOP antagonists, such as naloxone and naltrexone, block E_2 negative-feedback inhibition of LH in ovx animals (Van Vugt *et al.*, 1982), greatly increase the LH release in ovx, E_2-primed rats (Van Vugt, Aylsworth, Sylvester, Leung, & Meites, 1981), and advance the time of occurrence of the preovulatory LH surge (Lustig, Fishman, & Pfaff, 1989). That the EOP's effect on LH involves suppression of NE neurons is indicated by the elimination of naloxone's facilitation of LH release when NE synthesis is prevented by α-MPT or α-adrenergic receptors are blocked by phenoxybenzamine (Van Vugt *et al.*, 1981).

Barraclough *et al.* (1984) posit EOP restraint on NE activation of GnRH neurons throughout the entire GnRH system including the ME. Other investigators suggest that NE acts on the more medial components of the GnRH system (Kalra, 1986; Ramirez, Feder, & Sawyer, 1984). The MPN and ME, according to Akabori and Barraclough (1986), are particular locations where EOP modulation of NE release could initiate processes that eventuate in the LH surge. They found that morphine suppresses both the pulsatile release of LH and the increment in NE turnover that usually occurs in the MPN and ME during the afternoon in ovx, E_2-implanted animals. Luteinizing hormone can be released in morphine-treated rats by intraventricular administration of NE, EPI, or clonidine, an α_2-adrenergic agonist (Gallo & Kalra, 1983). This finding indicates that EOPs most likely inhibit release of NE from terminals impinging on GnRH neurons. Additional identification of EOP presynaptic modulation of NE neurons is provided by Diez-Guerra, Augood, Emson, and Dyer (1986), who found that NE released by electrical field stimulation of MPOA slices is inhibited by morphine placed in the perfusion chambers. Because no NE cell bodies are present in the MPOA, in this preparation the morphine must be preventing release of NE from fibers or their terminals.

Impairing either EOP neurons' response to E_2 or their secretory capabilities could lead to loss of cyclic LH surges. Any change that elevates EOP tone would be expected to interfere with normal estrous cycles or LH pulsatile activity. On the other hand, unusually low EOP tone would permit higher levels of NE and GnRH release that could be translated into a constant high level of ovarian E_2 secretion.

AGE-RELATED MODIFICATIONS

Both aging and neonatal androgen influence EOP neuronal function, but whether they do so in a similar manner remains to be determined. The aging CE female rat shares with the neonatally androgenized female and normal male diminished LH responsiveness to naloxone. Field and Kuhn (1988) found that 8- to 11-month-old intact CE animals exhibited significantly smaller rises in serum LH 30 minutes after injections of naloxone than similar 8- to 11-month-old but cycling animals or 6-month or younger females. Increasing the dosage of naloxone did not elevate the level of LH released. These results are viewed by the investigators as reflecting an age-related decrease rather than increase in EOP inhibition. The ineffectiveness of naloxone was attributed to age-related dysfunction in the LH system.

Only a few studies have related EOP brain concentration with aging in the female rat. Met-enkephalin concentrations in the whole hypothalamus were found not to differ between young cycling and older CE animals (Kumar, Chen, & Huang, 1980; Missale *et al.*, 1983). In one study, β-endorphin concentrations did not differ in whole hypothalamus of young and 18- to 20-month-old CE animals (Forman, Sonntag, Hylka, & Meites, 1983), but in another, it was lower in 14- and 25- than in 3-month-old females (Missale *et al.*, 1983). Old rats had significantly lower concentrations of β-endorphin in most of the major hypothalamic nuclei than young animals (Barden & Dupont, 1982). No difference was found between old and young animals in β-endorphin in the ME. The most common finding, therefore, is that hypothalamic EOP concentrations either do not change or decrease as a function of age. Accordingly, the CE state characteristic of senescent rats is not likely to be caused by an increase in opioid tone.

NEONATAL ANDROGEN MODIFICATIONS

Neonatal androgen administered up to day 6 after birth influences the density of opiate receptors in the MPN as measured by [^{3}H]naloxone binding (Hammer, 1984, 1985). The lower density of EOP binding characteristic of males can be duplicated in females by neonatally injecting the latter with TP, and the higher density pattern of females by neonatally castrating the males (Hammer, 1984, 1985). Using [^{3}H]etorphine, a μ-receptor agonist, as the ligand, Bicknell, Dyer, Mansfield, and Marsh (1986) found lower levels of binding in the MPOA, MBH, and stalk ME in males and females exposed to early androgen than in normal females and neonatally castrated males. Thus, the concentration of the μ-receptor subtype is sexually dimorphic and alterable by neonatal androgen manipulations. A lower number is present in the male than the female hypothalamus (Limonta, Maggi, Dondi, & Piva, 1989).

EOP neurons are distributed differently in the MPN of the sexes, each following the cytoarchitectonically established sexually dimorphic volume differences in this region (Simerly, McCall, & Watson, 1988). The volume of the MPNl is larger and contains more proenkephalin-immunoreactive fibers in female than in male rats. The reverse is true for the MPNc, the SDN of Gorski *et al.* (1980), which is larger and contains more proenkephalin-immunoreactive fibers in the male than does the corresponding region of the female.

An impressive difference in EOP distribution between the sexes is seen in the intensity of immunoreactivity to antibodies that bind Leu- and Met-enkephalin in the region of the AVPv. Using a protein E (PE) antibody that recognizes a fragment of the proenkephalin precursor to label both Leu- and Met-enkephalin, Simerly *et al.* (1988) found that males have seven times the number of PE-immunoreactive cell bodies as females in the AVPv. The difference in number of visualizable cell bodies and intensity of staining between the sexes can be reduced by neonatally administering androgen to the female and removing it from the male. Males also had two to three times the number of immunoreactive Leu-enkephalin cell bodies as females. One interpretation of these data provided by the investigators is that neonatal androgen might reduce the rate at which proenkephalin is processed into protein E and smaller, more releasable forms of enkephalins. Within the enkephalinergic neurons of the AVPv, therefore, the female rat may release enkephalins faster or store less than the male.

Interpretation of the EOP binding and immunocytochemical data is complicated by the finding that Met-enkephalin immunoreactivity in the rostral periventricular region, encompassing part of the AVPv, is greater in female than male rats, a difference that is reversible by perinatal treatments (Watson, Hoffman, & Wiegand, 1986). Thus, within a relatively restricted area involved in ovulation, two similar EOPs may be serving different functions. Evidence is available that Leu- and Met-enkephalins affect LH release differently: Met-enkephalin and its analogues inhibit, and Leu-enkephelin stimulates LH release in ovx rats (Leadem & Kalra, 1985).

β-Endorphin content in the MPOA and MBH is greater in female than in male rats. Again, this EOP condition is reversed by neonatal androgen manipulations (Diez-Guerra, Bicknell, Mansfield, Emson, & Dyer, 1987). The authors suggest that neonatal androgen increases the utilization of β-endorphins. Increased EOP release would be correlated with less EOP retained in neurons and compensatory downregulation of EOP receptors. This interpretation is compatible with the general finding of lower EOP binding in males' hypothalamic areas. Also, it suggests that neonatal androgen elevates adult EOP tone, a condition that makes it less likely for a given change in E_2 to increase the amplitude of LH pulses. However, at this time generalizations about overall EOP tone drawn from observations of EOP activity in one part of the brain must be made with reservation.

Higher EOP tone associated with neonatal androgen is congruent with the finding that before 35 days of age naloxone does not induce LH release in intact male and neonatally androgenized female rats but does so in normal females and neonatally castrated males (Sylvester, Sarkar, Briski, & Meites, 1985). Naloxone can increase LH levels in adult intact males as well as in females. However, neonatal androgen in doses of 50 μg or more reliably induces a recalcitrance to naloxone's LH-enhancing effects. Limonta, Dondi, Maggi, Martini, and Piva (1989) report that 1.25 mg TP given on day 2 after birth blocks naloxone's stimulatory LH action in 16-, 26-, and 60-day-old females. Peterson and Barraclough (1986) found that 50 μg of TP given on day 5 eliminates naloxone's potentiation of spontaneous LH pulses exhibited by adult ovx, E_2-implanted females. To distinguish whether NE or EOP neurons were altered by the neonatal androgen treatment, they compared morphine's stimulatory effect on prolactin release in neonatally androgenized and control animals. Morphine is believed to suppress tuberoinfundibular dopaminergic neurons that inhibit prolactin. In this study, however, morphine increased prolactin secretion to the same extent in the neonatally androgenized and control animals. Hence, they conclude that neonatal androgen alters the NE rather than the EOP neuronal system.

Results demonstrating that neonatal androgen modifies the functional capability of EOP neurons have been provided by Grossmann, Diez-Guerra, Mansfield, and Dyer (1987). In one study, serum LH released in response to electrical stimulation of the ventral NE tract was measured in normal and neonatally androgenized animals. Pretreatment with naloxone significantly increased LH release in the stimulated ovx, E_2-implanted, androgenized but not control animals. Because the NE axons were equally activated in the two groups, the enhanced LH is attributed to lessening of opioid tone.

In their other study, hypothalamic slices excised from neonatally androgenized and control animals were preincubated with [^{3}H]NE and subjected to electrical field stimulation. Naloxone added to the superfusion fluid reversed morphine inhibition of electrical-field-induced release of NE. Also, it enhanced to a greater extent NE release from hypothalamic slices taken from androgenized than from control animals. Because the slices contain no NE cell bodies, possible effects of neonatal androgen on NE perikarya or on neurons outside of the MPOA that could affect LH release, such as SCN temporal integrating neurons, were bypassed, and these data reveal direct modification of opioid tone by exogenous neonatal androgen.

In summary, the available data demonstrate that aging and neonatal androgen are associated with functional modifications in the EOP systems in the female rat. However, aging appears to be more related to decreased EOP tone and neonatal androgen with increased EOP tone in hypothalamic areas involved in LH release. Therefore, different factors may underlie naloxone's failure to elevate serum LH levels in aging normal female and young neonatally androgenized rats. In aging animals, the cause may be related more to alterations in the NE and LH systems than in the EOP neurons that act to modulate them. Although the functional significance of the prominent sexual dimorphisms in EOP neurons remains to be clarified, it is unlikely that with the passing of time the female brain is converging toward the masculine pattern of hypothalamic EOP receptive fields and neuronal distribution.

THE SCN AND SEROTONIN INFLUENCE ON LH RELEASE

THE SCN—CIRCADIAN PACEMAKER

The laboratory rat is representative of species whose reproductive physiology and behavior are temporally entrained to environmental photoperiodicities. Expression of the entrainment is dependent on the integrity of an endogenous circadian pacemaker or oscillator that provides baseline signals for coordinating physiological events (Moore, 1979; Zucker, 1976). The preovulatory LH surge is illustrative of this coordination. When the daily light/dark ratio is 14:10, the LH surge is timed to occur once every 4–5 days within a brief 2- to 4-hour window. Altering the light/dark ratio, for example, by maintaining animals in constant light, or disrupting the endogenous timing mechanisms interferes with ovulation.

In addition to the occurrence of the preovulatory LH surge, the expression of most behaviors is controlled by circadian factors. In the female rat, general activity and consummatory and sexual behaviors occur almost exclusively during evening hours. Using measures of locomotion, usually in activity wheels, experimenters have measured the period and other temporal characteristics of endogenous circadian periodicities in free-running situations in which environmental cues are minimized and in entrainted rhythms in which external circadian cues are present. In comparison to all other days of the cycle, not only is the amount of running higher but its onset is earlier on the evening of estrus (Albers, Gerall, & Axelson, 1981; Zucker, 1976). High levels of E_2 advance the onset of activity in both visually entrained and blinded free-running female rats and hamsters. This phase-shortening action of E_2 is not evident in male or in neonatally androgenized female rats (Albers, 1981; Morin, Fitzgerald, & Zucker, 1977). Thus, at least one aspect of the endogenous circadian system is sexually dimorphic. Neonatal androgen, in fostering development of the masculine characteristics in the circadian apparatus, including low sensitivity to E_2, simulates what would occur over time in the female circadian system.

The SCN is known to be the neural region in which the endogenous circadian signals are generated (Zucker, 1976; Rusak & Zucker, 1975). As noted earlier, lesions placed in the SCN produce the CE condition (Brown-Grant & Raisman, 1977). However, it is unlikely that all of the observed physiological changes involved in the CE condition associated with aging and neonatal androgenization are caused solely by modifications in the SCN. One reason, as noted earlier, is that animals with SCN lesions, although no longer spontaneously ovulating, retain the capability of responding to exogenous P with an LH surge (Wiegand et al., 1978). Their positive-feedback capability is intact, whereas it is impaired in aging animals and those neonatally exposed to androgen. It appears that the SCN acts on or serves as a gating mechanism for the separate neural circuitry involving the GnRH system and the neurotransmitters that modulate it (Walker, 1984; Wiegand & Terasawa, 1982). In this capacity it could be a participant in time- and neonatal-androgen-associated effects on reproduction.

Viewing endogenous circadian signals as acting on the GnRH system, Walker (1984) has presented a cogent theoretical formulation of the role of the SCN and serotonin (5-HT), its major neurotransmitter, in the young and aging female rat's ovulatory process. As a guiding tenet, he posits that evolution would not favor senescence resulting from independently occurring alterations or derangements in individual organs or components of physiological systems. Rather, he proposed that senescence most likely results from degenerative modifications in integrative centers that govern the "entire soma" (Walker, 1984). One such center could be the SCN, whose function is to integrate temporally the multiplicity of physiological and behavioral processes necessary for the survival of the individual and the species.

Serotonin: Age-Related Modifications

Serotonin is viewed as the primary neurotransmitter governing the expression of endogenous circadian rhythms, and as such, through its actions on the SCN rather than on the GnRH system, it indirectly influences LH release. By increasing the amplitude of the neurogenic temporal signals emitted from the SCN, 5-HT can facilitate the LH surge through enhanced synchronization of release of NE and other transmitters (Walker, 1984). However, when released in excessive amounts or at unusual rates, 5-HT disrupts the expression of SCN circadian rhythms. Thus, when 5-HT secretion is either above or below certain amplitude and rate values, normal timing of ovulation can be compromised.

Serotonin released in the MPOA and SCN originates from cell bodies located primarily in the medial and dorsal raphe nuclei of the brainstem (Azmita & Segal, 1978; Steinbusch, 1984; van der Kar & Lorens, 1979). The medial and dorsal raphe also supply the 5-HT innervation of the SCN and AN. The dorsal raphe cell bodies are the origin of 5-HT terminals in the ME.

Serotonin can directly inhibit or facilitate LH release (Lenahen, Seibel, & Johnson, 1986; Kalra & Kalra, 1983). Injecting 5-HT into the third ventricle decreases plasma LH in untreated ovx rats but does not alter LH levels on any day of the estrous cycle (Schneider & McCann, 1970). Serotonin might mediate the inhibition of pulsatile LH produced by electrical stimulation of the AN in ovx rats because the inhibition does not occur when 5-HT is depleted and is reinstituted when 5-HT levels are restored to normal levels (Gallo & Moberg, 1977).

A large number of studies have reported that 5-HT facilitates LH release. Blocking 5-HT synthesis by administering *para*-chlorophenylalanine (PCPA) reduces basal LH levels by 50% and eliminates episodic LH release in E_2-treated ovx rats (Héry, Laplante, & Kordon, 1976). Selective depletion of 5-HT in the MPOA lowers basal levels of LH (Johnson & Crowley, 1983). The initiation of LH pulses is preceded by increases in 5-HT turnover rate in the ME in E_2-treated ovx animals (Cohen & Wise, 1988b).

Time-dependent changes in medial raphe 5-HT neurons innervating the SCN could contribute to the increasing variability in the cyclical vaginal smear pattern and the eventual loss of spontaneous ovulation in the aging rat. A CE condition similar to that manifested by older anovulatory females can be induced in young females both by lowering the 5-HT content in the SCN with the 5-HT synthesis blocker PCPA (Walker, Cooper, & Timiras, 1980) and by increasing hypothalamic 5-HT content with the 5-HT precursor 5-hydroxytryptophan. Ovulation can be reinstated in CE animals by pharmacological alteration of 5-HT levels or 5-HT receptor sensitivity (Walker, 1982).

Evidence that can be interpreted to support the hypothesis that changes in the SCN are causally related to some aging effects has been provided by Cohen and Wise (1988b). They found a diurnal rhythm in the 5-HT turnover rate in the SCN that could be reversed by E_2 in young but not in middle-aged ovx rats. At the age when the amplitude of the serum LH is diminished and its onset is delayed, some aspect of the SCN circuitry is losing its reactivity to photoperiodic input and to circulating E_2.

Reservations have been expressed about the role of 5-HT as the dominant neuroactive substance mediating the age-related changes in the female rat (Wise, 1984a). In particular, the timing of 5-HT activity in the SCN is not what would be expected if it were involved in initiating the preovulatory LH surge. Neither the concentration of 5-HT nor that of its metabolite, 5-hydroxyindole acetic acid, increased during the critical time on the afternoon of the LH surge (Cohen & Wise, 1988a). Also, in young ovx females, SCN 5-HT turnover rates were at their highest values during the light phase (Cohen & Wise, 1988a). Administering E_2 to these females reversed the diurnal pattern so that the lowest 5-HT turnover rate was at 1800 hours, which, according to the light–dark schedule used in this study, was after the initiation of the LH surge. Thus, additional information is needed to define how 5-HT activity participates in providing a neurogenic signal through the SCN for phasic LH release.

NOREPINEPHRINE: AGE-RELATED AND NEONATAL ANDROGEN MODIFICATIONS

In addition to 5-HT, NE may influence SCN changes in the expression of circadian rhythms over time. Wise (1982b) found that NE turnover rate in the SCN significantly increased on the afternoon of proestrus from its morning level in young but not middle-aged females. In ovx females the morning-to-afternoon increase in NE turnover rate was present in young animals 2–4 days after implantation of an E_2 capsule, but in middle-aged females it was diminished on the fourth day (Wise, 1984b). Also, the concentration of SCN NE increased over time to a greater extent in the E_2-implanted young than middle-aged females. Thus, the time course of the SCN NE turnover rate indicates that diurnal variations are present in the SCN and that changes occurring in this measure of NE activity in middle-aged females are correlated with alterations in LH surge parameters.

Another manifestation of age-associated altered SCN expression of diurnal rhythmicity is the utilization of glucose, a measure of neuronal activity, in the SCN of ovx, E_2-treated females (Wise, Cohen, Weiland, & London, 1988). In young animals glucose utilization varies diurnally; it starts to rise with the onset of light and shows a significantly sharper increase between 1000 and 1200 hours, the time when SCN NE turnover rate increases. These late morning changes in the SCN, which occur just before the usual time of the preovulatory LH surge, are absent in males and untreated ovx females. Middle-aged females also exhibit a diurnal cycle of glucose utilization, but the increase between 1000 and 1200 hours is not statistically significant. Also, utilization values diminish rapidly after 1200 hours so that no difference is present between the end of the light and the beginning of the dark period. The investigators interpret these modifications in the rhythm and degree of SCN glucose utilization in the middle-aged female as indicating a change in the efficacy of the neural signal for ovulation.

Although few direct empirical comparisons have been made to determine whether neonatal androgen acts in a similar manner as aging on the SCN to produce phenotypically similar losses in reproductive competence, circumstantial physiological and morphological data are available that make it probable. Data obtained by Barraclough and associates (Lookingland & Barraclough, 1982; Lookingland et al., 1982; Wise, 1982b, 1984b) indicate that both neonatal androgenization and aging lower the morning-to-afternoon increase in SCN NE turnover rate in ovx, E_2-treated animals. The SCN NE concentration induced by E_2 is lowered in androgenized females in direct relation to the dose of neonatal TP (Lookingland & Barraclough, 1982).

Neonatal hormonal influences on SCN concentrations of NE have also been examined by Crowley, O'Donohue, and Jacobowitz (1978), who found the concentration to be higher in males than in females. Neonatally androgenized females had NE concentrations intermediate between those of males and normal females, and the NE levels in neonatally castrated males approached those of normal females. However, the hormonal status of the various groups at the time of NE measurement were different from each other. Gonads were not removed in all groups, and identical hormone replacement was not provided. Neonatal androgen in the studies by Lookingland et al. (1982) lowered, and in the study by Crowley et al. (1978) raised, SCN NE levels. However, measurements were made in the former studies on ovx, E_2-primed females and in the latter study on intact females with polyfollicular ovaries. The higher SCN NE levels observed in neonatally androgenized intact animals may have resulted from the higher levels of circulating E_2 in these animals compared to the controls.

In overview, the hypothesis elaborated by Walker (1984) that processes leading to a lessening or dampening in the amplitude or regularity of endogenous circadian signals account for some of the age-related deteriorations in the female rat's ovulatory capacity and receptive behaviors has received considerable empirical support. Although the SCN must be involved in these modifications, additional research is required to discern the relative contribution of 5-HT and NE and of the 10 or so other neuroactive substances localized in the SCN. Although limited, the available data indicate that neonatal androgen modifies the SCN's function in a manner that simulates some of its age-related effects. It is not known whether the aging and neonatal androgen consequences are caused by structural alteration of SCN itself or of cell body regions located elsewhere whose processes innervate it. Because of the paucity of SCN E_2 receptors, it is plausible to consider sites of structural change in areas outside the SCN.

Regulation of LH

Because the preovulatory LH surge occurs in response to rising plasma E_2 levels, its functional integrity should be related to the distribution and affinity of the E_2 receptor for both the hormone and chromatin responsive elements in critical neurons. McGinnis, Krey, MacLusky, and McEwen (1981) reported that the variations in nuclear E_2 receptor levels in neural and hypophyseal tissues parallel those of ovarian E_2 secretion throughout the estrous cycle: maximal levels of nuclear E_2 receptors are found on proestrus, the time of highest circulating E_2 levels, and both E_2 and its nuclear receptors are low on diestrus 1 and estrus. Using these results, the authors developed a protocol of low, medium, and high physiological E_2 replacement in ovx females. The ability of exogenous E_2 to induce an LH surge depended not only on the incremental circulating E_2 levels but also on a concurrent increase in nuclear E_2 receptor levels in neural tissue along with the temporal characteristics of nuclear retention of the E_2–receptor (ER) complex. Using a model similar to that of McGinnis *et al.* (1981), Camp and Barraclough (1985) also found that increasing serum E_2 from low to medium or to relatively high physiological levels in ovx rats significantly elevated nuclear E_2 receptor concentration in the MBH and POA and that these increases correlated highly with E_2- and P-induced amplification of LH release.

The absence of the spontaneous LH surge and many other reproductive dysfunctions shared by aging and neonatally androgenized rats might have as one of their common bases reduced nuclear E_2 receptor availability, decreased affinity of hormone for its receptor, or impaired interaction of the ER complex with chromatin. These types of relationships were suggested before much was known about steroid receptors in the often-stated view that early steroid exposure or aging altered tissue responsiveness to E_2 (Barraclough, 1966; Gerall & Kenney, 1970; Phoenix *et al.*, 1959).

Conceptual and Methodological Considerations

Advances in defining the mechanisms of action of steroid receptors necessitate reliance on a contemporary theory of their temporal interaction with subcellular components. In particular, the model of a cytoplasmic receptor binding E_2 to form an ER complex that is translocated to the cell's nucleus has been modified. Rather, it has been proposed that *in vivo* occupied and unoccupied receptors are localized to the nucleus (King & Greene, 1984; Walters, 1985). On E_2 binding, the ER complex undergoes alterations resulting in binding to chromatin responsive elements, followed by the initiation of transcription. The existence of unoccupied E_2 receptors in the cytoplasm is now believed to be a methodological artifact. Procedural variables, such as the temperature at which assays are run and the volume and salinity of buffers used in extraction and properties inherent to different forms the receptor takes in the presence of ligand along with the strength of ER–chromatin binding are now known to alter the subcellular distribution of the receptor when assayed. Thus, measures of cytosolic unoccupied E_2 receptors may result in a variety of possible interpretations when different parameters of the E_2 receptor are estimated under various hormonal conditions (Walters, 1985).

The underlying assumption guiding researchers in the past was that responsiveness of neural or somatic tissue to E_2 is determined by the number of its receptors available for hormone binding and subsequent genomic activation. Numerous attempts have been made to correlate differences in the ability of E_2 to release LH or induce behavior with measures of E_2 binding capacity and affinity in animals of various ages as well as in those subjected to neonatal manipulations. In general, results from the earliest studies do not offer a consistent set of findings with regard to either aging or sex differences and neonatal treatment, probably because of low assay sensitivity, the large blocks of tissue examined, and the methodological artifacts discussed above. More recent studies utilizing improved methodology and discrete regions of the brain for measuring E_2 receptor binding capacity, distribution, and affinity encourage the acceptance of the hypothesis that aging and neonatal androgen exposure have common subcellular consequences in various neuron groups.

When assessing differences in age groups or in neonatal exposure to steroid hormones, caution must be taken in interpreting results of studies measuring cytosolic E_2 receptors. Based on the methodology used, several interpretations may be possible, including differential strength of binding of the ER complex to chromatin responsive elements (i.e., nuclear retention).

AGE-RELATED MODIFICATIONS

Wise, McEwen, Parsons, and Rainbow (1984) investigated age-related alterations in cytosolic E_2 receptor concentrations in various microdissected brain areas. Cytoplasmic receptors were measured in young and two groups of middle-aged rats 7–14 days after ovx. Compared to young rats, 7- to 8-month-old rats did not differ in cytoplasmic E_2 receptor concentrations, whereas 10- to 11-month-old animals had a significantly lower cytosolic receptor concentration in the SCN–POA, MPN, and VMN. Although not statistically significant, the bed nucleus of the stria terminalis and POP had lower concentrations of cytosolic E_2 receptors in the 10- to 11-month-old group as well.

Using an exchange assay, Wise and Camp (1984) investigated nuclear E_2 receptor concentrations in ovx animals injected with E_2 1 hour before sacrifice. Middle-aged rats had a lower nuclear E_2 receptor concentration in the MPN, whereas the older rats had an additional decrease in receptor concentration in the MBH and pituitary. No change in affinity of the ligand for the receptor was evident with increased age, suggesting that the number of E_2 receptors diminished in the aging animals.

Cytosolic and nuclear E_2 receptor binding were assessed in young and middle-aged long-term ovx rats by Rubin, Fox, and Bridges (1986). The entire hypothalamus–preoptic area (HPOA), amygdala, and pituitary were dissected 1 hour after an injection of E_2. The HPOA nuclear E_2 receptor binding was significantly lower, and cytosolic E_2 receptor binding was significantly higher, in the middle-aged animals. Similar results were obtained in anterior pituitary extracts. Binding levels of either nuclear or cytosolic receptors in amygdala extracts were not different between the two age groups. Using DNA affinity chromatography and LH-20 column chromatography, the authors investigated the capacity of the cytosolic E_2 receptors to bind to DNA. No age-related differences were found in HPOA or pituitary in the capacity of cytosolic receptors to bind to DNA cellulose. Because the affinity of E_2 for its receptors does not vary as a function of age, the authors speculate that the efficiency with which the chromatin binds the ER complex may change with age (Rubin *et al.*, 1986). Similar findings and interpretations were reported previously in mice by Belisle, Bellabarba, and Lehoux (1985). The results of these studies support a hypothesis that the binding capacity of nuclear E_2 receptors in hypothalamic areas involved in ovulation decreases with age.

Jiang and Peng (1981) examined *in vivo* and *in vitro* uptake of $[^3H]E_2$ in the brains of young and old female rats. In the *in vivo* study, 48-hour ovx subjects were sacrificed 0.5, 1, or 3 hours after an IV injection of $[^3H]E_2$. Nuclear binding in the POA at 0.5 and 3 hours and in the MBH at 1 hour were significantly lower in the old than in the young rats. Assays of cytoplasmic and nuclear uptake of $[^3H]E_2$ in pooled hypothalamus and amygdala *in vitro* revealed what the authors interpreted to be impaired translocation of the ER from cytosol to nucleus in old CE rats. However, these results can also be interpreted as indicating a reduction in the strength of ER-complex–chromatin binding in the old CE animals.

NEONATAL ANDROGEN MODIFICATIONS

No consistent differences between the sexes were found in cytosolic E_2 binding (Korach & Muldoon, 1974) or in nuclear binding (Ogren, Vertes, & Woolley, 1976) or as measured by E_2 exchange assays in immature rats (Clark, Campbell, & Peck, 1972). However, when measured within 60 minutes after injecting $[^3H]E_2$, Whalen and Olsen (1978) found significantly less nuclear chromatin binding in whole hypothalamic blocks in males than in females. Examination of the temporal characteristics revealed that the sex difference was in the nuclear retention of the ER complex. During the first 1–2 hours after injection, $[^3H]E_2$ nuclear retention was significantly higher in the female than in the male rats. When injected ip, $[^3H]E_2$ accumulated in the nuclear fraction of the hypothalamic samples from females throughout the 2-hour sampling period, whereas in males, $[^3H]E_2$ nuclear concentration increased in the first hour and then progressively declined (Whalen & Massicci, 1975). These results also point to sex differences in the effectiveness of the ER-complex–chromatin binding. Classical data obtained using uterine tissue indicate that nuclear retention time and not absolute amount of E_2 bound to its receptor determines the maximal uterotrophic actions of E_2 (Anderson, Clark, & Peck, 1972).

Additional evidence for the importance of temporal binding characteristics in determining sex differences in E_2 receptor activity was provided by Nordeen and Yahr (1983), who were among the first researchers to use temporal parameters in analyzing sex differences in nuclear E_2 binding in separate regions of the hypothalamus. They found reliably less nuclear binding in the MBH of males than female following $[^3H]E_2$ injection. These results were independent of the time of sacrifice. Nuclear binding of $[^3H]E_2$ in the POA was lower in the male at 60 but not at 30 minutes after injection. They interpret the temporal difference in binding as reflecting characteristics of the discrete functions of the male's POA. At 30 minutes after E_2 injection, the increased binding level may be associated with the circuitry involved in male copulatory behavior, and at 60 minutes decreased binding may correlate with a defeminized gonadotropin secretory pattern.

Using an E_2 exchange assay on microdissected brain areas, Brown, Hochberg, Zielinski, and MacLusky (1988) examined $[^3H]E_2$ binding in cell nuclear extracts of adrenalectomized–gonadectomized male and female rats administered a near-saturating dose of E_2 iv 1 hour before sacrifice. Females had significantly higher $[^3H]E_2$ binding in nuclear extracts from the periventricular area of the MPOA, the

MPOA itself, and VMN but not from the ARC–ME region. Males had from 52% to 66% of the E_2 binding of females. Using autoradiography with $[^{125}I]MIE_2$ as the probe, which has a lower affinity for the E_2 receptor but higher specific activity than $[^3H]E_2$, they confirmed the sex differences found in the *in vitro* biochemical study.

Decreases in hypothalamic E_2 uptake or binding in animals exposed perinatally to exogenous androgens have been reported frequently. Anderson and Greenwald (1969) found fewer $[^3H]E_2$-labeled cells in the hypothalamus of neonatally androgenized than control female rats. By 60 days of age, females injected neonatally with 1 mg TP had lower $[^3H]E_2$ in cytosolic and nuclear hypothalamic extracts than controls (Vertes & King, 1971). Conflicting data were obtained by Maurer and Woolley (1974), who found no difference in $[^3H]E_2$ uptake in the POA among control females and those injected with 100 μg TP as well as males. However, they used whole tissue, and Whalen and Massicci (1975) found that separation of cytosol and nuclear fractions was necessary to detect sex differences in $[^3H]E_2$ binding. Olsen and Whalen (1980), using the same paradigm as Whalen and Olsen (1978), found less nuclear binding in neonatally androgenized than normal female rats. Nordeen and Yahr (1983) found that females manifesting defeminized gonadotropic and behavioral systems, because they were experimentally exposed to neonatal steroid, had significantly less $[^3H]E_2$ nuclear binding than control female peers in the POA and MBH when examined at 60 minutes after $[^3H]E_2$ injection. Therefore, in light of the cautionary statements regarding interpretation of cytosolic E_2 studies, the results of these studies lend credence to the hypothesis that neonatal androgen modifies the strength of the ER–chromatin interaction.

It can be concluded from the present evidence taken together that neither aging nor neonatal androgen alters the affinity of E_2 for its receptor. Instead, the reproductive dysfunctions associated with aging and neonatal androgen exposure appear to result from an alteration in the strength of the ER-complex–chromatin binding. Estradiol is capable of binding its receptor; however, the nuclear retention time necessary for full estrogenic effects is diminished in hypothalamic areas involved in ovulation and reproductive behavior. If the common locus of action of these two processes is the nuclear chromatin, a large proportion of the variance associated with measures of reproductive capability is being attributed to genomic factors. This does not rule out the possibility that neonatal steroids, particularly when available in high amounts, permanently alter cytoplasmic organelles. Last, it is likely that in the rat the estrogen receptor mediates both organizational and activational effects of hormone action. Thus, any natural or external scripting of this receptor's dynamics must have pervasive physiological and behavioral consequences.

CONCLUSIONS

It is concluded that experimental data support the supposition that age-related modifications in the female rat's reproductive capability can be partially attributed to the steroid hormones active early during fetal and neonatal periods. Steroid hormones derived from the gonads and adrenal glands and perhaps from neighboring fetuses act at the cellular level to set the maximum capacity with which various organs and brain structures will respond to estrogen. The rate of decline from the maximum capacity might be largely determined by other more general intrinsic cellular processes.

The specific intracellular processes influenced permanently by perinatal steroids are not known. One hypothesis that seems to conform conceptually and empirically to the available data is that early hormones modify how nuclear chromatin binds the estrogen–receptor complex and initiates transcription. Endogenous androgens prevailing throughout the perinatal period in the male diminish the capability of chromatin binding or retention of the estrogen–receptor complex to such an extent that brain regions containing them are not responsive to circulating estrogen from the onset of puberty. Thus, when E_2 is available, various organs and neuronal aggregates can not initiate changes in serum LH or receptive behaviors. The same result occurs in the female rat exposed to relatively high levels of exogenous androgen. At lower levels of androgen exposure the chromatin capability is modified less, resulting in the cell's continued responsiveness to E_2. But how this graded effect might be mediated by changes in chromatin binding characteristics is not known. One view is that the lower chromatin binding characteristic affects transcription and, as a consequence, reduces the number of estrogen receptors the cell can synthesize at any given time.

As indicated in the general review, many of the functions of neural regions and neuroactive substances are altered by neonatal androgen and aging. With regard to the critical spontaneous ovulatory mechanism, it is unlikely that modifications in GnRH neurons mediate either age-related or neonatal-androgen-induced diminution or delays in LH release in response to E_2. A convincing case has been made for their modifying activity in the NE and EPI catecholaminergic systems. The site of neonatal steroids and aging's genomic influence probably would be the A1–A2/C1–C2 neuronal groups in the brainstem because they concentrate estrogen and innervate hypothalamic areas implicated in phasic LH release.

β-Endorphin neurons in the AN and enkephalin neurons in the AVPv and other periventricular regions of the hypothalamus concentrate E_2. Prominent sexual dimorphisms in the distribution of EOP fibers and receptive fields have been described in these regions. Also, EOPs can be ascribed a role in the development of sexually dimorphic behaviors (see O. B. Ward Chapter 6). However, the proposed modification of the organizational hypothesis does not readily fit with the functioning of the EOP system. Facilitation of the LH surge involves E_2 acting to decrease EOP release (Lustig *et al.*, 1989). Thus, processes that decrease effectiveness of estrogen receptors in EOP neurons should be associated with increased EOP tone, which is the case with neonatal androgen but apparently not with aging.

Age-related modifications in the functioning of the endogenous pacemaker account for some of the age-related changes in the female rat's reproductive processes. Because neonatal androgen does not appear to alter the functioning of the circadian oscillator itself but its coupling to other physiological systems, and estrogen receptors do not appear to be in abundance in the SCN, it is concluded that aging and neonatal androgen consequences are not the result of a common set of SCN-mediated events.

Probably only a few of the sexually dimorphic areas in the brain are involved in ovulatory processes or receptive behavior. It seems reasonable to assume that sexually dimorphic areas relevant to mechanisms underlying phasic LH release and to receptive behaviors should contain estrogen receptors and should be altered by perinatal hormones and aging in a corresponding manner. Current results from studies involving neonatal hormonal manipulations, estrogen receptor analyses, and physiological measures of neural activity seem to localize the brain regions involved in phasic LH release in basal and medial regions bordering the third ventricle.

Although these regions are limited in number, studies seem to indicate that both aging and neonatal androgen have similar effects on their activity.

Perhaps the VMN is the most appropriate brain site in which to examine the usefulness of this extension of the organizational hypothesis. It is singularly important for the occurrence of receptive behaviors (Edwards & Mathews, 1977; Pfaff & Modianos, 1985), and its concentration of estrogen is sexually dimorphic and influenced by perinatal steroid hormones. The expression of receptive behaviors appears to involve fewer interrelated hypothalamic areas than the ovulatory system. Thus, quantitative expectations of the hypothesis concerning the influence of neonatal androgen and aging on the maximum level and rate of decline in reproductive processes that can be shown in receptive behaviors might also be examined at the physiological and estrogen-receptor level in the VMN.

Acknowledgments

Appreciation is expressed to Ms. Roxanne L. Reger for her constructive input to all phases of this effort and to Dr. Alison Hartman and Tulane Computing Services for indispensable aid in preparing the manuscript for publication.

REFERENCES

Akabori, A., & Barraclough, C. A. (1986). Gonadotropin responses to naloxone may depend upon spontaneous activity in noradrenergic neurons at the time of treatment. *Brain Research, 362*, 55–62.

Albers, H. E. (1981). Gonadal hormones organize and modulate the circadian system of the rat. *American Journal of Physiology, 241*, R62–R66.

Albers, H. E., Gerall, A. A., & Axelson, J. F. (1981). Effect of reproductive state on circadian periodicity in the rat. *Physiology and Behavior, 26*, 21–25.

Anderson, C. H., & Greenwald, G. S. (1969). Autoradiographic analysis of estradiol uptake in the brain and pituitary of the female rat. *Endocrinology, 85*, 1160–1165.

Anderson, J. N., Clark, J. H., & Peck, E. J. (1972). The relationship between nuclear receptor-estrogen binding and uterotrophic responses. *Biochemical and Biophysical Research Communications, 48*, 1460–1468.

Arai, Y. (1971). Some aspects of the mechanisms involved in steroid-induced sterility. In C. H. Sawyer & R. A. Gorski (Eds.), *Steroid hormones and brain function* (pp. 185–191). Los Angeles: University of California Press.

Arnold, A. P., & Breedlove, S. M. (1985). Organizational and activational effects of sex steroids on brain and behavior: A reanalysis. *Hormones and Behavior, 19*, 469–498.

Aschheim, P. (1976). Aging in the hypothalamic–hypophyseal–ovarian axis in the rat. In A. V. Everitt & J. A. Burgess (Eds.), *Hypothalamus, pituitary and aging* (pp. 376–416). Springfield, IL: Charles C. Thomas.

Aschheim, P. (1983). Relation of neuroendocrine system to reproductive decline in female rats. In J. Meites (Ed.), *Neuroendocrinology of aging* (pp. 73–101). New York: Plenum Press.

Asdell, S. A., Bogart, R., & Sperling, G. (1941). The influence of age and breeding upon the ability of the female rat to reproduce and raise young. *Memoirs, Cornell University Agriculture Experiment Station, 238*, 1–26.

Azmita, E. C., & Segal, M. (1978). An autoradiographic analysis of the differential ascending projections of the dorsal and median raphe nuclei in the rat. *Journal of Comparative Neurology, 179*, 641–668.

Barden, N., & Dupont, A. (1982). Distribution and changes in peptides in the brain. In K. McKerns & V. Pantic (Eds.), *Hormonally active brain peptides* (pp. 307–326). New York: Plenum Press.

Barnea, A., Cho, G., & Porter, J. C. (1980). Effects of aging on the subneuronal distribution of luteinizing hormone-releasing hormone in the hypothalamus. *Endocrinology, 106*, 1980–1988.

Barraclough, C. A. (1955). Influence of age on the response of preweaning female mice to testosterone propionate. *American Journal of Anatomy, 97*, 493–522.

Barraclough, C. A. (1961). Production of anovulatory sterile rats by single injections of testosterone propionate. *Endocrinology, 68*, 61–67.

Barraclough, C. A. (1966). Modifications in the CNS regulation of reproduction after exposure to steroid hormones. *Recent Progress in Hormone Research, 22*, 503–539.

Barraclough, C. A. (1973). Sex steroid regulation of reproductive neuroendocrine process. In R. O. Greep (Ed.), *Handbook of physiology: Endocrinology 11, Part 1* (pp. 29–56). Washington, DC: American Physiological Society.

Barraclough, C. A., & Gorski, R. A. (1961). Evidence that the hypothalamus is responsible for androgen-induced sterility in the female rat. *Endocrinology, 68*, 68–79.

Barraclough, C. A., & Leathem, J. H. (1954). Infertility induced in mice by a single injection of testosterone propionate. *Proceedings of the Society of Experimental Biology and Medicine, 85*, 673–674.

Barraclough, C. A., & Wise, P. M. (1982). The role of catecholamines in the regulation of pituitary luteinizing hormone and follicle-stimulating hormone secretion. *Endocrine Reviews, 3*, 91–119.

Barraclough, C. A., Wise, P. M., & Selmanoff, M. K. (1984). A role for hypothalamic catecholamines in the regulation of gonadotropin secretion. *Recent Progress in Hormone Research, 40*, 487–529.

Barry, J., Hoffman, G. E., & Wray, S. (1985). LHRH-containing systems. In A. Björklund & T. Hökfelt (Eds.), *Handbook of chemical neuroanatomy: Vol. 4. GABA and neuropeptides in the CNS: Part I* (pp. 166–215). New York: Elsevier.

Belisle, S., Bellabarba, D., & Lehoux, J.-G. (1985). Age-dependent, ovary-independent decrease in the nuclear binding kinetics of estrogen receptors in the brain of the C57BL/6J mouse. *American Journal of Obstetrics and Gynecology, 153*, 394–401.

Bicknell, R. J., Dyer, R. G., Mansfield, S., & Marsh, M. I. C. (1986). Opioid binding in the rat hypothalamus is sexually differentiated. *Journal of Physiology, 371*, 212P.

Bleier, R., Byne, W., & Siggelkow, I. (1982). Cytoarchitectonic sexual dimorphism of the medial preoptic and anterior hypothalamic areas in guinea pig, rat, hamster, and mouse. *Journal of Comparative Neurology, 212*, 118–130.

Bogdanove, E. M. (1954). Location of hypothalamic lesions affecting the adenohypophysis. *Anatomical Record, 118*, 282–283.

Brown, T. J., Hochberg, R. B., Zielinski, J. E., & MacLusky, N. J. (1988). Regional sex differences in cell nuclear estrogen-binding capacity in the rat hypothalamus and preoptic area. *Endocrinology, 123*, 1761–1770.

Brown-Grant, K., & Raisman, G. (1977). Abnormalities in reproductive function associated with the destruction of the suprachiasmatic nuclei in female rats. *Proceedings of the Royal Society of London, 198*, 279–296.

Camp, P., & Barraclough, C. A. (1985). Correlation of luteinizing hormone surges with estrogen nuclear and progestin cytosol receptors in the hypothalamus and pituitary gland. I. Estradiol dose response effects. *Neuroendocrinology, 40*, 45–53.

Chen, W.-P., Witkin, J. W., & Silverman, A.-J. (1989). Beta-endorphin and gonadotropin-releasing hormone synaptic input to gonadotropin-releasing hormone neurosecretory cells in the male rat. *Journal of Comparative Neurology, 286*, 85–95.

Cintra, A., Fuxe, K., Härfstrand, A., Agnati, L. F., Miller, L. S., Greene, J. L., & Gustafsson, J.-Å. (1986). On the cellular localization and distribution of estrogen receptors in the rat tel- and diencephalon using monoclonal antibodies to human estrogen receptor. *Neurochemistry International, 8*, 587–595.

Clark, J. H., Campbell, P. S., & Peck, E. J., Jr. (1972). Receptor estrogen complex in the nuclear fraction of the pituitary and hypothalamus of male and female immature rats. *Neuroendocrinology, 10*, 218–228.

Clemens, J. A., & Meites, J. (1971). Neuroendocrine status of old constant-estrous rats. *Neuroendocrinology, 7*, 249–256.

Clemens, J. A., Smalstig, E. B., & Sawyer, B. D. (1976). Studies on the role of the preoptic area in the control of reproductive function in the rat. *Endocrinology, 99*, 728–735.

Cohen, I. R., & Wise, P. M. (1988a). Effects of estradiol on the diurnal rhythm of serotonin activity in microdissected brain areas of ovariectomized rats. *Endocrinology, 122*, 2619–2625.

Cohen, I. R., & Wise, P. M. (1988b). Age-related changes in the diurnal rhythm of serotonin turnover in microdissected brain areas of estradiol-treated ovariectomized rats. *Endocrinology, 122*, 2626–2633.

Cooper, R. L., Conn, P. M., & Walker, R. F. (1980). Characterization of the LH surge in middle-aged female rats. *Biology of Reproduction, 23*, 611–615.

Crowley, W. R. (1988). Role of exogenous opioid neuropeptides in the physiological regulation of luteinizing hormone and prolactin secretion. In A. Negro-Vilar & P. M. Conn (Eds.), *Peptide hormones III: Effects and mechanisms of action* (pp. 79–115). Boca Raton: CRC Press.

Crowley, W. R., & Terry, L. C. (1981). Effects of an epinephrine synthesis inhibitor, SKF 64139, on the secretion of luteinizing hormone in ovariectomized female rat. *Brain Research, 204,* 231–235.

Crowley, W. R., O'Donohue, T. L., & Jacobowitz, D. M. (1978). Sex differences in catecholamine content in discrete brain nuclei of the rat: Effects of neonatal castration or testosterone treatment. *Acta Endocrinologica, 89,* 20–28.

Crowley, W. R., Terry, L. C., & Johnson, M. D. (1982). Evidence for the involvement of central epinephrine systems in the regulation of luteinizing hormone, prolactin, and growth hormone release in the rat. *Endocrinology, 110,* 1102–1107.

Dahlström, A., & Fuxe, K. (1964). Evidence for the existence of monoamine-containing neurons in the central nervous system. I. Demonstration of monoamines in the cell bodies of the brainstem neurons. *Acta Physiologica Scandinavica, 62* (Suppl. 232), 1–155.

Diez-Guerra, F. J., Augood, S., Emson, P. C., & Dyer, R. G. (1986). Morphine inhibits electrically stimulated noradrenaline release from slices of rat medial preoptic area. *Neuroendocrinology, 43,* 89–91.

Diez-Guerra, F. J., Bicknell, R. J., Mansfield, S., Emson, P. C., & Dyer, R. G. (1987). Effect of neonatal testosterone upon opioid receptors and the content of β-endorphin, neuropeptide Y and neurotensin in the medial preoptic and the mediobasal hypothalamic areas of the rat brain. *Brain Research, 424,* 225–230.

Dörner, G., & Staudt, J. (1968). Structural changes in the preoptic anterior hypothalamic area of the male rat following neonatal castration and androgen substitution. *Neuroendocrinology, 3,* 136–140.

Dunlap, J. L., Gerall, A. A., & Hendricks, S. E. (1972). Female receptivity in neonatally castrated males as a function of age and experience. *Physiology and Behavior, 8,* 21–23.

Dunlap, J. L., Gerall, A. A., & Carlton, S. R. (1978). Evaluation of prenatal androgen and ovarian secretions on receptivity in female and male rats. *Journal of Comparative and Physiological Psychology, 92,* 280–288.

Edwards, D. A., & Mathews, D. (1977). The ventromedial nucleus of the hypothalamus and the hormonal arousal of sexual behaviors in the female rat. *Physiology and Behavior, 14,* 439–453.

Estes, K. S., & Simpkins, J. W. (1982). Resumption of pulsatile luteinizing hormone release after alpha-adrenergic stimulation in aging constant estrous rats. *Endocrinology, 111,* 1778–1784.

Estes, K. S., & Simpkins, J. W. (1984a). Age-related alterations in dopamine and norepinephrine activity within microdissected brain regions of ovariectomized Long Evans rats. *Brain Research, 298,* 209–218.

Estes, K. S., & Simpkins, J. W. (1984b). Age-related alteration in catecholamine activity within microdissected rat brain regions of ovariectomized Fischer 344 rats. *Journal of Neuroscience Research, 11,* 405–417.

Everett, J. W., Sawyer, C. H., & Markee, J. E. (1949). A neurogenic timing factor in control of the ovulatory discharge of luteinizing hormone in the cyclic rat. *Endocrinology, 44,* 234–250.

Field, E., & Kuhn, C. M. (1988). Opioid inhibition of luteinizing hormone release declines with age and acyclicity in female rats. *Endocrinology, 123,* 2626–2631.

Finch, C. E. (1987). Neural and endocrine determinants of senescence: Investigation of causality and reversibility by laboratory and clinical interventions. In H. R. Warner, R. N. Butler, R. L. Sprott, & E. L. Schneider (Eds.), *Modern biological theories of aging* (pp. 261–308). New York: Raven Press.

Finch, C. E., & Langfield, P. W. (1985). Neuroendocrinology of aging. In C. E. Finch & E. L. Schneider (Eds.), *Handbook of the biology of aging* (2nd ed., pp. 79–90). New York: Van Nostrand Reinhold.

Finley, J., Lindstrom, P., & Petrusz, P. (1981). Immunocytochemical localization of β-endorphin-containing neurons in the hypothalamus. *Neuroendocrinology, 33,* 28–42.

Finley, J. W., Maderdut, J., & Petrusz, P. (1981). The immunocytochemical localization of enkephalin in the central nervous system of the rat. *Journal of Comparative Neurology, 198,* 541–565.

Flerkó, B., Petrusz, P., & Tima, L. (1967). On the mechanism of sexual differentiation of the hypothalamus: Factors influencing the "critical period" of the rat. *Acta Biologica Academiae Scientiarum, Hungary, 18,* 27–36.

Forman, L. J., Sonntag, W. E., Hylka, V. W., & Meites, J. (1983). Immunoreactive beta-endorphin in the plasma, pituitary and hypothalamus of young female rats on the day of estrus and intact and chronically castrated old constant estrous female rats. *Life Sciences, 33,* 993–999.

Gallo, R. V. (1980). Neuroendocrine regulation of pulsatile luteinizing hormone release in the rat. *Neuroendocrinology, 30,* 122–131.

Gallo, R. V. (1981). Pulsatile LH release during the ovulatory LH surge on proestrus in the rat. *Biology of Reproduction, 24,* 100–104.

Gallo, R. V., & Kalra, P. S. (1983). Pulsatile LH release on diestrus I in the rat estrous cycle: Relation to brain catecholamines and ovarian steroid secretion. *Neuroendocrinology, 37,* 91–97.

Gallo, R. V., & Moberg, G. P. (1977). Serotonin-mediated inhibition of episodic luteinizing hormone release during electrical stimulation of the arcuate nucleus in ovariectomized rats. *Endocrinology, 100,* 945–954.

Gerall, A. A., & Kenney, A. McM. (1970). Neonatally androgenized females' responsiveness to estrogen and progesterone. *Endocrinology, 87,* 560–566.

Gerall, A. A., & Ward, I. L. (1966). Effects of prenatal exogenous androgen on the sexual behavior of the female albino rat. *Journal of Comparative and Physiological Psychology, 62,* 370–375.

Gerall, A. A., Hendricks, S. E., Johnson, L. L., & Bounds, T. W. (1967). Effects of early castration in male rats on adult sexual behavior. *Journal of Comparative and Physiological Psychology, 64,* 206–212.

Gerall, A. A., Dunlap, J. L., & Sonntag, W. E. (1980). Reproduction in aging normal and neonatally androgenized female rats. *Journal of Comparative and Physiological Psychology, 94,* 556–563.

Gitler, M. S., & Barraclough, C. A. (1988). Stimulation of the medullary A1 noradrenergic system augments luteinizing hormone release induced by medial preoptic nucleus stimulation. *Neuroendocrinology, 48,* 351–359.

Gorski, R. A. (1968). Influence of age on the response to perinatal administration of a low dose of androgen. *Endocrinology, 82,* 1001–1004.

Gorski, R. A. (1973). Mechanisms of androgen induced masculine differentiation of the rat brain. In K. Lissak (Ed.), *Hormones and brain function* (pp. 27–45). New York: Plenum Press.

Gorski, R. A. (1979). The neuroendocrinology of reproduction: An overview. *Biology of Reproduction, 20,* 111–127.

Gorski, R. A. (1984). Critical role for the medial preoptic area in the sexual differentiation of the brain. In G. J. De Vries, J. P. C. DeBruin, H. B. N. Uylings, & M. A. Corner (Eds.), *Progress in brain research, Vol. 61: Sex differences in the brain* (pp. 129–146). Amsterdam: Elsevier.

Gorski, R. A., & Barraclough, C. A. (1963). Effects of low dosages of androgen on the differentiation of hypothalamic regulatory control of ovulation in the rat. *Endocrinology, 73,* 210–216.

Gorski, R. A., Harlan, R. E., Jacobson, C. D., Shryne, J., & Southam, A. (1980). Evidence for the existence of a sexually dimorphic nucleus in the preoptic area of the rat. *Journal of Comparative Neurology, 193,* 529–539.

Goy, R. W., & McEwen, B. S. (1980). *Sexual differentiation of the brain.* Cambridge, MA: MIT Press.

Goy, R. W., Bridson, W. E., & Young, W. C. (1964). Period of maximal susceptibility of the prenatal guinea pig to masculinizing actions of testosterone propionate. *Journal of Comparative and Physiological Psychology, 57,* 166–174.

Grady, K. L., Phoenix, G. H., & Young, W. C. (1965). Role of the developing rat testis in differentiation of the neural tissues mediating behavior. *Journal of Comparative and Physiological Psychology, 59,* 176–182.

Gray, G. D., Tennent, B., Smith, E. R., & Davidson, J. M. (1980). Luteinizing hormone regulation and sexual behavior in middle-aged female rats. *Endocrinology, 107,* 187–194.

Grossman, R., Diez-Guerra, F. J., Mansfield, S., & Dyer, R. G. (1987). Neonatal testosterone modifies LH secretion in the adult female rat by altering the opioid–noradrenergic interaction in the medial preoptic area. *Brain Research, 415,* 205–210.

Hammer, R. P., Jr. (1984). The sexually dimorphic region of the preoptic area in rats contains denser opiate receptor binding sites in females. *Brain Research, 308,* 172–176.

Hammer, R. P., Jr. (1985). The sex hormone-dependent development of opiate receptors in the rat medial preoptic area. *Brain Research, 360,* 65–74.

Handa, R. J., Condon, T. P., Whitmoyer, D. I., & Gorski, R. A. (1986). Gonadotropin secretion following intraventricular norepinephrine infusion into neonatally androgenized female rats. *Neuroendocrinology, 43,* 269–272.

Harlan, R. E., & Gorski, R. A. (1977). Steroid regulation of luteinizing hormone secretion in normal and androgenized rats at different ages. *Endocrinology, 101,* 741–749.

Harris, G. W. (1964). Sex hormones, brain development and brain function. *Endocrinology, 75,* 627–648.

Héry, M., Laplante, E., & Kordon, C. (1976). Participation of serotonin in the phasic release of LH. I. Evidence from pharmacological experiments. *Endocrinology, 99,* 496–503.

Hoffman, G. E., & Gibbs, F. P. (1982). LHRH pathways in the rat brain: Deafferentation spares a sub-chiasmatic LHRH projection to the median eminence. *Neuroscience, 7,* 1979–1993.

Hoffman, G. E., Fitzsimmons, M. D., & Watson, R. E. (1989). Relationship of endogenous opioid peptide axons to GnRH neurons in the rat. In R. G. Dyer & R. J. Bicknell (Eds.), *Brain opioids in reproduction* (pp. 125–134). Oxford: Oxford University Press.

Huang, H. H., & Meites, J. (1975). Patterns of reproductive function in aging female rats and effects of L-dopa administration. *Neuroendocrinology, 17,* 289–295.

Huang, H., Marshal, S., & Meites, J. (1976). Capacity of old versus young female rats to secrete LH, FSH, and prolactin. *Biology of Reproduction, 14,* 538–543.

Huang, H. H., Steger, R. W., Sonntag, W. E., & Meites, J. (1980). Positive feedback by ovarian hormones on prolactin and LH in old versus young female rats. *Neurobiology of Aging, 1,* 141–143.

Ibata, Y., Watanabe, K., Kinoshita, H., Kubo, S., & Sano, Y. (1979). The location of LH-RH neurons in the rat hypothalamus and their pathways to the median eminence. *Cell and Tissue Research, 198,* 381–395.

Ingram, D. L., Mandl, A. M., & Zuckerman, S. (1958). The influence of age on litter-size. *Journal of Endocrinology, 17,* 280–285.

Jennes, L., Beckman, W. C., Stumpf, W. E., & Grzanna, R. (1982). Anatomical relationship of serotonergic and noradrenergic projections with the GnRH system in septum and hypothalamus. *Experimental Brain Research, 46,* 331–338.

Jiang, M.-J., & Peng, M.-T. (1981). Cytoplasmic and nuclear binding of estradiol in the brain and pituitary of old female rats. *Gerontology, 27,* 51–57.

Johnson, M. D., & Crowley, W. R. (1983). Acute effects of estradiol on circulating luteinizing hormone and prolactin concentrations and on serotonin turnover in individual brain nuclei. *Endocrinology, 113,* 1935–1941.

Kalra, S. P. (1983). Inhibitory neuronal systems in regulation of gonadotropin secretion. In S. M. McCann & D. S. Dhindsa (Eds.), *Role of peptides and proteins in control of reproduction* (pp. 63–87). New York: Elsevier.

Kalra, S. P. (1986). Neural circuitry involved in control of LHRH secretion: Model for the preovulatory release. In W. F. Ganong & L. Martini (Eds.), *Frontiers in neuroendocrinology* (pp. 31–75). New York: Raven Press.

Kalra, S. P., & Kalra, P. S. (1983). Neural regulation of luteinizing hormone secretion in the rat. *Endocrine Reviews, 4,* 311–351.

Kim, Y. I., Dudley, C. A., & Moss, R. L. (1987). A1 noradrenergic input to medial preoptic–medial septal area: An electrophysiological study. *Neuroendocrinology, 45,* 77–85.

King, J. C., Tobet, S. A., Snavely, F. L., & Arimura, A. A. (1980). The LHRH system in normal and neonatally androgenized female rats. *Peptides, 1,* 85–100.

King, J., Tobet, S., Snavely, F., & Arimura, A. (1982). LHRH immunopositive cells and their projections to the median eminence and organum vasculosum of the lamina terminalis. *Journal of Comparative Neurology, 209,* 287–300.

King, W. L., & Greene, G. L. (1984). Monoclonal antibodies localize oestrogen receptor in the nuclei of target cells. *Nature, 307,* 745–747.

Kinoshita, F., Nakai, Y., Katakami, H., Kato, Y., Yajima, H., & Imura, H. (1980). Effect of β-endorphin on pulsatile luteinizing hormone release in conscious castrated rats. *Life Sciences, 27,* 843–846.

Korach, K. S., & Muldoon, T. G. (1974). Studies on the nature of the hypothalamic estradiol-concentrating mechanism in the male and female rat. *Endocrinology, 94,* 785–793.

Köves, K., Marton, J., Molnár, J., & Halász, B. (1981). (D-Met^2, Pro^5)-Enkephalinamide-induced blockade of ovulation and its reversal by naloxone in the rat. *Neuroendocrinology, 32,* 82–86.

Kumar, M. S. A., Chen, C. L., & Huang, H. H. (1980). Pituitary and hypothalamic concentrations of Met-enkephalin in young and old rats. *Neurobiology of Aging, 1,* 153–155.

LaPolt, P. S., Matt, D. W., Judd, H. L., & Lu, J. K. H. (1986). The relation of ovarian steroid levels in young female rats to subsequent estrous cyclicity and reproductive function during aging. *Biology of Reproduction, 35,* 1131–1139.

Leadem, C. A., & Kalra, S. P. (1985). The effects of various endogenous opioid peptides and opiates on luteinizing hormone (LH) and prolactin (PRL) secretion in ovariectomized rats. *Neuroendocrinology, 41,* 342–352.

LeBlond, C. B., Morris, S., Karakiulakis, G., Powell, R., & Thomas, P. J. (1982). Development of sexual dimorphism in the suprachiasmatic nucleus of the rat. *Journal of Endocrinology, 95,* 137–145.

Lenahen, S. E., Seibel, H. R., & Johnson, J. H. (1986). Evidence for multiple serotonergic influences on LH release in ovariectomized rats and for modulation of their relative effectiveness. *Neuroendocrinology, 44,* 89–94.

Léránth, C., Palkovits, M., MacLusky, N. J., Shanabrough, M., & Naftolin, F. (1986). Reproductive failure due to experimentally induced constant estrus does not alter the LH-RH fiber density in the median eminence of the rat. *Neuroendocrinology, 43,* 526–532.

Leung, P. C., Arendash, G. W., Whitmoyer, D. I., Gorski, R. A., & Sawyer, C. H. (1982). Differential effects of central adrenoreceptor agonists on luteinizing hormone release. *Neuroendocrinology, 34,* 207–214.

Limonta, P., Maggi, R., Dondi, D., & Piva, F. (1989). Endocrine regulation of brain opioid receptors. In R. G. Dyer & R. J. Bicknell (Eds.), *Brain opioid systems in reproduction* (pp. 27–38). Oxford: Oxford University Press.

Limonta, P., Dondi, D., Maggi, R., Martini, L., & Piva, F. (1989). Neonatal organization of the brain opioid systems controlling prolactin and luteinizing hormone secretion. *Endocrinology, 124,* 681–686.

Linnoila, M., & Cooper, R. L. (1976). Reinstatement of vaginal cycles in aged female rats. *Journal of Pharmacology and Experimental Therapeutics, 199,* 477–482.

Lookingland, K. J., & Barraclough, C. A. (1982). Changes in plasma hormone profiles and in hypothalamic catacholamine turnover rates in neonatally androgenized rats during the transition phase from cyclicity to persistent estrus (delayed anovulatory syndrome). *Biology of Reproduction, 27,* 282–299.

Lookingland, K. J., Wise, P. M., & Barraclough, C. A. (1982). Failure of the hypothalamic noradrenergic system to function in adult androgen-sterilized rats. *Biology of Reproduction, 27,* 268–281.

Lustig, R. H., Fishman, J., & Pfaff, D. W. (1989). Ovarian steroids and endogenous opioid peptide action in control of the rat LH surge. In R. G. Dyer & R. J. Bicknell (Eds.), *Brain opioid systems in reproduction* (pp. 3–26). Oxford: Oxford University Press.

Matsumoto, A., & Arai, Y. (1981). Effect of androgen on sexual differentiation of synaptic organization in the hypothalamic arcuate nucleus: An ontogenetic study. *Neuroendocrinology, 33,* 166–169.

Maurer, R. A., & Woolley, D. E. (1974). ^{3}H-Estradiol distribution in normal and androgenized female rats using an improved hypothalamic dissection procedure. *Neuroendocrinology, 14,* 87–94.

McGinnis, M. Y., Krey, L. C., MacLusky, N. J., & McEwen, B. S. (1981). Steroid receptor levels in intact and ovariectomized estrogen-treated rats: An examination of quantitative, temporal and endocrine factors influencing the efficacy of an estradiol stimulus. *Neuroendocrinology, 33,* 158–165.

Meites, J., Huang, H. H., & Riegle, G. D. (1976). Relation of the hypothalamo–pituitary–gonadal system to decline of reproductive functions in aging female rats. In F. Labrie, J. Meites, & G. Pelletier (Eds.), *Hypothalamus and endocrine functions* (pp. 3–20). New York: Plenum Press.

Meites, J., Huang, H. H., & Simpkins, J. W. (1978). Recent studies on neuroendocrine control of reproductive senescence in rats. In E. L. Schneider (Ed.), *The aging reproductive system* (pp. 213–235). New York: Raven Press.

Merchenthaler, I., Kovács, G., Lovász, G., & Sétáló, G. (1980). The preoptico–infundibular LH-RH tract of the rat. *Brain Research, 198,* 63–74.

Merchenthaler, I., Lengvári, I., Horváth, J., & Sétáló, G. (1980). Immunocytochemical study of the LHRH-synthesizing neuron system of aged female rats. *Cell and Tissue Research, 209,* 497–505.

Missale, C., Govoni, S., Croce, L., Bosio, A., Spano, P. F., & Trabucchi, M. (1983). Changes of β-endorphin and Met-enkephalin content in the hypothalamus–pituitary axis induced by aging. *Journal of Neurochemistry, 40,* 20–24.

Moore, R. Y. (1979). The retinohypothalamic tract, suprachiasmatic hypothalamic nucleus and central neural mechanisms of circadian rhythm regulation. In M. Suda, O. Hayaishi, & H. Nakagawa (Eds.), *Biological rhythms and their central mechanisms* (pp. 343–354). Amsterdam: Elsevier.

Moore, R. Y., & Card, P. (1984). Noradrenaline-containing neuron systems. In A. Björklund & T. Hökfelt (Eds.), *Handbook of chemical neuroanatomy: Vol. 2. Classical transmitters in the CNS: Part I* (pp. 123–156). New York: Elsevier.

Morin, L. P., Fitzgerald, K. M., & Zucker, I. (1977). Estradiol shortens the period of hamster circadian rhythms. *Science, 196,* 305–307.

Morley, J. E., Levine, A. S., Yim, G. K., & Lowy, M. T. (1983). Opioid modulation of appetite. *Biobehavioral Reviews, 7,* 281–305.

Morrell, J. I., McGinty, J. F., & Pfaff, D. W. (1985). A subset of β-endorphin- or dynorphin-containing neurons in the medial basal hypothalamus accumulates estradiol. *Neuroendocrinology, 41,* 417–426.

Naftolin, F. K., Ryan, K. J., Davies, I. J., Reedy, V. V., Flores, F., Petro, Z., Kuhn, M., White, R. J., Takaoka, Y., & Wolin, L. (1975). The formation of estrogen by central nervous system tissues. *Recent Progress in Hormone Research, 31,* 295–319.

Napoli, A. M., & Gerall, A. A. (1970). Effect of estrogen and anti-estrogen on reproductive function in neonatally androgenized female rats. *Endocrinology, 87,* 1330–1337.

Nordeen, E. J., & Yahr, P. (1983). A regional analysis of estrogen binding to hypothalamic cell nuclei in relation to masculinization and defeminization. *Journal of Neuroscience, 3,* 933–941.

Ogren, L., Vertes, M., & Woolley, D. (1976). *In vivo* nuclear ³H-estradiol binding in brain areas of the rat: Reduction by endogenous and exogenous androgens. *Neuroendocrinology, 21,* 350–365.

Olsen, K. L., & Whalen, R. E. (1980). Sexual differentiation of the brain: Effects on mating behavior and [³H]-estradiol binding by hypothalamic chromatin in rats. *Biology of Reproduction, 22,* 1068–1072.

Palkovits, M., & Brownstein, M. J. (1988). *Maps and guide to microdissection of the rat brain.* New York: Elsevier.

Pang, C. N., Zimmermann, E., & Sawyer, C. H. (1977). Morphine inhibition of the preovulatory surges of luteinizing hormone and follicle stimulating hormone in the rat. *Endocrinology, 101,* 1726–1736.

Peng, M. T., & Huang, H. H. (1972). Aging of hypothalamic–pituitary–ovarian function in the rat. *Fertility and Sterility, 23,* 535–542.

Peterson, S. L., & Barraclough, C. A. (1986). Effect of naloxone and morphine on LH and prolactin release in androgen-sterilized rats. *Neuroendocrinology, 44,* 84–88.

Pfaff, D., & Keiner, M. (1973). Atlas of estradiol-concentrating cells in the central nervous system of the female rat. *Journal of Comparative Neurology, 151,* 121–158.

Pfaff, D., & Modianos, D. (1985). Neural mechanisms of female reproductive behavior. In N. Adler, D. Pfaff, & R. W. Goy (Eds.), *Handbook of behavioral neurobiology: Vol. 7: Reproduction* (pp. 423–493). New York: Plenum Press.

Pfeiffer, C. A. (1936). Sexual differences of the hypophysis and their determination by the gonads. *American Journal of Anatomy, 58,* 195–225.

Phoenix, C. H., Goy, R. W., Gerall, A. A., & Young, W. C. (1959). Organizing action of prenatally administered testosterone propionate on the tissues mediating mating behavior in the female guinea pig. *Endocrinology, 65,* 369–382.

Popolow, H. B., King, J. C., & Gerall, A. A. (1981). Rostral medial preoptic area lesions' influence on female estrous processes and LHRH distribution. *Physiology and Behavior, 27,* 855–861.

Quadri, S. K., Kledzik, G. S., & Meites, J. (1973). Reinitiation of estrous cycles in old constant-estrous rats by central-acting drugs. *Neuroendocrinology, 11,* 248–255.

Raisman, G., & Field, P. M. (1973). Sexual dimorphism in the neuropil of the preoptic area of the rat and its dependence on neonatal androgen. *Brain Research, 54,* 1–29.

Ramirez, V. D., Feder, H. H., & Sawyer, C. H. (1984). The role of brain catecholamines in the regulation of LH secretion: A critical inquiry. In L. Martini & W. F. Ganong (Eds.), *Frontiers in neuroendocrinology* (pp. 27–84). New York: Raven Press.

Rance, N., Wise, P. M., Selmanoff, M. K., & Barraclough, C. A. (1981). Catecholamine turnover rates in discrete hypothalamic areas and associated changes in median eminence luteinizing hormone-releasing hormone and serum gonadotropins on proestrus and diestrus day 1. *Endocrinology, 108,* 1795–1802.

Robinson, S. M., Fox, T. O., Dikkes, P., & Pearlstein, R. A. (1986). Sex differences in the shape of the sexually dimorphic nucleus of the preoptic area and suprachiasmatic nucleus of the rat: 3-D computer reconstructions and morphometrics. *Brain Research, 371,* 380–384.

Rubin, B. S., King, J. C., & Bridges, R. S. (1984). Immunoreactive forms of LHRH in the brains of aging rats exhibiting persistent vaginal estrus. *Biology of Reproduction, 31,* 343–351.

Rubin, B. S., Fox, T. O., & Bridges, R. S. (1986). Estrogen binding in nuclear and cytosolic extracts from brain and pituitary of middle-aged female rats. *Brain Research, 383,* 60–67.

Rubinstein, L., & Sawyer, C. H. (1970). Role of catecholamine in stimulating the release of pituitary ovulating hormone in rats. *Endocrinology, 86,* 988–995.

Rusak, B., & Zucker, I. (1975). Biological rhythms and animal behavior. *Annual Review of Psychology, 26,* 137–171.

Sar, M., & Stumpf, W. E. (1981). Central noradrenergic neurones concentrate ³H-oestradiol. *Nature, 289,* 500–501.

Sar, M., Stumpf, W., Miller, R., Chang, K., & Cuatrecasas, P. (1978). Immunohistochemical localization of enkephalin in the rat brain and spinal cord. *Journal of Comparative Neurology, 182,* 17–37.

Schneider, H. P. G., & McCann, S. M. (1970). Mono- and indolamines and control of LH secretion. *Endocrinology, 86,* 1127–1133.

Shivers, B. D., Harlan, R. E., Morrell, J. I., & Pfaff, D. W. (1983). Absence of oestradiol concentration in cell nuclei of LHRH-immunoreactive neurones. *Nature, 304,* 345–347.

Simerly, R. B., & Swanson, L. W. (1986). The organization of neural inputs to the medial preoptic nucleus of the rat. *Journal of Comparative Neurology, 246,* 312–342.

Simerly, R. B., & Swanson, L. W. (1987). The distribution of neurotransmitter-specific cells and fibers in the anteroventral periventricular nucleus: Implications for the control of gonadotropin secretion in the rat. *Brain Research, 400,* 11–34.

Simerly, R. B., & Swanson, L. W. (1988). Projections of the medial preoptic nucleus: A *Phaseolus vulgaris* leucoagglutinin anterograde tract-tracing study in the rat. *Journal of Comparative Neurology, 270,* 209–242.

Simerly, R. B., Swanson, L. W., & Gorski, R. A. (1984). Demonstration of a sexual dimorphism in the distribution of serotonin-immunoreactive fibers in the medial preoptic nucleus in the rat. *Journal of Comparative Neurology, 225,* 151–166.

Simerly, R. B., Gorski, R. A., & Swanson, L. W. (1986). Neurotransmitter specificity of cells and fibers in the medial preoptic nucleus: An immunohistochemical study in the rat. *Journal of Comparative Neurology, 246,* 343–363.

Simerly, R. B., McCall, L. D., & Watson, S. J. (1988). Distribution of opioid peptides in the preoptic region: Immunohistochemical evidence for a steroid-sensitive enkephalin sexual dimorphism. *Journal of Comparative Neurology, 276,* 442–459.

Simerly, R. B., Chang, C., Muramatsu, M., & Swanson, L. W. (1990). Distribution of androgen and estrogen receptor mRNA-containing cells in the rat brain: An *in situ* hybridization study. *Journal of Comparative Neurology, 294,* 76–95.

Soper, B. D., & Weick, R. F. (1980). Hypothalamus and extrahypothalamic mediation of pulsatile discharges of luteinizing hormone in the ovariectomized rat. *Endocrinology, 106,* 348–355.

Steinbusch, H. W. M. (1984). Serotonin-immunoreactive neurons and their projections in the CNS. In A. Björklund, T. Hökfelt, & M. J. Kuhar (Eds.), *Handbook of chemical neuroanatomy: Vol. 3. Classical transmitters and transmitter receptors in the CNS: Part II* (pp. 68–125). New York: Elsevier.

Stumpf, W. E., Sar, M., & Keefer, S. A. (1975). Atlas of estrogen target cells in the rat brain. In W. E. Stumpf & L. D. Grant (Eds.), *Anatomical neuroendocrinology* (pp. 104–119). Basel: S. Karger.

Swanson, H. E., & van der Werff ten Bosch, J. J. (1964). The early androgen syndrome: Its development and the response to hemispaying. *Acta Endocrinologica, 45,* 1–12.

Sylvester, P. W., Sarkar, D. K., Briski, K. P., & Meites, J. (1985). Relation of gonadal hormones to differential LH response to naloxone in prepubertal male and female rats. *Neuroendocrinology, 40,* 165–170.

van der Kar, L. D., & Lorens, S. A. (1979). Differential serotonergic innervation of individual hypothalamic nuclei and other forebrain regions by the dorsal and median midbrain raphe nuclei. *Brain Research, 162,* 45–54.

van der Schoot, P. (1976). Changing pro-oestrous surges of luteinizing hormone in aging 5-day cyclic rats. *Journal of Endocrinology, 69,* 287–288.

Van Vugt, D. A., Aylsworth, C. F., Sylvester, P. W. Leung, F. C., & Meites, J. (1981). Evidence for hypothalamic noradrenergic involvement in naloxone-induced stimulation of luteinizing hormone release. *Neuroendocrinology, 33,* 261–264.

Van Vugt, D. A., Sylvester, P. W., Aylsworth, C. F., & Meites, L. (1982). Counteraction of gonadal steroid inhibition of luteinizing hormone release by naloxone. *Neuroendocrinology, 34,* 274–278.

Vertes, M., & King, R. J. B. (1971). The mechanism of oestradiol binding in rat hypothalamus: Effect of androgenization. *Journal of Endocrinology, 51,* 271–282.

Walker, R. F. (1982). Reinstatement of LH surges by serotonin neuroleptics in aging constant estrous rats. *Neurobiology of Aging, 3,* 253–257.

Walker, R. F. (1984). Impact of age-related changes in serotonin and norepinephrine metabolism on reproductive function in female rats: An analytical review. *Neurobiology of Aging, 5,* 121–139.

Walker, R. F., Cooper, R. L., & Timiras, P. S. (1980). Constant estrus: Role of rostral hypothalamic monoamines in development of reproductive dysfunction in aging rats. *Endocrinology, 170,* 249–255.

Walters, M. R. (1985). Steroid hormone receptors and the nucleus. *Endocrine Reviews, 6,* 512–543.

Ward, I. L. (1969). Differential effect of pre- and postnatal androgen on the sexual behavior of intact and spayed female rats. *Hormones and Behavior, 1,* 25–36.

Watson, R. E., Hoffman, G. E., & Wiegand, S. J. (1986). Sexually dimorphic opioid distribution in the preoptic area: Manipulation by gonadal steroids. *Brain Research, 398,* 157–163.

Watts, A. G., & Fink, G. (1984). Pulsatile luteinizing hormone release, and the inhibitory effect of estradiol-17β in gonadectomized male and female rats: Effects of neonatal androgen or exposure to constant light. *Endocrinology, 115,* 2251–2259.

Watts, A. G., Swanson, L. W., & Sanchez-Watts, G. (1987). Efferent projections of the suprachiasmatic nucleus: Studies using anterograde transport of *Phaseolus vulgaris* leucoagglutinin in the rat. *Journal of Comparative Neurology, 258,* 204–229.

Whalen, R. E., & Massicci, J. (1975). Subcellular analysis of the accumulation of estrogen by the brain of male and female rats. *Brain Research, 89,* 255–264.

Whalen, R. E., & Olsen, K. L. (1978). Chromatin binding of estradiol in the hypothalamus and cortex of male and female rats. *Brain Research, 152,* 121–131.

Wiegand, S., & Terasawa, E. (1982). Discrete lesions reveal functional heterogeneity of suprachiasmatic structures in regulation of gonadotropin secretion in the female rat. *Neuroendocrinology, 34,* 395–404.

Wiegand, S. J., Terasawa, E., & Bridson, W. E. (1978). Persistent estrus and blockade of progesterone-induced LH release follows lesions which do not damage the suprachiasmatic nucleus. *Endocrinology, 102,* 1645–1648.

Wilkes, M. M., & Yen, S. S. C. (1981). Attenuation during aging of the postovariectomy rise in median eminence catecholamines. *Neuroendocrinology, 33,* 144–147.

Wilson, J. G. (1943). Reproductive capacity of adult female rats treated prepuberally with estrogenic hormone. *Anatomical Record, 86,* 341–363.

Wise, P. M. (1982a). Alteration in proestrus LH, FSH and prolactin surges in middle-aged rats. *Proceedings of the Society of Experimental Biology and Medicine, 169,* 348–354.

Wise, P. M. (1982b). Norepinephrine and dopamine activity in microdissected brain areas of the middle-aged and young rat on proestrus. *Biology of Reproduction, 27,* 562–574.

Wise, P. M. (1984a). Comments on the role of monoamines, particularly serotonin and aging. *Neurobiology of Aging, 5,* 142–143.

Wise, P. M. (1984b). Estradiol-induced daily luteinizing hormone and prolactin surges in young and middle-aged rats: Correlations with age-related changes in pituitary responsiveness and catecholamine turnover rates in microdissected brain areas. *Endocrinology, 115,* 801–809.

Wise, P. M., & Camp, P. (1984). Changes in concentrations of estradiol nuclear receptors in the preoptic area, medial basal hypothalamus, amygdala, and pituitary gland of middle-aged and old cycling rats. *Endocrinology, 114,* 92–98.

Wise, P. M., & Ratner, A. (1980). LHRH-induced LH and FSH responses in the aged female rat. *Journal of Gerontology, 35,* 506–511.

Wise, P. M., McEwen, B. S., Parsons, B., & Rainbow, T. C. (1984). Age-related changes in cytoplasmic estradiol receptor concentrations in microdissected brain nuclei: Correlations with changes in steroid-induced sexual behavior. *Brain Research, 321,* 119–126.

Wise, P. M., Cohen, I. R., Weiland, N. G., & London, E. D. (1988). Aging alters the circadian rhythm of glucose utilization in the suprachiasmatic nucleus. *Proceedings of the National Academy of Science U.S.A., 85,* 5305–5309.

Wise, P. M., Dueker, E., & Wuttke, W. (1988). Age-related alterations in pulsatile luteinizing hormone release: Effects of long-term ovariectomy, repeated pregnancies and naloxone. *Biology of Reproduction, 39,* 1060–1066.

Zucker, I. (1976). Light, behavior, and biological rhythms. *Hospital Practice, 11,* 83–91.

Index